HEADACHE AND MIGRAINE
BIOLOGY AND MANAGEMENT

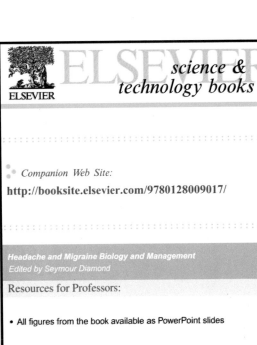

ELSEVIER *science & technology books*

Companion Web Site:

http://booksite.elsevier.com/9780128009017/

Headache and Migraine Biology and Management
Edited by Seymour Diamond

Resources for Professors:

- All figures from the book available as PowerPoint slides

ELSEVIER

**ACADEMIC
PRESS**

HEADACHE AND MIGRAINE BIOLOGY AND MANAGEMENT

Editor-in-Chief:

SEYMOUR DIAMOND
National Headache Foundation, Chicago, Illinois, USA

Associate Editors:

ROGER K. CADY

MERLE L. DIAMOND

MARK W. GREEN

VINCENT T. MARTIN

AMSTERDAM • BOSTON • HEIDELBERG • LONDON
NEW YORK • OXFORD • PARIS • SAN DIEGO
SAN FRANCISCO • SINGAPORE • SYDNEY • TOKYO
Academic Press is an imprint of Elsevier

Academic Press is an imprint of Elsevier
The Boulevard, Langford Lane, Kidlington, Oxford OX5 1GB
225 Wyman Street, Waltham MA 02451

Notices
Knowledge and best practice in this field are constantly changing. As new research and experience broaden our understanding, changes in research methods, professional practices, or medical treatment may become necessary.

Practitioners and researchers may always rely on their own experience and knowledge in evaluating and using any information, methods, compounds, or experiments described herein. In using such information or methods they should be mindful of their own safety and the safety of others, including parties for whom they have a professional responsibility.

To the fullest extent of the law, neither the Publisher nor the authors, contributors, or editors, assume any liability for any injury and/or damage to persons or property as a matter of products liability, negligence or otherwise, or from any use or operation of any methods, products, instructions, or ideas contained in the material herein.

Library of Congress Cataloging-in-Publication Data
A catalog record for this book is available from the Library of Congress

British Library Cataloguing-in-Publication Data
A catalogue record for this book is available from the British Library

ISBN: 978-0-12-800901-7

For information on all Academic Press publications
visit our website at http://store.elsevier.com

  Working together
to grow libraries in
developing countries

www.elsevier.com • www.bookaid.org

Publisher: Mica Haley
Acquisition Editor: Natalie Farra
Editorial Project Manager: Kristi Anderson
Production Project Manager: Melissa Read
Designer: Alan Studholme

Printed and bound in the United States of America

Dedication

To my wife of 66 years, Elaine; my three daughters — Judi Diamond-Falk, Merle Diamond, and Amy Diamond; my grandchildren — Brian Diamond-Falk, Emily Horowitz, Max Barack, Michael Barack, Jacob Barack, and Katelyn Barack; and my great-grandchildren — Zevan and Oliver Diamond-Falk; all of whom bore with me during this momentous task.

Contents

Preface

It is not enough to know what to say — one must know how to say it

Aristotle's Injunction

At this juncture, why did we see a need for a new book on headache? First, because of the proliferation in research, both biochemical and genetic, a review of the current status of headache was obvious. We also determined that solving the mysteries of headache, primarily migraine, indicated the need for an update in a simple correlation which is understandable to the researcher as well as the clinician. It is our goal to clarify, as simply as possible, the understanding of this conundrum.

Physicians representing various specialties and subspecialties, as well as alternative medicine practitioners, have confused some of the basics of diagnosis and treatment of headache and its management. It is a fact that a multitude of headache patients are treated symptomatically without regard to diagnosis. Many headache patients have never undergone a careful history or adequate physical and neurological examinations.

Our book will stress the importance of diagnosis and then appropriate choice of treatment. For the past 50 years, I have always stressed in my many lectures on headache that continuity of care is imperative to headache treatment. It is my hope that in this text, the importance of a continuum of treatment will be recognized. In my many years of treating patients, I have observed that those patients who are continually followed by their primary treating physician — whether a neurologist, internist, or family practitioner — will demonstrate the greatest improvement.

Headache should never be treated casually or with intolerance, especially in those with persistent complaints. I, and my associate editors, have carefully chosen each chapter for its relevance and significance in understanding cephalalgia. We have chosen the contributing authors thoughtfully, not only for their scientific acumen but also for their practical knowledge in treating headache patients. Too often, texts neglect these factors. It is my hope that this text will be used both as a reference and for methodology in treatment, and that it will be considered invaluable.

Seymour Diamond, MD
Chicago, Illinois, USA

About the Editor

Dr. Seymour Diamond is the National Headache Foundation's Executive Chairperson and Founder. Dr. Diamond is Director Emeritus and Founder of the Diamond Headache Clinic, Chicago, Illinois, and previously served as Adjunct Professor, Department of Cellular and Molecular Pharmacology, as well as Clinical Professor, Department of Family Medicine of The Chicago Medical School at Rosalind Franklin University of Medicine and Science, North Chicago, Illinois. From 2009 through 2010, he was also a Lecturer, Department of Family Medicine (Neurology), Loyola University Chicago/Stritch School of Medicine, Maywood, Illinois.

Dr. Diamond received his medical degree from the Chicago Medical School, from which he received the Distinguished Alumni Award in 1977. In November 2002, he received the President's Award from The Chicago Medical School Alumni Association.

Dr. Diamond has served as editor for 16 publications. In addition, he has published over 500 articles in the professional literature. Dr. Diamond has authored or co-authored over 63 books, including *Diagnosing and Managing Migraine*, 7th edition (2008); *Headache Through the Ages* (with Mary A. Franklin, 2005); *Conquering Your Migraine* (with Mary A. Franklin, 2001); *Headache and Your Child* (with Amy Diamond, 2001); and *The Headache Godfather* (with Charlie Morey, 2015). With Donald J. Dalessio, he has co-edited five editions of *The Practicing Physician's Approach to Headache*.

Dr. Diamond has lectured throughout the United States, Europe, and Asia, and has held more than 30 professional association offices in the past. In 1988 he was the first recipient of the Migraine Trust Lectureship, from the British Migraine Trust. He received the Lifetime Achievement Award from the American Association for the Study of Headache, in 1999.

For 66 years, Dr. Diamond has been married to Elaine, and they have three daughters: Judi Diamond-Falk, Merle Diamond, and Amy Diamond. Their family includes six grandchildren and two great-grandchildren.

Dr. Diamond is a life-long Chicago White Sox fan, and both he and his wife have achieved Gold Life Master levels in duplicate bridge.

List of Contributors

Andrew H. Ahn University of Florida College of Medicine, Gainsville, Florida, USA

Steven M. Baskin New England Institute for Neurology and Headache, Stamford, Connecticut, USA

José Biller Department of Neurology, Loyola University Chicago Stritch School of Medicine, Chicago, Illinois, USA

Jan Lewis Brandes Nashville Neuroscience Group, and Vanderbilt University School of Medicine, Nashville, Tennessee, USA

Roger K. Cady Headache Care Center, Springfield, Missouri, USA

Rachel Colman Department of Neurology, Icahn School of Medicine at Mt Sinai, New York, New York, USA

Wade M. Cooper Department of Neurology, University of Michigan, Ann Arbor, Michigan, USA

Merle L. Diamond Diamond Headache Clinic, Chicago, Illinois, USA

Seymour Diamond Diamond Fellowship and Educational Foundation, and Diamond Headache Clinic, Chicago, Illinois; National Headache Foundation, Chicago, Illinois, USA

Paul L. Durham Center for Biomedical & Life Sciences, Missouri State University, Springfield, Missouri, USA

Kathleen Farmer Headache Care Center, Springfield, Missouri, USA

Alexander Feoktistov Diamond Headache Clinic, Chicago, Illinois, USA

Johnathon Florczak Medical College of Wisconsin, Milwaukee, Wisconsin, USA

Mary A. Franklin National Headache Foundation, Chicago, Illinois, USA

Frederick G. Freitag Medical College of Wisconsin, Milwaukee, Wisconsin, USA

Benjamin W. Friedman Albert Einstein College of Medicine, Bronx, New York, USA

Benjamin Frishberg UCSD School of Medicine, The Headache Center of Southern California, Encinitas, California, USA

Jack Gladstein Department of Pediatric Neurology, and Headache Program, University of Maryland School of Medicine, Baltimore, Maryland, USA

Mark W. Green Department of Neurology, Icahn School of Medicine at Mt Sinai, New York, New York, USA

Howard S. Jacobs Department of Neurology, Nationwide Children's Hospital, and Ohio State University, Columbus, Ohio, USA

Robert G. Kaniecki Department of Neurology, University of Pittsburgh Medical Center, Pittsburgh, Pennsylvania, USA

Sylvia Lucas Departments of Neurology and Neurological Surgery, University of Washington Medical Center, Seattle, Washington, USA

Vincent T. Martin Department of Internal Medicine, University of Cincinnati, Cincinnati, Ohio, USA

Amit K. Masih Department of Neurology, Michigan State University, East Lansing, Michigan, USA

Edmund Messina The Michigan Headache Clinic, East Lansing, Michigan, and College of Human Medicine, Michigan State University, Michigan, USA

George R. Nissan Baylor University Medical Center, Baylor Headache Center, Dallas, Texas, and Texas A&M Health Science Center College of Medicine, Dallas, Texas, USA

Duren Michael Ready Headache Clinic, Baylor Scott & White, Temple, Texas, USA

A. David Rothner Center for Pediatric Neurology, Cleveland Clinic Main Campus, Cleveland, Ohio, USA

Elizabeth K. Seng Ferkauf Graduate School of Psychology of Yeshiva University, and Albert Einstein College of Medicine of Yeshiva University, New York, New York, USA

Robert B. Shulman Department of Psychiatry, Rush University Medical Center, and Rush Medical College, Chicago, Illinois, USA

Michael Star Department of Neurology, Loyola University Chicago Stritch School of Medicine, Chicago, Illinois, USA

Acknowledgments

I would like to thank my associate editors — Roger Cady, Merle Diamond, Mark Green, and Vincent Martin — whose knowledge and experience have contributed greatly to this volume.

Also, a note of gratitude to my colleagues — Kathleen Farmer, PsyD and Robert Daroff, MD, for their assistance in the review of specialized topics.

I am most appreciative of the support and encouragement of Natalie Farra, PhD, and Kristi Anderson of Elsevier. Their interest and expertise have been instrumental in the production of this book.

Finally, special thanks to Mary Franklin who has assisted me in most of my major editorial projects for 44 years. She has facilitated the completion of this text. I could not have undertaken this project without Mary's assistance and editorial abilities.

Seymour Diamond, MD

1

Introduction — The History of Headache

Seymour Diamond[1,2] and Mary A. Franklin[2]

[1]Diamond Fellowship and Educational Foundation, and Diamond Headache Clinic, Chicago, Illinois
[2]National Headache Foundation, Chicago, Illinois, USA

INTRODUCTION

In a previous monograph with my editorial collaborator, Mary Franklin, we reviewed the history of headache through the ages — in the arts and literature.[1] In this comprehensive work on headache, I would be remiss to not update the history of the advances in headache medicine during the 20th and early 21st centuries.

The history of headache treatment did not start with the discovery of the triptans. The approval of propranolol for the indication of migraine prophylaxis was not the nascent event for migraine prevention; neither was the introduction of dihydroergotamine into the migraine armamentarium. When Bayer started manufacturing acetylsalicylic acid for pain prevention, that was just one step in the long struggle for effective migraine and headache treatment, which has blossomed in recent years.

THE ANCIENTS

The earliest mention of headache can be found in Mesopotamia (modern-day Iraq), dating from 4000 BC. When the Ancients experienced headache, they blamed their affliction on Tiu, the evil spirit of headache. Our knowledge of the ancient Egyptians' headache management is found in the Ebers Papyrus, a collection of medical texts, named for the German Egyptologist George Ebers (1837–1898) who had acquired it. This papyrus contains the earliest written reference to the central nervous system and brain. For headache, the recommended treatment includes a combination of frankincense, cumin, ulan berry, and goose grease, to be boiled together and applied externally to the head.

The Egyptians also attributed the cause of headache as the work of an evil spirit. For those experiencing a "warmth in the head," the application of moistened mortar to the head was suggested. Another therapy was derived from Egyptian mythology — a combination of coriander, wormwood, juniper, honey, and opium. For joint pain, the Egyptians recommended a mixture of myrtle and willow leaves. The use of willow leaves is cited in treatment for an inflammatory condition: "... you must make cooling substances for him to draw the heat out ... leaves of the willow." Salicylic acid is derived from willow bark, and its use led to the discovery of aspirin. Later, the Assyrians, using stone tablets, recommended the use of willow leaves for treating inflammatory rheumatoid disorders, such as arthritis.[2]

The Greeks were the next to espouse willow bark as a treatment for pain. Hippocrates (4th or 5th century BC) recommended the extract of willow bark for headache pain. As we know, the teachings of Hippocrates formed the basis of medicine for centuries in the Greek and then the Roman Empires.

At Alexandria, Egypt, the Greeks established a center for medical education and practice. Once the Romans conquered this area, they maintained the center. Aretaeus of Cappodocia (AD 81–138) was probably educated at Alexandria and practiced medicine in Rome. He was the first to distinguish migraine from general headache, noting migraine's unilaterality, periodicity, and the associated symptom of nausea.[3] Aretaeus divided all diseases into acute and chronic. For headache, he described headaches of short duration, lasting a few days, as *cephalalgia*. The term "cephalea" referred to headaches which lasted longer. Because of migraine's one-sided occurrence, Aretaeus named it *Heterocrania*, meaning "half-a-head." The recommended treatment for headache by this ancient physician was counter-irritation in the form of application of blisters to the affected area, which had been shaved.

DOI: http://dx.doi.org/10.1016/B978-0-12-800901-7.00001-X

In Aretaeus' repertoire of blister agents were pitch, peilitory, euphorbium, lemnestis, or the juice of the thapsia.

During the 2nd century AD, Galen (131–201) gained prominence in Rome. Like Aretaeus, he was trained in Hippocratic medicine and became the court physician to Commodus, the heir of Marcus Aurelius. He is credited with describing migraine as *Hemicrania*.[4] Galen further advanced counter-irritation as a treatment for headache when he proposed the use of the electric torpedo fish applied to the forehead. This form of therapy foreshadowed the use of electrotherapy by Duchenne (1806–1875) and the transcutaneous electric stimulator (TENS) introduced in the late 20th century for all types of chronic pain.

THE MIDDLE AGES

The use of trephination for headache treatment was described by Paul of Aegina (625–690), who practiced in ancient Alexandria. The procedure, removing a circular portion of the skull, was believed to disturb the evil spirits which were causing the headache pain and allow them to escape through the wound (Figures 1.1, 1.2).

The fall of the Roman Empire did not mean the end of ancient Greek and Roman medicine. Those early texts on medicine influenced Arab physicians throughout the Islamic world from the 7th century and beyond. One of the most prominent of these physicians was Avicenna (980–1037). A native of Persia (modern day Iran), his textbook, *The Book of Healing*, was used by his contemporary Islamic physicians but was also available as a Latin translation for the scientists in Europe. Avicenna noted that headache location could vary between frontal, occipital, or generalized, and that one-sided headaches could be provoked by smells. He used cashews as a remedy for headache as well as other neurological and psychiatric disorders. Other Arab physicians wrote of treating headache, epilepsy, and syncope with *anomum nelegueta*, an African ginger.

In Cordoba, Spain, Abulcasis (935–1013) was physician to the Spanish caliph and was considered the greatest of Islamic medieval surgeons. His book, *Kitab al-Tasrif*, remained the leading textbook on surgery for the next five centuries in Europe and the Middle East. Abulcasis recognized the importance of the physician–patient relationship. Also, he advised his students to observe individuals closely in order to establish the appropriate diagnosis and select the most effective therapy. His recommended therapy for headache was extreme – applying a hot iron to the head of the individual with headaches. Another headache intervention

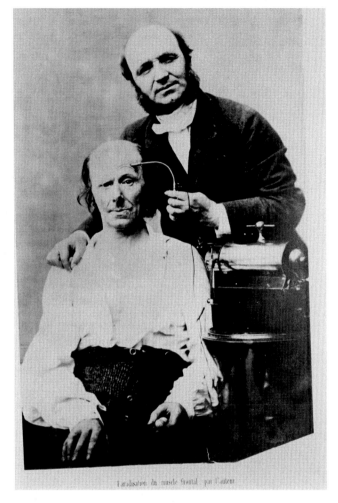

FIGURE 1.1 **Electrotherapy.** Guillaume-Benjamin Duchenne demonstrates electric stimulation therapy on a patient by holding an electric apparatus to the patient's head. ©*CORBIS.*

that he suggested was an incision made to the temple, and application of garlic to the wound.

In addition to its prominence in the Islamic world, Cordoba was also known as the birthplace of the medieval Jewish scholar and physician Maimonides (1135–1204). He studied medicine at Fez, Morocco, and later settled in Egypt, serving as court physician to the Sultan, Saladin, during the first crusade.[5] Maimonides' works on medicine continue to be studied, and it is apparent that he was influenced by Hippocrates and Galen. In his work on headache, Maimonides recognized various triggers of headache, including extremes of cold and heat, caused by changes in barometric pressure.

For headache treatment, Maimonides recommended that those suffering from a "strong midline headache, secondary to thick blood or internal coldness" could benefit from consumption of undiluted wine either during or after a meal. The warming effect of the

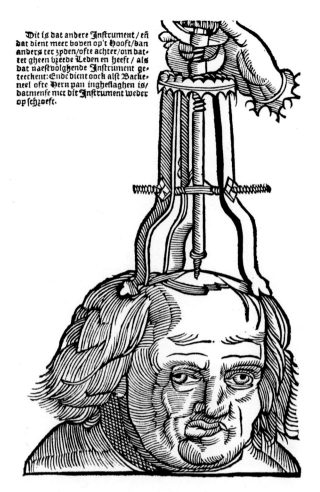

Dit is dat andere Instrument / en
dat dient meer boven op 't Hooft / van
anders ter zyden / ofte achter / om dat-
tet gheen breede Leden en heeft / als
dat naestvolghende Instrument ge-
teeckent: Ende dient oock alst Backe-
neel ofte Beyn van inghestaghen is /
dat men se met dit Instrument weder
op schroeft.

FIGURE 1.2 **Trephination, 1593.** Use of an elevator to remove a piece of bone from the skull. *Reproduced from the* Veldt Boeck van den Chirugia Scheel-Hans, *by Hans von Gersdorf (Amsterdam, 1593). Oxford Science Archive.*

FIGURE 1.3 "Scivias" (Know the ways of the Lord) by the German nun and mystic Hildegard von Bingen (1098–1179). The book, Codex Rupertsberg, disappeared during WW II. *Reproduced from a facsimile. Image reference: ART170094; Erich Lessing/Art Resource, New York, NY.*

wine would help, and also would thin the blood. Maimonides also instructed individuals with headache to refrain from physical exertion and other activities until their headache resolved. He cautioned that certain foods which were "rich in moisture" should be avoided, including melons, peaches, apricots, mulberries, fresh dates, etc.[6] For milder headache, Maimonides did not believe medication was appropriate, believing nature could relieve this pain without assistance.

During the same period, in what is now modern Germany, a remarkably intelligent and creative nun, the Abbess Hildegard of Bingen (1098–1179), became prominent in the Church because of her preaching. She is also remembered for her religious music and several texts that she wrote on a variety of subjects. In the world of headache medicine, she is known for the illuminated manuscripts that she created from her "visions," but which have been described as excellent depictions of migraine auras.[7] Hildegard lived in a area of Germany near the Rhine river. On the

opposite bank there was an outbreak of St Anthony's fire, ergotism, which was attributed to spoiled rye. On Hildegard's side of the river, it has been presumed that the spoiled crop maybe triggered an increase in migraine symptoms in those susceptible to migraine. In those exposed to ergot, hallucinations have been reported (Figure 1.3).

Throughout the Middle Ages in northern Europe, including the British Isles, herbal remedies were frequently mentioned. Elder seed (presumably from the alder tree) and willow bark are included in some recommended remedies. The popularity of willow bark for the creation of baskets pre-empted its use in pain management, and it is not seen again in herbal medicine until the late 18th century. Mugwort (*A. Vulgaris*) and wormwood (*A. Absinthium*) are also noted to have curative powers in relieving headaches. Bartholomew of

England, who lived during the 13th century, recommended scarification of the shin bones (making a series of cuts, scratches, incisions, etc., in the skin) in order to transfer whatever humor was causing the headache and send it to the lower extremities.

During the Renaissance, Galen's theories of medicine fell into disfavor as new principles of anatomy were established due to the dissection of human bodies. Barbers were still considered surgeons who did minor surgeries, treated trauma, and performed bloodletting. Bloodletting, which remained a common procedure through the 18th century, was believed to cure diseases, including headache.

THE 16TH TO 19TH CENTURIES

The monograph *Hemicrania* was written by Charles Lepois (1563–1633). This work focuses on the author's own headaches, which were only relieved by sleep or vomiting. He found relief with consumption of large amounts of fluids. In this treatise, Lepois notes that his headaches are triggered by changes in weather. He also described an association between one-sided headaches and epilepsy.

During the 17th century, in England, Nicholas Culpeper (1616–1654) created a furor in the "medical" community by publishing an English translation of the *Pharmacoporia* from the Latin. His colleagues felt betrayed, as they maintained secrecy about their methods by only using the "dead" language. Valerian, a perennial herb, was used as a remedy for colic or flatulence, and for "nervous headache," because of its sedative effects. Valerian's use in headache therapy continued through the 19th century. The active principle of Meglin's pills, used during the Victorian era, was valerian.[8] During this same period, the use of atropa belladonna was recommended for headache relief and insomnia, but to be administered on the scalp, not as an oral remedy.[9]

As explorers returned from the New World, they brought herbs and plants never before seen in Europe. The natives of the Americas used tobacco (*Nicotiana tabacum*) for the symptoms of asthma, for labor pains, and for headache. They also utilized the testes of beavers to create a spirit to relieve nervous headaches. This curative was not popularized in Europe. It is believed that the beaver's secretions contained a salicylate-like agent from the bark of trees consumed by the animal.

One of the founders of the Royal Society, Thomas Willis (1621–1675), significantly impacted medicine in the 17th century and beyond. As a proponent of the "New" or experimental science, he did not adhere to Galenic or Aristotelian dogma, which was taught at Oxford at that time. Willis belonged to the "iatrochemistry" school, which combined the study of medicine and chemistry, and which existed in England from 1525 through 1660. Of course, he is best remembered for identifying the *Cerebri anatome*, or Circle of Willis — the ring of communicating vessels under the base of the brain. His accounts of his treatment of Lady Conway from age of 12 through to her death at 48 are highly descriptive of his patient's battles with migraine.[10] His attempts at finding a cure for Lady Conway's attacks were fruitless.

The 18th century saw the widespread use of the "blister" for a variety of disorders, including headache. In the text *Bibliotheca Anatomica, Medica, Chirurgica, etc.*, published in 1712 in London, the blister is described and its ingredients itemized.[11] The blister consisted of an agent prepared from Spanish flies (*Cantharides*), which was placed on the skin, and "its crimony raises Blisters or Bladders." *Cantharis vesicatoria* — the blister beetle — was the agent, and the flies were dried and powdered. The compound was then mixed with "Leaven and Vinegar," and was to remain on the skin for at least 4 to 5 hours. Once the blisters had developed, they were lanced in order to release the "serous Humour." The blisters could remain on the patient for 1 to 2 days if the symptoms did not dissipate.[8] As late as 1911, Oppenheimer mentioned the use of a blister inserted into the tissues at the nape of the neck, and maintained there for several days.[12]

William Heberden (1710–1801), was Cambridge educated, and authored *Commentaries on the History and Cure of Diseases*. It is there that we see migraine with aura described as "hemicrania." Heberden also notes the accompanying gastrointestinal symptoms of a migraine attack. In addition to recommending the use of a blister and bloodletting, he suggests a pill made from aloe (*Mauritus hemp*) in combination with the African columbo root (*Jatorbiza columba*) or the East Indian columba wood (*Coscinium fenestratum*). Both of the latter contain columbin, which is an extremely bitter substance. He also suggests a nightly dose of pulverized myrrh, a bitter resin which derives from *Courmiphora aybssinica* — a tree found in Africa and the Arabian peninsula. Heberden also recognizes the need for general measures in treating headache, including the avoidance of noise, crowds, fatigue, and anxiety. He suggests the benefits of rest, quiet, warmth, and fresh air.

And now we return to the illusive bark of the willow tree. In a 1763 letter to the Earl of Macclesfield, the Reverend Edward Stone reported on his successful treatment of fever (probably related to malaria). The minister used 20 grains of the powdered willow bark in a dram of water every 4 hours.[13] It would be over 100 years before the salicylate powers of willow bark would evolve into the drug, aspirin.

In 1758 John Fordyce published *De Hemicrania*, a series of clinical observations based on the author's own migraine attacks.[14] For relief, Fordyce used *Valeriana sylvestris*, which he found very beneficial. One food item, hot buttered toast, was suggested as a headache trigger by Fordyce. This was also mentioned a century later by Edward Liveing (1843–1907), who noted that abstinence from butter could reduce the number of headache attacks.[15] For headache treatment, Liveing suggested the use of a combination of belladonna and the herb *Hyoscyamus niger* (henbane).

Fordyce's contemporary, John Fothergill (1712–1780), described his personal headache struggle in a series of reports. His focus was on the dietary triggers of headache, and he wrote the initial description of chocolate as precipitating headache. Although he did not relate heavy drinking as the cause of headaches, he felt that immoderate eating played a role. Recognizing the gastrointestinal symptoms associated with migraine, many 18th century physicians based treatment on purging. A common agent used for this process was calomel, which was later used in combination with soap containing aloe. Tissot, in 1790, suggested the use of vegetable bitters. Migraine and other disorders were treated by regular purgation as late as the 20th century.

T. Buzzard (1831–1919) believed that consuming cod liver oil would prevent migraine.[8] In 1858, J. Addington Symonds, Sr (1807–1871) presented a series of Goulstonian Lectures on migraine to the Royal College of Surgeons. He recommended the consumption of a glass of salty water 1 hour before breakfast.[16]

Emotional stress and intellectual strain were frequently cited triggers during the 19th century. William Osler (1849–1919) observed that excessive bookwork should be avoided in order to prevent migraine in children.[17] To treat emotional stress, a number of remedies were used, including the bromides, henbane, zinc, and valerian, alone or in combination.

A number of treatments were used throughout the 19th century. In 1858, Robert Bentley Todd recommended the use of potassium iodide.[8] *Cannabis indica* was suggested by Osler.[16] Gower's mixture, described by William Gowers (1845–1915), was a combination of liquid trinitrin, liquid strychnine, tincture of gelsemia, sodil bromide, hydrobrom dil, and aqueous chloroform.[18] This mixture was to be used three times daily, and its use continued into the 1940s. Patent medicines were very popular during this period. Bequerel's pills contained quinine sulfate, extract of digitalis, and extract of colchicum. A similar preparation was Debout's pills.

The efficacy of caffeine added to migraine preparations has been recognized.[19] During the 19th century *guarana* (Brazilian cocoa) was used, and was derived from the Amazonian climbing plant *Paulinne cupana* or *sorbilis*. *Guarana* contained three- to five-fold as much caffeine as ordinary coffee beans. Caffeine's efficacy in migraine treatment was questioned by Gowers. During that period, a proprietary preparation, Migrainin, was a popular headache remedy that contained caffeine, antipyrin, and citric acid. Oppenheimer, in his 1911 text, doubted its efficacy.[12] Bromocaffeine (guarana, hydrobromic acid, caffeine) was another remedy used at the end of the 19th century, but C.K. Mills, in 1898, warned of the risk of abuse of this agent.[20]

From the Middle Ages through the 20th century, the liver and the bile duct were thought to be the source of migraine. This led, in 1925, to the use of sodium taurocholate and sodium glycocholate (cholagoguic preparations) in migraine therapy. Their utilization was pre-empted by the introduction of phenobarbitone (Luminal®), which was recommended by Wilfred Harris.[8]

The search for an effective agent to produce analgesia was continuous. The Blue Pill (*pil. Hydrarg*) contained mercury, was given in large doses, and may have been continued after the acute attack had ended.[12] The severe and dangerous side effects of mercury ended its use for headache. As early as the 1st century AD, opium was given for pain relief, first noted by Dioscorides. It was utilized until the mid-19th century as laudanum for relief of migraine and other pain conditions. By that time, more effective synthetic analgesics were available. In 1805, morphine and codeine were isolated from opium. Morphine was used as a treatment for opium addiction, but its addictive qualities were then recognized. In 1874, heroin was synthesized from morphine by the English chemist C.R. Alder Wright, but the agent was not produced commercially until 1898 by the Bayer Pharmaceutical Company. The hope that heroin would be a remedy for morphine abuse was quickly dashed, as heroin addiction became one of society's greatest ills.

By the end of the 19th century analgesics were being marketed, including phenacetin, phenatone, antipyrine, antifebrin, and – the most efficient – amidopyrine. Because of its association with leukocytopenia, amidopyrine fell out of favor. At the same time Bayer was working on production of heroin, one of its chemists was trying to create an analgesic which would help his father with arthritic pain.

In 1828, a professor of pharmacology in Munich, Johann Andreas Buchner, isolated *salacin* from the extract of willow bark in water, and removed the tanins and other impurities.[21] In 1838 Rafaelle Piria, an Italian chemist working at the Sorbonne in Paris, divided salicin into a sugar component and an aromatic component. Piria then took the aromatic portion, which was the active substance, and transformed it into an acid – salicyclic acid. Two other researchers, Charles

Frederic Gerhardt and H. Von Gilm, both synthesized the product and produced a form of acetylsalicylic acid. Unfortunately, neither of their products were chemically pure, rendering the compound unstable. Industrial production of salicylic acid began in 1874, which reduced the price of the drug, thus making it widely available for treatment of rheumatic pain, headaches, and migraine.

The quest for a more refined version of salicylic acid was underway, with chemists recognizing the analgesic and antipyretic effects of the agent. In 1892, Wilhelm Siebel, at Bayer Pharmaceutical, worked on developing a form of salicylic acid with fewer side effects. His work produced Salol, which released salicyclic acid slowly and was associated with fewer side effects. However, in high doses Salol, which contained high levels of phenol, produced symptoms of poisoning. Searching for a non-toxic compoent, Siebel created Salophen, which resembled small crystalline flakes and was tasteless, odorless, and less toxic, with the same analgesic effect. Unfortunately, Salophen irritated mucous membranes.

During this period, Felix Hoffman (1868–1946), another Bayer chemist, hoped to produce an analgesic for his father who was suffering debilitating rheumatoid arthritis. The elder Hoffman had been prescribed sodium salicylate, which had a terrible taste and caused gastrointestinal complaints. On August 10, 1897, Hoffman was able to produce acetylsalicylic acid in a stable and 100% chemical form, as it did not contain any free salicylic acid. The compound developed by Hoffman maintained its therapeutic effect longer and was well tolerated. In 1898, Bayer marketed the new drug under a new name – aspirin. Bayer received its US patent on February 27, 1900, and a new age in pain relief was underway. Hoffman's aspirin discovery, at the time, was overshadowed by his synthesization of heroin on August 21, 1897. Bayer marketed heroin first. When physicians kept asking for more samples of heroin, the addiction issues became evident.

THE 20TH CENTURY ONWARDS

Headache medicine in the 20th century, and medicine in general, saw great progression. This was characterized by many physicians and scientists who fostered the growth of headache therapies. Thyroid extract in small doses was recommended by Kinnier Wilson in lieu of Gower's mixture.[22] For neuralgia, ammonium chloride was prescribed. This agent, which is a component of *Ens veneris*, was advocated for migraine by both Liveing and Symonds.[8]

United States of America

Critchley noted that the "modern treatment [of migraine] began with the introduction of ergot, and more particularly, of an ergotamine tartrate that could be taken orally".[8] In 1934, three separate articles described the use of ergotamine in migraine.[23–25] Research in ergotamine was also undertaken by one the most pre-eminent headache researchers, Harold G. Wolff (1892–1962). He authored a prominent textbook, *Headache and Other Head Pains*, in 1948, of which there have been eight editions.[26] His experiments in headache were conducted during the 1930s while he was Professor of Medicine (Neurology and Psychiatry) at Cornell University Medical College, and as a physician at New York Hospital, New York. Wolff's experiments focused on migraine mechanisms, clinical aspects, psychological aspects, and personality features. He conducted research that today would never be approved by Institutional Review Boards (IRBs), including microscopic observation of conjunctival vessels during the course of migraine episodes. During experiments, he invoked edema of the temporal artery during an attack in 12 patients. In other experiments, Wolff extracted fluid from tissues adjacent to the temporal arteries and then injected the fluid into the forearm skin, producing erythema with decrease of the pain threshold. This led to our understanding that a sterile inflammation occurred around the vessels involved in migraine. Prolonged attacks of migraine were due to this inflammation. Later researchers, including this author, deduced that the use of corticosteroids to reduce the inflammation had therapeutic value in the management of prolonged migraine.

For an acute headache, Wolff recommended the use of intramuscular ergotamine tartrate. The oral administration of this agent was not as effective. If ergotamine was not effective, Wolff suggested the use of aspirin, amidopyrine, and codeine, with or without a barbiturate or caffeine. The use of morphine was to be avoided because of the repetitive nature of migraine.

In the 1940s, a contemporary of Wolff, Arnold P. Friedman (1909–1990), founded the first headache clinic in the United States, at Montefiore Medical Center in the Bronx. He remained Director of the Montefiore Unit until the 1970s, when he moved to Tucson and became a Professor of Neurology at the University of Tucson. His work during the 1940s with returning soldiers who had sustained head injuries forms the basis of our knowledge of post-traumatic headaches.[27]

During the 1950s, Friedman helped with the development of the drug Fiorinal, which is a combination of aspirin, caffeine, and butalbital. Because butalbital is a barbiturate with addictive properties, this combined agent is not recommended for use in chronic headaches, including chronic migraine. In a 1958 interview, Dr Friedman indicated that for an ordinary headache he would recommend "coffee, an aspirin, and fresh air".[28] He served as chairman of the World Commission for the Study of Headache in 1958, was chairman of the headache sections of the American Medical Association and American Neurological Association, and served as President of the American Board of Psychiatrists and Neurologists, the American Association for the Study of Headache (now the American Headache Society), and the National Migraine Foundation (now the National Headache Foundation). His service to headache medicine included chairing the Ad Hoc Committee on the Classification of Headache, published in 1962, and the criteria for headache classification until the work of the International Headache Society in 1988.[29,30] The members of the Ad Hoc Committee included Knox H. Finley, John R. Graham, E. Charles Kunkle, Adrian M. Ostfeld, and Harold G. Wolff.

John R. Graham (1909−1990) worked as a Fellow under Harold G. Wolff, and became prominent in his own right as Chief of Medicine at Faulkner Hospital in 1950. At Faulkner, he founded the Headache Research Foundation, and remained its Director until 1986. With Wolff, he published on the mechanisms of ergotamine tartrate in migraine.[31] Other notable works included the use of corticosteroids in headache, and the role of methysergide (Sansert®) in headache. Methysergide remains, today, probably the most effective drug to use in episodic cluster headache if managed carefully. Its availability is limited in many countries.

Doctor Graham also worked on the development of a computer-based headache interview, and describing the headache that occurs in patients on renal dialysis.[32] His descriptions of the facial characteristics of cluster headache patients have been utilized in many textbooks on headache.[33] He wrote a series of headache profiles, which have been used for decades by physicians managing headache patients. Dr Graham served as President of the American Association for the Study of Headache, and was on the editorial boards of the journals *Headache* and *Cephalalgia*.

At the Mayo Clinic, while Wolff and Graham were conducting their research at Cornell, Bayard T. Horton (1895−1980) was treating patients with a variety of headaches. Horton is internationally recognized for his identification of two headache types: histaminic cephalgia (Horton's headache) and temporal arteritis (giant cell arteritis). E. Charles Kunkle renamed the former type "cluster headaches" in 1952.[34] Horton's work with histamine treatment dates to the early 1930s. Horton reported on 84 patients with "Bouts of pain from 2 to 20 times a week … They (the headaches) appeared and disappeared very quickly" and "65 patients obtained definite permanent relief for periods of from 2 to 18 months via treatment with twice-daily injection of histamine in increasing doses for 2 to 3 weeks".[35]

On June 8, 1959, at the annual meeting of the American Medical Association, a small group of interested physicians met and founded the American Association for the Study of Headache. The organization's purpose was delineated in an editorial in the first issue of the journal *Headache* in 1961:[36]

> to bring together men practicing in the various fields of medicine so that they may express their ideas and beliefs about various forms of headache and head pain. It is evident that one medical specialty will advocate a certain form of treatment for a particular type of headache. Another group of specialists, however, may stress an entirely different approach to the same type of problem. The American Association for the Study of Headache (AASH) provided a media to discuss such differences of opinion. The medical literature is cluttered with such variances of opinion. This is especially true in the case of the many forms of vascular headache.

The officers of the organization at the first annual meeting in St Louis in 1960 were Henry Ogden, President; Walter Alvarez, Vice President; Bayard T. Horton, Secretary; Robert E. Ryan, Sr, Treasurer; and George Waldbott, Moderator. At that meeting, they recommended the publication of their society's journal, *Headache*. Its initial co-editors were Doctors Ogden and Ryan. In the early years of the organization, through the late 1970s, the annual meetings of the AASH were held in conjunction with the annual convention of the American Medical Association. In 2000, the organization changed its name to the American Headache Society.

In 1963, the current author presented a poster, *The Masks of Depression*, at the annual meeting of the American Medical Association. I was approached by Lester Blumenthal, an officer of the AASH at that time, who asked me about depression and headache. The question intrigued me, and I went home and prepared a paper on "Depressive Headaches." My paper was accepted for presentation at the 1964 meeting of the AASH.[37] Before leaving that meeting, I was asked to serve on the Program Committee, and in June, 1965, was elected Secretary of the AASH, and remained on the Executive Board for the next 21 years. At the 1964 meeting, I met Donald Dalessio, of the Scripps Clinic in La Jolla, California. He was a fellow presenter, and in January 1965 he became the Editor of *Headache*. We would continue our professional collaboration for the next 35 years, serving as co-chairs of various headache courses, as well as co-writers/editors of *The Practicing Physician's Approach to the Headache Patient*, through its 5th edition, published in 1992.[38]

In 1970, I recognized the need for an organization for patients, and established the National Migraine Foundation (now the National Headache Foundation).

Since its founding, the NHF has been the premier organization for patients, their families, their healthcare providers, and the public. In addition to its educational activities, the NHF advocates for the headache patient with governmental agencies, insurance companies, and the pharmaceutical industry.

My initial interest in headache research was through my work on the tricyclic antidepressant, amitriptyline, in depressive comorbid conditions and, eventually, headache. Today, amitriptyline is the drug most frequently used in chronic pain, including headache. I had introduced clinical research into my practice early in the 1950s. Throughout my professional career I conducted research, including early studies on the use of propranolol.[39] With Jose Medina, we reported on the efficacy of indomethacin on cluster headache variant.[40]

The discovery of propranolol's efficacy in migraine was incidental. Rabkin and his group were using propranolol in a patient with cardiac disease who also had migraine headaches.[41] The patient reported a reduction in migraine attacks while being on propranolol. In 1977, I had the opportunity to testify at the FDA on the efficacy of this drug in the prophylaxis of migraine. The number of patients reviewed by the FDA's neurological committee was only 86, which included patients in my study and a small cohort in a study by Dr John Graham. On that small number, propranol received approval. The success of propranolol led to the research of other beta blockers in migraine prevention, and timolol became the second drug of this class to receive FDA approval for the indication of migraine prophylaxis.

Studies were conducted on the calcium channel blockers for migraine prophylaxis throughout the 1980s. Glen Solomon and his group reported on the efficacy of verapamil in migraine.[42] Studies have been conducted on several of the calcium channel blockers, including flunarizine, nimodipine, nifedipine, and diltiazem.

The 1990s and beyond saw the use of the anti-seizure medications in migraine prophylaxis. In 1988, the efficacy of divalproex sodium was reported by Sorenson.[43] During the 1990s, valproate was the only drug to receive FDA approval for migraine prophylaxis. Because of its success, investigations were conducted with topiramate, which eventually earned FDA approval.[44]

United Kingdom

In the United Kingdom, a number of physicians have impacted our knowledge of headache diagnosis and treatment. A contemporary of Wolff, Friedman, and Graham, Professor Macdonald Critchley (1900–1997) was a prominent neurologist, serving as President of the World Federation of Neurology. Headache was just one of his many interests, including the parietal lobes and aphasiology. He established a headache clinic at King's College Hospital in London, and was one of the founders of the British Migraine Trust. Professor Critchley was a prolific writer, and his latter two books are greatly treasured by this author: *The Divine Banquet of the Brain* and *The Citadel of the Senses and Other Essays*.[45,46] His work in neurology is edified by the following terms: Adie-Critchley syndrome – forced grasping and groping; and Klein-Levine-Critchley syndrome – hypersomnia and hyperphagia.

Edwin Bickerstaff (1920–2007) is forever linked to his description of basilar migraine, which is also known as Bickerstaff's syndrome or Bickerstaff's migraine.[47] Basilar migraine refers to a series of symptoms that typify a dysfunction of the posterior portion of the brain which is supplied by the vertebrobasilar artery system. The symptoms, which usually occur prior to the acute headache, include visual disturbances, tinnitus, hearing loss, diplopia, and vertigo. Ataxia is described as the most common symptom, and the patient may appear confused or disoriented. Slurring of speech may occur and the patient may be considered to be under the influence of a drug or alcohol.

Joseph N. Blau (1928–2010) was a consultant neurologist at the National Hospital for Neurology and Neurosurgery in London, and he later co-founded, with Marcia Wilkinson, the City of London Migraine Clinic. At this clinic, Dr Blau volunteered for 1 day weekly, as a consultant neurologist, for 30 years. Dr Blau was a prolific writer and a critical reviewer. He firmly believed that listening to the patient would provide the diagnosis, and was greatly focused on migraine precipitants. His textbook, *Migraine: Clinical, Therapeutic, Conceptual and Research Aspects*, was published in 1987.[48] He also wrote a book for patients, *The Headache and Migraine Handbook*.[49] He served as the honorary medical adviser for Migraine Action (formerly the British Migraine Association) from 1980 to 2007.

Marcia Wilkinson (1919–2013), similar to many physicians in headache medicine, was a migraine sufferer herself. In order to help her fellow migraine sufferers, she founded a migraine clinic at the Elizabeth Garrett Anderson Hospital in 1963. It was there that Dr Wilkinson focused on her early work on dietary migraine and the role of tyramine and other vasoactive amines, with her colleagues Dr Edda Hanington and Professor Eleanor Zeimis. She and her colleagues also reported on the use of clonidine in migraine patients who are sensitive to tyramine-containing foods.[50]

In 1970, the City Migraine Clinic was founded and Dr Wilkinson became its Medical Director. It was the first facility in the world where a patient could be evaluated and treated during an acute headache, and afforded the opportunity to study the efficacy of metoclopramide in providing pain relief, the promotion of gastric motility, and facilitation of the analgesic absorption. Because of the ever-increasing number of patients, the clinic was moved to Charterhouse Square and became the Princess Margaret Migraine Clinic in 1973. The Migraine Trust ended its management of the Clinic in 1979, and moved it to Charing Cross Hospital. At that time, with her colleague Dr J.N. Blau, Dr Wilkinson established the City of London Migraine Clinic in 1980 as an independent medical charity. In addition to their medical practice, Doctors Blau and Wilkinson conducted fundraising to keep the Clinic's doors open.

Dr Wilkinson was a founding member of the International Headache Society, and served as editor-in-chief of Cephalalgia from 1989 to 1992. In 2000, at the Migraine Trust meeting in London, she was the first recipient of the Elizabeth Garrett Anderson Award, which was presented to "a woman whose work has made an extraordinary contribution to relieving the burden of those afflicted by headache".[51]

Frank Clifford Rose (1926–2013) served for many years as Chairman of the Migraine Trust, and in 1965 was the first consultant neurologist at Charing Cross Hospital. Later he was affiliated with its Migraine Clinic, which was founded in 1974. He was a founding member of the European and International Headache Societies, and was Chair of the Research Committee on Migraine and Headache of the World Federation of Neurology. Under the patronage of Princess Margaret, who was a migraine sufferer, Dr Rose organized the Migraine Trust meetings, which occur every 2 years in London. He also served on the first Headache Classification committee. Twice, Dr Rose and his colleagues won the Wolff Award from the American Headache Society (1981, 1984). He edited over 70 books on headache, many based on symposia that he had organized.

Sir John Vane (1927–2004), a pharmacologist, was awarded the Nobel Prize in Physiology or Medicine in 1982 for his work on the mechanism of aspirin, and helped expand its use in cardiology and other diseases. His work included the development of two other drugs: the COX-2 inhibitors, which are used in pain and inflammation, and the ACE inhibitors used in hypertension and cardiac failure. He discovered, while working at the Wellcome Foundation, that aspirin blocks cyclooxygenase, which blocks production of the prostaglandins, which contribute to pain, swelling, and fever.

Australia

The use of antidepressants in migraine has been reported since the 1960s. James W. Lance, of Australia, with his colleague, Michael Anthony, reported on the use of monoamine oxidase inhibitors (MAOIs) for treatment of intractable migraine in 1969.[52] James Lance (1926–) has been a pre-eminent headache researcher. After attending medical school at the University of Sydney, he trained as a neurologist at the National Hospital in Queen Square in 1954, and then returned to Sydney. Taking the opportunity to pursue his research efforts, he worked at Massachusetts General Hospital in Boston in 1960. When he returned to Australia, he founded the Department of Neurology at the University of New South Wales. About this time, with D.A. Curran, he published an article on the treatment of chronic tension headache.[53]

Doctor Lance is the author of the seminal textbook Mechanisms and Management of Headache, which continued publishing through seven editions.[54] He served as President of the International Headache Society, and has received numerous awards from the American Headache Society, including the Wolff Award, the John R. Graham Lectureship, and the Arnold Friedman Lectureship.

Italy

The identification, isolation, and synthesis of the serum vasoconstrictor, serotonin, occurred between 1948 and 1953, and established the groundwork for the research and discovery of several agents in migraine therapy.[55] Serotonin was found to be released from the platelets during blood clotting. Its role in migraine was researched further during the 1950s and 1960s, including by Wolff.[56] During the search for a serotonin antagonist, methysergide (derived from lysergic acid) was synthesized. In 1959, the Italian researcher Federigo Sicuteri (1920–2003), published his findings on the use of methysergide in migraine.[57] Other studies corroborated his findings and noted the efficacy of methysergide over placebo in migraine.[58,59] Methysergide use in the prevention of both migraine and cluster headaches continued to increase until severe adverse effects were noted about 5 years after the initial efficacy reports. Continued use of methysergide was associated with fibrotic changes, including retroperitoneal fibrosis.[60]

In 1954, Professor Sicuteri became Director of the first active headache center in Europe, at the University of Florence. He and his colleagues conducted research on a large scale, and published many articles in professional journals. He founded the Italian Headache Society in 1976, and served as President of the International Headache Society. For his research, he was awarded the Wolff award. His work on serotonin receptors and methysergide presaged the development of the triptans.[61]

Scandinavia

The Scandinavians have contributed greatly to headache medicine. Karl Axel Ekbom (1907–1997), following training in Stockholm and Gothenburg, worked at the Serafimer Hospital in Stockholm, which for many years was the only neurology clinic in Sweden. Throughout his long career at Serafimer (1937–1958), Dr Ekbom was able to follow patients for extended periods. In 1945, in his doctoral thesis, he described restless legs syndrome (Willis-Ekbom syndrome). His clinical studies on cluster headache are appreciated by all physicians who have managed this difficult disorder. In 1947, he reported on the successful use of ergotamine in cluster headaches.[62]

His son, Karl Ekbom, continued his father's research into cluster headaches. In 1974, he published an article on the use of lithium in the management of chronic cluster headaches.[63] He was prompted to investigate this treatment method when Dr Ekbom noted an improvement in cluster headaches in a patient with manic-depressive psychosis who was treated with lithium.

Research into headache in children was greatly impacted by another Swedish physician, Bo Bille (1919–2001). As a young pediatrician, he began his investigation of recurrent headaches in children and early onset migraine in preschoolers. His epidemiological study on the subject began in 1955 in Uppsala, Sweden, and he published his findings in 1962.[64] Through questionnaires and parent participation, he was able to determine the rates of headache frequency in a normal population of school-aged children. These data have been confirmed in numerous follow-up studies. Doctor Bille also conducted a 6-year follow-up investigation of children presenting with migraine or migraine-like symptoms.[65] His final report of 40 years of following the original group of 73 children was published in 1997.[66] Doctor Bille stressed the differences in children and adults experiencing migraine, and the need to seek distinct treatments for each group.

Professor Ottar Sjaastad (1928–) was one of the founding members of the International Headache Society in 1981, and was the original editor of the journal *Cephalalgia*. Based in Trondheim, Norway, he has identified a number of headache entities, most prominently chronic paroxysmal hemicranias,[67] and other headache types responsive to indomethacin. He has also contributed to headache medicine through his descriptions of SUNCT.[68]

The contribution of Danish physicians to headache research has been enormous. In my role at the American Association for the Study of Headache, I was able to organize the first international headache conference with the help of Dr Donald Dalessio and a Danish physician, T. Dalsgaard-Nielsen (1896–1975), at Elsinore, Denmark, in June 1971. An international monograph was published, based on the proceedings.[69] Doctor Dalsgaard-Nielsen's work began during the 1940s [70] when he described the percutaneous nitroglycerin test for migraine. His work continued through his exceptional studies on the epidemiology of headache.[71]

The Copenhagen Acute Headache Clinic was opened in 1976 by Jes Olesen (1941–). Olesen and his colleagues, including Martin Lauritzen, are noted for their work on describing the spreading oligemia of migraine, as demonstrated with carotid angiography.[72] Doctor Olesen also headed the Headache Classification Committee of the International Headache Society, which formulated an extensive headache classification system.[73] With his colleagues, Peer Tfelt-Hansen and Michael Welch, Olesen has edited the extensive treatise, *The Headaches*.[74]

Recent Advances

In the 1970s, the scientists at Glaxo in the United Kingdom began investigation on trying to identifying the serotonin receptor type responsible for the beneficial effect of serotonin. The research team, led by Patrick P.A. Humphrey, discovered the serotonin receptor type 5-HT1B, which is mainly found in cranial rather than peripheral blood vessels. This discovery enabled them to design novel agonists which specifically stimulated these receptors, in order to produce selective vasoconstriction of the cranial vessels, which can become distended and inflamed. In 1988, the prototypical triptan, sumatriptan, was administered parenterally; it demonstrated efficacy and was well-tolerated in most patients.[75] In the US it became available in 1992, after publication of the results of two parallel-group trials in the acute treatment of migraine.[76] Originally available only for subcutaneous administration, other forms were eventually approved, including oral, intranasal, and intradermal.

At Merck's Neuroscience Center in the UK, Richard Hargreaves and his team discovered the triptan, rizatriptan.[77] Other triptans followed, including zolmitriptan and eleptriptan. The 1990s have been called the decade of the "triptan wars".[78]

With the newest classification of headache, an addition – chronic migraine – has expanded our nomenclature. Chronic migraine is defined as migraine headache occurring on 15 or more days per month for more than 3 months in the absence of medication overuse.[79] The use of onabotulinumtoxin A (Botox®) has been approved for use in the treatment of chronic migraine.[80]

CONCLUSION

The second half of the 20th century and the beginning of the 21st century have seen an enormous increase in knowledge and treatment modalities for migraine and headache. Starting in 1992, epidemological studies by Richard Lipton, Walter Stewart, and their colleagues have tremendously expanded our understanding of headaches and the recognition of the impact on patients' lives and society in general.[81] The National Headache Foundation is the sponsor of the American Migraine Prevalence and Prevention (AMPP) study, a longitudinal study of individuals with migraine. Data collection began in 2004 and continued through 2009. The survey instrument includes patient demographic characteristics, as well as symptom pattern and frequency data, so that cases can be classified into ICHD-III beta diagnostic categories. The AMPP study has provided many answers about headache patients, including their treatments, lives, and problems, and continues to offer insight into the management of the headache patient. It has generated dozens of articles and has been utilized by researchers to extend the frontiers of headache medicine.[82]

The understanding of headache has changed radically during this period. It is the hope of this author and editor-in-chief of this book that in the next 100 years, we will fulfill the vision of the National Headache Foundation of creating "A World Without Headache."

References

1. Diamond S, Franklin MA. *Headache Through the Ages*. Caddo, OK: Professional Communications; 2005.
2. Jack DB. One hundred years of aspirin. *Lancet*. 1997; 350:4327−4439.
3. Aretaeus. In: Adams F, ed. *The Extant Works of Aretaeus of Cappadocia*. London, UK: New Sydenham Society; 1856.
4. Critchley M. From Cappadocia to Queen Square. In: Smith R, ed. *Background to Migraine*. London, UK: Heinemann; 1967: 28−38.
5. Rosner F. Headache in the writings of Moses Maimonides and other Hebrew sages. *Headache*. 1993;33:315−319.
6. Rosner F. *The Medical Aphorisms of Moses Maimonides*. Haifa, Israel: Maimonides Research Institute; 1989:111
7. Fox M. *Illuminations of Hildegard of Bingen*. Santa Fe, NM: Bear & Company; 1985.
8. Critchley M. Byegone remedies for migraine. *Headache Q*. 1991; 2:171−176.
9. Knapp Jr RD. Reports from the past 2. *Headache*. 1963;3:112−122.
10. Critchley M. *Migraine: from Cappadocia to Queen Square. Background to Migraine*. London, UK: Heinemann; 1969.
11. *Bibliotheca, Anatomica, Medica, Chirurgica, Etc.*, Vol. 2. London; 1712:351−369.
12. Oppenheimer H. *Textbook of Nervous Diseases* [Bruce A, Trans.]. Vol. 2, 5th ed. Edinburgh, UK: Otto Schulze & Co; 1911.
13. Jack DB. One hundred years of aspirin. *Lancet*. 1997; 350:437−439.
14. Fordyce J, Balguy C. *De Hemicrania*. London, UK: *apud* D. Wilson & T. Durham; 1758.
15. Liveing E. *On Megrim, Sich Headache and Some Allied Disorders*. London, UK: J & A Churchill; 1873.
16. Symonds JA. Goulstonian lectures on "headache." *Med Times Gas*. 1858;498.
17. Osler W. *Principles and Practice of Medicine*. New York, NY: D. Appleton & Co.; 1912.
18. Gowers WR. *A Manual of Diseases of the Nervous System*. Vol. 2. London, UK: D. J&A. Churchill; 1888.
19. Diamond S, Migliardi JR, Armellino JJ, Friedman M, Gillings DB, Beaver WT. Caffeine in tension headache. *Clin Pharmacol Ther*. 1998;64:465−466. Letter to the Editor − Correction of: Diamond S, Migliardi JR, Armellino JJ, Friedman M, Gillings DB, Beaver WT. Caffeine as an analgesic adjuvant in tension headache. *Clin Pharmacol Ther*. 1994;56:576−586.
20. Mills CK. *The Nervous System and Its Diseases*. Philadelphia, PA: J.R. Lippincott Co.; 1898.
21. Zundorf U. *100 Years of Aspirin*. Leverkuesen, Germany: Bayer AG; 1998.
22. Wilson SAK. *Neurology*. Vol. 2. London, UK: Edward Arnold & Co.; 1940.
23. Brock S, O'Sullivan M, Young D. The effect of non-sedative drugs and other measures in migraine, with special reference to ergotamine tartrate. *Am J Med Sci*. 1934;188:253−260.
24. Lennox WG. The use of ergotamine tartrate in migraine. *N Engl J Med*. 1934;210:1061−1065.
25. Logan AH, Allen EV. The treatment of migraine with ergotamine tartrate. *Proc Mayo Clin*. 1934;9:585−588.
26. Wolff HG. *Headache and Other Head Pains*. New York, NY: Oxford University Press; 1948.
27. Brenner C, Friedman AP, Merritt HH, Denny-Brown D. Post-traumatic headache. *J Neurosurg*. 1944;1:379−392.
28. Narvaez AA. Dr Arnold P. Friedman, 81, dies; authority on migraine headaches. <www.nytimes.com/1990/09/20/obituaries/dr-arnold-p-friedman-81-dies-authority-on-migraine-headaches.html?pagewanted=print>; Accessed 20.06.14.
29. Ad Hoc Committee on Classification of Headache. Classification of headache. *JAMA*. 1962;179:717−718.
30. International Headache Society. Classification and diagnostic criteria of headache disorders, cranial neuralgias and facial pain. *Cephalalgia*. 1988;8(suppl 7):4−96.
31. Graham JR, Wolff HG. Mechanism of migraine headache and action of ergotamine tartrate. *Arch Neurol Psychiatry*. 1938;39:737−763.
32. Dalessio DJ. Obituary: John R. Graham. *Cephalalgia*. 1990;10:156.
33. Graham JR. Cluster headache. *Postgrad Med*. 1974;56:181−185.
34. Kunkle EC, Pfieffer JB, Wilhoit WM, Hamrick LW. Recurrent brief headaches in "cluster" pattern. *Trans Am Neurol Assoc*. 1952;77:240−241.
35. Horton BT, MacLean AR, Craig WM. A new syndrome of vascular headache: results of treatment with histamine: preliminary report. *Proc Staff Meet Mayo Clin*. 1939;14:257−260.
36. Editorial. *Headache* 1961;1:1.
37. Diamond S. Depressive headaches. *Headache*. 1964;4:255−259.
38. Diamond S, Dalessio DJ. *The Practicing Physician's Approach to the Headache Patient*. 5th ed. Baltimore, MD: Williams & Wilkins; 1992.
39. Diamond S, Medina JL. Double-blind trials of propranolol for migraine prophylaxis. *Headache*. 1976;16:24−27.

40. Diamond S, Medina JL. Cluster headache variant: the spectrum of a new syndrome and its response to indomethacin. *Arch Neurol.* 1981;38:705–709.
41. Rabkin R, Stables DP, Levin NW, Suzman MM, et al. The prophylactic value of propranolol in angina pectoris. *Am J Cardiol.* 1966;18:370–383.
42. Solomon GD, Griffith Steel MC, Spaccavento LJ. Verapamil prophylaxis of migraine. *JAMA.* 1983;250:2500–2502.
43. Sorenson KV. Valproate: a new drug in migraine prophylaxis. *Acta Neurol Scand.* 1988;78:345–348.
44. Brandes JL, Saper JR, Diamond M, Couch J, Lewis DW, Schmitt J, et al., for the MIGR-002 Study Group. Topiramate for migraine prevention: a randomized controlled trial. *JAMA.* 2004;291:965–973.
45. Critchley M. *The Divine Banquet of the Brain.* New York, NY: Raven Press; 1979.
46. Critchley M. *The Citadel of the Senses and Other Essays.* New York, NY: Raven Press; 1986.
47. Bickerstaff ER. Basilar artery migraine. *Lancet.* 1961;1:15–17.
48. Blau JN, ed. *Migraine: Clinical, Therapeutic, Conceptual and Research Aspects.* London, UK: Chapman and Hall; 1987.
49. Blau JN. *The Headache and Migraine Handbook.* London, UK: Corgi; 1986.
50. Wilkinson M, Neylan C, Rowsell AR. Clonidine in the treatment of migraine at the City Migraine Clinic in patients selected with tyramine. In: Dalessio DJ, et al., eds. *Proceedings, International Headache Symposium.* Copenhagen, Denmark: Sandoz; 1971:219–221.
51. The Migraine Trust. Obituary: Dr Marcia Wilkinson. <http://www.migrainetrust.org/news-obituary-dr-marcia-wilkinson-16585>; Accessed 20.06.14.
52. Anthony M, Lance JW. Monoamine oxidase inhibition in the treatment of migraine. *Arch Neurol.* 1969;21:263–268.
53. Lance JW, Curran DA. Treatment of chronic tension headache. *Lancet.* 1964;1:1235–1239.
54. Lance JW, Goadsby PJ. *Mechanism and Management of Headache.* 7th ed. Philadelphia, PA: Elsevier; 2005.
55. Rapport MM, Green AA, Page H. Partial purification of the vasoconstrictor in beef serum. *J Biol Chem.* 1948;174:735–738.
56. Wolff HG, Ostfeld AM, Chapman LF, Goodell H. Studies in headache: a summary of evidence implicating a locally active chemical agent in migraine. *Trans Am Neurol Assoc.* 1956:35–36 [81st meeting].
57. Sicuteri F. Prophylactic and therapeutic properties of 1-methyllysergic acid butanolamide in migraine. *Int Arch Allergy Appl Immunol.* 1959;15:300–307.
58. Southwell N, Williams JD, MacKenzie I. Methysergide in the prophylaxis of migraine. *Lancet.* 1964;1:523–524.
59. Pedersen E, Møller CE. Methysergide in migraine prophylaxis. *Clin Pharmacol Ther.* 1966;7:520–526.
60. Graham JR, Suby H, LeCompte PR, Sadowsky NL. Fibrotic disorders associated with methysergide therapy for headache. *N Engl J Med.* 1966;274:359–368.
61. Fanciullacci M. In memory of Federigo Sicuteri (1920–2003), a headache medicine pioneer. *Cephalalgia.* 2004;24:1090–1091.
62. Ekbom KA. Ergotamine tartrate orally in Horton's "Histaminic cephalgia" (also called Harris's "ciliary neuralgia"). *Acta Psychiatr Scand.* 1947;45:106–113.
63. Ekbom K. Liyium vid kroniska symptom av cluster headache. *Opusc Med.* 1974;19:148–156.
64. Bille B. Migraine in schoolchildren. *Acta Paed Scand.* 1962;51:1–151.
65. Bille B. Migraine in childhood and its prognosis. *Cephalalgia.* 1981;1:71–75.
66. Bille B. A 40-year follow-up of school children with migraine. *Cephalalgia.* 1997;17:488–491 [discussion 487].
67. Sjaastad O. Chronic paroxysmal hemicrania. *Ups J Med Sci.* 1980;32(suppl):27–33.
68. Sjaastad O, Zhao JM, Kruszewski P, Stovner LJ. Short-lasting unilateral neuralgiform headache attacks with conjunctival injection, tearing, etc. (SUNCT): III. Another Norwegian case. *Headache.* 1991;31:175–177.
69. Dalessio DJ, Dalsgaard-Nielsen T, Diamond S, eds. *Proceedings of the International Headache Symposium, Elsinore, Denmark.* Copenhagen, Denmark: Sandoz; 1971.
70. Dalsgaard-Nielsen T. Results of some diagnostic tests in different forms of headache. *Acta Psychiatr Neurol.* 1949;24:391–402.
71. Dalsgaard-Nielsen T, Ulrich J. Prevalence and heredity of migraine and migrainoid headaches among 461 Danish doctors. *Headache.* 1973;12:168–172.
72. Olesen J, Larsen B, Lauritzen M. Focal hperemia followed by spreading oligemia and impaired activation of rCBF in classic migraine. *Ann Neurol.* 1981;9:344–352.
73. Headache Classification Committee of the International Headache Society. Classification and diagnostic criteria for headache disorders, cranial neuralgias and facial pain. *Cephalalgia.* 1988;8(suppl 7):1–96.
74. Olesen J, Tfelt-Hansen P, Welch KMA, eds. *The Headaches.* New York, NY: Raven Press; 1993.
75. Humphrey PPA, Feniuk W, Perren MJ, Connor HE. GR43175, a selective agonist for the 5-HT1-like receptor in dog isolated saphenous vein. *Br J Pharmacol.* 1988;94:1123–1132.
76. Cady RK, Wendt JK, Kirchner JR, Sargent JD, Rothrock JF, Skaggs Jr H. Treatment of acute migraine with subcutaneous sumatriptan. *JAMA.* 1991;256:2831–2835.
77. Longmore J, Hargreaves RJ, Boulanger CM, Brown MJ, Desta B, Ferro A, et al. Comparison of the vasoconstrictor properties of the 5-HT1D-receptor agonists rizatriptan (MK-462) and sumatriptan in human isolated coronary artery: outcome of two independent studies using different experimental protocols. *Funct Neurol.* 1997;12:3–9.
78. Solomon S, Diamond S, Mathew N, Loder E. American headache through the decades: 1950–2008. *Headache.* 2008;48:671–677.
79. IHS Classification ICHD-II. <http://ihs-classification.org/en/02_klassifikation/02_teil1/01.05.01_migraine.html>; Accessed 23.09.14.
80. Cady RK. OnabotulinumtoxinA (botulinum toxin type-A) in the prevention of migraine. *Expert Opin Biol Ther.* 2010;10:289–298.
81. Stewart WF, Lipton RB, Celentano DD, Reed ML. Prevalence of migraine headache in the United States. Relation to age, income, race, and other sociodemographic factors. *JAMA.* 1992;267:64–69.
82. Lipton RB, Serrano D, Pavlovic JM, Manack AN, Reed ML, Turkel CC, et al. Improving the classification of migraine subtypes: an empirical approach based on factor mixture models in the American Migraine Prevalence and Prevention (AMPP) study. *Headache.* 2014;54:830–849.

2

Classification, Mechanism, Biochemistry, and Genetics of Headache

Andrew H. Ahn

University of Florida College of Medicine, Gainsville, Florida, USA

CLASSIFICATION

Various headache classifications have been created over the past five decades. In 2013 the International Headache Society (IHS) updated its classification, which is particularly useful for the researcher.[1] This classification is used throughout the world. For the practicing clinician, a simpler classification has been established (Box 2.1).[2] Although it does not have the scientific validity of the IHS classification, it is very practical for utilization in a busy clinical practice.

The descriptive *sensory qualities* of headache pain are elemental features of the diagnosis and classification of the headache disorders. The throbbing, pounding quality of migraine pain, the sustained, sharp ice-pick quality of cluster headache, and even the paroxysmal and explosive electrical qualities of trigeminal neuralgia pain all bear striking witness to how these descriptive pain sensory qualities diagnose and even define the disorder. That is, in addition to being essential features of the diagnosis of these individual disorders, these qualities are fundamental, even axiomatic, features of their respective theories of headache pathophysiology.

One such axiomatic view is the widely held notion that the throbbing pain of migraine arises from the sensory experience of pain-sensory afferents innervating blood vessels. This view is the basis for references to migraine as a so-called "vascular headache," and to explicit references to a "vascular theory" of migraine. In fact, this view towards the basis of throbbing pain is an ancient one, embraced even by Aristotle,[3] and likely endures until today because of our nearly universal subjective experiences of throbbing pain, which reinforce the conviction that pulsatile, rhythmic qualities of throbbing pain must somehow be related to the rhythmic flow of blood through the affected painful area.

One prominent feature of this theory was pioneered by Wolff to associate the vasoconstrictor action of ergotamine with its migraine-abortive properties.[4] Another feature is the descriptions by Penfield and McNaughton,[5] who described the referral patterns of head pain from stimulation of the blood vessels associated with the dura. The vascular theory was thus a central preoccupation of thinking on migraine for the remainder of the 20th century, revealing further distinctions between extracranial versus intracranial arteries and venous sinuses as a potential source of that pain.[6–8]

The view that pain elicited by the dilation of cranial vessels was the source of migraine pain, and that vasoconstriction was the key to the reversal of migraine, led to the discovery that the selective constriction of cranial vessels (and not those of the cardiac arteries) by the serotonin sub-selective agonist sumatriptan at serotonin 1B, 1D, and 1F receptors had selective anti-migraine properties. However, over the years, subsequent experiments and a growing body of evidence for the anti-migraine actions of the triptans at multiple sites within the central nervous system [9] left the craniovascular theory at odds with the prevailing view.[10] In addition, Strassman and Levy reviewed the available literature and concluded that there is in fact no evidence supporting that physiological changes in vessel caliber activate pain-sensory afferents in a manner consistent with the rhythmic or phasic activation of pain-sensory afferents.[11]

Furthermore, two recent observations leave little doubt that the original vascular theory has serious flaws. The fundamental premise of this theory is that craniovascular dilation is the cause of migraine pain, so it implies that cranial vessels are dilated during a migraine attack. Amin and colleagues, using high

DOI: http://dx.doi.org/10.1016/B978-0-12-800901-7.00002-1

BOX 2.1

Vascular Headache

Migraine
    With aura (classic)
    Without aura (common)
    Complicated
        Hemiplegic
        Ophthalmoplegic
        Basilar (migraine with brainstem aura)
    Chronic migraine
Cluster
Toxic vascular
Hypertensive

Tension-Type (Muscle Contraction) Headache

Depressive equivalents and conversion reactions
Chronic anxiety states
Cervical osteoarthritis
Chronic myositis
Chronic daily headache

Traction and Inflammatory Headaches

Mass lesions
Diseases of the eye, ear, nose, throat, and teeth
Arteritis, phlebitis, and cranial neuralgias
Occlusive vascular disease
Atypical facial pain
Temporomandibular disease

Adapted from Solomon.[2]

---

resolution MRI imaging of cranial blood vessels, could not demonstrate these changes in blood-vessel caliber during the ictal migraine period.[12] The study probed this theory thoroughly, looking at a range of intracranial and extracranial vessels, further testing elaborations of the craniovascular theory that hypothesized a selective dilation of craniovascular afferents and constriction by migraine drugs, such as sumatriptan.[13] One limitation of this approach is that MRI can only determine the caliber of blood flow, but does not as easily determine the thickness of the vessel wall or the forces on the vessel, and of course has limited temporal resolution for the amplitude of pulsatile flow.

Examining the vascular theory explicitly and at a higher resolution temporal scale, the theory also predicts that the *rate and rhythm* of throbbing pain, directly or with a delay, is related to arterial pulsations. Direct observations of this relationship were examined in a systematic way for migraine pain[14] and acute dental pain.[15] In both cases, the general psychophysical characteristic of the throbbing rhythm was too slow to be related to systole, at around 40–50 beats per minute, whereas arterial pulse in these subjects was in the usual physiological range, from 70 to 80 beats per minute.[14] Moreover, these "bedside" observations of people with migraine pain were confirmed and extended in a population of subjects with dental pain, whose simultaneous recordings of arterial pulse and psychophysical reports of their throbbing rhythm were further examined to formally exclude any possible relationship between these two rhythms.[15]

A further explanation of the throbbing percept still remains to be elucidated. However, the exclusion of heart rate as a correlate of throbbing events also excludes the pulsations of cerebrospinal fluid and intracranial pressure, as these pulsations do correspond to heart rate, and the statistical properties of throbbing of both migraine and dental pain both exclude the possibility that throbbing refers to pulsations of CSF pressure. A more plausible explanation would be that the throbbing rhythm refers to an inherent property of pain processing within the brain, such as was observed in a further patient with a persistent throbbing experience whose rhythm corresponded to the fluctuations of alpha power.[16]

One feature in the classification of migraine that will require further elucidation is the progression from episodic migraine to chronic migraine. The classification embraced by the International Headache Society is based simply on the number of headache days per month,[17] which does not capture the many biological considerations that are clearly a part of a continuum of progression from episodic to chronic, and also includes a large degree of natural history and variation from month to month.[18] The reader is urged to become familiar with the considerations that are relevant to this debate.[19]

MECHANISMS OF MIGRAINE-ASSOCIATED SYMPTOMS

An inclusive understanding of the pathophysiology of migraine should account for the associated symptoms such as photophobia. In fact, there has been remarkable progress in understanding mechanisms underlying photophobia, though these discoveries raise as many new questions as they would seem to answer.[20,21]

The term "photophobia," from a linguistic point of view, implies a fear or aversion to light. However, the clinical expression of photophobia in migraine generally emphasizes an intolerance to light,[20] such that even modest or low ambient illumination levels produce an uncomfortable, even painful experience that can also exacerbate migraine pain. Some patients with migraine, though, will deny pain plays any part in the experience at all,[22] stating that they merely prefer to be in a darkened room during a migraine attack. However, this preference for the dark during a migraine attack is also recognized as a form of photophobia.

It is important to bear in mind that photophobia is also a term associated with a range of primary ophthalmologic disorders, and is present in a range of medical conditions other than migraine.[20] Especially in ophthalmologic disorders, pain is an important part of the clinical condition to which the term refers, where a range of refractive and ocular pressure disorders, and even some medications, can underlie exquisite light-induced pain.[20] Recent work to elucidate these mechanisms has gone far to describe these mechanisms as being perhaps altogether distinct or complementary to those active in migraine.

Through an elegant series of electrophysiological and pharmacological experiments performed in anesthetized rat, Bereiter and colleagues showed that very bright light can evoke pain-associated activity of trigeminal afferents arising from within the orbit and passing through pain-responsive areas of the trigeminal nucleus.[23] These responses appear to be physiologically and pharmacologically related to the autonomic regulation of blood flow within the orbit, and mediated by sympathetic regulation through the posterior hypothalamus.[24]

Functional MRI recordings of a patient with a corneal abrasion and photophobia also point to the activation of primary somatosensory areas of the brain by light, and resonate extremely well with these findings.[25] These sensory responses offer a plausible explanation for how disorders within the orbit are associated with the disruption of the tight regulation of light, and can produce a prompt and prominent experience of pain. This close connection between light and primary trigeminal sensory activation among pre-chiasmal disorders was suggested by Digre and Brennan to be best referred to as "photo-oculodynia".[20]

On the other hand, recent work by Burstein and colleagues showed that painful stimuli to the head can also interact with exposure to light — independently of the above trigeminal circuit — through a purely visual sensory pathway, by retinal ganglion cell input to the pulvinar nucleus of the thalamus.[26] In these experiments, Noseda and colleagues suggested that the neurochemical basis of this pain-responsive circuit lies within multimodal sensory inputs within the thalamus and implicated a role for the non-image-forming light sensors within the retina mediated by melanopsin. To add to complexity, Matynia and Gorin identified a further non-melanopsin-dependent pathway.[27]

Long before the present work, Drummond used a classical psychophysical approach to quantify these experiences of photophobia in migraine, and demonstrated that a range of light intensities can elicit unpleasant perceptions. He was thus able to operationally define the concept of a "pain threshold" for light-induced migraine pain,[28] and to demonstrate a bidirectional interaction between pain and photophobia: light can induce sensations of pain, but painful stimulation of the forehead can also reduce the threshold for light sensitivity.[29] The discovery of a purely visual pathway to pain modulation thus offers a critical physiological explanation for Drummond's observed psychophysical interactions between light and pain, and also offers new insights into the processing of migraine-related pain in the brain.[26] Even in the absence of headache, migraineurs can have neurophysiological changes in visual processing that suggest an overall hypersensitivity, or lack of normal suppression of cortical activity.[30] Friedman suggested the term "photoallodynia" could be used in the specific case of migraine-induced sensitivity to light (setting apart the ophthalmologic causes of photophobia), which thus also captures these new insights on the convergence of light and the experience of pain from non-noxious stimuli (allodynia) through these central nociceptive pathways.[31]

Another perspective on photophobia highlights its behavioral and neurochemical basis through the signaling of the neuropeptide calcitonin gene-related peptide (CGRP). Recober, Russo, and colleagues showed that mutant mice with enhanced signaling by CGRP were found to have a greatly enhanced preference to avoid light.[32] However, because CGRP is also intimately associated with the signaling of pain by peripheral sensory afferents, nociceptive reflexes were also enhanced in this mouse.[33] Thus, the mutant mice in this experimental paradigm do not present the opportunity to clearly distinguish between the somatosensory and visual sensory aspects of these light-avoidant behaviors. However, these experiments clearly indicate the eloquent role that CGRP plays in the full range of sensory and behavior changes associated with migraine, and provide a valuable animal model of the pathophysiology of migraine.[34]

Yet another line of work yields further consideration as to whether photophobia could represent the unmasking of a primitive, instinctual behavioral reflex.

In 1927, Crozier and Pincus observed that neonatal rats, even prior to the opening of their eyes, turn away from a localized light source, which he termed "negative phototaxis."[35] Interestingly, this behavior occurs at a time when there are no image-forming photoreceptors functioning within the retina, which, Copenhagen and colleagues inferred from mutant mouse lacking a melanopsin gene, depends critically on the existing retinal ganglion cells expressing the non-image-forming photodetector melanopsin.[36]

Further work revealed that these stereotypical light-avoidant behaviors are not associated with the activation of the pain sensory pathways within the trigeminal nucleus; neither did it elicit the broadband vocalizations associated with other forms of pain, and thus did not show evidence for overt nociceptive activation.[37] However, light did activate ultrasonic vocalizations resembling distress calls, and is associated with the phosphorylation of extracellular related kinase (ERK) in the central amygdala[37] — a biochemical signal associated with pain-related behaviors in neurons within the central amygdala, and whose activity is associated with the negative and aversive aspects of pain.[38–40] In other words, neurochemical responses to light in neonatal mice suggest a functional distinction between the detection of pain within primary somatosensory areas and the processing of aversive affective experiences within the amygdala.

How the significance of these responses in neonatal mice should be extended to humans with migraine is at this point a matter of speculation. However, from a neurological perspective, degenerative disorders and lesions within the brain are often seen to unmask latent or evolutionarily conserved reflexes (such as the Babinski, grasp, and rooting reflexes). The expression of photophobia in adults with migraine could thus represent a transient neurologic dysfunction of normal, positive inclinations toward light, resulting in the unmasking of more primitive avoidant behavioral reflexes that, from an evolutionary viewpoint, could have conferred a behavioral advantage for the newborn.

The clinical expression of photophobia in migraine has the potential to correspond to any combination of these neurobiological mechanisms. Although ocular disorders undoubtedly activate local trigeminal nociceptive responses, they will clearly also be processed within the brain as an aversive experience. Moreover, these recent discoveries reinforce the sensible clinical view that disability and discomfort produced by migrainous photophobia can also be ameliorated in part by addressing one of the many potential comorbid ocular or medical conditions that may also be present, for which the reader is urged to refer to the review by Digre and Brennan.[20]

MECHANISMS OF MIGRAINE TRIGGERS AND RISK FACTORS

Recent brain-imaging data suggest that the hypothalamus may have a critical role in the early pathophysiology of migraine.[41] This study enlisted people with migraine who also have a premonition, or early warning symptom, prior to the onset of their attacks. Those with these so-called "premonitory symptoms" often note changes in mood or wakefulness, increased urination, yawning, or other changes that again imply the role of the hypothalamus in the earliest phase of the migraine attack. In order to approach the premonitory phase experimentally, the migraine was triggered with nitroglycerin.[42] The nitric oxide donor nitroglycerin has over the years gained wide acceptance as an experimental model for triggering migraine-like headache in humans,[43] and is broadly recognized for its ability to trigger a broad range of migraine-like phenomena in animal models of headache.[44] Here again, the specific activation of blood flow in the hypothalamus during this early phase of the migraine attack, before the onset of headache, implicates this important self-regulatory region of the brain for the initiation of the earliest aspects of the migraine attack.

In fact, other well-established trigger factors for migraine also implicate a role for the hypothalamus in initiating a migraine attack, such as acute psychological stress, a skipped meal, overexertion with overheating, and lack of sleep. For many with migraine, these factors are not absolute triggers (meaning that a migraine follows them every time) but are merely associations. More often than not, however, these factors can be thought of as cumulative physiological challenges that require activity by the hypothalamus to normalize the physiological stresses that they represent.[45,46]

This association between migraine triggers and the hypothalamus is further supported by recent studies suggesting that the hypothalamus can be a key neurochemical regulator of pain.[46] Among many examples, some of which may have implications for therapy, one recent line of evidence has shown that an antagonist to orexin, a peptide hormone associated with the regulation of feeding behaviors, can also suppress physiological responses to sensory stimulation to the head.[47]

In addition, oxytocin, a neuropeptide associated with prosocial and nurturing behaviors in mammals,[48,49] is thought to act on another hypothalamic regulator of blood pressure: the vasopressin receptor, which geneticist Jeffrey Mogil and colleagues linked to a genetic variant in the human population that can change responses to pain in men presented with a high-stress situation.[50,51] Again, interestingly, early testing shows that oxytocin could abort a migraine attack.[52]

Among other neurotransmitters of the hypothalamus that have also caught the attention of those who study migraine are the modulation of pain-related responses in the trigeminal nucleus and the thalamus from the dopaminergic cells group A11/A13 of the posterior hypothalamus,[53,54] the sleep-related hormone melatonin,[55] and the pituitary adenylate cyclase-activating polypeptide (PACAP).[56]

There are several lines of evidence that support an association between migraine and yet another important risk factor — exercise.[57] These lines of evidence are indirect, but they do offer convergent views linking migraine to a modifiable risk factor. First, migraine is indirectly associated with a sedentary lifestyle in epidemiological studies of populations with migraine, such as in the American Migraine Prevalence and Prevention (AMPP) study, a longitudinal population-based study that surveyed a population of 120,000 US households and followed >10,000 migraine sufferers annually;[58] the frequency of common comorbidities for adults with chronic migraine (CM) and episodic migraine (EM) in this large population-based sample indicated a strong association of a range of comorbid conditions in those with CM compared to EM.

Another supportive line of evidence derives from an epidemiological study, but goes one step further. In this cross-sectional survey of 92,566 of the adult inhabitants of Nord-Trøndelag county in Norway — HUNT2 [59] — 46,648 answered further questions about the frequency, duration, and intensity of their physical activity. The study showed, remarkably, a "dose-dependent" association between low physical activity and increasing risk and greater frequency of both migrainous and non-migrainous headache. Varkey and colleagues then created an exercise intervention and implemented a prospective randomized controlled trial of adults with two to eight migraine attacks per month, among those who were not already exercising regularly.[60] These 91 subjects were randomized into three arms of a study, consisting of home relaxation, exercise, or medical therapy with topiramate at up to 100 mg twice a day. All three arms of the study produced a modest but statistically significant reduction in headache frequency, though limitations of the study included the fact that it was in essence an "open-label" study that did not control for subject expectation and observer bias, and that exercise was not shown to separate itself from a control arm that was predicted to have no benefit.

The available data are thus limited, showing little evidence for a significant reduction in attack frequency or duration due to exercise, but in aggregate may support an interesting hypothesis that exercise (or possibly just the weight loss) may produce an overall reduction in pain intensity.[61,62] This finding, that exercise may preferentially reduce migraine pain intensity more effectively than migraine frequency or duration, could more precisely focus our attention on the mechanisms of the modulation of pain by exercise. From a neurochemical perspective, exercise would be well poised to be an effective modulator of the neural systems engaged by pain, including both the somatosensory detection of tissue injury— nociception — as well as the affective, cognitive, and motivational systems that are engaged by the injury or potential injury — the experience.

On the one hand, the data regarding the effects of exercise on nociceptive processing are complex and conflicting. Indeed, there is a range of exercise-induced headaches, including migraine. Staud, as well as several other studies, has shown that strenuous exercise has an anti-nociceptive effect in otherwise healthy subjects, but that it may enhance peripheral and central pain in those with fibromyalgia.[63] However, in those with fibromyalgia, which is often comorbid with migraine, graded exercise therapy, involving increasing periods of regular cardiovascular exercise, is broadly accepted as the most effective treatment for this condition. One large hurdle to its implementation is that the fear of pain and fear of movement in some patients is a major barrier in participating.[64]

On the other hand, recent studies suggest that exercise can indeed have an important role in the modulation of pain processing from an affective-motivational perspective though the activation of endogenous cannabinoid signaling [65] in an intensity-dependent manner.[66] The initial studies on this subject focused on the "runner's high," referring to the sense of well-being and perhaps even euphoria associated with prolonged cardiovascular activity. Traditionally thought to be due to activation of endogenous opioids,[67] the activation of opioid receptor binding activity within the brain has been recently demonstrated, showing that there are likely complex changes in opioid signaling in the brain that take place with both pain and exercise.[68]

However, another recent line of investigation explored an alternate hypothesis: that human endurance exercise has intrinsically rewarding properties, possibly even hedonic properties, through the production of natural ligands of the endogenous cannabinoid (CB1 and CB2) receptors. The two best-recognized so-called endocannabinoids (eCBs) are anandamide (AEA) and 2-arachidonylglycerol (2-AG).

Because moderate-intensity exercise increases plasma levels of highly lipophilic eCBs [69] that readily cross the blood–brain barrier,[70] they are well suited to the modulation of brain reward centers through the inhibition of GABA-ergic terminals in the mesolimbic dopamine system via CB1 receptors.[71] The resulting increase in dopamine signaling to reward centers such as the nucleus accumbens[72] is sufficient to stimulate reward-seeking behaviors in animal models.[73]

Cannabinoids are thought to exert analgesic effects in the nervous system, both by activating a descending modulation of pain,[74] and by the inhibition of sensory afferents via CB1 receptors.[75] Accordingly, the modulation of pain transmission by activated CB1 receptors in animal models of trigeminal nociception also indicates a potential role for the selective cannabinoid receptor, not only through the activation of descending modulators of C-fiber activation[76] but also through the selective activation of CB1 receptors in dural trigeminovascular nociceptive neurons.[77] There is recent evidence that activation of the CB2 receptor in the periphery may also be part of an effective anti-nociceptive pathway.[78] It is thus possible that exercise-induced release of endogenous cannabinoids may help to suppress exercise-related discomfort and sustain greater endurance and higher performance during exercise, and perhaps may even help to alleviate pain symptoms for those with migraine. However, the development of selective cannabinoid receptor agonists in the management of migraine is likely to be stonewalled by their hypnotic side effects and their likely abuse potential.

GENETICS AND HYPOTHALAMIC REGULATION OF SLEEP

Another predominant and important modifiable risk factor for migraine is sleep.[79] In fact, people with migraine can often identify disturbances of their normal sleep routine as a trigger of migraine: staying up too late, getting up earlier than usual, or even oversleeping. Accordingly, any persistent disturbance of regular sleep, such as an erratic sleep schedule, frequent changes in activity due to shift work, frequent interruptions of sleep through the night, or a chronic condition that disturbs sleep quality (such as obstructive sleep apnea), is well recognized as an important risk factor for chronic migraine.

Brennan and colleagues recently provided important genetic and neurochemical insight into the connection between migraine and those with deep disturbances in their circadian determinants of sleep, again highlighting the role of the hypothalamus in regulating migraine.[80] The circadian rhythm is a biological clock whose rhythm is regulated by daylight conditions and whose functions are to control many physiological signals, such as a morning surge of activity-promoting corticosteroids from the adrenal glands, and a night time peak of sleep-promoting melatonin from the pineal gland. This clock implicates the role of the hypothalamus in regulating a range of critical physiological functions, such as body temperature, blood pressure, feeding and satiety, blood glucose, and the regulation of sex hormones.

Nearly all organisms have evolved an internal circadian rhythm that corresponds closely to the 24-hour day cycle, but in humans a normal variability in individual circadian length underlies differences in the population between "morning larks" and "night owls." People with shorter circadian rhythms arrive sooner at the end of their physiological day, and these morning larks are said to have advanced sleep phase (ASP). Their internal rhythm both drives them to bed early and wakes them bright and early after a full night's sleep. The recognition of extreme familial forms of ASP (FASP) presented researchers with the opportunity to dissect the molecular pathways related to circadian rhythm in humans, leading to the discovery that a mutation of the *hPer2* gene — the human homolog of the period gene that regulates the daily rhythm of activity in fruit flies — is a key player in how the hypothalamus keeps track of the daily rhythm that drives our innate pattern of wakefulness and activity.[81] The extreme forms of the disorder also permitted the identification of kindreds with FASP associated with alterations of the CK1delta gene,[82] encoding an enzyme that is also implicated in the regulation of the *hPer2* gene in the hypothalamus, and whose altered forms showed reduced enzymatic activity.[83] A mouse model of one such human mutation in the CK1delta gene, from a kindred with FASP, then helped to draw inferences between mouse models of migraine susceptibility[84] and the expression of the CK1delta mutation.[80]

However, an incomplete picture emerges from genetic studies of migraine, which is detailed elsewhere in this book. In general, mutations of individual families and genes associated with rare subtypes of migraine (such as those above) have produced little generalizability to the rest of the population with migraine. To address this more general question, a completely different approach to migraine genetics would appear to be required, such as through genome-wide association studies of migraine. Through the combined efforts of many genome scientists (such as the International Headache Genetics Consortium), large collections of upwards from 23,285 individuals with migraine have identified 12 or more loci associated with susceptibility to migraine.[85] An optimistic point of view would be that further hypothesis-driven research still lies ahead to substantiate these genome-driven discoveries, and will reveal their exciting mysteries.

References

1. Headache Classification Committee of the International Headache Society (IHS). The International Classification of Headache Disorders, 3rd ed. (beta version). *Cephalalgia*. 2013;33:629—808.

2. Solomon GD. Classification and mechanism of headache. In: Diamond ML, Solomon GD, eds. *Diamond and Dalessio's The Practicing Physician's Approach to Headache.* 6th ed. Philadelphia, PA: W.B. Saunders; 1999:9.

3. Aristotle. Parva Naturalia: On Youth & Old Age, Life & Death and Respiration [Ogle W, Trans.]. London: Longmans & Co.; 322 BC.

4. Tunis MM, Wolff HG. Studies on headache; long-term observation of alterations in function of cranial arteries in subjects with vascular headache of the migraine type. *Trans Am Neurol Assoc.* 1953;3(78th Meeting):121–123.

5. Feindel W, Penfield W, McNaughton F. The tentorial nerves and localization of intracranial pain in man. *Neurology.* 1960;10:555–563.

6. Drummond PD, Lance JW. Extracranial vascular changes and the source of pain in migraine headache. *Ann Neurol.* 1983;13(1):32–37.

7. Shevel E, Spierings EH. Role of the extracranial arteries in migraine headache: a review. *Cranio.* 2004;22(2):132–136.

8. Shevel E. Confirmation of the extracranial site of action of sumatriptan. *Ann Neurol.* 2011;70(5):862:author reply 862–863.

9. Ahn AH, Basbaum AI. Where do triptans act in the treatment of migraine? *Pain.* 2005;115(1–2):1–4.

10. Goadsby PJ. The vascular theory of migraine – a great story wrecked by the facts. *Brain.* 2009;132(Pt 1):6–7.

11. Strassman AM, Levy D. Response properties of dural nociceptors in relation to headache. *J Neurophysiol.* 2006;95(3):1298–1306.

12. Amin FM, Asghar MS, Hougaard A, Hansen AE, Larsen VA, de Koning PJ, et al. Magnetic resonance angiography of intracranial and extracranial arteries in patients with spontaneous migraine without aura: a cross-sectional study. *Lancet Neurol.* 2013;12(5):454–461.

13. Amin FM, Asghar MS, Ravneberg JW, de Koning PJ, Larsson HB, Olesen J, et al. The effect of sumatriptan on cephalic arteries: a 3T MR-angiography study in healthy volunteers. *Cephalalgia.* 2013;33(12):1009–1116.

14. Ahn AH. On the temporal relationship between throbbing migraine pain and arterial pulse. *Headache.* 2010;50(9):1507–1510.

15. Mirza AF, Mo J, Holt JL, Kairalla JA, Heft MW, Ding M, et al. Is there a relationship between throbbing pain and arterial pulsations? *J Neurosci.* 2012;32(22):7572–7576.

16. Mo J, Maizels M, Ding M, Ahn AH. Does throbbing pain have a brain signature? *Pain.* 2013;54(7):1150–1155.

17. IHS CS. The International Classification of Headache Disorders: 2nd edition. *Cephalalgia.* 2004;24(Suppl 1):9–160.

18. Lipton RB, Penzien DB, Turner DP, Smitherman TA, Houle TT. Methodological issues in studying rates and predictors of migraine progression and remission. *Headache.* 2013;53(6):930–934.

19. Silberstein SD, Lipton RB, Dodick DW. Operational diagnostic criteria for chronic migraine: expert opinion. *Headache.* 2014;54(7):1258–1266.

20. Digre KB, Brennan KC. Shedding light on photophobia. *J Neuroophthalmol.* 2012;32(1):68–81.

21. Ahn AH, Brennan KC. Unanswered questions in headache: so what is photophobia, anyway? *Headache.* 2013;53(10):1673–1674.

22. Selby G, Lance JW. Observations on 500 cases of migraine and allied vascular headache. *J Neurol Neurosurg Psychiatry.* 1960;23:23–32.

23. Okamoto K, Tashiro A, Chang Z, Bereiter DA. Bright light activates a trigeminal nociceptive pathway. *Pain.* 2010;149(2):235–242.

24. Katagiri A, Okamoto K, Thompson R, Bereiter DA. Posterior hypothalamic modulation of light-evoked trigeminal neural activity and lacrimation. *Neuroscience.* 2013;246:133–141.

25. Moulton EA, Becerra L, Borsook D. An fMRI case report of photophobia: activation of the trigeminal nociceptive pathway. *Pain.* 2009;145(3):358–363.

26. Noseda R, Kainz V, Jakubowski M, Gooley JJ, Saper CB, Digre K, et al. A neural mechanism for exacerbation of headache by light. *Nat Neurosci.* 2010;13(2):239–245.

27. Matynia A, Parikh S, Chen B, Kim P, McNeill DS, Nusinowitz S, et al. Intrinsically photosensitive retinal ganglion cells are the primary but not exclusive circuit for light aversion. *Exp Eye Res.* 2012;105:60–69.

28. Drummond PD. A quantitative assessment of photophobia in migraine and tension headache. *Headache.* 1986;26(9):465–469.

29. Drummond PD, Woodhouse A. Painful stimulation of the forehead increases photophobia in migraine sufferers. *Cephalalgia.* 1993;13(5):321–324.

30. Coppola G, Schoenen J. Cortical excitability in chronic migraine. *Curr Pain Headache Rep.* 2012;16(1):93–100.

31. Friedman DI. Unanswered questions in headache: so what is photophobia, anyway? *Headache.* 2013;53(10):1675–1676.

32. Recober A, Kuburas A, Zhang Z, Wemmie JA, Anderson MG, Russo AF. Role of calcitonin gene-related peptide in light-aversive behavior: implications for migraine. *J Neurosci.* 2009;29(27):8798–8804.

33. Marquez de Prado B, Hammond DL, Russo AF. Genetic enhancement of calcitonin gene-related peptide-induced central sensitization to mechanical stimuli in mice. *J Pain.* 2009;10(9):992–1000.

34. Kaiser EA, Kuburas A, Recober A, Russo AF. Modulation of CGRP-induced light aversion in wild-type mice by a 5-HT(1B/D) agonist. *J Neurosci.* 2012;32(44):15439–15449.

35. Crozier WJ, Pincus G. Phototropism in young rats. *J Gen Physiol.* 1927;10(3):407–417.

36. Johnson J, Wu V, Donovan M, Majumdar S, Renteria RC, Porco T. Melanopsin-dependent light avoidance in neonatal mice. *Proc Natl Acad Sci USA.* 2010;107(40):17374–17378.

37. Delwig A, Logan AM, Copenhagen DR, Ahn AH. Light evokes melanopsin-dependent vocalization and neural activation associated with aversive experience in neonatal mice. *PLoS One.* 2012;7(9):e43787.

38. Carrasquillo Y, Gereau RWt. Activation of the extracellular signal-regulated kinase in the amygdala modulates pain perception. *J Neurosci.* 2007;27(7):1543–1551.

39. Carrasquillo Y, Gereau RWt. Hemispheric lateralization of a molecular signal for pain modulation in the amygdala. *Mol Pain.* 2008;4:24.

40. Ji RR, Gereau RWt, Malcangio M, Strichartz GR. MAP kinase and pain. *Brain Res Rev.* 2009;60(1):135–148.

41. Maniyar FH, Sprenger T, Monteith T, Schankin C, Goadsby PJ. Brain activations in the premonitory phase of nitroglycerin-triggered migraine attacks. *Brain.* 2013.

42. Afridi SK, Kaube H, Goadsby PJ. Glyceryl trinitrate triggers premonitory symptoms in migraineurs. *Pain.* 2004;110(3):675–680.

43. Magis D, Bendtsen L, Goadsby PJ, May A, Sanchez del Rio M, Sandor PS, et al. Evaluation and proposal for optimization of neurophysiological tests in migraine: part 2 – neuroimaging and the nitroglycerin test. *Cephalalgia.* 2007;27(12):1339–1359.

44. Tassorelli C, Joseph SA, Nappi G. Central effects of nitroglycerin in the rat: new perspectives in migraine research. *Funct Neurol.* 1996;11(5):219–223:227–235.

45. Alstadhaug KB. Migraine and the hypothalamus. *Cephalalgia.* 2009;29(8):809–817.

46. Noseda R, Kainz V, Borsook D, Burstein R. Neurochemical pathways that converge on thalamic trigeminovascular neurons: potential substrate for modulation of migraine by sleep, food intake, stress and anxiety. *PLoS ONE.* 2014;9(8):e103929.

47. Holland PR, Akerman S, Goadsby PJ. Modulation of nociceptive dural input to the trigeminal nucleus caudalis via activation of the orexin 1 receptor in the rat. *Eur J Neurosci.* 2006;24(10):2825–2833.

48. Donaldson ZR, Young LJ. Oxytocin, vasopressin, and the neurogenetics of sociality. *Science*. 2008;322(5903):900−904.

49. Meyer-Lindenberg A, Domes G, Kirsch P, Heinrichs M. Oxytocin and vasopressin in the human brain: social neuropeptides for translational medicine. *Nat Rev Neurosci*. 2011;12 (9):524−538.

50. Schorscher-Petcu A, Sotocinal S, Ciura S, Duprá A, Ritchie J, Sorge RE, et al. Oxytocin-induced analgesia and scratching are mediated by the vasopressin-1A receptor in the mouse. *J Neurosci*. 2010;30(24):8274−8284.

51. Mogil JS, Sorge RE, LaCroix-Fralish ML, Smith SB, Fortin A, Sotocinal SG, et al. Pain sensitivity and vasopressin analgesia are mediated by a gene−sex−environment interaction. *Nat Neurosci*. 2011;14(12):1569−1573.

52. Phillips WJ, Ostrovsky O, Galli RL, Dickey S. Relief of acute migraine headache with intravenous oxytocin: report of two cases. *J Pain Palliat Care Pharmacother*. 2006;20(3):25−28.

53. Charbit AR, Akerman S, Holland PR, Goadsby PJ. Neurons of the dopaminergic/calcitonin gene-related peptide A11 cell group modulate neuronal firing in the trigeminocervical complex: an electrophysiological and immunohistochemical study. *J Neurosci*. 2009;29(40):12532−12541.

54. Kagan R, Kainz V, Burstein R, Noseda R. Hypothalamic and basal ganglia projections to the posterior thalamus: possible role in modulation of migraine headache and photophobia. *Neuroscience*. 2013;248C:359−368.

55. Peres MF. Melatonin for migraine prevention. *Curr Pain Headache Rep*. 2011;15(5):334−335.

56. Kaiser EA, Russo AF. CGRP and migraine: could PACAP play a role too? *Neuropeptides*. 2013;47(6):451−461.

57. Ahn AH. Why does increased exercise decrease migraine? *Curr Pain Headache Rep*. 2013;17(12):379.

58. Buse DC, Manack A, Serrano D, Turkel C, Lipton RB. Sociodemographic and comorbidity profiles of chronic migraine and episodic migraine sufferers. *J Neurol Neurosurg Psychiatry*. 2010;81(4):428−432.

59. Varkey E, Hagen K, Zwart JA, Linde M. Physical activity and headache: results from the Nord-Trondelag Health Study (HUNT). *Cephalalgia*. 2008;28(12):1292−1297.

60. Varkey E, Cider A, Carlsson J, Linde M. Exercise as migraine prophylaxis: a randomized study using relaxation and topiramate as controls. *Cephalalgia*. 2011;31(14):1428−1438.

61. Dittrich SM, Gunther V, Franz G, Burtscher M, Holzner B, Kopp M. Aerobic exercise with relaxation: influence on pain and psychological well-being in female migraine patients. *Clin J Sport Med*. 2008;18(4):363−365.

62. Busch V, Gaul C. Exercise in migraine therapy − is there any evidence for efficacy? A critical review. *Headache*. 2008;48 (6):890−899.

63. Staud R, Robinson ME, Price DD. Isometric exercise has opposite effects on central pain mechanisms in fibromyalgia patients compared to normal controls. *Pain*. 2005;118 (1−2):176−184.

64. Nijs J, Roussel N, Van Oosterwijck J, De Kooning M, Ickmans K, Struyf F, et al. Fear of movement and avoidance behaviour toward physical activity in chronic-fatigue syndrome and fibromyalgia: state of the art and implications for clinical practice. *Clin Rheumatol*. 2013;32(8):1121−1129.

65. Raichlen DA, Foster AD, Gerdeman GL, Seillier A, Giuffrida A. Wired to run: exercise-induced endocannabinoid signaling in humans and cursorial mammals with implications for the "runner's high". *J Exp Biol*. 2012;215(Pt 8):1331−1336.

66. Raichlen DA, Foster AD, Seillier A, Giuffrida A, Gerdeman GL. Exercise-induced endocannabinoid signaling is modulated by intensity. *Eur J Appl Physiol*. 2013;113(4):869−875.

67. Morgan WP. Affective beneficence of vigorous physical activity. *Med Sci Sports Exerc*. 1985;17(1):94−100.

68. Boecker H, Sprenger T, Spilker ME, Henriksen G, Koppenhoefer M, Wagner KJ, et al. The runner's high: opioidergic mechanisms in the human brain. *Cereb Cortex*. 2008;18(11):2523−2531.

69. Sparling PB, Giuffrida A, Piomelli D, Rosskopf L, Dietrich A. Exercise activates the endocannabinoid system. *Neuroreport*. 2003;14(17):2209−2211.

70. Dietrich A, McDaniel WF. Endocannabinoids and exercise. *Br J Sports Med*. 2004;38(5):536−541.

71. Lupica CR, Riegel AC. Endocannabinoid release from midbrain dopamine neurons: a potential substrate for cannabinoid receptor antagonist treatment of addiction. *Neuropharmacology*. 2005;48 (8):1105−1116.

72. Mahler SV, Smith KS, Berridge KC. Endocannabinoid hedonic hotspot for sensory pleasure: anandamide in nucleus accumbens shell enhances "liking" of a sweet reward. *Neuropsychopharmacology*. 2007;32(11):2267−2278.

73. Justinova Z, Yasar S, Redhi GH, Goldberg SR. The endogenous cannabinoid 2-arachidonoylglycerol is intravenously self-administered by squirrel monkeys. *J Neurosci*. 2011;31(19):7043−7048.

74. Meng ID, Manning BH, Martin WJ, Fields HL. An analgesia circuit activated by cannabinoids. *Nature*. 1998;395(6700):381−383.

75. Agarwal N, Pacher P, Tegeder I, Amaya F, Constantin CE, Brenner GJ, et al. Cannabinoids mediate analgesia largely via peripheral type 1 cannabinoid receptors in nociceptors. *Nat Neurosci*. 2007;10(7):870−879.

76. Akerman S, Holland PR, Goadsby PJ. Cannabinoid (CB1) receptor activation inhibits trigeminovascular neurons. *J Pharmacol Exp Ther*. 2007;320(1):64−71.

77. Akerman S, Holland PR, Lasalandra MP, Goadsby PJ. Endocannabinoids in the brainstem modulate dural trigeminovascular nociceptive traffic via CB1 and "triptan" receptors: implications in migraine. *J Neurosci*. 2013;33(37):14869−14877.

78. Greco R, Mangione AS, Sandrini G, Nappi G, Tassorelli C. Activation of CB2 receptors as a potential therapeutic target for migraine: evaluation in an animal model. *J Headache Pain*. 2014;15:14.

79. Brennan KC, Charles A. Sleep and headache. *Semin Neurol*. 2009;29(4):406−418.

80. Brennan KC, Bates EA, Shapiro RE, Zyuzin J, Hallows WC, Huang Y, et al. Casein kinase idelta mutations in familial migraine and advanced sleep phase. *Sci Transl Med*. 2013;5 (183):183ra156, 181−111.

81. Toh KL, Jones CR, He Y, Eide EJ, Hinz WA, Virshup DM, et al. An hPer2 phosphorylation site mutation in familial advanced sleep phase syndrome. *Science*. 2001;291(5506):1040−1043.

82. Xu Y, Padiath QS, Shapiro RE, Jones CR, Wu SC, Saigoh N, et al. Functional consequences of a CKIdelta mutation causing familial advanced sleep phase syndrome. *Nature*. 2005;434(7033):640−644.

83. Xu Y, Toh KL, Jones CR, Shin JY, Fu YH, Ptacek LJ. Modeling of a human circadian mutation yields insights into clock regulation by PER2. *Cell*. 2007;128(1):59−70.

84. Bates EA, Nikai T, Brennan KC, Fu Y-H, Charles AC, Basbaum AI, et al. Sumatriptan alleviates nitroglycerin-induced mechanical and thermal allodynia in mice. *Cephalalgia*. 2010;30(2):170−178.

85. Anttila V, Winsvold BS, Gormley P, Kurth T, Bettella F, McMahon G, et al. Genome-wide meta-analysis identifies new susceptibility loci for migraine. *Nat Genet*. 2013;45(8):912−917.

3

Evaluation of the Headache Patient in the Computer Age

Edmund Messina

The Michigan Headache Clinic, East Lansing, Michigan, and
College of Human Medicine, Michigan State University, Michigan, USA

EVALUATION OF THE HEADACHE PATIENT

This chapter provides a practical approach to the evaluation of the headache patient. A good evaluation will determine the specific headache diagnosis (or diagnoses), determine the impact on patient quality of life, and identify comorbid conditions that can affect the diagnosis and treatment of a primary headache disorder.

Initial evaluation of the headache patient includes:

1. The history, including general history, headache history, and search for comorbidities.
2. The examination, including general examination, neurological examination, and headache examination.
3. Testing, when indicated by key elements in the history, according to current guidelines.

THE HEADACHE HISTORY

A patient may experience more than one type of headache, so the initial question should be: "How many different types of headache are you having?" A detailed history for each headache type should follow, guiding individuals to best express their symptoms (Table 3.1).

Structured Interview Versus Open Questioning

There is no substitute for a thorough medical history; there is an art to performing a good patient interview. The proper evaluation of a headache patient still requires a good amount of face-to-face time between the physician and the patient. In no other area of medicine is the art of the clinical history more significant than in evaluation of the headache patient, as the lines of inquiry cross many organ systems and disciplines. The history explores physical as well as behavioral symptoms.

Many aspects of the headache history are structured, naming dates, symptoms, treatment attempts, neurological review of systems, and general overview of systems. Time constraints are a major consideration in gathering structured information. There is no substitute for an experienced clinician who looks for inconsistencies in the history and observes the patient's body language and reactions to certain questions; a sideways glance to a spouse or parent may be very revealing. If personalized structured information can be gathered from patients before they meet with the clinician, there is more time for open conversation to explore subtleties that might lead to a more accurate diagnosis. Computer technology can help in acquiring structured information more efficiently.

A wise professor once said "Shut up and let the patient tell you what's wrong with them!" If we listen carefully, in a less structured conversation, patients may tell us things of which we didn't think to ask. During unstructured conversation some migraine patients may provide excessive detail, so we must guide them to reveal useful details and to clarify the inevitable ambiguities or inconsistencies that may have arisen from structured questioning.

For example, a new patient with chronic migraine said that she was unable to sleep more than an hour at a time; the structured history indicated

S. Diamond, R. K. Cady, M. L. Diamond & V. T. Martin (Eds):
Headache and Migraine Biology and Management.

DOI: http://dx.doi.org/10.1016/B978-0-12-800901-7.00003-3

TABLE 3.1   Headache History

1. How many types of headache do you experience?

2. When did you first experience a headache? Can you relate it to any event (head injury, infection, surgery, stress, or, for women, menses or pregnancy)?

3. How often do you experience a headache?

4. How long does your headache last? If you take a drug for your headache, how long will the headache last?

5. In what part of your head do you feel the pain? Does the pain move around your head? Is the headache always on one side?

6. During a headache, do you feel pain in your neck or shoulders?

7. How would you rate the severity of your headaches on a scale of 1 (least) to 10 (most) severe?

8. How would you describe the pain (throbbing, pulsating, tight head-band, deep boring sensation, ache, vise-like)?

9. Do you have any warning that a headache will soon start? Do you see flashing lights or different colors? Do you have problems with your vision (loss of visual field, images look larger or smaller, blurred vision)? Prior to the headache, do you feel hungry, fatigued, loss of appetite, burst of energy, quick tempered?

10. During a headache, do you have any associated symptoms (nausea, vomiting, dizziness, sensitivity to light or sound, facial flushing, nasal congestion, runny nose, tearing in either eye, eyelid droopy or swollen, ringing in the ears, blurred vision)?

11. Can you identify any factor that precipitates your headache? Are you affected by certain foods, alcoholic beverages, skipping meals, skipping caffeine beverages, lack of sleep, changes in weather, certain drugs (nitrates, indomethacin), exercise, stress, or, for women, menses?

12. How would you describe your sleep pattern? Do you experience difficulty falling asleep or staying asleep? Does the headache awaken you from a sound sleep? Do you awake and the headache has already started?

13. How would you describe your marital and family relationships? Are you under stress at home or at work? For children and adolescents, are you under stress at school? Have you missed work or school days because of the headaches?

14. What are your medical and surgical histories? Have you ever been hospitalized for headaches?

15. For women: What is your menstrual history? At what age did your periods start? Did you have headaches during your pregnancy? Are you on birth control pills or hormone supplements?

16. What medications do you use for your headaches? Are you on any preventive medications for headaches? What medications have you tried previously for your headaches?

17. Are you on medications for any other medical condition?

18. Are you allergic to any medications? Do you have seasonal allergies? Do you have any food allergies?

19. How many cups of coffee, tea, or caffeine-containing soda do you drink per day?

20. Do you smoke?

21. Do you drink alcoholic beverages on a daily basis? How often do you drink?

22. Do you use any recreational drugs?

23. Have you ever tried alternative therapies for your headaches (biofeedback, acupuncture, massage)?

24. Are you using any herbal remedies for your headaches?

25. Have you had any recent tests because of your headaches? Have you ever received neuroimaging (CT scan or MRI) because of your headaches?

maintenance insomnia. When asked why she could not stay asleep, she said that she needed to urinate every hour. Her review of systems revealed no symptoms of a urinary tract infection. Further questioning revealed that for the past 12 years, following a severe closed head injury, she had been drinking about 2 gallons of water each day. She was referred to an endocrinologist, who confirmed her post-traumatic diabetes insipidus, and treatment led to improved sleep and less frequent migraines.

Most experienced clinicians have similar anecdotes; sometimes we discover these factors incidentally and sometimes we find them through systematic interrogation. Since the diagnosis of primary headache depends mainly on history, we need to take a very accurate and thorough history in each patient.

The Special Challenge of Talking to the Headache Patient ... What Patients Tell Us

Since the headache history helps to differentiate benign primary headaches from potentially lethal secondary headaches, an accurate history is critical; however, unfortunately, patients may be imprecise unless we assist them. Medical terminology is often misused, and never assume that the patient has good healthcare literacy. Patients may list diagnoses that might be self-deduced or misdiagnosed by others, so it is important to not accept "legacy" diagnoses without confirming them with specific questions.

"Sinus" Headaches

Migraine is commonly misdiagnosed as "sinus headaches" by patients or their healthcare provider.[1] Misdiagnosis is often due to the misinterpretation of migraine symptoms based on the periorbital or frontal location, lacrimation, rhinorrhea, or seasonal/weather-related exacerbations. The clinician's job is to interpret these symptoms. Patients might say "sinus pressure" when they try to describe a frontal headache due to other causes.

"I Have a Pinched Nerve in My Neck"

Patients may use the term "pinched nerve" when describing posterior head pain. It is important to distinguish the common occurrence of pain in the posterior cranium and upper cervical muscles from a migraine, tension-type headache, occipital neuralgia, chronic cervical pain syndrome, or a cervical radiculopathy. It is helpful to ask if the upper neck is painful only when there is a headache.

"I can't Stay Asleep"

Disturbed sleep is an important comorbidity, so this is a significant part of the general history. When patients say they have trouble staying asleep, clarify whether there is a normal awakening and quick return to sleep versus awakening with the inability to fall back asleep. Some patients can sleep off a headache and others cannot sleep if their headaches are severe. Some individuals experience headaches that are triggered by specific stages of sleep, such as cluster or hypnic headaches.

Migraine Aura Versus Other Conditions

Ask the patient specifically about symptoms that occur before the pain of a headache begins, as many patients do not think of mentioning these unless they are specifically asked. The first migraine aura may be frightening, and many patients think they are having a stroke or other serious disorder. *All* symptoms that occur with the migraine aura need to be clarified.

The migraine aura is defined as "Recurrent attacks, lasting minutes, of unilateral fully reversible visual, sensory or other central nervous system symptoms that usually develop gradually and are usually followed by headache and associated migraine symptoms".[3] The aura can be one or more neurological symptoms which, in general, precede a migraine attack, sometimes occurring in the absence of a headache.

The specific criteria for migraine and migraine aura are discussed in Chapter 2. The clinical history needs to identify visual phenomena, sensory changes, altered speech, motor weakness, brainstem symptoms, and any other neurological symptoms that occur before or during a headache.

The duration of an aura lasts usually less than 60 minutes per type of neurological symptom; sometimes each element can continue for 60 minutes if the patient experiences more than one kind of aura.[2] If the symptoms deviate from this, more detail is necessary and they must be differentiated from other conditions. It is helpful to ask: "What do you notice before a headache?" It is important to differentiate a migraine aura from stroke, seizure, or other neurological disorders.

Most aura symptoms are visual, generally a scotoma, scintillation, or the so-called "fortification spectrum," which is a zigzag figure near the center of vision that may gradually spread right or left and have a scintillating edge. This usually leaves a scotoma in its wake.

A frequent type of aura is a sensory disturbance, usually described as "numbness" or "pins and needles" going up one side of the body or face, or even the tongue. Speech disturbances are less frequent. A motor aura is more concerning, since it may be mistaken for a stroke; however, it is helpful to know that this distribution may not respect vascular territories. When the symptoms are primarily referable to the brainstem, they can include slurred speech, vertigo, tinnitus, hyperacusis, diplopia, ataxia, and even a decreased level of consciousness.

"Blurred Vision"

Patients may use the term "blurred vision" when they experience a visual aura. They may tell you about symptoms in their "right eye" when they may actually be experiencing homonymous right-sided visual symptoms. This distinction is important, as a true monocular visual phenomenon is suspicious for other conditions such as detached retina, retinal vein occlusion, or amaurosis fugax.

"Numbness" or "Heaviness"

The terms "numbness" or "heaviness" must be clarified to determine whether a patient is experiencing sensory symptoms, motor weakness, or both. A patient may use the term "numbness" to describe motor weakness. It is helpful to ask if the symptom is a heaviness (weakness) versus lack of feeling, with or without tingling (paresthesia). Some individuals describe both symptoms. It is unlikely that a sudden loss of feeling will call attention to itself but individuals are more likely to become aware of tingling sensations, often mistaken for "lack of circulation."

If the patient describes weakness, ask functional questions such as "Were you able to bear weight on the leg?" or "Could you lift your arm without help?" Ask how quickly the weakness developed (seconds versus minutes or hours), its initial location, and where it may have spread. In general, brain ischemia can develop quickly and hemiplegic migraine has a slower march, often across vascular territories such as the middle cerebral and anterior cerebral arteries. Ask about the duration of weakness; hemiplegic migraine may last from minutes to days and TIAs are brief. Because there is an overlap, ask enough details to differentiate stroke from hemiplegic migraine.

When the patient describes numbness or paresthesia, ask about the location and whether there is a "march" up or down the face to upper and lower extremities. The objective is to differentiate a benign migraine sensory aura from a sensory stroke or non-convulsive seizure. Most common sensory auras of migraine and cheiro-oral syndrome, such as the mouth/lips and hand, are involved.

"Trouble Talking"

Speech problems preceding a severe headache may be due to aura, presenting as dysarthria, word-finding dysphasia, or even aphasia. Duration of symptoms and presence of a subsequent headache are important points in the history to help differentiate migraine from cerebral ischemia.

"Dizziness"

The term "dizziness" must further be characterized as vertigo, pre-syncopal lightheadedness, episodic disequilibrium, etc. Patients may describe their ataxia or even their parkinsonian gait instability as "dizziness."

An important red flag is the patient describing a combination of symptoms such as dysarthria, vertigo, tinnitus, hyperacusis, diplopia, ataxia, or decreased level of conscious associated with his or her headaches. It is important to determine the diagnosis of migraine with brainstem aura (previously known as basilar migraine) versus brainstem ischemia.

An accurate history requires constant clarification and verification. When symptoms do not make sense, probe deeper to reveal what the patient is actually experiencing. Patients may not be able to express their symptoms clearly, or to recognize the relationships of multiple symptoms. A skilled clinician can group seemingly unrelated symptoms into a syndrome.

Screening for Secondary Headaches

The first priority in making a headache diagnosis is to separate primary headache[2] from secondary headache. Secondary headaches can arise from localized or systemic illnesses,[3] including:

- Space-occupying lesions such as brain tumors, brain abscesses, or hematomas
- Infectious disorders such as meningitis or encephalitis
- Vascular disorders such as hemorrhages, infarctions, vasculitis, vascular dissections, venous sinus thrombosis, or various vasoconstriction syndromes
- Structural abnormalities such as hydrocephalus, Chiari malformations, vascular malformations, aneurysms, or low or high CSF pressure.

Primary headaches must be differentiated from secondary headaches through clinical history, physical examination, and testing. A patient with primary headaches may also develop secondary headache of more ominous origin, so a careful history is essential (see also *Chapter 4*, on screening and testing of the headache patient).

Clinical History – What We Need to Ask the Patient with Primary Headaches

After ruling out secondary headaches, we can attend to the primary headache(s) experienced by the patient. The clinical history process asks specific questions to identify the type of headaches the patient is experiencing. This process acquires information leading to a practical treatment plan, including identification of comorbidities that could exacerbate the headaches or impact upon the treatment plan itself. It is important to understand the impact of headaches on the patient's quality of life, since this will determine management.

Headache Diary

A headache diary can be very helpful. Keep it simple; a single piece of paper with 31 lines can record a month's headaches (Figure 3.1). Patients should track the presence of a headache, the severity, known triggers, and whether this interfered with their activities. Women should track when menses start and stop. Patients should keep track of nights when they had insomnia, oversleeping, or sleep problems like nightmares. A good headache diary should be easy for

FIGURE 3.1 Headache diary.

| Headache Diary for _____ | | Month _____ | | | | |

Day of month	When headache started	Duration in hrs	Grade (1-5)	Triggered by	Relieved by med, etc.	Menses start/stop
1						
2						
3						
4						
5						
6						
7						
8						
9						
10						
11						
12						
13						
14						
15						
16						
17						
18						
19						
20						
21						
22						
23						
24						
25						
26						
27						
28						
29						
30						
31						

the patient to keep, avoiding excessive input, and making it easy to observe patterns, as a linear record is preferable to one which looks like a grid.

The analog 10-point pain scale used in general pain clinics is less helpful than a functional scale for headache patients, such as the following:

Grade 1: Mild pain is present but does not need treatment
Grade 2: Pain requires intervention but does not curtail activities
Grade 3: Pain is disruptive, not working at peak ability but functioning
Grade 4: Unable to carry on daily activity because of pain and/or nausea
Grade 5: Unable to do anything but lie in bed

What Type of Headache is it?

The most important question is: "How many different types of headache are you having?"

It is common for patients to experience more than one type of headache, and the patient's description of different headache types will help to classify these headaches. For example, since tension-type headaches are very common, it is not unusual for a migraine person to also experience tension-type headaches. Some cluster patients may also have migraine and often can differentiate them.

Ask the patient what the pain feels like and whether there are accompaniments such as lacrimation, congestion, nausea/vomiting, photo/phonophobia, etc. Ask about time of the day or day of the week, and whether there is a relationship to menses. Divide the headache questions into broad categories, asking specific questions that might rule out certain headache types. For example, if the headaches are asymmetrical, and produce vomiting and light or sound sensitivity, it is not a tension-type headache. Likewise, if intense headaches are of short duration without treatment, migraine is less likely than cluster.

Ask about headaches associated with exertion, to separate exertional exacerbation of pre-existing migraines from primary exertional headaches which are suddenly triggered by exertion. These headaches need to be differentiated from primary cough headaches. This distinction is important, as exertion-related headaches may also be due to increased intracranial pressure or vascular malformations such as arteriovenous malformations or aneurysms.

How Often are Your Headaches Occurring?

This helps to establish diagnosis as well as treatment strategies. This question should be asked for each type of headache the patient experiences. It is not unusual for a patient with low-grade daily headaches to have exacerbations every few days or weeks. Chronic daily headaches may be a clue to medication overuse headache. It is important to know whether the headaches occur in cycles or "clusters." Are the headaches escalating, especially in recent weeks?

When Did These Headaches Begin?

Did these headaches begin after head trauma? Migraine commonly begins at puberty, but it may be quiescent until major life stressors occur. If daily headaches, when did they become daily? Did they begin suddenly one day and never go away, as we might see in new daily persistent headache (NDPH)? When did the patient begin taking daily headache-relieving medications?

What Do You Do When You Get a Headache?

Ask the patient: Do you lie down because of light and sound aversion? Do you have to leave work or school? Do you retreat to a dark and quiet place? Can you sleep off a headache?

How are you Treating your Headaches?

Questions include: Are you using heat or cold packs? Are you doing meditation or relaxation exercises? Do you reach for medication every time you have a headache? What do you take, and how often do you take it? How effective is your headache medication? Can you depend upon the medication to help relieve your headache? How are you trying to prevent your headaches? Are you using non-pharmacologic techniques or daily preventive medicines? How well is your plan working?

Daily headaches are commonly caused by overuse of headache-relieving medications, and it is critical to ask how often the patient is medicating. Individuals often underreport the use of over-the-counter medications, so it helps to ask, "Are you using non-prescription medicines?"

What Precedes your Headaches?

Symptoms occurring before a headache can help with determining the diagnosis as well as a treatment plan.

- Many headache sufferers experience a prodrome of certain symptoms (sleepiness, surges of energy, salt craving, or thirst) prior to a migraine. The prodrome may occur many hours before the attack. Sometimes, the prodrome can be used to trigger a proactive treatment plan.
- The aura is important, as discussed above.

What Triggers the Headaches?

Many patients are already aware of specific triggers, but a systematic history may reveal others. Ask: "What seems to give you a headache?"

- *Stress.* Inquire about stressors in daily life, especially in the workplace/school and in personal relationships. Migraine patients can be stressed by the fear of being fired due to absenteeism. How does the spouse or parent deal with the patient's headaches? Are they understanding and supportive? In adolescents and younger children: Are the parents separated? Is bullying causing school avoidance, which is being blamed on headaches? People will report that a migraine is triggered by crying. Sometimes, resolution of a stressful period of time will provoke a migraine the following day ("letdown headaches").[4] Stress sometimes worsens anxiety, a common comorbidity to be discussed later.
- *Dietary factors.* Food and beverage triggers are very individualized and it is helpful to track them in the headache diary, including the amount ingested, to find the headache threshold.
- *Sleep disturbances.* Undersleeping and oversleeping can make headaches more frequent. Headaches are common in adolescents who try to "catch up" by sleeping later on weekends.
- *Hormones.* Hormones may be a contributing factor. Starting or stopping estrogens can provoke migraine, as can certain phases in the menstrual cycle. Ask women about the relationship of headaches to menses.

What are the Accompaniments?

Associated symptoms such as nausea and vomiting are important not only to help differentiate the headache type but also to plan the approach when a headache occurs. For example, nausea and vomiting occurring early in a migraine could make an oral treatment plan impractical. Is the patient able to continue functioning, even if he or she does not treat the headache? Aside from the pain and nausea, migraine patients are often unable to think clearly and function in the workplace.

Inquire About Migraine Comorbidities

A comorbidity is defined as an illness that occurs more frequently in association with a specific disorder than would be found as a coincidental association in the general population.[5] Conditions that frequently coexist with migraine include depression, anxiety, sleep disorders, epilepsy, stroke, and other painful conditions. Identifying these comorbidities can help formulate a more inclusive treatment plan and avoid conflicting strategies. The clinician needs to treat the comorbidities as part of the headache plan or work in close cooperation with the patient's other clinicians.

For the sake of this headache discussion, it is helpful to divide these comorbidities into four overlapping categories.

1. *Conditions that may worsen headaches.* These are conditions which can exacerbate or trigger a headache attack, especially migraine. Headaches may be associated with stress caused by a hostile workplace or an unhappy marriage. Headaches may be triggered by anxiety attacks, and occur more commonly in patients with sleep disorders. Medications used to treat other conditions, such as the nitrates, can worsen migraine.
2. *Conditions that may worsen with the headaches or their treatment.* Migraine or cluster attacks can disrupt nighttime sleep. Chronic constipation can be worsened by certain migraine preventive medications, or by opioids being used inappropriately to treat migraine. Esophageal reflux or ulcer conditions can worsen when patients treat headaches with non-steroidal anti-inflammatories. Comorbidities can be bidirectional: chronic headache can exacerbate depression, and major depression is more common in migraine patients.[6,7]
3. *Conditions that may mask the headache diagnosis.* Medication overuse headaches can coexist with daily headaches of other causes. When evaluating migraine patients who are receiving opioids for other painful conditions, it is important to determine better solutions for their other painful problems and discontinue opioid use.
4. *Conditions that can be co-managed with the headache disorder.* Because of the comorbidity between migraine, fibromyalgia, anxiety, depression, insomnia, and other conditions, it is rational to consider overlapping treatment plans. For example, if the patient is on a beta-blocker for hypertension, perhaps the medication can be converted to one that more readily crosses the blood—brain barrier, such as propranolol, to help prevent migraine. Likewise, in patients with tremor problems, propranolol can be used for both migraine and tremor.

Many antidepressants can be useful when anxiety and depression coexist with migraine. Some of these agents are useful for other chronic pain states, such as fibromyalgia. Some of the tricyclic antidepressants are useful in migraine, as well as irritable bowel syndrome or enuresis, and possibly for some of the primary insomnias.

Asking About the Common Migraine Comorbidities

Medical assessment and psychiatric screening are essential to the evaluation and treatment of headache disorders and associated coexisting conditions that might complicate or contraindicate specific headache treatments.

Depression and Anxiety

Depression and anxiety are key factors in headache syndromes. A careful psychiatric and social history is important, inquiring about marriage, job, and symptoms of depression or anxiety. Validated scales, such as the PHQ-2[8] and PHQ-9,[9] are very helpful for uncovering depression, or at least in initiating the conversation. Anxiety screening tools include the Beck Anxiety Inventory.[10,11]

Evaluation instruments and assessment scales are helpful, but they do not establish a diagnosis. Paper forms and simple computer-aided forms are commonly inaccurate, so informal questioning remains important — especially in the elderly, in whom depression symptoms are harder to elicit.

Sleep Disorders

A careful sleep history is essential when managing headache patients because of the high incidence of sleep disorders. Identifying these issues can lead to better treatment strategies. Migraineurs report difficulty initiating and maintaining sleep, and migraines are commonly triggered by sleep disturbance. Sleep complaints occur with greater frequency among chronic than among episodic migraineurs. Ironically, migraineurs try to seek refuge in sleep but frequently experience insomnia.[12,13] Patients with cluster and hypnic headaches experience fragmented sleep because of awakenings.

Excessive daytime sleepiness may be due to insomnia or to sleep apnea, and both can exacerbate migraine. In addition to asking about insomnia or daytime sleepiness, the Epworth Sleepiness Scale[14] and the Insomnia Severity Index[15] can provide more precision.

Be suspicious of sleep apnea when daytime sleepiness is not readily explained by insomnia, under-sleeping, or medication side effects. Apnea causes disruption of normal sleep patterns and reduces REM sleep, making headache control more difficult. Obstructive apnea may cause disruptive snoring, but central apnea is silent. Patients using chronic opioids have a higher incidence of central and obstructive apnea.[16]

Insomnia is defined as difficulty in falling asleep or in maintaining sleep. Many patients with stressful lives or anxiety will say, "I am unable to turn off my brain." The International Classification of Sleep Disorders[17] defines this as psychophysiological insomnia, when the patient experiences heightened arousal, has excessive focus, and worries about sleep because of increased levels of cognitive and somatic arousal at bedtime. This is common in migraine.

Sleep-related bruxism[17] is caused by repetitive involuntary contractions of jaw muscles during sleep, commonly causing disrupted sleep and chronic temporal headaches. Adolescents commonly have a delayed sleep phase because their preferred bedtime is much later at night and they are in a constant state of fatigue due to inadequate nighttime sleep duration. They need to awaken earlier than they should because of school schedules. Managing their sleep phase may significantly reduce the number of migraines in many adolescents.

Fibromyalgia

Fibromyalgia is common in patients with migraine, especially in chronic daily headaches.[18,19] Fibromyalgia is a type of centralized pain state, not unlike migraine, and it should be suspected in patients with multifocal pain not explained by inflammation or injury.[20] These shared features have many therapeutic options in common.

Restless Legs Syndrome

Restless legs syndrome[17] can lead to disrupted sleep and subsequent headaches. It is important that this condition is recognized as some medications used by headache patients, such as certain anxiolytics and antidepressants, can, paradoxically, exacerbate this problem.

Other Somatic Complaints

Patients with chronic headache will commonly experience a variety of other somatic complaints, such as back pain, joint pain, dizziness, constipation, diarrhea, etc.[21] These symptoms may influence therapeutic choices.

Other Systemic Complaints

To treat a patient with a headache disorder appropriately, a general medical history will identify potential constraints to your treatment strategy. Conditions such as liver disease, renal disease, hypertension, epilepsy, stroke, or heart disease may influence the choice of preventive and acute medications.

Substance Abuse

Substance abuse is important at any age. It is vital to ask about ethanol and street drug use, since this may well influence the headache treatment plan. Many states have an online registry, which tracks individual patients' prescription history of controlled substances.

Cardiovascular Comorbidity

Migraine is associated with an increased risk of cardiovascular disease,[22] so it is reasonable to obtain a history of prior cardiovascular disease and traditional cardiovascular disease risk factors before prescribing triptans or ergotamines. In patients with a history of cardiovascular disease, an EKG may be helpful before starting tricyclic antidepressants.[23]

Quality of Life

A good headache history includes questions about quality of life. A patient with frequent but mild headaches may require a less aggressive treatment plan than someone who is unable to function during attacks.

The impact of headaches on social life, family time, and relationships is important. It is essential to know if the spouse or parent reacts to a patient's headaches with empathy, anger, or indifference. Ask about the impact of headaches on work attendance and quality of work, and about the reactions of coworkers. In children, question the influence of headaches on schoolwork. Is the child absent on a regular basis? How are the parents reacting to these headaches? Is the school system being cooperative? The MIDAS (Migraine Disability Assessment Scale)[24] is a useful way to determine the impact of headaches on the adult patient's quality of life.

Previous Treatment Attempts

It is important to identify previous treatment attempts. Which preventive medications have been tried? In what doses? Were there any side effects? Was the medication effective, and why was it discontinued? Which over-the-counter medications have been used, currently or in the past? The frequency of use is an important question, since over-the-counter combination medications are common causes of medication overuse headache. Has the patient tried any acute medications for migraine? Were they effective, and were there any side effects? Which rescue medications or painkillers have been tried? Are they in current use?

Have any "alternative" or "complementary" strategies been used? Has the patient tried biofeedback, yoga, relaxation training, or mindfulness techniques? Has the patient ever been treated with nerve blocks, stimulators, or surgery to treat migraine? Have they tried acupuncture or counseling? Have they ever received injections of onabotulinum toxin?

EXAMINATION OF A HEADACHE PATIENT

A general physical examination is important, especially to determine headaches related to systemic illness. Measure vital signs, height, and weight. Postural blood pressure measurements are helpful if the patient has complained of lightheadedness, for the possible use of medications that could cause drops in blood pressure. Listen to the lungs for wheezing, the heart for arrhythmia, and observe the general habitus and posture. Note grooming and behaviors during the examination, pain provoked by moving onto the examination table, and any peculiarities about clothing, piercings, and other body modification art.

A careful neurological examination is important to document any abnormalities, which may alter your diagnosis of a primary headache disorder. It will be useful for comparison if the patient should later develop neurological symptoms. Examine carefully the cranial nerves, motor, sensory, coordination, and gait functions, as well as the patient's affect, speech, and appropriateness.

The eyes should be examined[25] for extraocular movement abnormalities and the presence of nystagmus. Attention should be paid to pupillary symmetry and reactivity. The fundoscopic examination is important to observe for papilledema. Confrontation visual field testing is valuable; test each eye independently. Visual acuity testing is also important.

It is important to auscultate the carotid arteries for bruits (at any age) and to palpate the superficial temporal arteries for induration, reduced pulsations, and tenderness. Examine the temporalis muscles for tenderness or asymmetry.

Examination of the head and neck includes observation of cervical range of motion to rotation, flexion, and extension. The cervical muscles and trapezius muscles should be palpated for tenderness and trigger points (hard, tender areas). The posterior cranium should be palpated for tenderness, especially over the upper cervical muscles and over the greater occipital nerves. Observe the patient's posture and symmetry of shoulder height. If the patient describes a specific location of pain on the head or face, gently palpate it. This is particularly true when pain is located in specific nerve territories such as the greater occipital or supraorbital nerves.

TESTING

For patients with longstanding intermittent classical headache presentations such as recurrent migraine or cluster occurring in classical patterns, scanning has a very low yield. In a new patient with headache, the various causes of secondary headaches must be considered. The American Academy of Neurology (AAN) and the American Headache Society participate in the Choosing Wisely® movement started by the American Board of Internal Medicine (ABIM Foundation) to reduce unnecessary testing.

The use of electroencephalography (EEG) in headaches has not been justifiable since the advent of the CT scanner during the 1970s. One of the top recommendations of the AAN is *not* to use EEG to evaluate headache disorders.[26]

The American Headache Society Choosing Wisely Task Force[27] recommends against performing neuroimaging studies in patients with stable headaches that meet typical criteria for migraine. If scanning is needed, magnetic resonance imaging (MRI) is recommended instead of computed tomography (CT) scanning for headache, except in emergency settings.

When meeting a headache patient for the first time, the current author obtains routine labs, including complete blood count (CBC), liver enzymes, blood urea nitrogen (BUN), and creatinine, to determine adverse effects of self-medication and confirm that any planned medications will not cause problems. If not recently tested, look at TSH and B12 levels because of their effect on the nervous system.

COMPUTER-ASSISTED HISTORY TAKING

Many clinics obtain clinical information from a patient prior to the visit, gleaning parts of the history from paper questionnaires, computer programs, or by using an assistant. These methods may not be as accurate as a personal interview with a seasoned clinician.

It has been shown that one of the most *inaccurate* methods of obtaining a medical history is with a written questionnaire. Unless a clinician reviews each point individually, there is a large possibility of error.

A report from the Institute of Medicine[28] stated that about 90 million Americans have difficulty understanding and using written health information. Many people cannot read well enough to understand a simple questionnaire. It has been shown that self-administered questionnaires are unsatisfactory in diagnosing headache,[29] although some questionnaires have been useful in epidemiological studies.[30,31]

This author has been using computer-assisted history taking[32] since the 1980s, and application of the technique has evolved as technology has improved. Surprisingly, little has been written about the use of computers in taking medical histories, compared with the volumes written about decision support and electronic medical records. An excellent review appeared several years ago, and the points remain valid.[33]

It has been the author's experience that expert systems, which simulate human reasoning, are helpful but still do not replace the flexible mind of an experienced clinician. Such systems, however, are extremely useful in gathering structured information, validating it by seeking additional information, and clarifying inconsistencies. This is true in the general medical history and in screening for specific problems with sleep and mood. The author has provided a version of this expert system, available as a public service for patients, at www.arbormedicus.com. Remember, the computerized medical history does not replace interaction with a skilled clinician; it just makes it more efficient and thorough. It is also important not to depend entirely on the results of a computerized history. Critical points must be verified in a patient interview.

TELEMEDICINE

In the context of this chapter, telemedicine is defined as the act of delivering headache care to a patient at a remote location. Theoretically, this can be done by phone, email, or "snail mail," but the face-to-face encounter is essential when dealing with the complexities of the headache patient. This chapter has emphasized the behavioral comorbidities related to headache disorders and how the patient encounter should be as personal as possible. For this reason, video telemedicine would be essential to optimize patient bonding at a distance.

In this author's opinion, telemedicine should not replace the initial patient encounter because a physical examination is required and there needs to be an initial personal human contact. Telemedicine should not be thought of as video-on-demand medicine, but it can provide specialty care to patients in remote locations where such care would not otherwise be available.

The actual logistics of telemedicine can vary with location. Telemedicine centers may have a "set" where the clinician sits in front of a camera and accesses the patient's electronic record on another screen. Patients can be at home, if they have the video capabilities, or could go to a remote telemedicine center close to their home where an on-site assistant can monitor blood

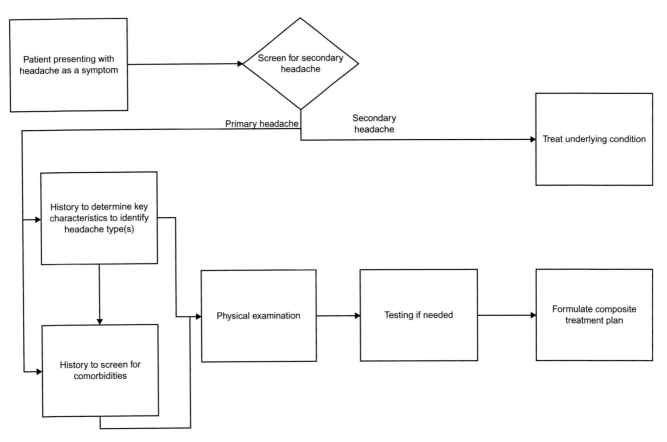

FIGURE 3.2   Overview of the headache evaluation process.

pressure or perform an examination as directed by the remote doctor. The future holds many interesting developments, such as eyeglass-mounted cameras and patient-mounted electronic monitoring devices.

This chapter describes the complexities of the headache patient and the advantages of computer-assisted history taking. Thus, it would be logical that expert history-taking systems are incorporated into the telemedicine process. Again, it is essential that the patients realize that they are ultimately dealing with a fellow human who considers them important enough to spend time to solve their problems, whether they are in the same room or many miles away.

SUMMARY

This chapter discusses the need to identify secondary headache and the different types of primary headache(s), and to identify comorbidities, in order to develop a comprehensive treatment plan (Figure 3.2). It should be clear to the reader that direct human interaction is key to diagnosing and managing the headache patient.

References

1. Cady RK, Schreiber CP, Billings C. Subjects with self-described "sinus" headache meet IHS diagnostic criteria for migraine. Poster presented at the 10th International Heacache Congress, June 29–July 2, 2001; New York, NY. Poster Session IA: Diagnosis and Clinical Features, P1–A17. (Abstract published in Cephalalgia, 2001; 21:241–245.)
2. The International Classification of Headache Disorders, 3rd ed (beta version). *Cephalalgia* 2013;33:629–808.
3. Dodick DW. Pearls: headache. *Semin Neurol.* 2010;30:74–81.
4. Lipton RB, Buse DC, Hall CB, Tennen H, DeFreitas TA, Borkowski TM, et al. Reduction in perceived stress as a migraine trigger. *Neurology.* 2014;82(16):1395–1401.
5. Lipton RB, Silberstein SD. Why study the comorbidity of migraine? *Neurology.* 1994;44(10, Suppl 7):S4–S5.
6. Breslau N, Lipton RB, Stewart WF, Schultz LR, Welch KMA. Comorbidity of migraine and depression: investigating potential etiology and prognosis. *Neurology.* 2003;60:1308–1312.
7. Lipton RB, Stewart WF. Migraine headaches: epidemiology and comorbidity. *Clin Neurosci.* 1998;5:2–9.
8. Kroenke K, Spitzer RL, Williams JB. The patient health questionnaire-2: validity of a two-item depression screener. *Med Care.* 2003;41:1284–1294.
9. Kroenke K, Spitzer R, Williams W. The PHQ-9: validity of a brief depression severity measure. *J Gen Intern Med.* 2001;16:606–616.
10. Beck AT, Epstein N, Brown G, Steer RA. An inventory for measuring clinical anxiety: psychometric properties. *J Consult Clin Psychol.* 1988;56(6):893–897.
11. Maizels M, Smitherman TA, Penzien DB. A review of screening tools for psychiatric comorbidity in headache patients. *Headache: J Head Face Pain.* 2006;46(Issue Supplement s3):S98–S109.
12. Kelman L, Rains JC. Headache and sleep: examination of sleep patterns and complaints in a large clinical sample of migraineurs. *Headache: J Head Face Pain.* 2005;45(7):904–910.
13. Rains JC, Poceta JS. Headache and sleep disorders: review and clinical implications for headache management. *Headache: J Head Face Pain.* 2006;46(9):1344–1363.
14. Johns MW. Reliability and factor analysis of the epworth sleepiness scale. *Sleep,* 15(4):376–381.
15. Bastien CH, Valliéres A, Morin CM. Validation of the insomnia severity index as an outcome measure for insomnia research. *Sleep Med.* 2001;2:297–307.
16. Webster LR, Choi Y, Desai H, Webster L, Grant BJ. Sleep disordered breathing and chronic opioid therapy. *Pain Med.* 2008;9(4):425–432.
17. International Classification of Sleep Disorders. 3rd ed. American Academy of Sleep Medicine, Darien, IL.
18. Peres MFP, Young WB, Kaup AO, Zukerman E, Silberstein SD. Fibromyalgia is common in patients with transformed migraine. *Neurology.* 2001;57:1326–1328.
19. Evans RW, de Tommaso M. Migraine and fibromyalgia. *Headache.* 2011;51:295–299.
20. Clauw DJ. Fibromyalgia a clinical review. *JAMA.* 2014;311 (15):1547–1555.
21. Tietjen GE, Brandes JL, Digre KB, Baggaley S, Martin V, Recober A, et al. High prevalence of somatic symptoms and depression in women with disabling chronic headache. *Neurology.* 2007;68 (2):134–140.
22. Bigal ME, Kurth T, Santanello N, Buse D, Golden W, Robbins M, et al. Migraine and cardiovascular disease: A population-based study. *Neurology.* 2010;74(8):628–635.
23. Kirkham KE, Colon RJ, Solomon GD. The role of cardiovascular screening in headache patients. *Headache: J Head Face Pain.* 2011;51(2):331–337.
24. Stewart WF, Lipton RB, Dowson AJ, Sawyer J. Development and testing of the migraine disability assessment (MIDAS) questionnaire to assess headache-related disability. *Neurology.* 2001;56 (suppl. 1):S20–S28.
25. Friedman DI, Digre KB. Headache medicine meets neuro-ophthalmology: exam techniques and challenging cases. *Headache: J Head Face Pain.* 2013;53(4):703–716.
26. Langer-Gould AM, Anderson WE, Armstrong MJ, Cohen AB, Eccher MA, Iverson DJ, et al. FAAN, The American Academy of Neurology's Top Five Choosing Wisely recommendations. *Neurology.* 2013;81(11):1004–1011.
27. Loder E, Weizenbaum E, Frishberg B, Silberstein S, American Headache Society Choosing Wisely Task Force. Choosing wisely in headache medicine: the American Headache Society's list of five things physicians and patients should question. *Headache: J Head Face Pain.* 2013;53(10):1651–1659.
28. Nielsen-Bohlman L, Panzer AM, Hamlin B, Kindig DA, eds. Health Literacy: A Prescription to End Confusion. Committee on Health Literacy Board on Neuroscience and Behavioral Health, April 8, 2004.
29. Rasmussen BK, Jensen R, Olesen J. Questionnaire versus clinical interview in the diagnosis of headache. *Headache: J Head Face Pain.* 1991;31(5):290–295.
30. Torelli P, Beghi E, Manzoni GC. Validation of a questionnaire for the detection of cluster headache. *Headache: J Head Face Pain.* 2005;45(6):644–652.
31. Abrignani G, Ferrante T, Castellini P, Lambru G, Beghi E, Manzoni GC, et al. Description and validation of an Italian ICHD-II-based questionnaire for use in epidemiological research. *Headache: J Head Face Pain.* 2012;52(8):1262–1282.
32. Messina E. Computerized history taking. In: Jonathan Javitt, ed. *Computers in Medicine: Applications and Possibilities.* Philadelphia, PA: WB Saunders Co; 1986:159–186.
33. Bachman JW. The patient–computer interview: a neglected tool that can aid the clinician. *Mayo Clin Proc.* 2003;78:67–78.

4

Screening and Testing of the Headache Patient

George R. Nissan

Baylor University Medical Center, Baylor Headache Center, Dallas, Texas, and Texas A&M Health Science Center College of Medicine, Dallas, Texas, USA

INTRODUCTION

The majority of patients who present with headache have normal physical and neurologic examinations and most headache disorders can be diagnosed based on history without supplemental testing. However, it is necessary to distinguish primary headache disorders from secondary headaches. Since headache is a common presentation in both the primary care office and emergency department settings, it is important to complete a thorough headache history and physical examination. The differential diagnosis for headache is extensive. This chapter discusses diagnostic testing in the headache patient, including the use of neuroimaging, lumbar puncture, electroencephalography (EEG), and laboratory studies.

NEUROIMAGING

Following completion of the history and physical examination, deciding which patients will need neuroimaging studies is determined by multiple factors. These factors include new onset headache in patients greater than 50 or under 5 years of age; the first or worst headache in the patient's life (including "thunderclap headache"); new daily persistent headache; a progression of a pre-existing headache; headache with seizure activity; headache associated with focal neurologic symptoms or signs; headache associated with cancer or infectious diseases such as HIV; pregnancy; and headache associated with papilledema, cognitive impairment, or personality change (Table 4.1).[1] Despite the higher cost, magnetic resonance imaging (MRI) of the brain is the imaging modality of choice, but non-contrast computed tomography (CT) of the brain can generally exclude space-occupying lesions. CT is

TABLE 4.1 Clinical Situations to Consider Neuroimaging for Headaches[a]

1. The first or worst headache/thunderclap headache
2. New or unexplained neurologic signs or symptoms
3. Recent significant change in the pattern, frequency or severity of headaches
4. Headache always on the same side
5. New daily persistent headache or chronic daily headache
6. Headache that does not respond to treatment
7. New-onset headaches after age 50
8. New-onset headaches in patients with cancer or HIV infection
9. Headaches associated with symptoms and signs such as fever, stiff neck, papilledema, cognitive impairment or personality change
10. Patients with headaches and seizure

[a]*Data from Evans,[1] Diagnostic testing for migraine and other primary headaches. Neurol Clin 2009; 27:393–415.*

generally preferred to MRI for the diagnosis of acute subarachnoid hemorrhage and of bony abnormalities, and for the evaluation of acute head trauma. However, some disorders may be overlooked on routine CT of the brain, including vascular disease, neoplastic disease, cervicomedullary lesions, and infections.

MRI of the brain is more sensitive than CT for the detection of ischemia, posterior fossa and cervicomedullary lesions, white matter abnormalities, cerebral venous thrombosis, subdural and epidural hematomas, meningeal disorders (including sarcoid), cerebritis, brain abscess, pituitary lesions, and neoplasms – especially when located in the posterior fossa. MRI should also be considered for headache precipitated by exertion, cough, or Valsalva, which may be due to Chiari malformation or other intracranial

S. Diamond, R. K. Cady, M. L. Diamond & V. T. Martin (Eds):
Headache and Migraine Biology and Management.

DOI: http://dx.doi.org/10.1016/B978-0-12-800901-7.00004-5

mass lesions. If the headache is worse when standing, an MRI with gadolinium may detect cerebrospinal fluid (CSF) leak. If the headache is worse when lying down, an MRI can be obtained to rule out cerebral venous thrombosis. In instances when an MRI is contraindicated due to a pacemaker or aneurysm clip, a CT scan may be ordered. In cases of patients with claustrophobia, an open MRI with improved image quality can be obtained.

In cases of chronic sinusitis, coronal CT scan of the sinuses without contrast is the preferred initial neuroimaging study, which adequately displays soft-tissue attenuation. However, CT imaging is less helpful for acute sinusitis, which is diagnosed primarily by clinical findings. According to the American College of Radiology (ACR) Appropriateness Criteria, most uncomplicated acute or subacute sinus disease is a clinical diagnosis, and a non-contrast CT scan is the examination of choice in recurrent or chronic sinus disease. CT scan findings should be interpreted in conjunction with clinical and endoscopic findings.[2] MRI of the sinuses is reserved for aggressive disease with ophthalmic or intracranial complications, especially for characterization of a sinus mass, or fungal disease in an immunocompromised patient.

Neuroimaging with MRI of the brain in spontaneous intracranial hypotension, which is often due to a spontaneous CSF leak and is an uncommon cause of new daily persistent headache, may be normal. Common imaging findings include subdural fluid collections, pachymeningeal enhancement, pituitary hyperemia, and brain sagging.[3] CT-myelography is the study of choice to identify the CSF leak, but is not always necessary to make the diagnosis. For unusually rapid CSF leaks, particularly those ventral to the spinal cord, digital subtraction myelography or dynamic CT-myelography are indicated.[4] Some patients with spontaneous intracranial hypotension verified by intracranial MRI are never found to have a CSF leak using current techniques.

The American Academy of Neurology guidelines have suggested that neuroimaging is not indicated if the patient presents with symptoms consistent with migraine, has a normal neurological exam, and has no red flags such as thunderclap onset, fever or other systemic complaints, stiff neck, or papilledema.[5]

Subarachnoid Hemorrhage

For the detection of an acute subarachnoid hemorrhage (SAH), a CT scan without contrast is the neuroimaging study of choice.[6] If the scan is performed within 24 hours after the bleed, the blood is demonstrated within the subarachnoid space and ventricles in up to 92% of cases.[7] The CT scan should be performed with thin cuts through the base of the brain to increase its sensitivity.[8] The ability of CT to detect SAH is highest within the first 6 to 12 hours after the hemorrhage, and is then nearly 100%.[9] The sensitivity progressively decreases over time to nearly 58% at day 5. Since the consequences of missing SAH are grave, most guidelines mandate a follow-up lumbar puncture if the CT scan is negative and the diagnosis strongly suspected.[10] Limited data suggest that proton density and flair sequences on brain MRI may be as sensitive as head CT for the acute detection of SAH.[11] Flair and T2 sequences on MRI have a higher sensitivity for subacute cases of SAH in patients that present more than 4 days after the bleed.[12] As is the case with a negative CT scan, a follow-up lumbar puncture should be obtained if the MRI is negative.[10]

Cerebral Venous Thrombosis

Cerebral venous thrombosis (CVT) is an uncommon but serious disorder in which headache is the most frequent symptom, often in association with stroke symptoms and seizures. Headaches associated with CVT are more frequent in women and young patients compared with men or older adults.[13] Headache is usually the first symptom of CVT, and can be the only symptom or can precede other symptoms and signs by days or weeks. The headache onset is usually gradual, increasing over several days.[14] However, some patients with CVT have sudden explosive onset of severe head pain (i.e., thunderclap headache) that mimics subarachnoid hemorrhage.[15] The headache may also resemble migraine with aura.[16] In suspected cases of CVT, an MRI using T2 susceptibility-weighted sequences in combination with two-dimensional time of flight MR venography is the most sensitive imaging method for demonstrating the thrombus and the occluded dural sinus or vein.[17]

White Matter Abnormalities

Most patients with migraine never develop relevant changes on MRI. However, white matter abnormalities (WMA) are often detected on MRI but are rarely seen on CT scan of the brain. White matter abnormalities are foci of hyperintensity on both proton density and T2-weighted images in the deep and periventricular white matter and regions of the posterior circulation. The percentage of WMA for all types of migraine ranges from 12% to 46%.[18] In patients less than 40 years of age, white matter lesions are more commonly located in the centrum semiovale and frontal subcortical white matter. After the age of 40, there is a predilection for involvement of the deep white matter at the level of the basal ganglia.[19] The lesions correlate with increasing age but not with migraine subtypes. These lesions are more common with an increasing number of aura attacks, increasing headache frequency, or longer duration of the attack.[18] White matter lesion incidence occurs at the same incidence in tension-type headache as in migraine. A certain degree of white matter change is expected with age, although it does not mean that these changes are completely benign. Increased white matter hyperintensities are associated with higher risk of stroke and dementia, as well as higher mortality in general, although the lesions themselves may not be causing problems. White matter hyperintensities have been associated with diabetes, hypertension, heavy alcohol use, obesity, and cigarette smoking.[18] No clear consensus is established regarding the need for follow-up imaging of white matter abnormalities in a patient with no other intracranial imaging abnormalities.

LUMBAR PUNCTURE

The decision to proceed with a lumbar puncture is undertaken before intracranial imaging only if acute meningitis is suspected. Otherwise, an MRI or CT scan of the brain is obtained initially. CSF opening pressure is most accurately measured while the patient is in a lateral decubitus position and relaxed, with at least partial extension of the head and legs. Lumbar puncture can be diagnostic for subarachnoid hemorrhage, high or low CSF pressure, meningitis, encephalitis, meningeal carcinomatosis, or lymphomatosis. Prior to obtaining a lumbar puncture, the platelet count should be at least 50,000 if the patient has a history of blood dyscrasias.

A lumbar puncture should strongly be considered in the following circumstances: in a patient who presents with fever and stiff neck or other symptoms or signs suggestive of an infectious cause; if the headache is the first or worst of the patient's life; if the headache is atypical of primary headache disorders; or when changes in CSF pressure are suspected, as in pseudotumor cerebri, assuming no mass lesions are identified.

Subarachnoid Hemorrhage

Lumbar puncture (LP) is mandatory in patients for whom there is a strong suspicion of subarachnoid hemorrhage (SAH) and a normal head CT.[6] Clinical prediction rules are not sufficiently sensitive to exclude SAH without a lumbar puncture.[20] When analyzing cerebrospinal fluid, red blood cells are present in almost all cases of SAH, although may take hours to develop. The classic LP findings of SAH are an elevated opening pressure and an elevated red blood cell (RBC) count that does not diminish from CSF tube 1 to tube 4. The differential of RBCs between tubes 1 and 4, and immediate centrifugation of the CSF, can help differentiate bleeding in SAH from that due to a traumatic lumbar puncture.[21] The presence of xanthochromia, a pink or yellow supernatant that usually but not invariably represents hemoglobin degradation products, helps to distinguish SAH from a traumatic lumbar puncture if bloody CSF is obtained. The visualization of xanthochromia can be deceptive, and is best determined by spectrophotometry when available. The red blood cells usually clear within 6 to 30 days.

Bacterial and Aseptic Meningitis

For suspected cases of bacterial meningitis, the CSF should be sent for Gram stain and bacterial cultures, in addition to determination of CSF glucose, protein, cell counts, and assays for specific bacterial antigens and endotoxins. In acute bacterial meningitis, CSF findings can include glucose <40 mg/dL, protein level >45 mg/dL, and a pleocytosis with cell counts greater than 1000/mm^3 (Table 4.2). During the early stages of infection, lymphocytes predominate.

In aseptic meningitis, enteroviruses such as echovirus and Coxsackie viruses A and B are most commonly involved. Other less common viruses include cytomegalovirus, herpes simplex, adenovirus, rubella, mumps, Epstein-Barr, herpes zoster, influenza, parainfluenza, and human immunodeficiency virus (HIV). In cases of viral meningitis, CSF findings reveal a mild pleocytosis with a predominance of polymorphonuclear cells early in the infection. CSF glucose is usually normal or slightly decreased, and CSF protein is normal or mildly elevated. After analyzing the CSF for viral antigens and cultures, the virus responsible is usually identified in only 12% of cases.[22]

TABLE 4.2 Common Cerebrospinal Fluid (CSF) Findings in Acute Bacterial Meningitis

1. Positive Gram stain and/or culture

2. Elevated protein (greater than 45 mg/dL and often 100–500 mg/dL)

3. Decreased glucose (less than 40 mg/dL)

4. White blood cell count of 1000–5000/μL

5. Neutrophils usually greater than 80%

6. Elevated lactate dehydrogenase (LDH)

Encephalitis

Encephalitis, which may be due to viral, bacterial, fungal, or parasitic agents, is an inflammation of the brain parenchyma (the encephalon). This condition is in contrast to meningitis, which involves the meninges and not the brain parenchyma itself. The presence or absence of normal brain function is an important distinguishing feature between encephalitis and meningitis, although many patients have a meningo-encephalitis. Patients with meningitis typically have relatively normal cerebral function despite severe headache and lethargy. Patients with encephalitis have alterations in brain function including altered mental status, but may include motor or sensory deficits, altered behavior and personality changes, and speech or movement disorders. Encephalitis usually consists of an acute febrile illness with mental status changes, in addition to possible seizures, aphasia, hemiparesis, ataxia, myoclonic jerks, nystagmus, reflex asymmetry, and cranial neuropathies. Common causes of viral encephalitis include herpes simplex virus, influenza, Epstein Barr, measles, mumps, and arbovirus. The initial examination of the CSF, although not diagnostic, will usually confirm the presence of inflammatory disease of the CNS. The CSF findings with aseptic meningitis and encephalitis are generally indistinguishable. A common and treatable form of encephalitis, herpes simplex encephalitis, causes hemorrhagic encephalitis, such that the lumbar puncture also shows a large number of RBCs and hemorrhage into the temporal lobes.

Idiopathic Intracranial Hypertension (Pseudotumor Cerebri)

Idiopathic intracranial hypertension (IIH), which usually occurs in women who are often obese and of childbearing years, is a disorder of generalized brain edema with subsequent elevated CSF pressure. It is also called "pseudotumor cerebri." The patient primarily presents with headache and has no localizing neurologic findings. The headaches are often lateralized and throbbing or pulsatile in character. The headache may be intermittent or persistent, and may occur daily or less frequently. Associated nausea and vomiting can occur, similar to that seen in migraine. Some patients describe headache exacerbation with changes in posture, and some may report that relief occurs with non-steroidal anti-inflammatory medications (NSAIDs) and/or rest.[23] No evidence of obstruction of the ventricular system or secondary cause of elevated CSF pressure will be observed. Visual findings can include diplopia due to a cranial nerve VI paresis, transient visual obscurations, and papilledema. Neuroimaging, preferably with MRI of the brain, is required to exclude secondary causes of intracranial hypertension. If the neuroimaging study reveals no structural etiology for intracranial hypertension, a lumbar puncture is performed to document an opening pressure and to exclude other conditions. Ophthalmologic evaluation, including visual field testing, is required to document the severity of optic nerve involvement and monitor response to treatment. Lumbar puncture findings reveal elevated CSF pressure (>200 mm of water in the non-obese and >250 mm of water in the obese patient).[24] Reduction of CSF pressure with serial lumbar punctures has only short-lived effects and is still of unproven efficacy. During pregnancy, serial lumbar punctures can be undertaken in order to avoid medical therapy[25] and as a temporary measure prior to surgery in non-pregnant patients.

Acetazolamide is often used as a first-line agent for lowering intracranial pressure. In the event of intolerance to acetazolamide, loop diuretics such as furosemide may be used as a replacement. Many patients experience relief of headache symptoms with these treatments; however, the improvement in headache with medication treatment does not necessarily parallel the improvement in papilledema.[26] CSF diversion procedures (ventricular shunting) are highly effective in lowering intracranial pressure. In some facilities, they remain the procedures of choice for treating patients who do not respond to maximum medical treatment. Shunts are also indicated for patients with intractable headaches, patients living in regions where there is no access to a surgeon comfortable with optic nerve sheath fenestration, and patients in whom optic nerve sheath fenestration has failed. Other treatments, such as weight reduction, may be more effective and may have less associated morbidity.[27] The ophthalmic surgical approach to managing patients with progressive vision loss and papilledema involves cutting slits or rectangular patches in the dura surrounding the optic nerve immediately behind the globe.[28] Optic nerve sheath fenestration has been demonstrated to reverse optic nerve edema and to bring about some recovery of optic nerve function. In addition, it may decrease headache in many patients. Despite the general lack of an intracranial pressure-lowering effect, unilateral surgery occasionally has a bilateral curative effect on the papilledema. However, if this is not the case, the opposite nerve must undergo the same procedure.[29]

ELECTROENCEPHALOGRAPHY

Prior to the era of CT and MRI technology, the electroencephalogram (EEG) was often considered in the evaluation of headache.[30] However, in recent years, practice parameter suggestions have refuted the need to obtain an EEG in the evaluation of a patient who presents with headache. The first report from the Quality Standards Subcommittee of the American Academy of Neurology (AAN), in 1995,[31] suggested that:

> The electroencephalogram (EEG) is not useful in the routine evaluation of patients with headache. This does not preclude the use of EEG to evaluate headache patients with associated symptoms suggesting a seizure disorder, such as atypical migrainous aura or episodic loss of consciousness. Assuming head imaging capabilities are readily available, EEG is not recommended to exclude a structural cause for headache.

In 2013, the AAN published an article regarding the "Top Five Choosing Wisely Recommendations".[32] The first of the five recommendations states "Don't perform EEGs for headaches. EEG has no advantage in diagnosing headache, does not improve outcomes and increases cost. Recurrent headache is the most common pain problem, affecting 15% to 20% of people."

For headaches associated with what appears to be seizure activity, routine EEG is often deceptive in differentiating epileptic seizures from psychogenic non-epileptic seizures.[33] A normal interictal EEG does not exclude the possibility of epilepsy or confirm the diagnosis of psychogenic non-epileptic seizures. Video-EEG monitoring combines extended EEG monitoring with time-locked video acquisition, allowing for analysis of clinical and electrographic features during a captured event. The yield of monitoring is high, and 73% to 96% of patients will have typical psychogenic non-epileptic seizures within the first 48 hours.[34] These patients often need neuropsychological evaluation to assess both cognitive functioning and psychological domains. Personality profile testing such as the Minnesota Multiphasic Personality Inventory (MMPI) can be helpful in supplementing formal psychiatric evaluations and highlighting comorbid psychiatric disturbances.[35] Patients with psychogenic non-epileptic seizures typically have high scores on somatization, hypochondriasis, and hysteria subscales.[36]

LABORATORY STUDIES

Clinical laboratory studies are generally not helpful in the diagnosis of common headache disorders. However, certain headache disorders warrant the use of laboratory studies. In suspected cases of giant cell arteritis, erythrocyte sedimentation rate (ESR) and C-reactive protein (CRP) are indicated. In patients with headaches and arthralgias, ESR, rheumatoid factor (RF), and anti-nuclear antibody (ANA) may be obtained to diagnose collagen vascular disease such as lupus.[37] Testing for mononucleosis should be undertaken in adolescents who present with headaches, sore throat, and cervical adenopathy. A complete blood count (CBC), liver function tests, Lyme antibody, and HIV testing should be utilized in the screening of the patient with a suspected infection or a refractory chronic daily headache pattern. A CBC can diagnose anemia since headache may be a symptom of anemia, especially if the hemoglobin concentration is reduced by one-half or more. CBC and platelet count can also be obtained to rule out thrombotic thrombocytopenic purpura (TTP) as a cause of headache.

Thyroid Function Studies

Headache can be a symptom in up to 14% of cases of hypothyroidism. Thyroid function testing, including thyroid stimulating hormone (TSH), free T4, and T3, is undertaken in the initial evaluation of the headache patient, especially in adolescents and young adults. However, the American Academy of Neurology has stated that there is inadequate documentation in the medical literature to support laboratory and diagnostic studies in the child and adolescent recurrent headache populations if the physical examination is normal.[38]

Other Laboratory Studies

Blood urea nitrogen (BUN) and creatinine may be obtained to exclude renal failure as a possible cause of headache. Serum calcium can exclude hypercalcemia, which can be associated with headache. If a pituitary lesion is suspected, prolactin and cortisol may be obtained, in addition to neuroimaging with MRI of the brain since MRI is the best imaging modality for detecting sellar and suprasellar masses. In headache associated with seizure activity, prolactin levels may also be elevated, but only immediately after an epileptic seizure – depending on the seizure type.[39] However, prolactin is unreliable as a seizure marker and is no longer used in this setting.

Medication Compliance Monitoring

For patients using migraine or cluster headache preventive medication therapy, clinical blood levels may be indicated to monitor medication compliance and potential toxicity. Some of these medications include lithium, valproic acid, carbamazepine, and tricyclic antidepressants. A urine drug screen can exclude potential addictive exogenous sources that may contribute to headache diagnosis and treatment.

The urine drug screen may be considered not only in the initial evaluation of the headache patient, but also in follow-up office visits for appropriate patients.

Genetic Testing

In recent years, investigations have shown that genetic factors may play an important role in the three most prevalent headache syndromes: migraine, cluster headache, and tension-type headache. However, the routine use of genetic testing in the screening of primary headache disorders is not generally recommended. Genetic testing is sometimes considered in cases of suspected familial hemiplegic migraine (FHM), which is an autosomal-dominant form of migraine with aura. In addition to symptoms characteristic of migraine with aura, patients with FHM may present with hemiparesis that can last from minutes to weeks in duration and is reversible, resolving without sequelae. In 1996, a voltage-gated P/Q-type calcium channel alpha 1A-subunit gene (CACNA1A) was identified on chromosome 19 in the FHM candidate region.[40] To date, the first three types of FHM, which are channelopathies that are numbered according to the gene involved, account for some but not all of the cases of FHM. FHM1 is caused by mutations in the CACNA1A gene, FHM2 is caused by mutations in the ATP1A2 gene, and FHM3 is caused by mutations in the SCN1A gene. The diagnosis of familial hemiplegic migraine is predominantly a clinical one. However, genetic testing for mutations involving the genes that cause familial hemiplegic migraine (CACNA1A, ATP1A2, and SCN1A) may be most useful for patients with early onset severe sporadic hemiplegic migraine associated with nystagmus, seizures, or other atypical neurologic manifestations.[41] In addition, genetic testing may be helpful for cases of familial hemiplegic migraine when the attack severity or permanent neurologic manifestations are dissimilar from those of affected relatives.[42] The yield of genetic testing is low in patients with adult onset hemiplegic migraine, particularly when there are no associated permanent neurologic features.

Migraine with aura occurs in about 30% of cerebral autosomal dominant arteriopathy with subcortical infarcts and leukoencephalopathy (CADASIL) cases, and is usually an early sign of the disease.[43] CADASIL is usually suspected when there is a positive family history for stroke or dementia in patients who have typical clinical signs and extensive white matter changes on brain MRI. It is an autosomal-dominant inherited angiopathy that, in the 1990s, was shown to be caused by mutations in the NOTCH3 gene on chromosome 19.[44] CADASIL is now recognized as an important cause of stroke in the young. Genetic screening does not detect all patients with CADASIL. In up to 4% of patients, sequencing of all exons fails to identify a mutation.[45] As a result, a skin biopsy is indicated if genetic testing is negative in patients in whom there is a high index of clinical suspicion for the diagnosis of CADASIL.

CONCLUSION

Although the history and physical examination remain the most important components in the evaluation of the headache patient, it is necessary to differentiate between primary and secondary headache disorders. This may include evaluation with neuroimaging, laboratory studies, and lumbar puncture. The routine use of electroencephalography (EEG) in the evaluation of the headache patient is generally inappropriate. The recommendations have provided a framework to limit the use of neuroimaging with CT or MRI of the brain to specific high-risk headache presentations. Despite recent advances in the understanding of the role of genetics in primary headache disorders, the use of genetic testing is not commonplace in the clinical setting.

References

1. Evans RW. Diagnostic testing for migraine and other primary headaches. *Neurol Clin.* 2009;27:393−415.
2. Cornelius RS, Martin J, Wippold FJ, Aiken AH, Angtuaco EJ, Berger KL, et al. ACR appropriateness criteria sinonasal disease. *J Am Coll Radiol.* 2013;10:241−246.
3. Scheivink WI. Spontaneous cerebrospinal fluid leaks. *Cephalalgia.* 2008;28:1345−1356.
4. Scheivink WI. Novel neuroimaging modalities in the evaluation of spontaneous cerebrospinal fluid links. *Curr Neurol Neurosci Rep.* 2013;13:358.
5. American Academy of Neurology. The utility of neuroimaging in the evaluation of headache in patients with normal neurological examination. *Neurology.* 1994;44:1353−1354.
6. Vermeulen M, van Gijn J. The diagnosis of subarachnoid haemorrhage. *J Neurol Neurosurg Psychiatry.* 1990;53:365−372.
7. Bederson JB, Connolly ES, Batjer HH, Dacey RG, Dion JE, Diringer MN, et al. Guidelines for the management of aneurysmal subarachnoid hemorrhage: a statement for healthcare professionals from a special writing group of the Stroke Council, American Heart Association. *Stroke.* 2009;40:994−1025.
8. Latchaw RE, Silva P, Falcone SF. The role of CT following aneurysmal rupture. *Neuroimaging Clin N Am.* 1997;7:693−708.
9. Perry JJ, Stiell IG, Sivilotti ML, Bullard MJ, Emond M, Symington C, et al. Sensitivity of computed tomography performed within six hours of onset of headache for diagnosis of subarachnoid haemorrhage: prospective cohort study. *BMJ.* 2011;343:d4277.
10. Connolly Jr ES, Rabinstein AA, Carhuapoma JR, Derdeyn CP, Dion J, Higashida RT, et al. Guidelines for the management of aneurysmal subarachnoid hemorrhage: a guideline for

healthcare professionals from the American Heart Association/ American Stroke Association. *Stroke.* 2012;43:1711–1737.

11. Wiesmann M, Mayer TE, Yousry I, Medele R, Hamann GF, Brückmann H. Detection of hyperacute subarachnoid hemorrhage of the brain by using magnetic resonance imaging. *J Neurosurg.* 2002;96:684–689.

12. Mitchell P, Wilkinson ID, Hoggard N, Paley MN, Jellinek DA, Powell T, et al. Detection of subarachnoid haemorrhage with magnetic resonance imaging. *J Neurol Neurosurg Psychiatry.* 2001;70:205–211.

13. Coutinho JM, Ferro JM, Canhão P, Barinagarrementeria F, Cantú C, Bousser MG, et al. Cerebral venous and sinus thrombosis in women. *Stroke.* 2009;40:2356–2361.

14. Stam J. Thrombosis of the cerebral veins and sinuses. *N Engl J Med.* 2005;352:1791–1798.

15. de Bruijn SF, Stam J, Kappelle LJ. Thunderclap headache as first symptom of cerebral venous sinus thrombosis. CVST Study Group. *Lancet.* 1996;348:1623–1625.

16. Slooter AJ, Ramos LM, Kappelle LJ. Migraine-like headache as the presenting symptom of cerebral venous sinus thrombosis. *J Neurol.* 2002;249:775–776.

17. Chu K, Kang DW, Yoon BW, Roh JK. Diffusion-weighted magnetic resonance in cerebral venous thrombosis. *Arch Neurol.* 2001;58:1569–1576.

18. Bashir A, Lipton RB, Ashina S, Ashina M. Migraine and structural changes in the brain: a systematic review and meta-analysis. *Neurology.* 2013;81:1260–1268.

19. Hamedani AG, Rose KM, Peterlin BL, Mosley TH, Coker LH, Jack CR, et al. Migraine and white matter hyperintensities: the ARIC MRI study. *Neurology.* 2013;81:1308–1315.

20. Mark DG, Hung YY, Offerman SR, Rauchwerger AS, Reed ME, Chettipally U, et al. Nontraumatic subarachnoid hemorrhage in the setting of negative cranial computed tomography results: external validation of a clinical and imaging prediction rule. *Ann Emerg Med.* 2013;62:1–10.

21. Wijdicks EF, Kallmes DF, Manno EM, Fulgham JR, Piepgras DG. Subarachnoid hemorrhage: neurointensive care and aneurysm repair. *Mayo Clin Proc.* 2005;80:550–559.

22. Kupila L, Vuorinen T, Vainionpää R, Hukkanen V, Marttila RJ, Kotilainen P. Etiology of aseptic meningitis and encephalitis in an adult population. *Neurology.* 2006;66:75–80.

23. Wall M. The headache profile of idiopathic intracranial hypertension. *Cephalalgia.* 1990;10:331–335.

24. Whiteley W, Al-Shahi R, Warlow CP, Zeidler M, Lueck CJ. CSF opening pressure: reference interval and the effect of body mass index. *Neurology.* 2006;67:1690–1691.

25. Evans RW, Friedman DI. Expert opinion: the management of pseudotumor cerebri during pregnancy. *Headache.* 2000; 40:495–497.

26. Friedman DI, Jacobson DM. Idiopathic intracranial hypertension. *J Neuroophthalmol.* 2004;24:138–145.

27. Brazis PW. Clinical review: the surgical treatment of idiopathic pseudotumour cerebri (idiopathic intracranial hypertension). *Cephalalgia.* 2008;28:1361–1373.

28. Sinclair AJ, Kuruvath S, Sen D, Nightingale PG, Burdon MA, Flint G. Is cerebrospinal fluid shunting in idiopathic intracranial hypertension worthwhile? A 10-year review. *Cephalalgia.* 2011;31:1627–1633.

29. Goh KY, Schatz NJ, Glaser JS. Optic nerve sheath fenestration for pseudotumor cerebri. *J Neuroophthalmol.* 1997;17:86–91.

30. Gronseth GS, Greenberg MK. The utility of electroencephalogram in the evaluation of patients presenting with headache: a review of the literature. *Neurology.* 1995;45:1263–1267.

31. American Academy of Neurology. Practice parameter: the electroencephalogram in the evaluation of headache. *Neurology.* 1995;45:1411–1413.

32. American Academy of Neurology. Five things patients and physicians should question. Choosing Wisely: An Initiative of the ABIM Foundation 2014; <www.choosingwisely.org/doctor-patient-lists/american-academy-of-neurology>.

33. Lesser RP. Psychogenic seizures. *Neurology.* 1996;46:1499–1507.

34. Cragar DE, Berry DT, Fakhoury TA, Cibula JE, Schmitt FA. A review of diagnostic techniques in the differential diagnosis of epileptic and nonepileptic seizures. *Neuropsychol Rev.* 2002;12:31–64.

35. Dikmen S, Hermann BP, Wilensky AJ, Rainwater G. Validity of the Minnesota Multiphasic Personality Inventory (MMPI) to psychopathology in patients with epilepsy. *J Nerv Ment Dis.* 1983;171:114–122.

36. Schramke CJ, Valeri A, Valeriano JP, Kelly KM. Using the Minnesota Multiphasic Inventory 2, EEGs, and clinical data to predict nonepileptic events. *Epilepsy Behav.* 2007;11:343–346.

37. Amit M, Molad Y, Levy O, Wysenbeek AJ. Headache in systemic lupus erythematosus and its relation to other disease manifestations. *Clin Exp Rheumatol.* 1999;17:467–470.

38. Lewis DW, Ashwal S, Dahl G, Dorbad D, Hirtz D, Prensky A, et al. Practice parameter: evaluation of children and adolescents with recurrent headaches: report of the Quality Standards Subcommittee of the American Academy of Neurology and the Practice Committee of the Child Neurology Society. *Neurology.* 2002;59:490–498.

39. Wyllie E, Lüders H, MacMillan JP, Gupta M. Serum prolactin levels after epileptic seizures. *Neurology.* 1984;34:1601–1604.

40. Ophoff RA, Terwindt GM, Vergouwe MN, van Eijk R, Oefner PJ, Hoffman SM, et al. Familial hemiplegic migraine and episodic ataxia are caused by mutations in the Ca2+ channel gene channel CACNL1A4. *Cell.* 1996;87:543–552.

41. Russell MB, Ducros A. Sporadic and familial hemiplegic migraine: pathophysiological mechanisms, clinical characteristics, diagnosis, and management. *Lancet Neurol.* 2011;10:457–470.

42. Riant F, Ducros A, Ploton C, Barbance C, Depienne C, Tournier-Lasserve E. De novo mutations in ATP1A2 and CACNA1A are frequent in early-onset sporadic hemiplegic migraine. *Neurology.* 2010;75:967–972.

43. Liem MK, Oberstein SA, van der Grond J, Ferrari MD, Haan J. CADASIL and migraine: a narrative review. *Cephalalgia.* 2010;30:1284–1289.

44. Joutel A, Corpechot C, Ducros A, Vahedi K, Chabriat H, Mouton P, et al. Notch3 mutations in CADASIL, a hereditary adult-onset condition causing stroke and dementia. *Nature.* 1996;383:707–710.

45. Peters N, Opherk C, Bergmann T, Castro M, Herzog J, Dichgans M. Spectrum of mutations in biopsy-proven CADASIL: implications for diagnostic strategies. *Arch Neurol.* 2005;62:1091–1094.

5

Overview of Migraine: Recognition, Diagnosis, and Pathophysiology

Mark W. Green

Department of Neurology, Icahn School of Medicine at Mt Sinai, New York, New York, USA

RECOGNITION OF MIGRAINE

Migraine is among the primary headache disorders — those without any underlying structural, infectious, or metabolic cause. Migraine is a common, debilitating, and recurring affliction affecting 18% of women and 6% of men.[1] In children the prevalence is higher in males.[2] It is best to understand migraine as a phenotype that is comprised of multiple genotypes, most of which have not yet been identified. The genotype confers sensitivity for a variety of triggers to precipitate an attack of migraine. There are multiple subtypes of migraine, but the most common are *migraine without aura* (75%) followed by *migraine with aura* (25%). The term "migraine without aura" refers to headache attacks that are not preceded by or accompanied by any focal neurological complaints. In many ways, migraine reflects a lowered threshold for the development of headache. Migraine attacks are often triggered events, yet the triggers of migraine are mundane. Most individuals might develop a headache should the trigger be severe enough. Surprisingly, headache is not a *sine qua non* for the diagnosis of migraine, not more important for diagnosis than symptoms like vomiting and aura. In most cases of migraine, sufferers feel normal between attacks but retain a low threshold for the development of subsequent attacks. Additionally, even interictally, migraineurs show a hypersensitivity to sensory stimuli and have abnormal processing of sensory information.[3,4]

Tension-type headache is four times more common than migraine,[5] but is rarely disabling and is self-treated. The Landmark Study evaluated a multinational cohort of patients consulting a primary care physician with the complaint of episodic headache.[6] In 94% of cases, these patients fulfilled the ICHD criteria for migraine or migrainous headache.[7] It further documents that although tension-type headache is the most common headache, the overwhelming majority of patients seeking medical care have migraine.

HISTORY OF MIGRAINE

The term "migraine" is derived from the Greek *hemi* and *kranion*, although it is now recognized that migraine is often not hemicranial. Hippocrates described prodromata and stated that combinations of humors gave rise to headache with vomiting and gastric upset. The Egyptians, in 3500 BC, described a remedy for "warmth in the head" composed of a moistened clay crocodile bound around the temples in an attempt to cool the head and obtain relief. By 1883, in Germany, Eulenburg used injections of ergot extract in five cases of headache, which he referred to as a "vasoparalytic form of hemicrania." In 1938, Wolff and Graham demonstrated that the amplitude of superficial temporal pulsations diminished with the successful amelioration of migraine pain using intravenous ergotamine, a vasoconstrictor.[8] This led to the theory that migraine attacks had two components. The aura phase, when present, was felt to be due to vasoconstriction and the headache phase, when present, was felt to be due to vasodilatation. The studies that led to this theory were often performed on medical residents, and likely would not be acceptable under the current system requiring IRB approvals. However, this vascular theory was taught for decades, as pharmaceutical companies rushed to develop even more powerful vasoconstrictors. In 1941, Lashley, a psychologist, observed his own visual aura and measured the progression of the event across his visual field. He calculated the speed of the event moving across his visual

S. Diamond, R. K. Cady, M. L. Diamond & V. T. Martin (Eds):
Headache and Migraine Biology and Management.

DOI: http://dx.doi.org/10.1016/B978-0-12-800901-7.00005-7

cortex at 3 mm/min. Because his migraine aura symptoms propagated, he argued that a neural disturbance causing it must have propagated as well.[9] Leão, 3 years later, upon applying potassium or glutamate to the cortex of rabbits while studying epilepsy, demonstrated that a wave of depolarization of neurons occurred. This was followed by lengthy suppression of neuronal activity progressing at a rate of 3 mm/min, in agreement with the calculation of a possible propagation of a wave at the same speed. This phenomenon was named "cortical spreading depression" (CSD).[10]

EPIDEMIOLOGY OF MIGRAINE

Migraine is highly prevalent in the population, with the cumulative prevalence being 43% in females and 18% in males. In young children migraine is more prevalent in males, but by puberty females predominate.[11]

It is likely that the expression of migraine emanates from those with genetic underpinnings who encounter internal and external triggers. Internal factors can include stress or the let-down following stress, menstruation or ovulation, and sleep — generally too little or too much. Examples of external factors include alcohol, weather changes, missing meals, dehydration, glare, and flickering lights. Changes in environment are the common denominator to most triggers. Charles Darwin, who suffered from migraine, had headaches precipitated by any change in his routine. On his impending wedding to Emma Wedgwood, he wrote: "My last two days in London, when I wanted to have the most leisure, were rendered very uncomfortable by a bad headache, which continued for two days and two nights, so that I doubted whether it ever meant to go and allow me to be married".[12] Those with migraine have a cerebral cortex that appears to be hyperexcitable. Triggers that are innocuous to most may not be innocuous to those with migraine. Most individuals without migraine can have several glasses of red wine and not suffer a headache, but those with migraine might suffer a severe attack with as little as one glass.

Two premonitory symptoms of migraine need to be distinguished: *prodrome* and *aura*. Most forms of migraine are preceded by often vague symptoms known as prodromes. Kellman reported that the average duration of a prodrome was 9.4 hours.[13] That study reported the most prevalent prodromes to be tiredness, mood change (which can be depression or euphoria), and gastrointestinal symptoms. Prodromes also include yawning, cold hands and feet, and food cravings. Often, a prodromal symptom is mistaken for a migraine trigger. There is little evidence that supports chocolate as a migraine trigger, yet a craving for chocolate can be a prodromal symptom. Those acting upon this craving will then blame chocolate for triggering an attack. In Blau's review on prodromes in 1980, he wrote that "George Elliot felt 'dangerously well before an attack' and Sir John Forbes had an 'irresistible and horrid drowsiness'; Lady Conway ate her supper with a 'greedy appetite' and DuBois Reymond's migraines were 'in general preceded by constipation'".[14]

Migraine is typically an episodic disorder with significant recovery between attacks. However, some individuals have chronic migraine, defined as 15 or more days monthly with headache. One-quarter of migraines occur with aura. The term "aura" is unfortunate in that auras can occur before an attack, during it, and occasionally in association with headache types other than migraine. Migraines are recurrent attacks that are associated with various autonomic phenomena such as photophobia, phonophobia, osmophobia, nausea, and vomiting. Often, a generalized sensory hypersensitivity is experienced. All of these features are not required in order to make a diagnosis of migraine.

The international classification of headache disorders was initially developed primarily for use in research studies. It is possible for the headache to be bilateral and non-pulsatile, and still be a migraine if the other features are present. The lack of recognition of this fact often leads to the misdiagnosis of migraine as tension-type headaches. Many individuals were taught that head pain which increases with leaning forward suggests sinusitis, but this is a very common migraine complaint.

Migraine without aura criteria is delineated (Box 5.1). Note that with a history of two or more attacks with aura, despite many without, the diagnosis remains migraine with aura. This diagnosis is clinical, and there are no objective markers of aura in these diagnostic criteria.

Most auras involve visual illusions in 99% of cases, but can involve sensory symptoms, motor symptoms, and speech symptoms (Box 5.2).[15] Furthermore, auras are not always followed by headache attacks, and multiple auras can occur in succession. Auras are stated to last less than hour, but Viana and colleagues found a significant number to have longer lasting auras.[16] Auras characteristically have positive features, such as photopsias and scintillations, which are then followed by negative features like scotomata. The most common visual aura begins as a small shimmering region (a positive neurological event) followed by a blind spot (a negative neurological event) expanding and moving to the periphery over about 20 minutes.

This aura may be comprised of colored or uncolored scintillations, or can have a zigzag outline (Figure 5.1).

BOX 5.1

MIGRAINE WITHOUT AURA DIAGNOSTIC CRITERIA

A. At least five attacks fulfilling criteria B–D

B. Headache attacks lasting 4–72 hours (untreated or unsuccssfully treated)

C. Headache has at least two of the following four characteristics:
1. Unilateral location
2. Pulsating quality
3. Moderate or severe pain intensity
4. Aggravation by or causing avoidance of routine physical activity (e.g., walking or climbing stairs)

D. During the headache, at least one of the following:
1. Nausea and/or vomiting
2. Photophobia and phonophobia

E. Not better accounted for by another ICHD-III diagnosis

Adapted from ICHD-III beta (www.ihs-classification.org/_downloads/ mixed/International-Headache-Classification-III-ICHD-III-2013-Beta.pdf).

BOX 5.2

MIGRAINE WITH AURA DIAGNOSTIC CRITERIA

A. At least two attacks fulfilling criteria B and C

B. One or more of the following fully reversible aura symptoms:
1. Visual
2. Sensory
3. Speech and or language
4. Motor
5. Brainstem
6. Retinal

C. At least two of the following four characteristics:
1. At least one aura symptom spreads gradually over at least 5 minutes, and/or two or more symptoms occur in succession

2. Each individual aura symptom lasts 5–60 minutes
3. At least one aura symptom is unilateral
4. The aura is accompanied, or followed within 60 minutes, by headache

D. Not better accounted for by another ICHD-III diagnosis, and transient ischemic attack has been excluded

Adapted from ICHD III-beta (www.ihs-classification.org/_downloads/ mixed/International-Headache-Classification-III-ICHD-III-2013-Beta.pdf).

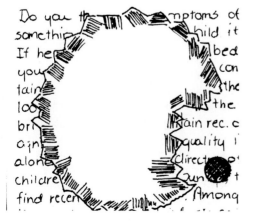

FIGURE 5.1 Migraine without aura diagnostic criteria.

Visual distortions, including change in the size and shape of objects (metamorphopsia) or mosaic vision, can occur. There can be a great deal of variation. It is not invariable that a headache follows each aura, and the aura does not have to precede the headache but rather can occur at any point in the attack. Even in those with migraine headaches with aura, 13% to 47% will experience some aura attacks without headache.[17,18] When an aura is present, it is assumed there is CSD. When there is no aura evident, it is possible that the aura is involving "silent" regions of the brain. It is often recognized that migraineurs frequently have cognitive changes surrounding their attack, and it is possible that these are due to unrecognized cortical spreading depression.

Most visual auras involve a homonomous visual field and are therefore binocular, although most patients will misinterpret this as involving the left or right eye. However, rarely will migraine be truly monocular. This phenomenon is often called "ocular migraine" or "retinal migraine" in the ICHD classification.[7] Despite the terms, migraines can involve the optic nerve or choroid, and would be better described as involving the anterior visual pathways. The diagnosis of ocular migraine can be strongly suspected with a painless monocular visual disturbance lasting less than 30 minutes and followed by a headache in individuals under the age of 40. This needs to be distinguished from amaurosis fugax, usually due to a carotid embolus, where the attacks are briefer — at 5 minutes or less — and often described as a shade in the eye. In the elderly, it also needs to be distinguished from giant cell arteritis. An evaluation for inflammatory disease and carotid stenosis is important in the appropriate age group.

When the aura is sensory, it often begins as a tingling (a positive neurological event), which is then replaced by numbness (a negative neurological event). Most sensory auras are in a cheiro-oral distribution, with paresthesias often beginning in the hand and then gradually ascending to the shoulder, ipsilateral face, and lips. This pattern might be pathognomic of migraine.

Should headache follow, it is most commonly a pressure-like or pulsatile pain felt over the temple or eye and often associated with nausea, photophobia, and phonophobia. Migraines are most commonly recognized when the pain is unilateral, yet it is commonly bilateral. The attack is generally accompanied by nausea, and light and sound sensitivity. The most sensitive migraine symptom is that it worsens with activity and therefore, during the disabling attack, migraineurs instinctively prefer to be in bed with the lights off and in a quiet environment.

ACEPHALGIC ATTACKS

In 1814, James Ware described cases of teichopsia without a subsequent headache in those with migraine.[19] It is known that in migraine, headache is not invariable and the aura symptoms can occur in isolation or with nausea, photophobia, or phonophobia. Lewis Carroll, in Chapter 4 of *Through a Looking Glass*, wrote: "at fifty years, 'tis said, afflicted citizens lose their sick headaches".[20] C. Miller Fisher described "late-life migraine accompaniments" in elderly patients presenting solely with migraine auras of various types.[21] However, this phenomenon can occur at any age. A personal or family history of migraine in association with aura always strongly suggests a diagnosis of migraine.

THE SPECTRUM OF MIGRAINE ATTACKS

Migraineurs typically experience multiple headache types. In the past, terms such as "mixed headaches" were used. In 1988, Raskin introduced the concept of the "spectrum of migraine" in that migraineurs could experience migraines with aura, migraines without aura, migraines minus a feature (migrainous), and tension-type headaches. He expressed that these syndromes were more similar than different.[22] Cady and Lipton's "spectrum study" queried migraineurs and most reported having migraines, migrainous, and tension-type headaches.[23] When administered sumatriptan, all headache types responded similarly to the same treatment. The mild ones may be reported as *tension-type headaches*. Those with significant autonomic phenomena may be reported as *sinusitis*, and the disabling ones as *migraines*. When the authors tried to identify those with disabling tension-type headaches, but without also having migraines, these patients were rare and did not respond to sumatriptan. This supports the notion that clinicians should be cautious of diagnosing disabling tension-type headache in their patients, and should query whether they also have migraines. The implication is that disabling tension-type headaches are generally a variant of migraine, and should be treated as such. Most likely these represent the spectrum of patients' migraine attacks, all a reflection of their hyperexcitable cortex, which predisposes them to attacks of head pain.

In the event that a migraine attack is not terminated and progresses, second- and third-order trigeminal neurons are activated. The clinical correlate to this activation is that the location of the pain expands, sometimes to involve the entire head, and the pain is more continuous and less pulsatile. Cutaneous allodynia is also a hallmark of this "central sensitization." Two-thirds of migraine patients will ultimately develop cutaneous allodynia with their attacks.[24,25] Often the pain that began as a periorbital or temporal pain then extends to extracephalic locations. Sufferers report that they need to remove their hats; they do not want to touch their hair or scalp, and remove jewelry, as there is amplification of trivial stimuli into pain. At that point, triptans tend to be relatively ineffective,[26] and patients may report that they now have a "tension headache" that evolved from their migraine as they have diffuse scalp pain. Dihydroergotamine or non-steroidal anti-inflammatory agents may still be effective at this state. As headaches become more chronic, there is more central sensitization and cutaneous allodynia, the nociceptive fields enlarge, and there is a more generalized and pervasive form of headache.

THE DIAGNOSIS OF MIGRAINE

The pain of migraine is commonly unilateral, but can be bilateral. In children, it is usually bilateral. Although the quality of pain is often pulsatile, it may be pressure-like or stabbing, even in cases where it is initially pulsatile. Light and sound sensitivity ensues, there may be intolerance to light and sound, or these stimuli might amplify the pain of the attack. A characteristic feature of a migraine is that it worsens with movement.

All of these factors shape the behavior of migraineurs during an attack. If permitted, they often prefer to lie in bed on a few pillows and remain still in a quiet, dark, cool environment.

Despite the high prevalence of migraine in the United States, the diagnosis rate remains poor. Two large epidemiological studies have been performed: American Migraine I and American Migraine II. The American Migraine I study reported a diagnosis rate of 38% in those with headaches fulfilling the IHS criteria for migraine.[27] When repeated 10 years later as the American Migraine II study, the diagnosis rate increased only to 48%.[28] It reported that 37% of migraineurs had been diagnosed as having tension-type headaches and 32% as having sinus headaches.[29]

It is striking that so many migraines are diagnosed as sinus headache in the United States, despite the fact that the International Headache Society has not recognized the validity of such a diagnosis. Much of this confusion arises from the common usage of this term by the population, and the frequent sales of "sinus headache" drugs. The trigeminal autonomic reflex consists of parasympathetic activation following trigeminal nerve activation via nociception. This stems from the polysynaptic connections between the trigeminal nucleus caudalis and the superior salivatory nucleus, which is parasympathetic. The parasympathetic innervation of the dura causes release of vasoactive intestinal peptide (VIP), as well as nitric oxide and acetylcholine. These events can lead to nasal congestion and eye-tearing, which is a reflection of parasympathetic activation. Barbanti showed that the autonomic symptoms are commonly present in migraines, with 41% having eye-tearing, 41% having nasal congestion, and 46% having both.[30] These findings demonstrate that eye-tearing and nasal congestion are more common as migraine symptoms than vomiting and aura. In addition to eye-tearing and nasal congestion, many migraineurs will report weather changes as a trigger of their attacks. In a study of self-diagnosed or physician-diagnosed headache, 80% fulfilled the IHS criteria for migraines or migrainous (missing one feature of migraine).[31] In this group, 84% had sinus pressure, 82% complained of sinus-located pain, and 63% had nasal congestion.

The misdiagnosis of migraine as tension-type headache is also common. The trigeminal nucleus caudalis extends caudally from C2–C4 although there is a significant variation in its caudal extent. Afferents from the upper cervical spine, in particular the C1 and C2 dorsal horn, as well as the trigeminal nerve, converge in the trigeminal nucleus caudalis and can be active during an attack, referring pain to the back of the neck. Since the occipital nerve, which is the medial branch of the dorsal primary ramus of C2, is often tender, blocking this nerve with local anesthetic affords relief for some migraines. Kaniecki showed that 75% of migraines were associated with neck pain, which often lead to the misdiagnosis of a migraine as a tension-type headache.[32] Similar to the autonomic symptoms of migraine of lacrimation and rhinorrhea, neck pain is far more common as a migraine symptom than are vomiting and aura.

The ICDH-3 beta criteria are used in research studies.[33] In practice, it is common that many attacks that vary from these descriptions are still best understood and managed as migraines.

PRECIPITATING FACTORS

As migraineurs have a hyperexcitable cerebral cortex, triggers that are benign in others can elicit migraine attacks in those genetically predisposed to migraine. An individual's triggers will not invariably produce an attack. Multiple triggers may be additive, and often operate to ultimately reduce the migraine threshold and trigger an attack. An example would be a woman who reports that she can tolerate one glass of any type of wine, but not two, except around the time of her menstruation when she cannot drink any alcohol.

Food triggers are commonly reported but are often undocumented. Andress-Rothrock, in analyzing triggers reported by migraineurs, found emotional stress reported in 59%, too much or too little sleep in 54%, and missing meals in 39%.[34] Most premenopausal women reported that menstruation was a trigger. It is rare for a single trigger to invariably trigger a migraine. Those with significant food triggers can usually identify them. Routinely mandating an extensive diet for those with migraine but without reported food triggers is generally unjustified. A regular eating schedule and adequate hydration is more relevant. Missing or delaying a meal is common as a migraine trigger.

Alcohol remains a common trigger of migraine, often in small amounts. Among products with alcohol, those that are heavily fermented, such as brandy, port, and red wines, are most likely to trigger attacks. Aside from alcohol, these drinks contain high levels of phenolic acid, tyramine, sulfites, as well as histamine.

However, many foods that also contain tyramine can trigger attacks in predisposed patients. These include hard cheeses, vinegar, yogurt, or any product that is highly fermented. Processed meats, including hot dogs, contain nitrites. Monosodium glutamate is often seen in high concentrations in certain Asian foods, as well as many snack foods, canned soups, and bottled salad dressings. Glutamate is an excitatory amino acid and therefore capable of triggering cortical spreading depression. Thus, it is understandable why monosodium glutamate can be a migraine trigger. This agent is an additive to many processed foods, and was originally reported as causing "Chinese restaurant syndrome." The artificial sweetener aspartame sometimes triggers attacks, and aspartate is also an excitatory amino acid.

Caffeine is a common ingredient in over-the-counter and combination prescription headache drugs. It is clear that many headaches can be aborted with the use of caffeine. However, many individuals consume large amounts of caffeine in their diet and through the use of medications. These individuals may develop caffeine withdrawal between doses, which by itself can become a migraine trigger. The chronic use of caffeine that exceeds 200 mg daily is not recommended, and the amounts and timing should be relatively stable.

Chocolate contains phenylethylamine, which is a potential migraine trigger. As previously stated, a craving for chocolate is, in many cases, a prodromal symptom of migraine and the actual ingestion of chocolate may not be the trigger. Head trauma can also trigger attacks, particularly in a known migraineur.[35] These post-traumatic headaches may seem relatively minor, like "footballer's migraine."

Odors, particularly pungent ones, are often reported to trigger attacks. Perfume intolerance is a common complaint among migraineurs.

Hougaard's study questioned the reliability of self-reported triggers.[36] Patients reported that exposure to bright light and exercise were triggers, although this was not replicated under controlled circumstances.

SECONDARY HEADACHES

Migraine defines an individual's threshold for headaches. Migraineurs are more likely to develop headache as a symptom of a brain tumor or other secondary headache causes.[37] Furthermore, the headache they subsequently develop is often similar to their original migraines, only more frequent and severe. Therefore, an individual with a pre-existing history of migraine who then develops a significant progression of these attacks should be re-evaluated to assure that they have not developed a secondary headache.

PATHOPHYSIOLOGY

For many years, migraine was thought to be a "vascular" disorder. Agents were believed to work if they were able to constrict cerebral vasculature. It is generally recognized that the slowly propagating wave of neuronal and glial depolarization, cortical spreading depression, is the physiologic correlate of aura. The wave of cortical spreading depression differs from epilepsy in many ways. It typically emanates from the occipital cortex radiating anteriorly at a rate of 2–3 mm/minute. It is unfortunate that this event has been labeled as "cortical spreading depression," since the initial wave is of cortical activation, and then depolarization of the cortex is followed by a long-term suppression of activity. Epilepsy is most likely to be localized in the temporal area, and the occipital cortex is relatively non-epileptogenic. The gray matter in a migraine appears to be activated and intensely depolarized. The observed blood flow changes of hyperemia followed by oligemia reflect blood flow requirements, given this physiologic state of neurons being in a refractory period. The vascular changes, which had been the focus of migraine research and drug development, probably represent blood flow serving the metabolic requirements during these neuronal and glial events. It has been shown, however, that this meningeal and extracranial vasodilation is neither necessary nor sufficient to explain the pain of migraine.[38] Extracranial structures are responsible for the production of head pain; in particular, the first division of the trigeminal nerve innervates dural vessels, extracranial arteries, venous sinuses, and the meninges. These connect to the gasserian ganglion, representing a first-order trigeminal neuron, then enter the pons and descend into the trigeminal nucleus caudalis. These also connect rostrally to the thalamus and from the thalamus into the somatosensory cortex where pain is appreciated and localized, and to the limbic system where there is affective processing.

It has been theorized that cortical spreading depression leads to a diffusion of a variety of chemicals that are released into the meninges, where they can activate trigeminal nociceptors with the subsequent development of headache. These agents include excitatory amino acids, nitric oxide, calcitonin gene-related peptide (CGRP), substance P, neurokinin A, potassium, and arachidonic acid.[39,40] It has been demonstrated that CSD increases the activity of dural nociceptors and central trigeminovascular neurons.[41] CSD leads to a sterile inflammatory problem within the meninges known as neurogenic inflammation, and to vasodilatation of meningeal arteries. Since the meninges' sensory innervation is via VI, it also explains why the pain is commonly referred to the eye and temple, as these first-order trigeminal neurons become activated and send a signal centrally into the thalamus.

It is not certain that CSD is the trigger for all migraine attacks, in particular, those without aura. Individuals who have migraine with aura also experience attacks of headache without aura. "Silent" CSD is difficult to document, as it requires imaging to be done prior to, or at least early in, attacks, in a disorder that is sporadic. It appears that the vascular changes, long the focus of migraine treatment research, focused on an epiphenomenon. This correlates well with the observable phenomena of scintillations, a positive neurological complaint followed by a scotoma — a negative neurological complaint. CSD also occurs in the cerebellum, and could explain some cases where dizziness accompanies an attack.[42] Functional imaging studies have documented that cortical spreading depression exists, particularly in a traumatized brain, using MRI, PET, and magnetoencephalography.[43–45] This is a convincing pathophysiological correlate of migraine with aura; it is more problematic to document this in migraine without aura since the actual onset of an attack is less clear. It is necessary to study multiple subjects without aura even before attacks are initiated, which is logistically difficult. There is evidence that CSD might occur even in migraine without aura. Woods studied a single case of migraine without aura that exhibited positron emission tomography evidence of CSD, although some argued that some visual blurring may have occurred.[46] Geraud studied seven cases of untriggered migraine without aura within 6 hours of the onset of the attacks, using positron emission tomography.[47] Cerebral hypoperfusion is associated with increased blood flow in the hypothalamus and brainstem.

Cortical spreading depression is a complex and poorly understood phenomenon. Whereas neurons are generally involved, astrocytes may play an important role as well. Astrocytes regulate the extracellular environment, normalizing regions of low magnesium or regions of abnormal quantities of glutamate and potassium. Therefore, astrocytic dysfunction can lead to regions of the brain being more hyperexcitable than usual.

It remains a mystery why these changes in blood flow that are seen in aura may also be seen in migraine without aura, and why the changes that occur during aura are often far more extensive than the clinical manifestations of the aura. During CSD there is an increase in intracellular calcium, and glial cells communicate via calcium waves, probably causing a phenomenon similar to that seen in neuronal CSD. One form of hemiplegic migraine involves abnormalities that are only expressed in astrocytes, and it is likely that a great deal of migraine pathology is astrocytic rather than neuronal.

However, the fundamental hyperexcitability of the cerebral cortex of migraineurs can be documented with a variety of studies. If a transcranial magnetic stimulator is applied over the occiput of individuals, most see phosphenes. Migraineurs see phosphenes at a very low threshold compared with non-migraineurs. In neurological testing, the use of an optokinetic nystagmus drum is common to evaluate visual pathways and nystagmus. However, most migraineurs find this to be adversative. In life, migraineurs often dislike stripes and checkerboard patterns, particularly if they are moving. A drive home at sundown, with the sun behind trees, can cause an unpleasant strobe effect. Transcranial magnetic stimulation activates linear detector neurons in the occiput, which explains why migraineurs often have the illusion of zigzag lines as part of their aura.

Migraine pain may be due to activation of the trigeminovascular pain pathway. This is comprised of trigeminal sensory afferent neurons that innervate cranial tissues, in particular the large cerebral arteries and the meninges. It is suggested that the neuronal influence leads to alterations in blood flow. Most of these blood flow changes described in migraine are probably secondary to many of the physiologic events that occur during a migraine attack. It is likely that cortical spreading depression is associated with the release of a variety of chemicals in the meninges, which activates cranial meningeal nociceptors of first-order trigeminal neurons. The subsequent pulsatile nature of the headache may be due to activated trigeminal nociceptors on extracranial arteries, or perhaps these activated trigeminal nociceptors in the meninges are reacting to normal CSF pulsations which would otherwise be insensate. This meningeal activation also leads to meningeal inflammation, which includes dural plasma protein extravasation and a sterile meningeal inflammatory response. The symptoms of a severe migraine attack are strikingly similar to the symptoms of an attack of meningitis, and this is understandable given the similarity of meningeal inflammation of migraine and meningeal inflammation of meningitis. Indeed, in both cases sufferers complain of a pulsatile headache, nausea, photophobia, neck stiffness, and pain upon eye movements. Vasodilatation alone will not account for the pain of migraine. Vasoactive intestinal peptide is a neurotransmitter involved in parasympathetic activation, but is not a trigger of a migraine attack. The blood flow changes observed in migraine are likely reflections of the metabolic changes that are a consequence of this cortical activation. In a migraine attack, there is an initial hyperemia of the brain, which is then followed by a prolonged period of oligemia.

Calcitonin gene-related peptide (CGRP) is widely expressed in multiple cell types, and certainly throughout the central and peripheral nervous system. Likely most relevant to migraine is its abundance in sensory nerve terminals of intracerebral and extracranial blood vessels. With trigeminal nerve activation there is an antidromic release of CGRP. When CGRP was infused into migraineurs, it produced a migraine-like headache.[48] Various CGRP antagonists and antibodies to the CGRP receptor are in development as potential migraine treatments.

Nitric oxide (NO) is also involved in migraine pathogenesis. Glyceryl trinitrate infusions reliably trigger migraine attacks in migraineurs.[49]

COMORBIDITIES OF MIGRAINE

It is important to recognize comorbidities of migraine, since, by definition, they will be encountered in this group more than by chance.[50] It is important to recognize that in comorbid conditions there is no implication that one causes the other. Psychiatric comorbidities include depression, bipolar disorder, social phobias, and anxiety. Treating both with a single agent is desirable but frequently not possible, as many antidepressants and anxiolytics can increase migraine and *vice versa*. Other comorbidities include Raynaud's phenomenon, asthma, irritable bowel syndrome, epilepsy, and stroke.

THE INHERITANCE OF MIGRAINE

There is little known about the genetics of migraine other than three forms of familial hemiplegic migraine, which are discussed in Chapter 6, on complicated migraine. Genes have not been identified for the most common forms of migraine, and the phenotype of migraine is likely due to a polygenic and multifactorial inheritance.

Migraines tend to run in families, which does not prove that it has a genetic basis. If one parent has migraine, there is a 50% chance of the offspring developing migraine.[51] Should both parents have migraine, the risk of migraine is 80%. The suspicion of a migraine diagnosis should be high when multiple family members have disabling episodic headaches.

PROGRESSION OF MIGRAINE

Obesity has been recognized to be a significant risk factor for the transformation of episodic attacks into a chronic form of migraine. Obesity induces a proinflammatory state and, as migraine is an inflammatory disorder, can become an important migraine trigger. It is well known that elevated levels of estrogen can trigger migraines in obese women even if postmenopausal, as these individuals may have extraovarian estrogen synthesis. Another mechanism by which obesity can become a migraine trigger is through the development of hypersomnia sleep apnea syndrome. Such patients often experience awakening in the morning with headache which is likely secondary to hypercapnia. Higher BMIs will increase the risk of all chronic headaches, particularly chronic tension-type headaches. Higher BMI is positively correlated with headache frequency and disability, and when overweight patients subsequently reduce their BMI it often leads to a reduction in headache frequency.[52] Bariatric surgery for those who are morbidly obese is often recommended.

COMPLICATIONS OF MIGRAINE

Kruit has documented a high frequency of white matter hyperintensities in the posterior circulation, and these might represent small infarcts.[53,54] Although more prevalent in migraine with aura, they were also seen in migraine without aura. Another complication of migraine is that of progression into chronic migraine. Migrainous stroke can occur, particularly in those with aura, although the strokes attributed to migraine are commonly remote from the actual migraine attacks. This will be discussed in detail in Chapter 6.

References

1. Stewart W, Lipton R, Celentano D, Reed M. Prevalence of migraine headache in the United States. Relation to age, income, race, and other sociodemographic factors. *JAMA*. 1992;267:64−69.
2. Bille B. Migraine in school children. A study of the incidence and short-term prognosis, and the clinical, psychological and electroencephalographic comparison between children with migraine and matched controls. *Acta Paediatr*. 1962;51:614−616.
3. Coppola G, Pierelli F, Schoenen J. Is the cerebral cortex hyperexcitable or hyperresponsive in migraine? *Cephalalgia*. 2007; 27:1427−1439.
4. Aurora S, Wilkinson F. The brain is hyperexcitable in migraine. *Cephalalgia*. 2007;27:1427−1439.
5. Schwartz B, Stewart W, Simon D, Lipton R. Epidemiology of tension-type headache. *JAMA*. 1998;279:381−383.
6. Tepper S, Dahlof C, Dowson A, Newman L, Mansbach H, Jones M, et al. Prevalence and diagnosis of migraine in patients consulting their physician with the complaint of headache: data from the Landmark study. *Headache*. 2004;44:856−864.

7. The International Headache Society (IHS). *IHS Classification ICHD-2*. London, UK: IHS. <http://ihs-classification.org/en/>.

8. Graham J, Wolff H. Mechanisms of migraine headache and action of ergotamine tartrate. *Arch Neurol Psych*. 1938;39:737−763.

9. Lashley K. Patterns of cerebral integration indicated by the scotomas of migraine. *Arch Neurol*. 1941;46:331−339.

10. Leão A. Spreading depression of activity in cerebral corte. *J Neurophysiol*. 1944;7:369−390.

11. Stewart W, Wood C, Reed M, Roy J, Lipton R, AAMP Advisory Group. Cumulative lifetime migraine incidence in women and men. *Cephalalgia*. 2008;28:1170−1178.

12. Saunders M. Review: Darwin's Victorian Malady − evidence for its medically induced origin. *West J Med*. 1973.

13. Kellman L. The premonitory symptoms (prodromes): a tertiary care study of 893 migraineurs. *Headache*. 2004;44:865−872.

14. Blau J. Migraine prodromes separated from the aura: complete migraine. *BMJ*. 1980;281:658−681.

15. Russell M, Olesen J. Nosographic analysis of the migraine aura in a general population. *Brain*. 1996;119:355−361.

16. Viana M, Spregner T, Andelova M, Goadsby P. The typical duration of a migraine aura: a systematic review. *Cephalalgia*. 2013;33:483−490.

17. Alvarez W. The migrainous scotoma as studied in 618 persons. *Am J Ophthalmol*. 1960;49:489−504.

18. Queiroz L, Rapoport A, Weeks R, Sheftell F, Siegel S, Baskin S. Characteristics of migraine visual aura. *Headache*. 1997;37:137−141.

19. Diamond S, Franklin M. *Headache Through the Ages*. New York, NY: Professional Communications; 2005.

20. Carroll L. *Through the Looking-Glass, and What Alice Found There*. 1871.

21. Fisher CM. Late-life (migrainous) scintillating zigzags without headache: one person's 27-year experience. *Headache*. 1999;39:391−397.

22. Raskin N. Tension headache. In: Raskin N, ed. *Headache*. 2nd ed. New York, NY: Churchill Livingstone; 1988.

23. Lipton R, Cady R, Steward W. Diagnostic lessons from the spectrum study. *Neurology*. 2002;58(suppl 6):S27−S31.

24. Burstein R, Yarnitsky D, Goor-Aryeh I, Ransil B, Bajwa Z. An association between migraine and cutaneous allodynia. *Ann Neurol*. 2000;47:614−624.

25. Lipton R, Bigal M, Ashina S, Burstein R, Silberstein S, Reed ML, et al. Cutaneous allodynia in the migraine population. *Ann Neurol*. 2008;63:148−158.

26. Burstein R, Jakubowsky M. Analgesic triptan action in an animal model on intracranial pain: a race against the development of central sensitization. *Ann Neurol*. 2004;55(1):27−36.

27. Lipton RB, Stewart WF, Simon D. Medical consultation for migraine: results from the American migraine study. *Headache*. 1998;38:87−96.

28. Lipton RB, Diamond S, Reed M, Diamond ML, Stewart WF. Migraine diagnosis and treatment results from the American migraine study II. *Headache*. 2001;41:638−645.

29. Schreiber C, Hutchinson S, Webster CJ, Ame M, Richardson MS, Powers C. Prevalence of migraine in patients with a history of self-reported or physician-diagnosed "sinus headache". *Arch Int Med*. 2004;164:1769−1772.

30. Barbanti P, Fabbrini G, Pesare M, Vanacore N, Cerbo R. Unilateral cranial autonomic symptoms in migraine. *Cephalalgia*. 2002;22:256−259.

31. Diamond M. The role of concomitant headache types and non-headache comorbidities in the under diagnosis of migraine: data from the American Migraine Study II. *Neurology*. 2002;58(suppl 6):S3−S9.

32. Kaniecki R. Migraine and tension-type headache: an assessment of challenges in diagnosis. *Neurology*. 2002;4:256−259.

33. The International Headache Society (IHS). *IHS classification ICHD-III*. London, UK: IHS. <www.ihs-classification.org/_downloads/mixed/International-Headache-Classification-III-ICHD-III-2013-Beta.pdf>.

34. Andress-Rothrock D, King W, Rothrock J. An analysis of migraine triggers in a clinic-based population. *Headache*. 2010;50:1366−1370.

35. Packard R, Ham L. Pathogenesis of posttraumatic headaches and migraine: a common headache pathway? *Headache*. 1997;37:142−152.

36. Hougaard A, Amin F, Hauge A, Ashina M, Olesen J. Provocation of migraine with aura using natural trigger factors. *Neurology*. 2013;80(5):428−431.

37. Forsyth P, Posner J. Headaches in patients with brain tumors: a study of 111 patients. *Neurology*. 1993;43:1678−1683.

38. Brennan K, Charles A. An update on the blood vessel in migraine. *Curr Opin Neurol*. 2010;23:266−277.

39. Bolay H, Moskowitz M. The emerging importance of cortical spreading depression in migraine headache. *Rev Neurol (Paris)*. 2005;161:655−657.

40. Dalkara T, Zervas N, Moskowitz M. From spreading depression to the trigeminovascular system. *Neurol Sci*. 2006; S86−S90.

41. Zhang X, Levy D, Noseda R, Kainz V, Jakubowski M, Burstein R. Activation of meningeal nociceptors by cortical spreading depression: implications for migraine with aura. *J Neurosci*. 2010;30:8807−8814.

42. Ebner T, Chen G. Spreading acidification and depression in the cerebellar cortex. *Neuroscientist*. 2003;3(9):37−45.

43. Fabricius M, Fuhr S, Bhatia R, Boutelle M, Hashemi P, Strong AJ, et al. Cortical spreading depression and peri-infarct depolarization in acutely injured human cerebral cortex. *Brain*. 2005;129:778−790.

44. Strong A, Fabricius M, Boutelle M. Spreading and synchronous depressions of cortical activity in acutely injured human brain. *Stroke*. 2002;33:2738−2743.

45. Hadjikhani N, Sanchez Del Rio M, Wu O. Mechanisms of migraine aura revealed by functional MRI in human visual cortex. *Proc Natl Acad Sci USA*. 2001;98:4687−4692.

46. Woods R, Iacoboni M, Mazziotta J. Brief report: bilateral spreading cerebral hypoperfusion during spontaneous migraine headache. *N Eng J Med*. 1994;331:1689−1692.

47. Geraud G, Denuelle M, Fabre N, Payoux P, Chollet F. Positron emission tomographic studies of migraine. *Rev Neurol (Paris)*. 2005;161:666−670.

48. Olessen J, Diener H, Husstedt W, Goadsby PJ, Hall D, Meier U, et al. BIBN 4096 BS Clinical Proof of Concept Study Group. Calcitonin gene-related peptide receptor antagonist BIBN 4096 BS for the acute treatment of migraine. *N Eng J Med*. 2004;350:1104−1110.

49. Afridi SK, Matharu MS, Lee L, Kaube H, Friston KJ, Frackowiak RSJ, et al. A PET study exploring the laterality of brainstem activation in migraine using glyceryl trinitrate. *Brain*. 2005;128:932−939.

50. Bigal ME, Lipton RB. The epidemiology, burden, and comorbidities of migraine. *Cephalalgia*. 2002;22:256−257.

51. Schurks M. Genetics of migraine in the age of genome-wide association studies. *J Headache Pain*. 2012;13(1):1−9.

52. Bigal M, Lipton R. Modifiable risk factors for migraine progress (or for chronic daily headaches) − clinical lessons. *Headache*. 2006;465:144−146.

53. Kruit M, van Buchem M, Hofman P, Bakkers J, Terwindt GM, Ferrari MD, et al. Migraine as a risk factor for subclinical brain lesions. *JAMA*. 2004;291:427−434.

54. Kruit M, Launer L, Ferrari M, van Buchem M. Small infarcts in the posterior circulation territory in migraine. The population-based CAMERA study. *Brain*. 2005;128:2068−2077.

6

Complicated Migraine

Mark W. Green and Rachel Colman

Department of Neurology, Icahn School of Medicine at Mt Sinai, New York, New York, USA

INTRODUCTION

Although the term "complicated migraine" is no longer used in the current headache classification, it remains in common use by clinicians.[1]

It generally refers to migraine with focal neurological complaints associated with migraine attacks, but excludes visual auras. Migraine auras are recurrent episodes of transient focal neurologic symptoms that can occur in association with migraine, before or during the headache. The International Headache Society recognizes three "typical" auras: visual, sensory, and language.[1] The pathophysiology of migraine aura is debated, but is most widely felt to be a manifestation of cortical spreading depression (CSD). The CSD typically spreads from the occipital pole, producing the more common visual aura, and can spread anteriorly to the primary sensory cortex, motor cortex, or language areas to produce sensory, motor, or language manifestations. Immediately before or at the onset of aura symptoms, regional cerebral blood flow is found to be decreased in the cortex corresponding to the clinically affected area, and often includes an even wider area. Blood-flow reduction usually starts posteriorly and spreads anteriorly, and is far below the ischemic threshold (except in the case of migrainous stroke). After one to several hours, a gradual transition into hyperemia occurs in the same region.

While migraine with visual aura is most common, less common types of aura often lead to misdiagnosis, extensive work-up, and confusion on the part of the patient and physician. Most migraineurs have exclusively migraine without aura. Many people who have attacks with aura also have attacks without aura. Often mistaken as aura, premonitory symptoms can occur hours to a day or more prior to a migraine attack (with or without aura). Prodromal symptoms include various combinations of fatigue, stiff neck, sensitivity to light or sounds, difficulty in concentrating, depression or euphoria, cold hands and feet, blurred vision, yawning, nausea, and pallor. Migraine auras are sometimes associated with a headache that does not fulfill criteria for migraine without aura. In other cases, migraine aura may occur without headache. Aura similar to those seen in migraine attacks has also been described with other distinct headache types, including cluster headache. The relationship between aura and these headaches has not been fully elucidated.

Many visual auras are associated with occasional paresthesias in the extremities. The paresthesias often culminate in numbness, reflecting the activation of neurons, followed by depression of cortical neurons as a reflection of CSD. Patients whose migraines are accompanied by symptoms in the extremities virtually always exhibit visual aura symptoms as well. Hence, the International Headache Society does not recognize a significant distinction between migraine with visual aura and hemiparasthetic migraine.[1]

It is known that Sigmund Freud suffered with migraine and described cognitive abnormalities as prodromal symptoms during his attacks, as he wrote[2]:

> The mild attacks of migraine from which I still suffer, usually announce themselves hours in advance by my forgetting names, and at the height of these attacks ... it frequently happens that all proper names go out of my head ... slips of the tongue do really occur with particular frequency when one is tired, have a headache or is threatened with migraine. In the same circumstances, proper names are easily forgotten. Some people are accustomed to recognize the approach of an attack of migraine when proper names escape them in this way.

TYPES OF MIGRAINE AURAS AND "COMPLICATED MIGRAINE"

The ICDH-III beta headache classification categorizes four subtypes of migraine with aura, with further subtypes in each.[1] These include: Migraine with

S. Diamond, R. K. Cady, M. L. Diamond & V. T. Martin (Eds):
Headache and Migraine Biology and Management.

DOI: http://dx.doi.org/10.1016/B978-0-12-800901-7.00006-9

aura, Migraine with brainstem aura, Hemiplegic migraine, and Retinal migraine. Those not listed under migraine with aura are separate complications of migraine, including status migrainosus, persistent aura without infarction, migrainous infarction, and migraine aura-triggered seizure.

Hemiplegic Migraine

Recurrent motor paralysis in migraine was first described in 1910, by Clarke, in the *British Medical Journal*.[3] During the 1950s, more cases were reported.[4,5] Although rare, hemiplegic migraines (HMs) are of particular interest as they are the only forms for which some of the genotypes have been identified. HM can be either sporadic or familial.[6] There are approximately 200 published cases of sporadic hemiplegic migraine (SHM), and 100–200 known families. A population-based epidemiological survey of sporadic and familial hemiplegic migraine (FHM) was performed in Denmark, and the data indicated that the prevalence of the sporadic form was at least 0.002%[6] while the prevalence of the familial form was at least 0.003%.[7] Isolated cases are diagnosed as having SHM, and those who have at least one affected first-degree or second-degree relative with motor symptoms are diagnosed as having FHM, although the manifestations and criteria for diagnosis are otherwise similar. Published data on affected families suggest that FHM has an autosomal dominant mode of inheritance.[8] This has led to the identification of many of the genes involved, and their relationship to chromosome 19p.[9]

Familial Hemiplegic Migraine

Mutations in the ion transportation genes CACNA1A, ATP1A2, and SCN1A coding for a (P/Q type) voltage-gated calcium channel, neuronal sodium–potassium pump, and neuronal voltage-gated sodium channel, respectively, all can cause the familial hemiplegic migraine phenotype.[10–13] These are referred to as FHM1, FHM2, and FHM3, respectively.

FHM1

The first type of familial hemiplegic migraine (FHM1) is categorized by the common gene involved: CACNA1A on chromosome 19p13. Different CACNA1A mutations cause episodic ataxia type 2 and spinocerebellar ataxia type 6. More than 30 FHM1 mutations have been identified in familial and sporadic cases, and most are of the missense type.[8] CACNA1A was first reported as the main familial hemiplegic migraine gene, and was initially described as mutated in one-half of the families affected, including all those with permanent cerebellar signs.[10,13] In affected families, permanent cerebellar signs are found in many but not all those with hemiplegic migraine. A common mutation present in 40% of unrelated families with FHM1 is a Thr666Met substitution.[6] Those with the Thr666Met substitution had the highest penetrance of hemiplegic migraine (98%), severe attacks with coma (50%), and nystagmus (86%).[13]

FHM2 & FHM3 and Other Familial Variants

The second form of familial hemiplegic migraine (FHM2) affects ATP1A2 gene at 1q23, which encodes the α_2 subunit of the A1A2 glial sodium–potassium ATPase pump.[9] This gene is expressed primarily in astrocytes, which differs from FHM1 and FHM3, and is thought to cause inefficient glutamate clearance by astrocytes and consequent increased cortical excitatory neurotransmission.[14] More than 60 FHM2 mutations have been identified in familial and sporadic cases, and most are missense mutations. Permanent cerebellar signs have also been seen with FHM2.[8] FHM3 involves a sodium channel SCN1A gene, which encodes the pore-forming subunit of neuronal Na v1.1 channels.[12] To date, only five FHM3 mutations have been reported in five families.[8] Other mutations have been found to cause the hemiplegic migraine phenotype, including a mutation in SLC1A3, encoding the glial glutamate transporter EAAT1, which was identified in a boy with pure hemiplegic migraine.[15] A homozygous deletion in SLC4A4, encoding the electrogenic sodium bicarbonate ($Na^+–HCO3^-$) co-transporter NBCe1, was associated with familial hemiplegic migraine as well. Notably, the two sisters in whom the deletion was found also had renal tubular acidosis and ocular abnormalities.[16] Hence, SLC1A3 and SLC4A4 might be the fourth and fifth genes to be implicated in familial hemiplegic migraine, but there have been an insufficient number of families elucidated.

The actual attacks are similar in SHM and FHM, although individuals present differently. Typical HM attacks begin in the first or second decade of life. The mean frequency of hemiplegic attacks is low, with an average of three attacks per year; however, this number is highly variable among individuals and families. In many cases, the frequency and the severity of the attacks decrease in adulthood.[8] A frequently reported trigger of a hemiplegic event is head trauma, as well as emotional stress and exertion. Most with hemiplegic migraine also have attacks of migraine with typical aura (without weakness), with a similar prevalence as reported in the general population. The hemiplegic attacks include gradually progressing visual, sensory, motor, aphasic, and basilar-type symptoms, accompanied by headache. The clinical presentation of sporadic and familial cases varies from pure HM type to severe early-onset forms that are seen with recurrent coma and cerebral edema, permanent cerebellar ataxia, and, rarely, epilepsy, elicited repetitive transient blindness, or mental retardation.[8]

To diagnose HM, motor symptoms must be present, and are usually accompanied by sensory symptoms.[7] The aura symptoms slowly progress over 20–30 minutes, and various aura symptoms occur successively. Generally, the first aura symptoms are visual, followed by sensory, motor and/or aphasic, and basilar symptoms.[17] Motor, sensory, visual, and aphasic symptoms are frequently all present during attacks, and usually at least four aura symptoms are experienced (Box 6.1).[8]

Sensory symptoms seen in HM combine positive phenomena, such as pain, cold, or paresthesias, and negative phenomena, such as numbness, and are usually described in a "cheiro-oral" distribution.[6,17] Many patients report positive phenomena beginning in one of their fingers and radiating up an arm to affect the face, tongue, trunk, and then a leg.[17] In some patients the negative sensory features are more prominent, including alien-limb syndrome or a substantial sensory loss.[8] Motor weakness involves areas affected by sensory symptoms, and can vary from mild clumsiness to hemiplegia. Sensory-motor symptoms usually begin in one hand and gradually radiate to the arm and the face. Most cases are unilateral; however, bilateral symptoms are seen in up to 35% of FHM1 cases. The bilateral symptoms may be simultaneous, or may start unilaterally and progress to bilateral symptoms.[6,17] Visual phenomena can be positive or negative, and are typical of migrainous visual auras, with reports of lights, zigzag lines, or scotomata, which can be colored or uncolored. In both sporadic and familial hemiplegic migraine, the aura typically starts peripherally and consists of a scotoma. In migraine with aura, the visual aura generally starts centrally and frequently consists of zigzag lines.[17] Aphasic symptoms usually present as difficulty with articulation, word-finding, and comprehension, and with prosodic language production. Basilar-type phenomena occur in HM, are often diverse, and include simultaneous bilateral paresis or paresthesia, simultaneous bilateral visual symptoms, dysarthria, vertigo, diplopia, tinnitus, reduced level of hearing, reduced level of consciousness, reduced ability to balance, drop attacks, crossed symptoms, and change of symptoms from side to side.[17] Many patients with HM fulfill criteria for basilar migraine.

In the vast majority of patients (up to 95% in a series), headache is present in all attacks.[8,17] For most patients, the pain starts during the aura after the onset of visual symptoms. The headache can be bilateral or unilateral, and can be ipsilateral or contralateral to the motor weakness.[8] There does not appear to be a link between the side of the headache and contralateral symptoms, as ipsilateral aura symptoms are as common as contralateral symptoms to the headache.[8] A small percentage of HM sufferers do not experience headache.[17]

The criteria for duration of the HM aura symptoms are controversial, with many case reports and case series of attacks lasting longer than the ICHD-III beta criteria (Box 6.1) which state the motor symptoms should resolve within 72 hours.[1] Numerous published cases of motor auras reported symptoms lasting longer than 72 hours, and up to 4 weeks.[8] It has been hypothesized that the reason for the long duration of persistent negative features is a more pronounced neuronal depolarization in hemiplegic migraine than that seen in typical migraine aura.[7,8] The mechanism of initiation of cortical spreading depression has been suggested as different in HM and migraine with or without aura.

BOX 6.1

HEMIPLEGIC MIGRAINE DIAGNOSTIC CRITERIA

A. At least two attacks fulfilling criteria B and C

B. Aura consisting of both of the following:
1. Fully reversible motor weakness
2. Fully reversible visual, sensory and/or speech/language symptoms

C. At least two of the following four characteristics:
1. At least one aura symptom spreads gradually over ≥5 minutes, and/or two or more symptoms occur in succession
2. Each individual non-motor aura symptom lasts 5–60 minutes, and motor symptoms last <72 hours

3. At least one aura symptom is unilateral*
4. The aura is accompanied, or followed within 60 minutes, by headache

D. Not better accounted for by another ICHD-III diagnosis, and transient ischemic attack and stroke have been excluded

*Aphasia is always regarded as a unilateral symptom; dysarthria may or may not be.
Adapted from ICHD-III beta.[1]

The common migraine triggers, calcitonin gene-related peptide, and nitric oxide are not believed to trigger hemiplegic migraines.[10–20]

Differential diagnosis and extensive work-up are critical for a patient who presents with symptoms suggestive of hemiplegic migraine. It must remain a diagnosis of exclusion. A first episode requires an urgent work-up to exclude stroke, mass lesion, epilepsy, and infectious or inflammatory diseases. The diagnosis of HM cannot be established with certainty after a single attack. Angiography is believed to worsen or trigger hemiplegic attacks and is generally not helpful diagnostically, since cases have shown either vasoconstriction or vasodilatation.[8] During the attacks, abnormalities in objective testing can be variable. CT or MRI of the brain can be abnormal, demonstrating areas of cerebral edema and swelling of the cortical ribbon. Transcranial Doppler of the cerebral arteries has shown diffuse or localized increase in intracranial arterial velocities. EEG findings can be variable, with diffuse slow waves contralateral to the motor deficit that can persist for several weeks. Sharp waves have been noted as well; however, seizures or status epilepticus are rare. CSF studies can be grossly abnormal, with elevated white blood cell counts usually in the range of 12–290 white blood cells per mm^3, with lymphocytic pleocytosis, but occasionally neutrophils and granulocytes are seen. Protein can be elevated up to 1 g/L, although glucose is usually normal. The key feature is that all of these objective tests resolve between attacks.[8]

Genetic testing for the three genes associated with FHM is most useful in early-onset sporadic cases that present with typical associated neurological signs, and in familial cases when the severity of attacks or permanent neurological features are different from those of affected relatives.[8] Genetic testing is costly, and may not be necessary for those with known family members or with a negative secondary work-up in sporadic cases.

Isolated cases have been reported in the literature in which recurrent hemiplegic migraine attacks were seen in association with a meningioma, meningitis, encephalitis, Sturge-Weber syndrome, and various other inflammatory or metabolic disorders.[21,22] There are also case reports of patients with associated hereditary cerebral angiopathies, including CADASIL (cerebral autosomal-dominant arteriopathy with subcortical infarcts and leukoencephalopathy), amyloid angiopathy, and MELAS (mitochondrial encephalomyopathy, lactic acidosis, and stroke-like episodes).[23–25]

Treatment

Treatment decisions are based on case reports, and some success has been reported with intranasal ketamine and verapamil as an abortive therapy. Prophylaxis with verapamil, lamotrigine, sodium valproate, and acetazolamide has demonstrated success in a small number of reported cases.[8] Triptans and other vasoconstricting agents, such as dihydroergotamine (DHE) or isometheptene, are avoided due to a historical clinical concern for worsening of aura in the setting of possible cerebral vasoconstriction.

Basilar Migraine

Basilar migraine or "migraine with brainstem aura" is a rare form of migraine which is accompanied by dysfunction of the brainstem (Box 6.2). Typically, sufferers report non-positional vertigo, diplopia, and dysarthria. Often, a mild or significant change in level of consciousness is noted. Other symptoms include

---

BOX 6.2

BASILAR MIGRAINE DIAGNOSTIC CRITERIA

A. At least two attacks fulfilling criteria B–D

B. Aura consisting of visual, sensory and/or speech/language symptoms, each fully reversible, but no motor or retinal symptoms

C. At least two of the following brainstem symptoms:
  1. Dysarthria
  2. Vertigo
  3. Tinnitus
  4. Hypacusis
  5. Diplopia
  6. Ataxia
  7. Decreased level of consciousness

D. At least two of the following four characteristics:
  1. At least one aura symptom spreads gradually over ≥ 5 minutes, and/or two or more symptoms occur in succession
  2. Each individual aura symptom lasts 5–60 minutes
  3. At least one aura symptom is unilateral
  4. The aura is accompanied, or followed within 60 minutes, by headache

E. Not better accounted for by another ICHD-III diagnosis, and transient ischemic attack has been excluded

Adapted from ICHD-III beta.[1]

hypacusis, tinnitus, and bilateral paresthesias. Headache commonly, but not invariably, follows. This condition is difficult to diagnose with certainty in the elderly, in whom brainstem strokes are far more common than migraines. It differs from sporadic and familial hemiplegic migraine, as there are no motor features. Acute confusional migraine should also be considered in the differential diagnosis. The etiology of coma in basilar migraine is not fully understood; theories include severe vasospasm of the basilar artery,[26] as well as dysfunction of the ascending reticular activating system via GABAergic mechanisms.[27] Case reports show promise with the use of lamotrigine for basilar migraine,[28,29] as well as for prevention of other "troubling" aura symptoms. For those with prolonged aura, or complicated features, prevention of the headache is not sufficient, and none of the typical headache prevention agents addresses the aura itself. Open label trials have shown benefit from lamotrigine with a significant ($>75\%$) reduction in frequency of aura in 21 of 36 patients responding to treatment.[29]

Retinal Migraine

Other uncommon visual migraine variants are "ocular," "retinal," or "anterior visual" migraine, since they affect the eye rather than the visual field. The pathophysiology of this disorder is unknown. When a patient presents with monocular visual loss, the leading differential diagnosis is amaurosis fugax due to infarct, embolus, or vasculitic phenomenon, and migraine is a diagnosis of exclusion. Most migraineurs describe their homonomous visual disturbance as emanating from an eye, rather than a visual field, so careful questioning is necessary.

Definite retinal migraine, as defined by the ICHD-III beta criteria, is a rare cause of transient monocular visual loss (Box 6.3).[1] The attacks include scintillations, scotomata, or blindness, and are associated with migraine headache and its associated features. The aura is fully reversible, and there are monocular positive and/or negative visual phenomena. To meet criteria, the diagnosis must be confirmed during an attack by either a formal visual field examination and/or a patient's drawing of a monocular field defect. The aura generally spreads gradually over 5 minutes and lasts from 5 to 60 minutes. The aura is accompanied or followed within an hour by headache.[30] Most cases of transient monocular visual loss diagnosed as "retinal migraine" would more appropriately be diagnosed as "presumed retinal vasospasm".[30] Cases of permanent monocular vision loss have been reported in migraine, but evaluation for other causes is necessary. A case report addressed the difference between embolic and non-embolic monocular visual loss.[31] Embolic visual loss usually presents as a blackout of visual symptoms, with a "curtain"-like phenomenon or altitudinal field loss. In contrast to non-embolic causes, TIAs are often short-lived, with 1–5 minutes of vision loss and occasional positive phenomena. With non-embolic monocular visual loss, the recovery usually occurs in reverse order. The authors note that headache is an uncommon feature of non-embolic monocular vision loss, but this disorder, like migraine, is more common in women than men. The causes are controversial, with cortical spreading depolarization and retinal vasospasm being the leading theories. In consideration of the concern for retinal vasospasm and the risk of ischemia, many physicians suggest avoiding vasoconstrictive agents, such as triptans and ergots, in such cases. Petzold's team and other researchers have reported some success with nifedipine as a preventive agent.[31]

BOX 6.3

RETINAL MIGRAINE DIAGNOSTIC CRITERIA

A. At least two attacks fulfilling criteria B and C

B. Aura consisting of fully reversible monocular positive and/or negative visual phenomena (e.g., scintillations, scotomata, or blindness) confirmed during an attack by either or both of the following:
  1. Clinical visual field examination
  2. The patient's drawing (made after clear instruction) of a monocular field defect

C. At least two of the following three characteristics:
  1. The aura spreads gradually over ≥ 5 minutes
  2. Aura symptoms last 5–60 minutes
  3. The aura is accompanied, or followed within 60 minutes, by headache

D. Not better accounted for by another ICHD-III diagnosis, and other causes of amaurosis fugax have been excluded

Adapted from ICHD-III beta.[1]

Migraine with Prolonged Aura

Migraine with persistent or prolonged aura is rare and difficult to treat. The medical literature on the topic is limited to a small number of case reports.[1] The ICHD III beta defines the disorder as aura persisting for one week or more without evidence of infarction on neuroimaging, in a patient who has previous attacks of migraine with aura.[1] A case series describes the phenomenon as involving the entire visual field and usually consists of diffuse small particles such as TV static, snow, lines of ants, dots, and rain.[32] Multiple case reports also show more classic prolonged aura, with geometric shapes and scintillating scotomata.[33] A magnetoencephalographic study showed that the visual cortex in patients with persistent visual aura maintains a steady-state hyperexcitability. The steady-state of excitability supports persistent visual aura as migraine spectrum disorder, suggesting sustained excitatory effects are related to a reverberating cortical spreading depression.[32] However, sustained visual aura differs from migraine with aura in that it is often refractory to treatment with migraine preventive or acute agents. Case series do suggest success with valproic acid, lamotrigine, and furosemide.[31] There is some literature *in vivo* and *in vitro* supporting the use of nimodipine.[34] A paper by Schankin and colleagues suggested that visual snow is not a persistent aura but rather another phenomenon entirely, and further evaluation is needed.[35]

Ophthalmoplegic Migraine

Ophthalmoplegic migraine (OM) may be a misnomer. ICHD-III beta classifies this disorder under "cranial neuralgias and central causes of facial pain".[1] The prior version of the guide listed OM as a migraine variant. OM is generally thought of as a recurrent childhood syndrome whose symptoms will fully resolve. It presents with a migraine-like headache, which follows (within 4 days) an episode of paresis of either cranial nerve III, IV, or VI. Most reported cases fully resolve in days to months, although there are case reports of longer lasting symptoms following multiple attacks.[36]

OM creates much conflict in the literature. A recent review[37] found that in up to one-third of cases the associated head pain was not migrainous in quality and neither were there associated migrainous symptoms, such as nausea or vomiting. The symptoms were overwhelmingly side-locked; a marked time lag was noted between headache onset and ophthalmoplegia, extending up to 14 days. CSF studies, when performed, were overwhelmingly negative. MRI with contrast performed during an attack in children showed a reversible focal thickening and enhancement of the cisternal tract of the involved nerve, usually cranial nerve III, at the root exit zone.[38] Repeat MRI following resolution of symptoms shows reduction or even complete resolution of the enhancement. However, the thickening of the nerve can persist for weeks, months, or even years.[39]

Adults present differently. In a case series published by Lal and colleagues in 2009, the majority of cases reported a single attack which does not recur.[40] Cranial nerve VI is more commonly affected in adults as compared with children. Most adults experienced antecedent worsening in the severity of their migraines prior to developing ophthalmoplegia, either during or within 24 hours of a severe migraine attack. In adults, cranial neuroimaging is normal.[38] Lal's series of 62 patients has been questioned as non-generalizable, as it only evaluated the Indian subcontinent, and may not in fact be representative of OM.[39] In this series, the majority of cases reported a single episode of ophthalmoplegia. Some believe OM is in fact a migrainous attack, while others believe it is a neuropathy with varying causes.

According to the ICHD-III beta,[1] many attacks previously considered to be OM in adults are better categorized as migraine with brainstem aura. The childhood variant, however, may be a cranial neuropathy. Ambrosetta and colleagues[36] hypothesized that these attacks in adults and children are the same disorder. They believed OM is migrainous and caused by ischemic reversible breakdown of the blood−nerve barrier due to vasospasm during the attack. They suggested that the blood−nerve barrier in adults is more mature and effective than in children. In adults, there is less cerebral edema and no MRI findings. They suggested that in children with OM, the frequent pupillary involvement could be due to thickening and enlargement of the third nerve causing a compression of the fibers from inside. In adults, the absence of the enlargement of the nerve could account for sparing of pupillary fibers.[34]

No treatment trials for ophthalmoplegic migraine have been published. Oral corticosteroids may be of benefit in treating acute exacerbations, based on available case series.[37]

VISUAL DISTURBANCES IN MIGRAINE

A number of visual phenomena have been reported in the literature and have been categorized within the migraine family. Barriga[41] presented a case series of seven patients who presented during a hemicranial migraine attack with ipsilateral mydriasis. In each of these patients, a cholinergic supersensitivity in the symptomatic pupil was demonstrated, pointing to a dysfunction of the ipsilateral ganglionic parasympathetic fibers. The synchronous co-localization of the features suggests a pathogenic link between the pupillary dysfunction and migraine. The authors postulate that the likely explanation includes a latent Adie's pupil (postganglionic parasympathetic paresis) triggered during a particular migraine attack versus a ciliary ganglionic lesion/dysfunction produced by the migrainous process. Alternatively, it could be ophthalmoplegic migraine with selective parasympathetic paresis or an episodic ciliary ganglionitis with migrainous features.

TRANSIENT GLOBAL AMNESIA

Transient amnesias have been reported as migraine auras. Some have speculated that transient global amnesia (TGA) is a variant of migraine.[42] In TGA, the retrograde and anterograde amnesia tends to last less than 24 hours, which is consistent with migraine auras. However, TGA attacks, unlike migraines, tend not to be recurrent. Any treatment, even anti-migraine agents, does not seem to prevent TGA. TGA has a predilection for the middle-aged and elderly — not the usual demographics of migraine, which tends to affect younger individuals. One author believed that cortical spreading depression is a feasible explanation for TGA.[43] The pathophysiology of TGA is even more elusive than that of migraine, so any relationship is highly speculative.

CADASIL

Cerebral autosomal dominant arteriopathy with subcortical infarcts and leukoencephalopathy (CADASIL) is caused by mutations in the NOTCH3 gene located on chromosome 19.[44] NOTCH3 gene encodes for a transmembrane receptor which is solely expressed in vascular muscle cells in humans. Histopathological findings show a degeneration of vascular smooth muscle cells with adjacent deposits of granular osmophilic material (GOM) and fibrous thickening of the arterial walls.[45] CADASIL is characterized by cerebrovascular disease that often progresses to dementia and generally begins in middle age.[46] About 30% of these patients are affected by migraine attacks, the majority having aura, which is often the first symptom of the disease.[45] Brain MRIs reveal white matter hyperintensities in the anterior temporal lobe, and periventricular white matter, with or without lacunar infarctions and microbleeds, in both symptomatic as well as asymptomatic adult carriers of the NOTCH3 mutation.[47] The lacunar infarcts are associated with cognitive dysfunction.[48] The CADASIL literature

reports the pooled prevalence of migraine as 43%, when excluding Asian countries, in which migraine prevalence in CADASIL is extremely low.[45] The data are unclear, but migraine with aura can be inferred in most of these cases. The exact pathway through which the NOTCH3 mutation in CADASIL leads to an increased prevalence of migraine with aura is unknown. Studies show that CSD is enhanced in mice expressing a vascular NOTCH3 CADASIL mutation or a NOTCH3 knockout mutation. The authors postulate that the vascular smooth muscle defect, due to NOTCH3 mutation, leads to an enhanced spreading depression susceptibility.[49]

Conversely, migraineurs have a high prevalence of non-specific white matter lesions, and occasionally these represent a secondary cause for headache such as CADASIL. CADASIL is frequently under-recognized and under-diagnosed. It should be considered in cases in which a patient presents with: (1) one or more recurrent subcortical ischemic strokes (especially before age 60 and in the absence of known vascular risk factors), or migraine (especially with aura, including atypical or prolonged auras), and/or early cognitive decline or subcortical dementia; (2) bilateral, multifocal, T2/FLAIR hyperintensities in the deep white matter and periventricular white matter with lesions in the anterior temporal pole, external capsule, basal ganglia, and/or pons; and (3) an autosomal-dominant family history of migraine, early-onset stroke, or dementia.[50]

HaNDL SYNDROME

HaNDL, or Headache with Neurological Deficits and cerebrospinal fluid Lymphocytosis, is a rare disorder. It was first described in the 1980s and was called "pseudo-migraine," with temporary neurological symptoms and lymphocytic pleocytosis (PMP). It is a self-limiting disorder, with multiple episodes that recur (1–20 times). The condition usually resolves without any long-term sequelae within a few months.[51] It is more common in adults aged 30–40 years, although cases in children have been reported.[52,53] Its features include severe deficits involving differing vascular territories, with concomitant CSF pleocytosis and no MRI evidence of infarction. Attacks can range from a few hours to 2–3 days. Frequently, symptoms begin with a fever and symptoms of a viral illness followed by headache. Often there is no history of migraine, and more men are reported to suffer from HaNDL than women. The neurological symptoms can develop at any time before, during, or following the headache. The most common focal symptoms are sensory, followed by aphasia, motor deficits, and more unusual visual phenomena. CSF studies in a report of 50 cases[54] showed an elevated opening pressure (100−400 mmH$_2$O) in most cases. There is elevated protein (20−250 mg/dL), lymphocytosis (10−760 cells), normal glucose, and no oligoclonal bands. Due to the concern about symptoms for encephalitis, viral and microbiological studies are appropriate. Invariably, in this disorder, these studies will be negative. It is essential to rule out neoplastic and granulomatous disorders as well as HIV and neurosyphilis, mycoplasma, and neuro-brucellosis within the appropriate clinical context.

A seizure disorder is part of the differential of such cases, and the EEG frequently shows focal slowing. SPECT studies performed on patients with HaNDL while symptomatic showed focal areas of decreased uptake consistent with their clinical symptoms.[54] It is not likely to be a true migraine variant since it is generally a monophasic disorder and the duration of symptoms and prevalence of infrequent visual symptoms separates it from the other "complicated" migraine variants. HaNDL is most likely triggered by a viral illness causing an aseptic inflammation of the leptomeninges and vasculature, leading to its common signs and symptoms. Considering the transient nature and quick return to baseline, TIA is often diagnosed. TIA uncommonly causes headache, can last up to 24 hours, and is far more common in older adults.

MIGRALEPSY AND OCCIPITAL SEIZURES

Migraine can be comorbid with epilepsy. Epilepsy can also imitate migraine − particularly occipital seizures, where both visual phenomena and headache can occur. Migraine auras typically last 5−30 minutes, with enlarging, generally uncolored scintillations that are replaced by scotomata, and move across half of a visual field over this time period. Zigzag lines are common. The visual disturbances of occipital seizures are brief, usually colored, generally lasting 1−3 minutes, and start in the periphery and move across the entire visual field.[55]

A variety of abnormalities, which can include epileptiform discharges, can occur in migraine. The diagnosis of occipital epilepsy, in what otherwise appears to be migraine, must be established with extreme care. Routine EEG is not justified in cases of migraine with aura.[56]

UNUSUAL SENSORY COMPLICATIONS OF MIGRAINE

The most common sensory disturbances of migraine are paresthesias, often replaced by numbness. However,

other symptoms can occur. Synesthesias occurred on several occasions in a patient with migraine with visual aura. Synesthesias refer to mixing senses such that one sensory disturbance is experienced as another. In this case, staring at bright lights caused her to experience an intense taste of lemon.[57] Many visual hallucinations are also associated with migraine. Metamorphopsia refers to distortions in the size and shape of objects, often faces and people. Macrosomatognosia refers to perceiving a body part as unusually large. In a paper on migraine hallucinations published by Lippman in 1952, a patient wrote[58]:

> my attacks used to be quite frequent, and about every 6 months I would have a major attack that lasted for weeks and required hospitalization. It was at these times that I experienced the sensation that my head had grown to tremendous proportions and was so light that it floated up to the ceiling, although I was sure it was still attached to my neck. I used to try to hold it down with my hands … this sensation would pass with the migraine but would leave me with a feeling that I was very tall. When walking down the street I would think I would be able to look down on the tops of others heads, and it was very frightening and annoying not to see as I was feeling. The sensation was so real when I would see myself in a window or full-length mirror, it was quite a shock to realize that I was still my normal height of under five feet. This happened quite often.

Disturbances of smell, which include hyperosmia and perversions of taste, have been reported. Blau reported a patient who during a migraine attack could smell a rose 20 feet away, and another who needed to dilute orange juice to one-third of its native concentration.[59] Various gustatory and olfactory hallucinations can also be seen with migraine attacks.

IS ANGIOGRAPHY SAFE IN MIGRAINE?

Since both small- and large-vessel diseases are in the differential diagnosis of "complicated migraine," cerebral angiography is frequently considered. Reports have suggested that catheter angiographic complications are increased in migraineurs. These include hemiplegia and hemisensory loss, confusion, angina, transient confusion, and transient amnesia.[60] It is unknown whether there are any additional risks to performing CT angiography in these patients.

TREATMENT OF COMPLICATED MIGRAINE

Few large treatment trials have been reported for complicated migraine attacks, and most recommendations are based on anecdotes and case reports.

Since these cases are rare and difficult to diagnose with certainty as migraine, large and numerically significant controlled trials are unlikely to be completed.

Triptans have been shown to be safe in attacks associated with visual and sensory auras, although they do not treat those phenomena. Triptans are contraindicated in hemiplegic migraine and basilar migraine. Although not recommended, the contraindication resulted from concern that a triptan might constrict an artery and cause the syndrome to become permanent. Another concern is that a hemiplegia and basilar syndrome could be a manifestation of stroke and misdiagnosed as a migraine, and then treated as a migraine. Klapper and colleagues studied 13 patients treated with triptans who had basilar migraine, familial hemiplegic migraine, and migraine with prominent and prolonged auras who did not show any adverse events.[61] Artto treated 79 patients with episodic and familial hemiplegic migraine with triptans, and reported no serious adverse events.[62] These numbers remain too small to suggest a change to the current recommendations.

OnabotulinumtoxinA, approved by the FDA only to treat chronic migraine, was used successfully in a single patient with hemiplegic migraine.[63] Others have been treated with valproate and lamotrigine.[64]

References

1. Headache Classification Committee of the International Headache Society. The International Classification of Headache Disorders, 3rd edition (beta version). *Cephalalgia*. 2013;33(9): 629–808.
2. Diamond S, Franklin MA. *Headache Throughout the Ages*. New York, NY: Professional Communications Inc; 2005.
3. Clarke JM. On recurrent motor paralysis in migraine, with report of a family in which recurrent hemiplegia accompanied the attacks. *Br Med J*. 1910;1:1534–1538.
4. Blau J, Whitty C. Familiar hemiplegic migraine. *Lancet*. 1955; ii:1115–1116.
5. Whitty C. Familial hemiplegic migraine. *J Neurol Neurosurg Psychiatry*. 1953;16:172–177.
6. Thomsen L, Ostergaard E, Olesen J, Russell M. Evidence for a separate type of migraine with aura: sporadic hemiplegic migraine. *Neurology*. 2003;60:595–601.
7. Thomsen L, Eriksen M, Roemer S, Andersen I, Olesen J, Russell M. A population-based study of familial hemiplegic migraine suggests revised diagnostic criteria. *Brain*. 2002;125: 1379–1391.
8. Russell M, Ducros A. Sporadic and familial hemiplegic migraine: pathophysiological mechanisms, clinical characteristics, diagnosis, and management. *Lancet Neurol*. 2011:457–470.
9. Joutel A, Bousser M, Biousse V, Labauge P, Chabriat H, Nibbio A, et al. A gene for familial hemiplegic migraine maps to chromosome 19. *Nat Genet*. 1993;5:40–45.
10. Ophoff R, Terwindt G, Vergouwe M, van Eijk R, Oefner PJ, Hoffman SM, et al. Familial hemiplegic migraine and episodic ataxia type-2 are caused by mutations in the Ca^{2+} channel gene CACNL1A4. *Cell*. 1996;87:543–552.
11. De Fusco M, Marconi R, Silvestri L, Atorino L, Rampoldi L, Morgante L, et al. Haploinsufficiency of ATP1A2 encoding the Na^+/K^+ pump alpha2 subunit associated with familial hemiplegic migraine type 2. *Nat Genet*. 2003;33:192–196.
12. Dichgans M, Freilinger T, Eckstein G, Babini E, Lorenz-Depiereux B, Biskup S, et al. Mutation in the neuronal voltage-gated sodium channel SCN1A in familial hemiplegic migraine. *Lancet*. 2005;366:371–377.
13. Ducros A, Denier C, Joutel A, Cecillon M, Lescoat C, Vahedi K, et al. The clinical spectrum of familial hemiplegic migraine associated with mutations in a neuronal calcium channel. *N Engl J Med*. 2001;345:17–24.
14. Leo L, Gherardini L, Barone V, De Fusco M, Pietrobon D, Pizzorusso T, et al. Increased susceptibility to cortical spreading depression in the mouse model of familial hemiplegic migraine type 2. *PLoS Genet*. 2011;7(6):e1002129.
15. Freilinger T, Koch J, Dichgans M, Mamsa H, Jen J. A novel mutation in SLC1A3 associated with pure hemiplegic migraine. *J Headache Pain*. 2010;11(Suppl. 1):90.
16. Suzuki M, Van Paesschen W, Stalmans I, Horita S, Yamada H, Bergmans BA, et al. Defective membrane expression of the Na (+)-HCO3(−) cotransporter NBCe1 is associated with familial migraine. *Proc Natl Acad Sci USA*. 2010;107:15963–15968.
17. Russell M, Iversen H, Olesen J. Improved description of the migraine aura by a diagnostic aura diary. *Cephalalgia*. 1994;14:107–117.
18. Hansen J, Thomsen L, Marconi R, Casari G, Olesen J, Ashina M. Familial hemiplegic migraine type 2 does not share hypersensitivity to nitric oxide with common types of migraine. *Cephalalgia*. 2008;28:367–375.
19. Hansen J, Thomsen L, Olesen J, Ashina M. Calcitonin gene-related peptide does not cause the familial hemiplegic migraine phenotype. *Neurology*. 2008;71:841–847.
20. Hansen J, Thomsen L, Olesen J, Ashina M. Familial hemiplegic migraine type 1 shows no hypersensitivity to nitric oxide. *Cephalalgia*. 2008;28:496–505.
21. Vetvik K, Dahl M, Russell M. Symptomatic sporadic hemiplegic migraine. *Cephalalgia*. 2005;25:1093–1095.
22. Thomsen L, Olesen J. Sporadic hemiplegic migraine. *Cephalalgia*. 2004;24:1016–1023.
23. Hutchinson M, O'Riordan J, Javed M, Quin E, Macerlaine D, Wilcox T, et al. Familial hemiplegic migraine and autosomal dominant arteriopathy with leukoencephalopathy (CADASIL). *Ann Neurol*. 1995;38:817–824.
24. Uitti R, Donat J, Rozdilsky B, Schneider R, Koeppen A. Familial oculoleptomeningeal amyloidosis. Report of a new family with unusual features. *Arch Neurol*. 1988;45:1118–1122.
25. Montagna P, Gallassi R, Medori R, Govoni E, Zeviani M, Di Mauro S, et al. MELAS syndrome: characteristic migrainous and epileptic features and maternal transmission. *Neurology*. 1988;38(5):751–754.
26. Frequin S, Linssen W, Pasman J, Hommes O, Merx H. Recurrent prolonged coma due to basilar artery migraine. A case report. *Headache*. 1991;31(2):75–81.
27. Requena I, Indakoetxea B, Lema C, Santos B, García-Castiñeira A, Arias M. Coma associated with migraine. *Rev Neurol*. 1999;29: 1048–1051.
28. Cologno D, d'Onofrio F, Castriota O, Petretta V, Casucci G, Russo A, et al. Basilar-type migraine patients responsive to lamotrigine: a 5-year follow-up. *Neurol Sci*. 2013;34(Suppl. 1):S165–S166.
29. Pascual J, Caminero A, Mateos V, Roig C, Leira R, García-Moncó C, et al. Preventing disturbing migraine aura with lamotrigine: an open study. *Headache*. 2004;44(10):1024–1028.
30. Hill D, Daroff R, Ducros A, Newman N, Biousse V. Most cases labeled as "retinal migraine" are not migraine. *J Neuroophthalmol*. 2007;27(1):3–8.

31. Petzold A, Islam N, Plant G. Patterns of non-embolic transient monocular visual field loss. *J Neurol.* 2013;260:1889–1900.

32. Chen W, Lin Y, Fuh J, Hämäläinen M, Ko Y, Wang S. Sustained visual cortex hyperexcitability in migraine with persistent visual aura. *Brain.* 2011;134(pt 8):2387–2395.

33. Liu G, Schatz N, Galetta S, Volpe N, Skobieranda F, Kosmorsky G. Persistent positive visual phenomena in migraine. *Neurology.* 1995;8:664–668.

34. San-Juan O, Zermeno P. Migraine with persistent aura in a Mexican patient: case report and review of the literature. *Cephalalgia.* 2007;27:456–460.

35. Schankin C, Maniyar F, Digre K, Goadsby PJ. "Visual snow" – a disorder distinct from persistent migraine aura. *Brain.* 2014;137:1419–1428.

36. Ambrosetto P, Nicolini F, Zoli M, Cirillo L, Feraco P, Bacci A. Ophthalmoplegic migraine: From questions to answers. *Cephalalgia.* 2014;34:914–919.

37. Gelfand A, Gelfand J, Prabakhar P, Goadsby P. Ophthalmoplegic "migraine" or recurrent ophthalmoplegic cranial neuropathy. *J Child Neurol.* 2012;27:759–766.

38. Miglio L, Feraco P, Tani G, Ambrosetto P. Computed tomography and magnetic resonance imaging findings in ophthalmoplegic migraine. *Pediatr Neurol.* 2010;42:434–436.

39. Mark A, Casselman J, Brown D, Sanchez J, Kolsky M, Larsen III TC, et al. Ophthalmoplegic migraine: reversible enhancement and thickening of the cisternal segment of the oculomotor nerve on contrast-enhanced MR images. *Am J Neuroradiol.* 1998;19(10):1887–1891.

40. Lal V, Sahota P, Singh P, Gupta A, Prabhakar S. Ophthalmoplegia with migraine in adults: is it ophthalmoplegic migraine? *Headache.* 2009;49(6):838–850.

41. Barriga F, López de Silanes C, Gili P, Pareja J. Ciliary ganglioplegic migraine: migraine-related prolonged mydriasis. *Cephalalgia.* 2011;31(3):291–295.

42. Caplan L, Chedru F, Lhermitte F, Mayman C. Transient global amnesia in migraine headache. *Neurology.* 1981;31:1167–1170.

43. Olesen J, Jorgensen M. Leão's spreading depression in the hippocampus explains transient global amnesia. A hypothesis. *Acta Neurol Scand.* 1986;73:219-210

44. Joutel A, Vahedi K, Corpechot C, Troesch A, Chabriat H, Vayssière C, et al. Strong clustering and stereotyped nature of Notch3 mutations in CADASIL patients. *Lancet.* 1997;350:1511–1515.

45. Liem M, Oberstein S, Van der Grond J, Ferrari M, Haan J. CADASIL and migraine: a narrative review. *Cephalalgia.* 2010;11:1284–1289.

46. Dichgans M, Mayer M, Uttner I, Brüning R, Müller-Höcker J, Rungger G, et al. The phenotypic spectrum of CADASIL: clinical findings in 102 cases. *Ann Neurol.* 1998;44:731–739.

47. van den Boom R, Lesnik Oberstein S, Ferrari M, Haan J, Van Buchem M. Cerebral autosomal dominant arteriopathy with subcortical infarcts and leukoencephalopathy: MR imaging findings at different ages – 3rd–6th decades. *Radiology.* 2003;229:683–690.

48. Liem M, van der Grond J, Haan J, van den Boom R, Ferrari MD, Knaap YM, et al. Lacunar infarcts are the main correlate with cognitive dysfunction in CADASIL. *Stroke.* 2007;38:923–928.

49. Eikermann-Haerter K, Yuzawa I, Dilekoz E, Joutel A, Moskowitz M, Ayata C. Cerebral autosomal dominant arteriopathy with subcortical infarcts and leukoencephalopathy syndrome mutations increase susceptibility to spreading depression. *Ann Neurol.* 2011;69(2):413–418.

50. Gladstone J, Dodick D. Migraine and cerebral white matter lesions: when to suspect cerebral autosomal dominant arteriopathy with subcortical infarcts and leukoencephalopathy (CADASIL). *Neurologist.* 2005;11(1):19–29.

51. Cifelli A, Vaithianathar L. Syndrome of transient Headache and Neurological Deficits with cerebrospinal fluid Lymphocytosis (HaNDL). *BMJ Case Reports.* 2011. Available from: http://dx.doi.org/10.1136/bcr.03.2010.2862.

52. Gonçalves D, Meireles J, Rocha R, Sampaio M, Leão M. Syndrome of transient headache and neurologic deficits with cerebrospinal fluid lymphocytosis (HaNDL): a pediatric case report. *J Child Neurol.* 2013;28(12):1661–1663.

53. Rossi L, Vassella F, Bajc O, Tönz O, Lütschg J, Mumenthaler M. Benign migriane like syndroe with CSF pleocytosis in children. *Dev Med Child Neurol.* 1985;49(9):1648–1654.

54. Gómez-Aranda F, Cañadillas F, Martí-Massó J, Díez-Tejedor E, Serrano PJ, Leira R, et al. Pseudomigraine with temporary neurological symptoms and lymphocytic pleocytosis. A report of 50 cases. *Brain.* 1997;120:1105–1113.

55. Queiroz L, Rapoport A, Weeks R, Sheftell F, Siegel S, Baskin S. Characteristics of migraine visual aura. *Headache.* 1997:137–141.

56. Seth N, Ulloa C, Solomon G, Lopez L. Diagnostic utility of routine EEG studies in identifying seizure as the etiology of the index event in patients referred with a diagnosis of migraine and not otherwise specified headache disorders. *Clin EEG Neurosci Soc.* 2012;43(4):323–325.

57. Alstadhaug K, Benjaminsen E. Synesthesia and migraine: case report. *BMJ Neurology.* 2010;10:121–123.

58. Lippman C. Certain hallucinations peculiar to migraine. *J Nervous Mental Disease.* 1952;116:346–351.

59. Blau N, Solomon F. Smell and other sensory disturbances in migraine. *J Neurol.* 1985;232:275–276.

60. Shuaid A, Hachinski V. Migraine and risks from angiography. *Arch Neurol.* 1988;45(8):911–912.

61. Klapper J, Mathew N, Nett R. Triptans in the treatment of basilar migraine and migraine with prolonged aura. *Headache.* 2001;10:981–984.

62. Artto V, Nissila M, Wessman M, Palotie A, Färkkilä M, Kallela M. Treatment of hemiplegic migraine with triptans. *Eur J Neurol.* 2007;14(9):1053–1056.

63. Dhawan P, Dhawan P. OnabotulinumtoxinA injections for the pain relief and long-term symptom control in a patient with hemiplegic migraine: case study. *J Headache Pain.* 2013;14(suppl. 1):188.

64. Pelzer N, Stam A, Carpay J, Vries BD, van den Maagdenberg AM, Ferrari MD, et al. Familial hemiplegic migraine treated by sodium valproate and lamotrigine. *Cephalalgia.* 2014;34:708–711.

Cerebrovascular Disease and Migraine

Michael Star and José Biller

Department of Neurology, Loyola University Chicago Stritch School of Medicine, Chicago, Illinois, USA

INTRODUCTION

It has been suspected for over a century that there is an association between migraine headaches and stroke.[1] Recognizing the extent and underlying etiology of that relationship has been studied more recently.

Multiple studies and meta-analyses have demonstrated that patients who have migraines are at higher risk of developing an ischemic stroke (Table 7.1[2–6]). Four recent meta-analyses[2–4,7] confirm that individuals suffering from migraine with aura are at roughly double the risk of having an ischemic stroke. Only one meta-analysis has demonstrated a statistically significant increase in the risk of stroke in individuals with migraine without aura.[2] Some of the relevant studies document that the correlation between migraine and stroke is seen only in younger patients.[3,8] The risk of stroke is more than tripled in migraineurs who smoke, and the combination of smoking and oral contraceptives increases the hazard ratio seven-fold.[7] Two recent studies have shown that migraines are also associated with other cardiovascular risk factors, including myocardial infarction and arterial claudication.[5,9] Many of these same studies suggest an increasing risk of stroke with increasing migraine frequency, though only in patients experiencing migraine with aura.[8,10] However, a similar relationship to migraine frequency does not exist with other cardiovascular events. A recent study[11] demonstrated that migraineurs have strokes with better functional outcomes when compared with strokes in non-migraineurs.

Recent literature also suggests there is a correlation between migraines and hemorrhagic strokes. A meta-analysis of eight case–control and cohort studies of association between migraine and hemorrhagic stroke demonstrated an overall pooled adjusted estimate of hemorrhagic stroke in subjects with any migraine versus control to be 1.48.[12] The risk was higher among women with migraines (1.55). This analysis did not demonstrate a specific association between hemorrhagic strokes with migraine with aura or without aura, and also did not demonstrate any difference between specific subtypes of hemorrhagic stroke. In one of the studies, the risk of fatal hemorrhagic strokes was greater among migraineurs compared with controls.[13] However, due to the very low absolute risk of hemorrhagic stroke in these studies, there is no indication to alter any of the current management guidelines for migraine treatment and prophylaxis.

THE BIOLOGY BEHIND THE RELATIONSHIP

The biology underlying the relationship between migraine and stroke is poorly defined (Figure 7.1).[14] One possibility is that although the majority of strokes in migraineurs do not occur in the height of a migraine attack, migraine and stroke have a direct cause-and-effect relationship. For example, an underlying causal agent of migraines with aura, such as cortical spreading depression,[15,16] might burden the subcortical white matter sufficiently to cause an ischemic stroke. Other researchers point to the evidence that migraineurs are more likely to have poor cardiovascular risk profiles,[5,9,17] including an association with increased c-reactive protein, non-HDL cholesterol, and Apo B100 levels,[18] and with endothelial dysfunction, as demonstrated by lower numbers of endothelial progenitor cells,[19] suggesting that strokes and migraines are separate disorders caused by a common set of environmental risk factors. Recent studies demonstrate the presence of specific genes that could predispose to both migraine and stroke,[20] suggesting a susceptibility to stroke and migraine that is present at birth. One particular study of the genetics of familial hemiplegic

S. Diamond, R. K. Cady, M. L. Diamond & V. T. Martin (Eds):
Headache and Migraine Biology and Management.

DOI: http://dx.doi.org/10.1016/B978-0-12-800901-7.00007-0

TABLE 7.1   Relative Risk (with 95% Confidence Interval) between Migraine and Ischemic Stroke as Demonstrated in Various Population-based Studies and Meta-analyses

	Bigal *et al.* [2] [a]	Spector *et al.* [3] [b]	Schurks *et al.* [4] [b]	Etminan *et al.* [5] [b]	Kurth *et al.* [6] [c]
Overall migraine	1.61 (1.19–2.18)	2.04 (1.72–2.43)	1.73 (1.31–2.29)	2.16 (1.89–2.48)	1.36 (0.97–1.92)
Migraine with aura	3.14 (2.25–4.38)	2.25 (1.53–3.33)	2.16 (1.53–3.03)	2.27 (1.61–3.19)	1.11 (0.69–1.78)
Migraine without aura	0.89 (0.61–1.30)	1.24 (0.86–1.79)	1.23 (0.90–1.69)	1.83 (1.06–3.15)	1.73 (1.10–2.71)

[a]*Case–control study.*
[b]*Meta-analysis.*
[c]*Cohort study.*
Table includes data from Kurth et al.[16] *Lancet Neurol. 2012;11:92–100.*

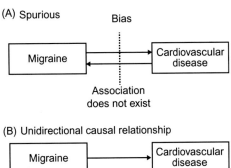

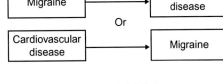

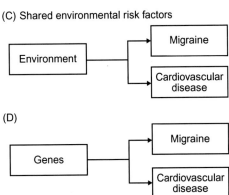

FIGURE 7.1   Four different theories on the underlying cause of the correlation between patients with migraines and stroke. *Reproduced from Bigal* et al.[14] *Neurology. 2009;72(21):1864–71, with permission.*

migraine suggests the possibility that migraineurs may have an increased susceptibility to mild ischemic events causing a stroke that would otherwise have been subclinical or a TIA.[21] Some researchers have implicated a potential association with the homozygous MTHFR C677T gene mutation, but further studies are needed.[22] Other investigators have suggested that the cause of this relationship is the risk of stroke associated with medications used to treat migraines, including triptans and ergotamine.[23–25] Taking all of

these possible explanations into account, the research may point to stroke and migraine sharing a reciprocal causal relationship. There is a significant amount of research attempting to further elucidate this multifaceted relationship.

NEUROIMAGING

The potential promise of using advanced neuroimaging — usually MRI — to better understand the relationship between stroke and migraine has produced a prodigious literature connecting migraineurs with subcortical white matter lesions.[26–29] A meta-analysis of these and similar studies[30] has confirmed this association. However, a recent large population-based study demonstrates that white matter hyperintensities are the same in migraineurs compared to individuals with non-migraine headache types,[31] which calls into question the direct relationship between migraines and these MRI findings. The CAMERA-2 study[29] was unable to demonstrate any change in cognitive function associated with the MRI findings. A recent study of white matter MRI changes in pediatric migraineurs also failed to demonstrate any correlation between migraine status and MRI findings.[32] The cumulative effect of this research suggests ascribing no clinical significance to subcortical white matter hyperintensities seen on MRI.

THE RELATIONSHIP BETWEEN MIGRAINES AND SECONDARY CAUSES OF STROKE

There have been many recent studies demonstrating a relationship between migraines and secondary causes of stroke. Cervicocephalic arterial dissection (CAD) is the most common single etiology of stroke in the young,[33] making up as much as 59% of strokes in the Helsinki Young Stroke Registry in patients aged

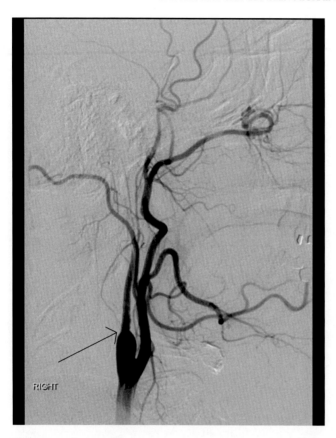

FIGURE 7.2 Catheter cerebral angiography demonstrating a cervical dissection of the right internal carotid artery. The arrow depicts the most proximal location of the dissection.

15 to 49 years (Figure 7.2).[34] A recent meta-analysis[35] demonstrated a two-fold increase in the risk of CAD among patients with migraines compared with controls. When comparing patients with CAD and stroke with patients with no CAD and stroke, those with CAD were 1.5 times more likely to have migraines than the controls.[33] The strong association between migraines and CAD is suggested to be a major contributor to the increased risk of stroke seen in migraineurs. Possible explanations for the association between migraines and CAD include underlying endothelial dysfunction in migraineurs,[19] a shared genetic predisposition,[36] and an increase in serum elastase demonstrated in patients with migraines.[37] CAD has been reported to present with migraine-like headaches.[38] Both extracranial and intracranial CAD can either cause steno-occlusive lesions or dissecting aneurysmal dilatations of the lesions. Patients with dissecting aneurysmal dilatation associated with CAD were 2.7 times more likely to have migraines versus CAD without dissecting aneurysm.[39] Multiple studies have demonstrated the long-term efficacy of conservative treatment of symptomatic extracranial CAD with dissecting aneurysm with a single antiplatelet

medication.[40,41] There is, however, some discussion[42] of the use of endovascular intervention in symptomatic intracranial vertebrobasilar dissecting aneurysms, though the current research suggests that perioperative complications outnumber the progression of these dissecting aneurysms to further ischemic damage, and, pending further study, we cannot recommend endovascular treatment at this time.

Multiple studies of the relationship between left-to-right shunts in patients with patent foramen ovale (PFO) and stroke demonstrate a statistically significant association between the two conditions.[43–46] Subsequent randomized controlled studies of individuals with migraines without a history of stroke did not demonstrate any association between PFO and migraines.[47,48] Due to three retrospective studies of patients undergoing percutaneous PFO closure demonstrating a subsequent decrease in the frequency and prevalence of migraines after the procedure,[49–51] the multi-center randomized control trial Migraine Intervention with STARFlex Technology (MIST) was initiated. The study failed to demonstrate an improvement in patients' migraines, though it demonstrated a high prevalence of PFOs in patients experiencing migraine with aura.[52] Due to these results and the failure of the CLOSURE I trial,[53] which failed to demonstrate superiority of PFO closure and medical therapy over medical therapy alone, the percutaneous closure of a PFO as a treatment cannot be recommended at the current time.

Cerebral autosomal dominant arteriopathy with subcortical infarcts and leukoencephalopathy (CADASIL) is a relatively well studied genetic disorder that demonstrates a correlation between stroke and migraine and is classified as 6.8.1 in the third edition of the International Classification of Headache Disorders (ICHD).[54] Approximately 20–40% of patients with CADASIL develop migraines with aura, and 60–85% of patients develop TIAs and strokes, which shows a strong overlap between the two diseases (Figure 7.3).[55] Other genetic disorders that can cause stroke and migraine headaches include familial hemiplegic migraine (FHM)[56]; hereditary endotheliopathy with retinopathy, nephropathy, and stroke (HERNS);[57] and hereditary angiopathy with nephropathy, aneurysm, and muscle cramps (HANAC)[58] (Table 7.2). FHM is an autosomal dominant channelopathy that causes migraine-like headaches accompanied by hemiplegia and an aura. The onset of symptoms is usually in the second decade, and it often resolves by age 50. Patients with a similar clinical picture but no family history are categorized as sporadic hemiplegic migraine. HERNS is a genetic mutation on chromosome 3p21 that causes diffuse endotheliopathy, progressive retinopathy and visual loss, stroke, and, eventually, dementia, usually in the third or fourth decade of life. HANAC is an

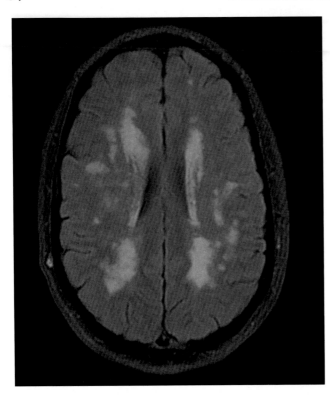

FIGURE 7.3   Axial T2 FLAIR MRI of a patient with CADASIL. Note the bilateral subcortical T2 hyperintensities that are characteristic in this disorder.

TABLE 7.2   Disorders Associated with Migraines and Stroke

Genetic	CADASIL
	CARASIL
	MELAS
	Hereditary hemorrhagic telangiectasia
	HERNS
	HANAC
Vascular	Cervical and intracranial arterial dissection
	Brain arteriovenous malformations
	Patent foramen ovale
	Sneddon's syndrome
	Moyamoya disease
Hematologic	Antiphospholipid antibody syndrome
	Essential thrombocytosis
	Polycythemia vera

Abbreviations: CADASIL, cerebral autosomal dominant arteriopathy with subcortical infarcts and leukoencephalopathy; CARASIL, cerebral autosomal recessive arteriopathy with subcortical infarcts and leukoencephalopathy; MELAS: mitochondrial encephalomyopathy, lactic acidosis, and stroke-like episodes; HERNS, hereditary endotheliopathy with retinopathy, nephropathy, and stroke; HANAC, hereditary angiopathy with nephropathy, aneurysm and muscle cramps.

autosomal dominant disorder associated with COL4A1 mutation that causes white matter changes, strokes, renal artery changes, and muscle cramping associated with creatinine kinase elevation, and is classified as ICHD 6.8.3.

THE DIAGNOSTIC CHALLENGE

The relationship between stroke and migraines is further complicated by the diagnostic challenge of differentiating between a migraine with a complex aura, a TIA or stroke with an accompanying headache, and other similar entities. Important questions to ask about a reported aura include the presence of positive or negative symptoms, chromaticity versus achromaticity, stereotyped aura or novel aura, the presence of precipitants, its evolution, and its duration. In a true migrainous aura, a patient's symptoms are usually described as having a progressive and marching quality, as opposed to the very acute onset in a stroke or TIA. The "march" experienced is either visual (fortification spectra) or sensory (perioral and/or arm tingling) symptoms which then proceed to dysphasia or motor symptoms. Migraines often have a prodrome consisting of blurred vision or flickering bright lights that would rarely be seen in a TIA or stroke. Though migraines often occur in younger patients, the initial onset can occur later in life as well, and, in older patients, can occur without a headache.[59] TIAs usually occur in the context of multiple vascular risk factors, have a sudden onset of focal neurological symptoms, and spontaneously resolve without a headache.

Strokes and TIAs that also cause headaches are another diagnostic consideration when treating a patient with acute onset of headaches and neurologic deficits. In the Lausanne Stroke Registry, 18% of patients with stroke reported a headache that started with the onset of symptoms.[60] Strokes and TIAs are more likely to be associated with headaches when they occur in the vertebrobasilar circulation, and rarely occur with lacunar infarcts. Thus, subtle findings on the neurological examination consistent with ischemia in the posterior circulation are important to note if there is concern for a basilar TIA. A migrainous infarction, classified as ICHD 1.4.3, occurs when a patient with pre-existing migraine with aura has his or her typical aura persisting longer than 60 minutes, and neuroimaging confirms the ischemic infarction in the relevant territory.[61] Migrainous infarctions are known to make up 0.5–1.5% of all ischemic strokes, and 13.7% of ischemic strokes in patients 45 years of age or younger,[62] and they have a predilection for posterior circulation infarctions on subsequent MRI.[61,63]

BOX 7.1

ETIOLOGIES OF THUNDERCLAP HEADACHE

With neck rigidity:
  Subarachnoid hemorrhage
  Meningitis
  Encephalitis
  Systemic infection
Without neck rigidity:
  Cervicocephalic artery dissection
  Acute ischemic stroke
  Intracerebral hemorrhage
  Expanding intracranial aneurysm
  Cerebral venous sinus thrombosis
  Pituitary apoplexy
  CSF hypotension syndrome
  Acute obstructive hydrocephalus
  Hypertensive crisis
  Posterior reversible encephalopathy syndrome
  Sphenoid sinusitis
  Primary thunderclap headache

Thunderclap headache, generally defined as a headache of sudden onset and great intensity, can be a presenting sign of an underlying cerebrovascular disease (Box 7.1). A sudden elevation in blood pressure can also cause a headache, referred to as a "hypertensive headache." A hypertensive headache is described as a headache occurring with elevation of the systolic blood pressure over 180 mmHg or the diastolic blood pressure rising over 120 mmHg.[64] These headaches are often relieved with treatment of the underlying hypertension, though not always. Migraine headaches also can cause a patient to have reactive hypertension, often confusing the clinical picture. Ischemic and hemorrhagic strokes can also cause reactive hypertension and headache along with focal neurological dysfunction, and should be ruled out in cases where all three are present.

Other underdiagnosed syndromes that may mimic migraine and stroke disorders include late-life migraine accompaniments,[65] headache with temporary neurologic deficits and cerebrospinal fluid lymphocytosis (HaNDL syndrome),[66] and stroke-like migraine attacks after radiation therapy (SMART syndrome).[67] Late-life migraine accompaniments, also known as acephalgic migraines, are visual symptoms that are usually positive in nature and are not accompanied by a typical migrainous headache. They usually occur in people over the age of 50, and are not rare, having

been noted to occur in 1.33% of women and 1.08% of men.[65] HaNDL syndrome is a benign and self-limited disorder whose presentation can be protean, ranging from focal weakness[68] to altered mental status, and is classified as ICHD 7.3.5.[69] The SMART syndrome is an idiopathic disorder that is often found in patients treated with focal brain radiation for brain tumors and who develop transient headaches and focal neurological deficits about 2–10 years after radiation therapy.[67]

SUMMARY

There is much to be studied concerning the relationship between migraine and stroke. Though there is a demonstrated association between the two, little is agreed upon regarding the source of that association. It is important to note the clinical implications of this association. There is no indication that patients with migraine, either with or without aura, and no other vascular risk factors should be started on an antiplatelet medication.[31] This is because the increased risk of major bleeding while taking an antiplatelet is higher than the potential benefit.[5,70] When it is unclear that a patient over 50 years old has experienced a late-onset migraine with aura versus TIA, it is important to proceed with an investigation of vascular risk factors and treat the patient as if he or she had experienced a TIA, including treatment with an antiplatelet. In younger patients with migraine with aura, modifiable vascular risk factors, such as smoking and oral contraceptive use, should be determined. Patients requiring birth control should be advised to use alternative means of contraception, and the risk of taking estrogen-progesterone combinations should be explained to them in detail. Though alternatives such as progesterone-only oral contraceptives have not demonstrated any increased risk of stroke, further research is warranted.[71] While triptan use in migraine patients has been shown to be generally safe,[23,25] ergotamine use should be avoided in patients with migraines, both as a treatment for migraine and for other disorders. Patients with migraines and stroke should avoid triptans as well. Generally, there is no need to further investigate patients diagnosed with migraines with aura, but if the migraines are atypical, with late onset, change in quality, or associated with unilateral symptoms with contralateral headaches, they should undergo further testing for possible underlying disorders, including hypercoagulable and genetic diseases. Similarly, in patients with stroke who have a history of migraines, the diagnostic workup and treatment for their stroke are the same as those without a history of migraines.

References

1. Fèrè C. Note sur un cas de migraine ophtalmique à accès rèpètès suivis de mort. *Rev Med (Paris)*. 1883;3:194−201.
2. Bigal M, Kurth T, Santanello N, Buse D, Golden W, Robbins M, et al. Migraine and cardiovascular disease a population-based study. *Neurology*. 2010;74(8):628−635.
3. Spector JT, Kahn SR, Jones MR, Jayakumar M, Dalal D, Nazarian S. Migraine headache and ischemic stroke risk: an updated meta-analysis. *Am J Med*. 2010;123(7):612−624.
4. Schurks M, Rist PM, Bigal ME, Buring JE, Lipton RB, Kurth T. Migraine and cardiovascular disease: systematic review and meta-analysis. *BMJ*. 2009;339:b3914.
5. Etminan M, Takkouche B, Isorna FC, Samii A. Risk of ischaemic stroke in people with migraine: systematic review and meta-analysis of observational studies. *BMJ*. 2005;330(7482):63.
6. Kurth T, Slomke MA, Kase CS, Cook NR, Lee IM, Gaziano JM, et al. Migraine, headache, and the risk of stroke in women: a prospective study. *Neurology*. 2005;64(6):1020−1026.
7. MacClellan LR, Giles W, Cole J, Wozniak M, Stern B, Mitchell BD, et al. Probable migraine with visual aura and risk of ischemic stroke: the stroke prevention in young women study. *Stroke*. 2007;38(9):2438−2445.
8. Kurth T, Schurks M, Logroscino G, Buring JE. Migraine frequency and risk of cardiovascular disease in women. *Neurology*. 2009;73(8):581−588.
9. Winsvold B, Hagen K, Aamodt A, Stovner L, Holmen J, Zwart J. Headache, migraine and cardiovascular risk factors: the HUNT study. *Eur J Neurol*. 2011;18(3):504−511.
10. Donaghy M, Chang CL, Poulter N; European Collaborators of The World Health Organisation Collaborative Study of Cardiovascular Disease and Steroid Hormone Contraception. Duration, frequency, recency, and type of migraine and the risk of ischaemic stroke in women of childbearing age. *J Neurol Neurosurg Psychiatry*. 2002;73(6):747−750.
11. Rist PM, Buring JE, Kase CS, Schurks M, Kurth T. Migraine and functional outcome from ischemic cerebral events in women. *Circulation*. 2010;122(24):2551−2557.
12. Sacco S, Ornello R, Ripa P, Pistoia F, Carolei A. Migraine and hemorrhagic stroke: a meta-analysis. *Stroke*. 2013;44(11):3032−3038.
13. Kurth T, Kase CS, Schurks M, Tzourio C, Buring JE. Migraine and risk of haemorrhagic stroke in women: prospective cohort study. *BMJ*. 2010;341:c3659.
14. Bigal ME, Kurth T, Hu H, Santanello N, Lipton RB. Migraine and cardiovascular disease: possible mechanisms of interaction. *Neurology*. 2009;72(21):1864−1871.
15. Ayata C. Cortical spreading depression triggers migraine attack: pro. *Headache*. 2010;50(4):725−730.
16. Kurth T, Chabriat H, Bousser M. Migraine and stroke: a complex association with clinical implications. *Lancet Neurol*. 2012;11(1):92−100.
17. Scher AI, Terwindt GM, Picavet HS, Verschuren WM, Ferrari MD, Launer LJ. Cardiovascular risk factors and migraine: the GEM population-based study. *Neurology*. 2005;64(4):614−620.
18. Kurth T, Ridker P, Buring J. Migraine and biomarkers of cardiovascular disease in women. *Cephalalgia*. 2008;28(1):49−56.
19. Rodriguez-Osorio X, Sobrino T, Brea D, Martinez F, Castillo J, Leira R. Endothelial progenitor cells: a new key for endothelial dysfunction in migraine. *Neurology*. 2012;79(5):474−479.
20. Anttila V, Winsvold BS, Gormley P, Kurth T, Bettella F, McMahon G, et al. Genome-wide meta-analysis identifies new susceptibility loci for migraine. *Nat Genet*. 2013;45(8):912−917.
21. Eikermann-Haerter K, Lee JH, Yuzawa I, Liu CH, Zhou Z, Shin HK, et al. Migraine mutations increase stroke vulnerability by facilitating ischemic depolarizations. *Circulation*. 2012;125(2):335−345.
22. Kowa H, Yasui K, Takeshima T, Urakami K, Sakai F, Nakashima K. The homozygous C677T mutation in the methylenetetrahydrofolate reductase gene is a genetic risk factor for migraine. *Am J Med Genet*. 2000,96(6):762−764.
23. Hall GC, Brown MM, Mo J, MacRae KD. Triptans in migraine: the risks of stroke, cardiovascular disease, and death in practice. *Neurology*. 2004;62(4):563−568.
24. Pezzini A, Del Zotto E, Giossi A, Volonghi I, Costa P, Dalla Volta V, et al. The migraine-ischemic stroke relation in young adults. *Stroke Res Treat*. 2010;2011:304921.
25. Wammes-van der Heijden EA, Rahimtoola H, Leufkens HG, Tijssen CC, Egberts AC. Risk of ischemic complications related to the intensity of triptan and ergotamine use. *Neurology*. 2006;67(7):1128−1134.
26. Rocca MA, Ceccarelli A, Falini A, Colombo B, Tortorella P, Bernasconi L, et al. Brain gray matter changes in migraine patients with T2-visible lesions: a 3-T MRI study. *Stroke*. 2006;37(7):1765−1770.
27. Cooney BS, Grossman RI, Farber RE, Goin JE, Galetta SL. Frequency of magnetic resonance imaging abnormalities in patients with migraine. *Headache*. 1996;36(10):616−621.
28. Fazekas F, Koch M, Schmidt R, Payer F, Freidl W, Lechner H. The prevalence of cerebral damage varies with migraine type: a MRI study. *Headache*. 1992;32(6):287−291.
29. Palm-Meinders IH, Koppen H, Terwindt GM, Launer LJ, Konishi J, Moonen JM, et al. Structural brain changes in migraine. *JAMA*. 2012;308(18):1889−1897.
30. Bashir A, Lipton RB, Ashina S, Ashina M. Migraine and structural changes in the brain: a systematic review and meta-analysis. *Neurology*. 2013;81(14):1260−1268.
31. Kurth T, Diener HC, Buring JE. Migraine and cardiovascular disease in women and the role of aspirin: subgroup analyses in the women's health study. *Cephalalgia*. 2011;31(10):1106−1115.
32. Mar S, Kelly JE, Isbell S, Aung WY, Lenox J, Prensky A. Prevalence of white matter lesions and stroke in children with migraine. *Neurology*. 2013;81(16):1387−1391.
33. Metso TM, Tatlisumak T, Debette S, Dallongeville J, Engelter ST, Lyrer PA, et al. Migraine in cervical artery dissection and ischemic stroke patients. *Neurology*. 2012;78(16):1221−1228.
34. Putaala J, Metso AJ, Metso TM, Konkola N, Kraemer Y, Haapaniemi E, et al. Analysis of 1008 consecutive patients aged 15 to 49 with first-ever ischemic stroke: the Helsinki young stroke registry. *Stroke*. 2009;40(4):1195−1203.
35. Rist PM, Diener HC, Kurth T, Schurks M. Migraine, migraine aura, and cervical artery dissection: a systematic review and meta-analysis. *Cephalalgia*. 2011;31(8):886−896.
36. Debette S, Markus HS. The genetics of cervical artery dissection: a systematic review. *Stroke*. 2009;40(6):e459−66.
37. Tzourio C, El Amrani M, Robert L, Alperovitch A. Serum elastase activity is elevated in migraine. *Ann Neurol*. 2000;47(5):648−651.
38. Haraguchi K, Toyama K, Ito T, Hasunuma M, Sakamoto Y. A case of posterior cerebral artery dissection presenting with migraine-like headache and visual field defect: usefulness of fast imaging employing steady-state acquisition (FIESTA) for diagnosis. *J Stroke Cerebrovasc Dis*. 2012;21(8):906.e5−906.e7.
39. Touze E, Randoux B, Meary E, Arquizan C, Meder JF, Mas JL. Aneurysmal forms of cervical artery dissection: associated factors and outcome. *Stroke*. 2001;32(2):418−423.
40. Benninger DH, Gandjour J, Georgiadis D, Stockli E, Arnold M, Baumgartner RW. Benign long-term outcome of conservatively treated cervical aneurysms due to carotid dissection. *Neurology*. 2007;69(5):486−487.
41. Kim BM, Kim SH, Kim DI, Shin YS, Suh SH, Kim DJ, et al. Outcomes and prognostic factors of intracranial unruputured vertebrobasilar artery dissection. *Neurology*. 2011;76(20):1735−1741.

42. Jin SC, Kwon DH, Choi CG, Ahn JS, Kwun BD. Endovascular strategies for vertebrobasilar dissecting aneurysms. *Am J Neuroradiol.* 2009;30(8):1518–1523.

43. Lamy C, Giannesini C, Zuber M, Arquizan C, Meder JF, Trystram D, et al. Clinical and imaging findings in cryptogenic stroke patients with and without patent foramen ovale: the PFO-ASA Study. Atrial septal aneurysm. *Stroke.* 2002;33 (3):706–711.

44. Wilmshurst PT, Nightingale S, Walsh KP, Morrison WL. Effect on migraine of closure of cardiac right-to-left shunts to prevent recurrence of decompression illness or stroke or for haemodynamic reasons. *Lancet.* 2000;356(9242):1648–1651.

45. Anzola GP, Magoni M, Guindani M, Rozzini L, Dalla Volta G. Potential source of cerebral embolism in migraine with aura: a transcranial doppler study. *Neurology.* 1999;52(8):1622–1625.

46. Schwedt TJ, Demaerschalk BM, Dodick DW. Patent foramen ovale and migraine: a quantitative systematic review. *Cephalalgia.* 2008;28(5):531–540.

47. Rundek T, Elkind MS, Di Tullio MR, Carrera E, Jin Z, Sacco RL, et al. Patent foramen ovale and migraine: a cross-sectional study from the Northern Manhattan Study (NOMAS). *Circulation.* 2008;118(14):1419–1424.

48. Garg P, Servoss SJ, Wu JC, Bajwa ZH, Selim MH, Dineen A, et al. Lack of association between migraine headache and patent foramen ovale: results of a case–control study. *Circulation.* 2010;121(12):1406–1412.

49. Post MC, Thijs V, Herroelen L, Budts WI. Closure of a patent foramen ovale is associated with a decrease in prevalence of migraine. *Neurology.* 2004;62(8):1439–1440.

50. Schwerzmann M, Wiher S, Nedeltchev K, Mattle HP, Wahl A, Seiler C. Percutaneous closure of patent foramen ovale reduces the frequency of migraine attacks. *Neurology.* 2004;62(8): 1399–1401.

51. Azarbal B, Tobis J, Suh W, Chan V, Dao C, Gaster R. Association of interatrial shunts and migraine headaches: impact of transcatheter closure. *J Am Coll Cardiol.* 2005;45(4):489–492.

52. Dowson A, Mullen MJ, Peatfield R, Muir K, Khan AA, Wells C, et al. Migraine intervention with STARFlex technology (MIST) trial: a prospective, multicenter, double-blind, sham-controlled trial to evaluate the effectiveness of patent foramen ovale closure with STARFlex septal repair implant to resolve refractory migraine headache. *Circulation.* 2008;117(11):1397–1404.

53. Furlan AJ, Reisman M, Massaro J, Mauri L, Adams H, Albers GW, et al. Closure or medical therapy for cryptogenic stroke with patent foramen ovale. *N Engl J Med.* 2012;366(11):991–999.

54. Headache Classification Committee of the International Headache Society (IHS). The International Classification of Headache Disorders, 3rd edition (beta version). *Cephalalgia.* 2013;33(9): 629–808.

55. Chabriat H, Joutel A, Dichgans M, Tournier-Lasserve E, Bousser MG. Cadasil. *Lancet Neurol.* 2009;8(7):643–653.

56. Russell MB, Ducros A. Sporadic and familial hemiplegic migraine: pathophysiological mechanisms, clinical characteristics, diagnosis, and management. *Lancet Neurol.* 2011;10(5): 457–470.

57. Ballabio E, Bersano A, Bresolin N, Candelise L. Monogenic vessel diseases related to ischemic stroke: a clinical approach. *J Cereb Blood Flow Metab.* 2007;27(10):1649–1662.

58. Alamowitch S, Plaisier E, Favrole P, Prost C, Chen Z, Van Agtmael T, et al. Cerebrovascular disease related to COL4A1 mutations in HANAC syndrome. *Neurology.* 2009;73(22):1873–1882.

59. Kurth T, Diener HC. Migraine and stroke: perspectives for stroke physicians. *Stroke.* 2012;43(12):3421–3426.

60. Kumral E, Bogousslavsky J, Van Melle G, Regli F, Pierre P. Headache at stroke onset: the lausanne stroke registry. *J Neurol Neurosurg Psychiatry.* 1995;58(4):490–492.

61. Wolf ME, Szabo K, Griebe M, Forster A, Gass A, Hennerici MG, et al. Clinical and MRI characteristics of acute migrainous infarction. *Neurology.* 2011;76(22):1911–1917.

62. Arboix A, Massons J, Garcia-Eroles L, Oliveres M, Balcells M, Targa C. Migrainous cerebral infarction in the Sagrat Cor Hospital of Barcelona stroke registry. *Cephalalgia.* 2003;23(5): 389–394.

63. Laurell K, Artto V, Bendtsen L, Hagen K, Kallela M, Meyer EL, et al. Migrainous infarction: a Nordic multicenter study. *Eur J Neurol.* 2011;18(10):1220–1226.

64. Friedman BW, Mistry B, West J, Wollowitz A. The association between headache and elevated blood pressure among patients presenting to an emergency department. *Am J Emerg Med.* 2014;.

65. Wijman CA, Wolf PA, Kase CS, Kelly-Hayes M, Beiser AS. Migrainous visual accompaniments are not rare in late life: the Framingham Study. *Stroke.* 1998;29(8):1539–1543.

66. Berg MJ, Williams LS. The transient syndrome of headache with neurologic deficits and CSF lymphocytosis. *Neurology.* 1995;45 (9):1648–1654.

67. Kerklaan JP, Lycklama a Nijeholt GJ, Wiggenraad RG, Berghuis B, Postma TJ, Taphoorn MJ. SMART syndrome: a late reversible complication after radiation therapy for brain tumours. *J Neurol.* 2011;258(6):1098–1104.

68. Goncalves D, Meireles J, Rocha R, Sampaio M, Leao M. Syndrome of transient headache and neurologic deficits with cerebrospinal fluid lymphocytosis (HaNDL): a pediatric case report. *J Child Neurol.* 2013;28(12):1661–1663.

69. Nelson S. Confusional state in HaNDL syndrome: case report and literature review. *Case Rep Neurol Med.* 2013;2013:317685.

70. De Berardis G, Lucisano G, D'Ettorre A, Pellegrini F, Lepore V, Tognoni G, et al. Association of aspirin use with major bleeding in patients with and without diabetes. *JAMA.* 2012;307 (21):2286–2294.

71. Chakhtoura Z, Canonico M, Gompel A, Thalabard JC, Scarabin PY, Plu-Bureau G. Progestogen-only contraceptives and the risk of stroke: a meta-analysis. *Stroke.* 2009;40(4):1059–1062.

8

Acute and Preventative Treatment of Episodic Migraine

Roger K. Cady and Kathleen Farmer

Headache Care Center, Springfield, Missouri, USA

PART 1

INTRODUCTION

Migraine is a leading cause of medical disability. Worldwide, it is the fifth leading cause of disability for women and in the top 20 for men.[1] Migraine is also the most prevalent neurological condition evaluated by healthcare practitioners in the world,[2,3] and is substantially more prevalent than Alzheimer's, Parkinson's disease, and epilepsy combined.[4] While it is estimated that 12% of the general population lives with migraine each year, its lifetime prevalence is estimated to be greater than 30%.[5] Historically, migraine has been considered an episodic pain syndrome. Individual attacks generally last from 4 to 72 hours and can recur at a frequency ranging from five lifetime attacks to an unrelenting daily experience. Migraine affects people of all ages, races, cultures, and socioeconomic status.

The pathophysiologic origins of migraine are thought to reside in the brain. People with migraine are considered to have a genetically hyperexcitable nervous system with an enduring propensity to generate the clinical event of migraine when exposed to various changes in the internal and/or external environment.[6,7] Technically, migraine is considered a syndrome defined by relatively few of the symptoms observed during the actual attack. In clinical practice migraine is the "great mimic," often generating numerous symptoms that are not part of the diagnostic criteria. This commonly creates diagnostic and therapeutic uncertainty for patients and healthcare professionals. Migraine can, however, be considered both an attack and a chronic disease; each perspective is important when constructing dynamic, enduring pharmacological treatment and management plans for patients.

Treatment of migraine is generally divided into acute and preventative. Acute treatments are used intermittently to reverse an attack of migraine after the attack has started, while preventative treatments are generally taken on a daily basis to protect the nervous system from generating future migraines. It is estimated that 97% of those with migraine use acute treatments while only 8–13% use preventative treatment.[8,9] These two treatment paradigms are frequently used in concert with each other. Despite a wide array of treatments being available, migraine continues to be underdiagnosed, inadequately treated, and a leading cause of disability.[10,11]

The primary goal of migraine pharmacology is to reduce the impact and disability of both the attack and the frequency of migraine. This is best accomplished when the patient is confident with the diagnosis of migraine, and has been educated to understand the nuances and complexities of treatment. In most instances, the patient will ultimately decide which headaches to treat, when to treat them, what medication to use, and what defines treatment success. Motivation and adherence to treatment are improved when the patient is included in the decision-making process and has reasonable expectations of treatment outcomes.

This chapter primarily explores pharmacological and non-pharmacological interventions for acute and preventative treatment of migraine. It is an attempt to integrate medical evidence with clinical factors that synergize and optimize therapeutic outcomes.

S. Diamond, R. K. Cady, M. L. Diamond & V. T. Martin (Eds):
Headache and Migraine Biology and Management.

DOI: http://dx.doi.org/10.1016/B978-0-12-800901-7.00008-2

ACUTE MEDICATIONS FOR MIGRAINE

The development of sumatriptan in the early 1990s ushered in a much heralded "new era" for the acute treatment for migraine, and with it came a promise from the medical community of improved migraine care. Pivotal clinical trials of sumatriptan published in 1991 demonstrated that 70% of subjects with moderate to severe migraine had relief of headache and associated symptoms within 60 minutes, and 50% of subjects were pain free.[12,13] The development of sumatriptan also was an important catalyst for serious scientific study of migraine which literally transformed migraine into a significant and important medical entity worthy of study and drug development.[14–16] In the decades that followed, subcutaneous sumatriptan, seven oral triptans, and two nasal sprays were approved and brought to market. Regulatory trials demonstrated a lower efficacy for oral formulations than observed with subcutaneous sumatriptan, specifically in terms of overall headache relief, pain–free efficacy, and time to migraine resolution. Despite these shortcomings, oral triptans rapidly became the standard acute migraine intervention prescribed by healthcare professionals.

A meta-analysis of regulatory trials of all oral triptans except frovatriptan revealed that 2-hour pain freedom occurred in approximately 25–35% of trial subjects, and 17% of these subjects sustained pain freedom from 2 to 24 hours when treating moderate to severe migraine.[17] Later studies utilizing early intervention (treating the migraine early after onset and when the headache was mild in intensity) demonstrated significant improvement in the efficacy of oral triptans, and was an important step forward in improving clinical outcomes with oral triptans.[18,19] However, later estimates suggested that less than 50% of patients can actually utilize this strategy effectively due to limitations imposed by pharmacy benefit policies as well as patient and provider behaviors.[20,21] This may in part account for assessments of recent studies suggesting that as many as 30% of patients prescribed an oral triptan are dissatisfied,[22] and 80% are willing to try new treatments.[23]

CLINICAL FACTORS

The most important skill required to manage and optimize treatment plans for migraine patients is an organized approach to obtaining and integrating key information about both migraine attacks and the individual patient experiencing them. A simple pneumonic to assist in this effort is called the "5 Ps": pattern, phenotype, patient, pharmacology, and precipitants. Attending to these five components of information can assist in creating individualized behavioral, pharmacological, and non-pharmacological interventions for most migraine patients.

Pattern

Assessment of the pattern of migraine should be carried out at each office visit, perhaps by suggesting "Explain to me how your headaches have changed over time or since we last visited." The answer allows for rapid differentiation of the pre-existing pattern of migraine from other primary or new-onset secondary headache disorders.[9,24,25] A pattern of new or different headaches suggests a possible secondary headache disorder, while a stable pattern of headache over many months is in all likelihood a primary headache disorder. A pattern of accelerating migraine frequency is a fundamental determination of the need for preventive pharmacology. Pattern, more than any other feature of migraine, is essential for the early detection of migraine progression. Repetitive headaches of shorter duration (less than 3 hours) can be differentiated from those of longer duration (4–72 hours), and a pattern of progressive headache symptoms associated with escalating use of acute medication can suggest medication overuse headache (MOH). Monitoring the pattern of headache over time can also help determine the benefit of both acute and preventative treatment.

Phenotype

Phenotype addresses the characteristics and associated symptoms of a specific headache event. This is not simply a checklist used to make a diagnosis, but rather an in-depth understanding of how the events of migraine unfold for a specific patient. Assessment of phenotype clarifies the diagnosis, but equally important is that it creates a rational approach to therapeutics.[26] Asking the question "What do you experience as your worst headache develops?" leads to understanding of the physiological potential of the nervous system to generate primary headaches. For example, if a headache phenotype has several migraine features, it will likely respond to migraine-specific medication.[27,28] If the phenotype is devoid of migraine features, other therapeutic interventions may be more appropriate. Understanding the phenotype of the headache permits rational, physiologically based choices of acute interventions. Common phenotypes described by patients are migraine, tension-type headache, "sinus migraine," migraine causing awakening from sleep, abrupt-onset migraine, migraine associated with early onset of nausea, and migraine associated with autonomic features. Exploring individual phenotypes of migraine and the relationship between them can clarify the spectrum of migraine being experienced by the patient.

Patients often consider each phenotype they experience as a unique or different headache diagnosis, and consequently develop unique treatment approaches for each. In reality, however, these phenotypes may be clinical variations of the same diagnosis, which is most often migraine. For example, physician- or patient-diagnosed "sinus headache" and tension-type headache phenotypes, when examined closely, are migraine.[28,29] Assisting patients to understand their spectrum of headache presentations often results in more effective and appropriate targeting of acute therapeutic interventions.

Patient

Determining the patient's function between episodes of headache is paramount in screening for comorbidity and the impact and disability of migraine. A simple question such as "What do you feel like between episodes of migraine?" will often lead to discussions of medical and psychological comorbidities, as well as the patient's fears and anxieties. Specific queries about sleep, mood, and changes in general health can target the appropriate selection of preventative interventions and treatment needs of specific comorbid diseases.[30–32] It is essential to treat comorbid diseases effectively rather than assume they will resolve as migraine improves. Many comorbidities, if untreated, can become independent risk factors for poor treatment outcomes and development of chronic migraine (CM).[33] There are several simple screening questionnaires that can assist in this effort, such as the Migraine Disability Assessment Screen (MIDAS) or Patient Health Questionnaire (PHQ-9).

Pharmacology

It is critical to have a complete understanding of all the medications a patient is using to treat migraine as well as other medical conditions. Asking the patient for a detailed description of their treatment strategy with regard to migraine assesses the risk of or presence of overuse and/or the ineffective use of acute medications.[34] Overuse of many migraine-specific and symptomatic medications can lead to MOH even if it is prescribed for another medical condition. In addition, understanding the dynamics of a patient's decision-making around medication usage assesses barriers to optimal use of acute treatment. Frequently patients develop complex schemes to determine the

"treatment worthiness" of a specific headache, and these dynamics may be important barriers to long-term successful treatment outcomes. Several common barriers to effective acute intervention are "waiting to see" if the headache becomes severe, "stockpiling," and quantity limitations.[35] This assessment also identifies barriers for preventative therapy where patients may believe that because migraine is less frequent, they no longer have a need for daily medication.

Precipitants

To elicit protective or provoking factors for migraine, ask a question such as "Are you aware of events or other factors that put you at risk or prevent you from having a migraine?" This explores the patient's understanding of risk factors, preventive factors, and triggers for migraine. Patients often take an overly simplistic view and believe the last "trigger" they experienced "caused" their migraine, thus missing the numerous other concurrent risk factors that set the stage for the migraine attack. For example, a woman dining at a Chinese restaurant may assume that the monosodium glutamate in her meal "caused" her migraine despite the fact she was also near her menses, struggling to meet major deadlines at work, and experiencing stress at home. Clearly, the stage for migraine was "set" before she ate Chinese food. This more global view of risk events is best evaluated by utilizing diaries that, over time, can assist in pinpointing both risk and protective behaviors associated with episodes of migraine.[36–39]

Important risk and protective factors should be identified and incorporated into a comprehensive treatment plan. An important point for consideration is that as the frequency of migraine increases, so too does a patient's susceptibility to most risk factors. Patients with very frequent migraine find that virtually anything can provoke their next migraine. This makes identification of "triggers" more challenging. When migraine is frequent it becomes increasingly important to assist patients to understand that multiple risk factors, rather than a single triggering event, lead to migraine. One method to assist patients to understand this distinction is to explain the concept of the "migraine threshold."

The *migraine threshold* is the physiological point at which the nervous system can no longer sustain its integrity and an attack of migraine is initiated. The migraine threshold is likely determined by genetics, and the interrelationship of multiple risk and protective factors impacting the nervous system at any point in time. This concept is worthy of discussion with most migraineurs. By suggesting self-nurturing activities that support and strengthen the migraine threshold, patients can engage in activities that compensate for their vulnerability to migraine. This empowers patients with a better understanding of migraine and a sense of control over individual migraine attacks. Ultimately, this transforms the migraineur's perception of the world from being hostile and unpredictable into being understandable and controllable.

ATTACK-BASED ACUTE TREATMENT OF MIGRAINE

Along with patient education, effective acute treatment is the cornerstone of migraine management. Effective acute care is a multidimensional complex endeavor that includes understanding the nature of the attack, the progression or stage of migraine disease, and the patient. Inherent in this premise is that acute treatment is not simply about terminating a single attack of migraine, but is also scalable to provide effective treatment that may be successfully implemented for multiple attacks over extended periods of time. Over the span of decades, rarely does a patient experience attacks of migraine that are symptomatically identical. Most patients experience a spectrum of migraine attacks, from "big" to "little," and these unique expressions of

migraine often require an individualized approach to acute treatment.

Further, successful acute treatment must ultimately be understood and directed by the patient, and it needs to be both obtainable and affordable. This is best accomplished through education conducted in a collaborative dynamic with patients that allows them to participate and understand treatment objectives and outcome goals.

PHASE-BASED ACUTE TREATMENT OF MIGRAINE ATTACKS

An attack of migraine is a dynamic process that evolves over the span of hours to days. The efficacy and benefit of acute treatment is significantly influenced by the timing of initiating pharmacological interventions during the development of the migraine attack (see Figure 8.4 and Table 8.1, below). At times, acute treatment can be successfully initiated before predictable attacks of migraine begin, such as attacks of migraine associated with menses. Guiding patients regarding when to initiate various acute treatments relative to the phase of migraine they are experiencing can improve efficacy, reduce headache recurrence, minimize drug-related adverse events,[40] and possibly reduce the risk of migraine chronification.[41]

TABLE 8.1   Efficacy of Acute Treatment Based on Headache Phase

Medication	Vulnerability phase	Premonitory phase	Mild headache phase	Moderate to severe headache phase	Rescue phase	Postdrome phase
Sumatriptan SC			++++	++++	+++	
Parenteral triptans; nasal and transdermal			++++	+++	++	
Oral triptans	Frovatriptan Sumatriptan Zolmitriptan	Frovatriptan Naratriptan	++++	++	+	Likely beneficial*
Sumatriptan/naproxen	Likely beneficial	Likely beneficial*	++++	+++	++	Likely beneficial*
Diclofenac powder	Likely beneficial	Likely beneficial*	++++	+++	++	Likely beneficial*
Anti-emetics alone or in combination				+++	+++	
Dihydroergotamine IV	+++	+++	+++	+++	+++	
Opioids				++	+++	
Phenothiazines			+++	+++	+++	
Metoclopramide			++	++	++	
Chlorpromazine				+++	+++	

*Adapted from Bedell AW, et al. Patient-centered strategies for effective management of migraine. Primary Care Network, 2000.
Level of efficacy: +, somewhat; ++, mild; +++, moderate; ++++, high.

The Therapeutic Phases of Migraine

An attack of migraine can be divided into six phases that can assist patients in determining appropriate options for acute therapeutic interventions. Not every patient experiences all six phases of migraine, and even in the same patient the migraine experience is not always the same. Thus, use of patient diaries to detect and assess attack variability and the effectiveness of acute intervention is of paramount importance.

1. Vulnerability Phase

Certain events carry a high probability of initiating migraine. The most notable and well-studied is the perimenstrual time period for some women. Several studies of triptans and NSAIDs have demonstrated efficacy at preventing or reducing anticipated attack severity when acute medication is administered based on the occurrence of a known risk event, but prior to the occurrence of recognized migraine symptomology. There is substantial evidence for frovatriptan as a "short-term" or "pre-emptive" prophylactic in menstrual migraine (Figure 8.1).[42] There is also evidence for sumatriptan,[43] zolmitriptan,[44] and dihydroergotamine[45] for short-term or pre-emptive prophylaxis in menstrual-related migraine. In fact, the published guidelines for preventative therapy by the American Academy of Neurology and American Headache Society suggest there is Class A evidence for frovatriptan as a short-term prophylactic medication, and Class B for sumatriptan and zolmitriptan.[46] Further evidence exists in smaller studies for naproxen and other non-steroidals utilized in the same fashion.[47] Whether other predictable attacks of migraine can be successfully prevented or attenuated with this strategy is less well studied. In one study, daily topiramate was compared with frovatriptan administered as an intermittent "pre-emptive" treatment at a patient's discretion, including prior to the onset of headache. In this study, both drugs were efficacious at reducing the frequency of migraine attacks over a 2-month time period.[48] Clearly, further study of acute treatment in the vulnerability phase of migraine is warranted, as it holds the potential of significantly reducing the impact and frequency of migraine events as well as the drug exposure of using daily preventative medications.

2. Premonitory Phase

The premonitory phase of migraine consists of non-headache symptoms beginning hours to days before the onset of headache. These include symptoms such as irritability, fatigue, food cravings, muscle tension/pain, and cognitive changes. It is estimated that 33−87% of migraine attacks are preceded by definable premonitory symptoms.[49] Despite the non-specific nature of premonitory symptoms, they often impose significant disability. In a study by Giffin and

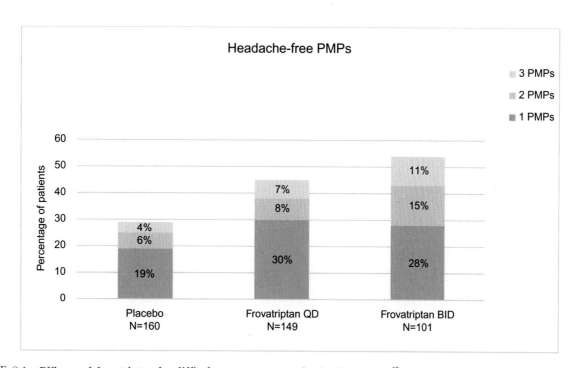

FIGURE 8.1 **Efficacy of frovatriptan for difficult to treat menstrual migraine attacks.**[42] Number of women who benefited 1 out of 3 attacks of menstrual migraine; number who benefited 2 out of 3 attacks; number who benefited 3 out of 3 attacks. Greater than 50% of women had 1 out of 3 attacks prevented by treating during the vulnerable phase of migraine. Over 40% of women prevented 2 out of 3 attacks. The women in the placebo group had a much lower response rate. PMP, perimenstrual periods.

coworkers, 70% of the disability of a migraine attack occurred prior to the onset of a migraine-associated headache.[50] Further studies by Farmer and Reeves have demonstrated significant cognitive impairment attributable to the premonitory phase of migraine,[51,52] and acute treatment utilized during this pre-headache phase has been shown to be effective at reversing predicted migraine attacks and symptoms.[53]

More recent studies of acute treatment initiated during the premonitory period have been conducted with both naratriptan (Figure 8.2)[54] and sumatriptan.[55] In these studies there was a reduction of expected attacks and attenuation of anticipated disability. Effective treatment delivered during the premonitory period has the potential to stop the attack before reaching moderate to severe pain and disability, for some attacks of migraine.[49,56]

3. Aura

The aura phase of migraine is estimated to occur in 20–25% of migraine attacks, and typically consists of fully reversible positive and negative neurological symptoms. The most common auras are visual, but can include sensory, speech, or motor symptoms.[57] Auras typically last from 5 to 60 minutes, and can be another source of significant disability for some patients. Migraine with aura is also associated with several infrequent but important health risks, including cerebrovascular and cardiovascular disease.[58] Vigilance is needed to reduce other known vascular risk factors, such as tobacco use, possibly birth control pills, and obesity, when managing patients with migraine with aura.

For many patients the occurrence of aura can be used to "time" the delivery of acute pharmacological interventions and thus reduce the intensity or occurrence of headache. However, to date there are no effective acute treatments to terminate auras. Specifically, triptans taken during the aura phase may prevent or attenuate headaches, but do not alter the aura.[59]

4. Headache

It is useful to divide the headache phase of migraine, based on headache intensity, into mild, moderate, and severe.[18] The intensity of the headache associated with migraine at the time that an acute intervention is initiated is a significant determinate of treatment outcome. In general, there is an inverse relationship between acute treatment efficacy and the intensity of headache when acute treatment is initiated, especially with oral triptan medications.[18] Providing diaries for patients that track the intensity of migraine headache when acute intervention is initiated is useful, as it provides the clinician with the opportunity to amplify education or consider formulation changes.

TREATMENT OF MIGRAINE WHEN THE HEADACHE IS MILD

Treatment of migraine attacks early after they begin, and while the headache is mild in intensity, is called "early intervention".[27,60,61] Acute treatment with an oral triptan during the mild headache phase of migraine is about twice as likely to result in a pain-free outcome as treatment initiated during severe headache.[62,63] Further clinical studies demonstrate that there is less recurrence of migraine and fewer medication-related adverse events when patients use an early intervention strategy. Clearly, initiating

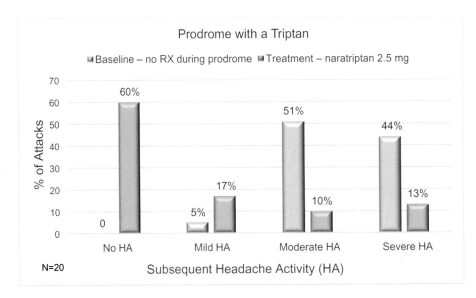

FIGURE 8.2 **Efficacy of naratriptan.**[54] The figure demonstrates that 60% of anticipated attacks did not occur due to pre-emptive treatment with naratriptan. No subjects in the placebo group avoided migraine with the use of placebo. Overall 77% using pre-emptive treatment had no or mild headaches, vs 65% of subjects in placebo group that developed moderate to severe headaches.

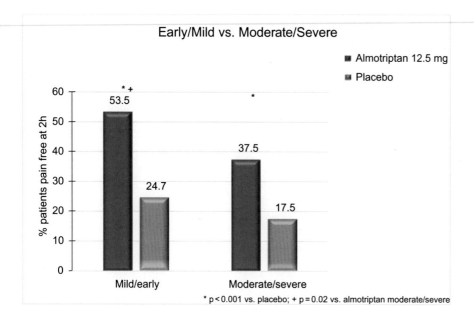

FIGURE 8.3 Almotriptan in early treatment in acute migraine, mild vs severe.[19]

treatment early, when the headache is mild, can reduce the time with attack-related disability. As the attack of migraine progresses from mild to moderate/severe, the effectiveness of tablets of acute medication lessens substantially. The optimal treatment strategy is to medicate early when the headache severity is mild. Numerous studies have supported the efficacy of early intervention (Figure 8.3).[19,27,64] However, technically, early migraine often does not always fulfill diagnostic criteria for migraine, and thus regulatory agencies have avoided support of this treatment paradigm.

TREATMENT OF MIGRAINE DURING MODERATE TO SEVERE HEADACHE

Initiating acute treatment during moderate to severe migraine often reduces the efficacy of an oral acute medication and increases the time the nervous system is in a state of migraine. Regulatory trials are based on treatment of moderate to severe headache to ensure that subjects in clinical trials treated an attack of migraine as defined by International Headache Society (IHS) criteria. While treatment utilizing this treatment paradigm may provide a degree of scientific and statistical clarity of the data, it is not necessarily an optimal treatment paradigm for patients seeking relief from migraine. As suggested earlier, the intensity of migraine headache is a significant driver of drug efficacy, especially for oral triptans. The more intense the headache, the less efficacious the oral medication, the more likely there will be recurrence of migraine and attack-related adverse events.[18]

Taking these important clinical observations forward, clinical data support that acute treatment initiated during moderate headache with oral triptans

provides better clinical outcomes than acute migraine treatment initiated during severe headache.[18]

The relation of drug efficacy to headache intensity is generally understood as being a consequence of the development of central sensitization.[65] As pain intensifies and becomes increasingly centralized (particularly with the occurrence of cutaneous allodynia), the efficacy of oral triptans is dramatically lessened.[66] Consequently, it is essential that patient-oriented healthcare professionals assist patients in understanding that initiating acute treatment early in the progression of a migraine attack provides optimal efficacy, minimizes attack disability, restores function, and reduces direct and indirect costs associated with migraine. This means acute intervention when the migraine headache is less severe improves clinical outcomes and reduces the time for which the nervous system endures migraine.

5. Resolution, Recovery, and the Postdrome Phase of Migraine

The physiology underpinning the resolution of migraine remains poorly understood. It generally occurs with sleep or with acute medications, but at times vomiting or significant psychological events have been reported to terminate a migraine attack.[67] Anecdotally, medications such as 5-HT1B/1D agonists, non-steroidal anti-inflammatories, anti-emetics, or analgesics have all been reported to successfully treat postdromal symptoms.[46] It is important that patients treating migraine establish complete and, ideally, prolonged recovery to their normal level of neurological baseline function between migraine events.[68] Giving the nervous system ample time for complete recovery between attacks may reduce the risk of future migraine and the risk of migraine transformation.[69]

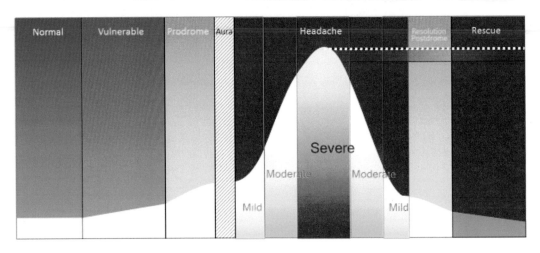

FIGURE 8.4   Phased-based pharmacological interventions for acute migraine. *Reproduced with permission from the Primary Care Network© 2014.*

6. Rescue Phase

The rescue phase of migraine refers to an attack that has not responded well to initial acute interventions and continues to produce significant disability. The pathophysiology of status migraine is poorly understood, but likely relates to central sensitization and continued peripheral and central drive of pain pathways.[70]

Consequently, in order to avoid the use of emergency or unscheduled medical services, many migraineurs need to have rescue medications on hand to use if acute medication fails to terminate an attack. Few large clinical trials have been conducted on rescue medications; however, parenteral triptans, dihydroergotamine, neuroleptics/anti-emetics, and analgesics have been employed in this role. Procedures such as occipital nerve blocks or sphenopalatine ganglion blockade may also be effective.[71] Studies have shown that sustained periods of migraine associated with poor or marginal acute treatment outcomes, and over time, may be a risk factor for CM.[72]

Figure 8.4 and Table 8.1 illustrate phase-based pharmacological interventions for acute migraine, and the efficacy of acute treatment based on the headache phase, respectively.

ACUTE MEDICATIONS FOR EPISODIC MIGRAINE

Goals

The International Headache Society states that the most appropriate goal for clinical trials of acute treatment is pain free within 2 hours of treatment. Patient-desired outcomes from acute treatment of migraine include rapid onset, complete headache resolution,

and no recurrence.[73] Sustained pain-free response for 2–24 hours reflects the desired patient goal for acute treatment.[74] Applying this standard to clinical trial data of acute pharmacological interventions suggests that less than one-fourth of migraines treated in clinical trials with oral triptans achieve these two goals. Translated into clinical terms, this suggests that an average migraineur with 3 attacks per month will endure 27 attacks per year that do not achieve this high but desired standard. Over the span of a decade, this translates to 270 migraine attacks with less than ideal clinical response. Clearly, there is a need to provide migraine patients with better outcomes with the use of acute treatment.

A more sensitive therapeutic alternative to optimize acute treatment is to define with individual patients the different therapeutic needs to manage their unique spectrum of migraine attacks. Patients are then provided the necessary acute treatments to match their treatment needs over multiple attacks of migraine. Often, patients may utilize an oral NSAID or oral triptan for migraines where they can intervene while the headache pain is mild if they are not having nausea. For more advanced attacks, especially those associated with nausea or symptoms of gastric stasis or atony, a parenteral triptan may be more effective. Additionally, a rescue therapy and strategy is made available for home use if the first intervention is ineffective. The strategy of providing multiple acute treatments to address the variability of acute treatment needs over multiple migraine attacks has been termed *attack-based care*, and is a departure from stratified and step-care models (see Figure 8.6, below).[75] Critical factors to successfully utilize attack-based care are patient education and collaborative tactical decision-making about different therapeutic needs based on individual attack characteristics. This includes decisions about timing of treatment (early), management of nausea, appropriate use of parenteral interventions, and providing rescue treatment to have on hand if an initial acute intervention is not effective.

While one might argue that 2-hour pain free and sustained pain free (2–24 hours) are too lofty a goals for currently available pharmacology, the counter argument might be that the acceptance of less than ideal outcomes represents a system failure of providers, patients, and third-party stakeholders to fully appreciate the importance of optimized acute treatment of migraine.

Medications Indicated for Treatment of Acute Migraine (Table 8.2)

Triptans

Triptans as a therapeutic category are frequently considered the gold standard for acute migraine therapy.[76,77] While likely true for subcutaneous sumatriptan, there is little comparative evidence to support this assertion for oral triptans. In fact, studies comparing oral sumatriptan to acetaminophen/aspirin/caffeine failed clearly to distinguish one product as being superior over the other,[78] and clinical trials of diclofenac demonstrated similar efficacy to that seen in trials with oral triptans.[79] On the other hand, subcutaneous sumatriptan demonstrates efficacy that is clearly superior to all other acute migraine interventions.

All triptans are serotonin receptor agonists with high affinity for the 5-HT1B/1D receptor. These receptors are localized to the trigeminal vasculature (5-HT1B) and trigeminal neurons (5-HT1D). Activation of the 5-HT1B receptors during migraine results in vasoconstriction (vascular normalization) of the dilated blood vessels. Activation of the 5-HT1D receptor inhibits the vesicular release of CGRP, substance P, and other inflammatory peptides from the trigeminal afferents.[80,81] This prevents plasma extravasation and elevates the sensory threshold for trigeminal afferents. There are also 5-HT1D receptors in the central nervous system, but it is unknown if triptans have clinically meaningful activity in the central nervous system. Calcitonin gene-related peptide (CGRP) is released during migraine, and is presumed responsible for vasodilation and initiation of pain transmission.[82] Both vasoconstriction and inhibition of neuropeptide release appear to be important mechanisms in how triptans abort migraine.[83]

The comparative efficacy of various oral triptans is difficult to ascertain, as few direct comparative studies have been conducted. In the clinical development programs of triptans, oral sumatriptan became the de facto comparative standard. However, over time, later clinical trials were able to enrich their study populations through exclusion of subjects with adverse effects from or non-response to earlier triptans. In addition, beyond trials with subcutaneous sumatriptan, no studies were restricted to subjects who were naïve to other triptan products. While all triptans have the same mechanism of action, there are unique pharmacological characteristics of different oral triptans that may be clinically meaningful – for example, frovatriptan has a much longer half-life than other triptans. Clinicians often report different patient responses to different triptans, and therefore it is worthwhile trying another triptan if one fails or has bothersome adverse events associated with its use. However, there is little evidence that multiple sequential trials of oral triptans will ultimately lead to the "right" oral triptan, or that there are specific patient populations for whom a specific oral triptan is indicated.

An important limitation to understanding the clinical value of oral triptans is that they were studied in regulatory trials for moderate to severe migraine headache. However, for a number of reasons, such as drug absorption when nausea is present or the increased likelihood of central sensitization as headache intensity increases, this treatment paradigm clearly underestimates the potential clinical value of oral triptans when applied appropriately in clinical practice.

Another limitation to the clinical use of triptans is exaggerated concerns of significant adverse events, especially catastrophic cardiovascular and cerebrovascular events. Initial concerns of vascular consequences with triptans were largely predicated on migraine being a vascular disease and prior knowledge of the vascular mechanisms of ergotamines as acute treatment of migraine. However, over the past two decades millions of patients have utilized hundreds of million doses of triptans, with little evidence that triptans are associated with significant cardiovascular or cerebrovascular events.[84–86] This is not to imply that there is no vascular risk with triptan use, and patients require careful cardiovascular screening before triptan therapy should be initiated or continued, but in general triptans are well tolerated and safe acute treatments for migraine.

An additional area of exaggerated safety concern with triptan therapy is serotonin syndrome. Except perhaps in rare instances, there is little evidence to support serotonin syndrome as a common adverse consequence to concomitant use of selective serotonin reuptake inhibitors or serotonin norepinephrine reuptake inhibitors (SSRIs/SNRIs) and a triptan. In fact, the American Headache Society has written a position paper questioning this association.[87] Their contention is supported by large claims databases, and the fact that serotonin syndrome is mediated through the

TABLE 8.2   Acute Medications for Migraine

Generic treatment	Doses
Aspirin tablets	325 mg to 650 mg
Acetaminophen tablets	325 mg to 1000 mg
Combination analgesics	
Aspirin plus acetaminophen plus caffeine tablets	250 mg plus 250 mg plus 65 mg
Isometheptene mucate plus acetaminophen plus dichloralphenazone tablets	65 mg plus 325 mg plus 100 mg
Butalbital plus aspirin plus caffeine tablets	50 mg plus 325 mg plus 40 mg
Butalbital plus acetaminophen plus caffeine tablets	50 mg plus 325 mg plus 40 mg
Ergotamine alkaloids	
Ergotamine tartrate plus caffeine tablet	1 mg plus 100 mg
Ergotamine tartrate plus caffeine suppository	2 mg plus 100 mg
DHE nasal spray	0.5 mg/nostril (repeat in 15 min 1 × for 2-mg total dose)
DHE IM or SC	1 mg
Neuroleptics and anti-emetics	
Chlorpromazine capsule	10 mg to 25 mg
Metoclopramide IV	10 mg+
Prochlorperazine	25 mg
NSAIDs	
Diclofenac K tablets	50 mg to 100 mg
Diclofenac potassium	50 mg
Flurbiprofen tablets	100 mg to 300 mg
Ibuprofen tablets	200 mg to 1200 mg
Naproxen sodium tablets	550 mg to 1100 mg
Naproxen tablets	250 mg to 500 mg
Piroxicam tablets	40 mg
Tolfenamic acid tablets	200 mg to 400 mg
Opiate analgesics	
Butorphanol nasal spray	1 to 2 mg
Triptans	
Almotriptan tablets	12.5-mg tablets
Naratriptan tablets	1 mg or 2.5 mg
Rizatriptan tablets	5 mg or 10 mg
Rizatriptan orally disintegrating tablets	5 mg or 10 mg
Sumatriptan tablets	25 mg, 50 mg or 100 mg
Sumatriptan nasal spray	5 mg or 20 mg
Sumatriptan SC self-injection	6 mg
Zolmitriptan tablets	2.5 mg or 5 mg
Zolmitriptan orally disintegrating tablets	2.5 mg or 5 mg

5-HT2 receptors for which triptans have minimal affinity.[88,89]

Finally, oral triptans, like most acute treatment medications, are incriminated in the etiology of MOH.[90,91] It may be that while this association is well documented, it is often the messenger rather than the message that is blamed. In reality, when a patient begins to require frequent doses of any acute medication for relief of severe migraine, treatment measures need to be advanced to include additional non-pharmacological and pharmacological interventions before, rather than after, MOH is established. In many instances, MOH is an iatrogenic consequence involving both provider and patient behaviors.

FORMULATIONS: THE KEY TO LONG-TERM SUCCESSFUL ACUTE INTERVENTION

Different formulations of triptans provide a unique opportunity for improving the clinical outcomes of acute interventions. A majority of patients (70%) experience nausea during migraine, especially if they treat a moderate to severe headache. In addition, gastric atony or gastric stasis is associated with migraine during and between attacks of migraine.[92] These factors strongly suggest that gastrointestinal absorption (GI) of oral medications is delayed or impaired during many attacks of migraine, and GI absorption may be impaired even before the symptom of nausea is present. Surrogate markers for impending nausea in migraine often include symptoms such as anorexia, bloating, belching, or queasiness. Absorption of oral acute medications can be inconsistent over the span of numerous migraine attacks. This may be observed clinically as an inconsistent response to the same oral acute intervention. In addition, nausea is a significant driver of migraine-related disability.[93] Careful clinical assessment of migraine-associated GI symptoms often dictates the inclusion of alternate formulations of triptans as tools to ensure drug absorption and consistent control of acute attacks.

SUBCUTANEOUS SUMATRIPTAN Subcutaneous sumatriptan is the fastest and most effective acute treatment for migraine. Initial studies demonstrated that at 1 hour, 70% of subjects with moderate to severe migraine experienced relief (moderate or severe headache becoming mild or no headache) and 50% were pain free. By 2 hours, 80% had headache relief. In a more recent open-label, multicenter Phase IV study of subcutaneous sumatriptan delivered by needleless technology (SUMAVEL® DosePro®) and involving 212 patients, 85.9% of subjects had pain relief and 60.7% were pain free at 1 hour while 35.4% of subjects sustained pain freedom without rescue interventions for 2−24 hours. In addition, the revised Patient

Perception of Migraine Questionnaire (PPMQ-R) and Overall Satisfaction and Confidence scores increased significantly from baseline to the end of treatment (mean ± SD 65.7 ± 19.8 vs 73.7 ± 29.1, $P = 0.0007$). Patients who were "satisfied" or "very satisfied" increased significantly from baseline (36.3%) to the end of treatment (64.0%) for all global satisfaction domains, including "overall satisfaction." Observed adverse events with the needleless injection system were primarily pain at injection site, local reactions, and triptan sensations.[94,95]

These studies underscore the value of providing access of subcutaneous sumatriptan (needle-based or needleless) for patients to use for attacks associated with GI symptoms, significant gastric atony or gastric stasis, rapid onset migraine, or attacks treated later in their evolution, such as early morning migraine. They can also be an effective rescue intervention.

NASAL FORMULATIONS

Nasal Sprays Nasal spray formulations of sumatriptan (5 and 20 mg) and zolmitriptan (2.5 and 5 mg) are approved for the treatment of acute migraine. Sumatriptan nasal spray, 20 mg, demonstrated an efficacy in between 55% and 64% of subjects for 2-hour headache relief vs placebo rates of between 25% and 36% in regulatory trials of migraine with headache of moderate to severe intensity. Pain-free efficacy was between 18% and 33%, and sustained pain-free efficacy from 2−24 hours was 20%. It was well tolerated, with "unpleasant taste" occurring in approximately 25% vs 1.7% for placebo.[96]

Zolmitriptan nasal spray, 5 mg, produced headache relief rates of approximately 70% at 2 hours vs 31% for placebo in clinical trials.[97] Pain free at 2 hours occurred in 35.5% of subjects with zolmitriptan nasal spray and 6.4% for placebo. It was well tolerated, with unusual taste experienced in 21% of subjects vs 3% for placebo. It had a rapid onset of action, and in one study statistical benefit was noted within 15 minutes by a small subset of subjects.

Dry Nasal Powder of Sumatriptan A closed-palate Breath Powered™ delivery system is used to deposit sumatriptan powder into the nasal cavity during oral exhalation. Exhalation causes closure of the nasopharynx, resulting in greater deposition of drug into the nasal cavity.[98] With this device 16 mg of sumatriptan is delivered as a powder, allowing it to adhere to the nasal mucosa where it is absorbed.

Pain relief at 2 hours was reported in 68% of subjects vs 45% with placebo, and pain free at 2 hours was reported in 34% of subjects vs 17% with placebo. Sustained pain-free rates were 28% vs 11.5%, and 48-hour sustained pain-free was reported in 20% vs 8.7%.[99] Adverse events were primarily bad or unpleasant taste (21% vs 4% for placebo) and nasal discomfort (13% vs 2% for placebo). There were few "triptan sensations" reported by subjects in this study, which is likely attributable to the smaller dose of triptan.

Nasally absorbed formulations as a whole are efficacious and well tolerated. When properly utilized, they are a vehicle to circumvent the impaired GI absorption associated with some migraine attacks and are easy to use, making them suitable for first-line interventions for many patients. There are few comparator data on nasal formulations and oral medications, but they appear equally or more efficacious and well tolerated.

IONTOPHORETIC TRANSDERMAL DELIVERY OF SUMATRIPTAN A new delivery system is the sumatriptan iontophoretic transdermal patch (Zecuity™), which is a single-use transdermal patch containing 86 mg of sumatriptan that delivers 6.5 mg of sumatriptan over 4 hours with iontophoresis. In a randomized placebo-controlled study of 469 subjects, this product was well tolerated and its efficacy was superior to placebo for both pain-free and pain-relief response at 2 hours (18% vs 9%, pain free; $P = 0.009$ and 52.9% vs 28.6%, headache relief; $P < 0.0001$). There were few triptan sensations reported with use of this technology, and adverse events consisted primarily of pain and erythema at the site of application. Additionally, there was significant reduction in migraine-associated symptoms of nausea (83.6% vs 63.2%; $P = 0.0028$), photophobia (51% vs 36%; $P = 0.0028$), and phonophobia (55% vs 39%; $P = 0.0002$) active vs placebo, respectively.[100] In a long-term safety study, transdermal sumatriptan maintained consistency and efficacy in up to 12 treated attacks. This delivery method may be useful for patients with migraine-related nausea, vomiting, or gastroparesis.[101] Zecuity™ may also be an option for patients with needle phobias, or with inconsistent, poor, or non-response to oral or nasal formulations.

ORAL FORMULATIONS Oral triptans are optimal medications for episodic migraine, especially when patients can consistently intervene during the mild headache of migraine. They should be prescribed with limitations on frequency of use at 2 days per week, and patients need to understand that using an oral triptan more than 2 days per week may not necessarily be a failure of medication but is an indication that it is time to use parenteral formulations of triptans, to change or add another therapeutic class of abortive medication, to initiate preventive medications, and/or to address behavioral and lifestyle issues. Oral triptans are often stratified by their half-life. In general, short half-life triptans are sumatriptan, zolmitriptan, rizatriptan, and almotriptan; intermediate half-life triptans

are eletriptan and naratriptan; and the long half-life triptan is frovatriptan. The clinical rationale is that short half-life triptans have a more rapid onset while longer duration triptans have a more sustained response. This differentiation has never been clearly documented in clinical studies.

Several oral triptans are also provided in unique and often beneficial oral formulations. Both zolmitriptan and rizatriptan are available as a rapidly dissolving oral wafer or tablet that does not require water. While these are convenient for many patients, this formulation remains an oral formulation, and the onset is not any faster than with tablets — in fact, there may be a slight delay due to transit time through the esophagus.

All triptans are available in two doses. In adults, it is usually best to initiate treatment at the higher dose. If unpleasant non-serious adverse events occur, the dose can be stepped down.

Sumatriptan 85 mg/Naproxen 500 mg Combination (Treximet®) A combination of sumatriptan, 85 mg, and naproxen sodium, 500 mg, is marketed as Treximet®. In two large clinical trials this combination was more efficacious than placebo. (Study 1: 65% vs 28%; $P < 0.001$ and Study 2: 57% vs 29%; $P < 0.001$). Further, there was a reduction of associated migraine symptoms of photophobia and phonophobia at 2 hours post-dose. Nausea reduction was significant in Study 1 but not in Study 2 (Study 1: 71% vs 65%, $P = 0.007$; Study 2: 65% vs 64%, $P = 0.71$). The 2- to 24-hour sustained pain-free efficacy was statistically superior to either component of the product used alone as well as placebo. The adverse event profile was similar to that of sumatriptan monotherapy.[102] Sumatriptan/naproxen may be useful as an early intervention, when migraine recurrence is common, or when migraine is associated with dysmenorrhea.

Non-steroidal Anti-inflammatory Drugs

Non-steroidal anti-inflammatory drugs (NSAIDs) are a diverse class of compounds that provide both analgesic and anti-inflammatory benefits. Most NSAIDs non-selectively inhibit the activity of cyclo-oxygenase 1 and 2 (COX-1 and COX-2), leading to inhibition of prostaglandin and thromboxane production. NSAIDs are a mainstay of acute therapy for migraine, used as monotherapy or in combination with triptans or other abortives. Earlier research on the selective COX-2 inhibitor rofecoxib demonstrated efficacy in migraine shortly before being taken off the market because of potential cardiovascular risks.[103] Interestingly, a study combining rizatriptan with low doses (25 mg) of rofecoxib showed promising efficacy and lower recurrence rates than with rizatriptan alone.[104] COX-2 inhibitors such as celecoxib preferentially inhibit the inducible COX-2 enzyme that is elaborated during attacks of migraine, but it is unclear if selective COX-2 inhibitors are more efficacious or safer than non-specific COX-1/COX-2 inhibitors.

While the onset of analgesic effects of NSAIDs is relatively rapid, the onset of the anti-inflammatory effects of NSAIDs generally takes many hours or days to develop.[105] However, recent animal studies suggest that prostaglandins in the trigeminal ganglion and CNS are critical to the recruitment by glial cells of neurons into the amplification of pain processing, and thus this may be an important component of their mechanism of action in migraine.[106,107]

Ibuprofen in both 200- and 400-mg doses has demonstrated short-term efficacy in reducing the headache of migraine, but sustained pain-free rates were similar to placebo. Higher dosage of 400 mg also was effective in relief of photophobia and phonophobia.[108] Naproxen sodium also demonstrated significant efficacy in moderate to severe headache over placebo at doses of 500 mg.[104,109] A combination of aspirin—acetaminophen—caffeine has also demonstrated efficacy as an abortive in acute migraine, and demonstrated efficacy similar to oral sumatriptan 50 mg.[110] Non-steroidal drugs are popular because they are effective, not a narcotic, not a steroid, and inexpensive. However, a majority of patients will have failed at least one of these agents before seeking medical consultation for migraine.

More recently, an oral solution of potassium diclofenac (Cambia®) received FDA approval for acute migraine.[111] In two multicenter trials, one European and the other US, the pain-free response for the oral solution of K-diclofenac was 24.7% and 18.5% for the diclofenac tablet, and 11.7% for placebo ($P < 0.0001$, 0.0035, 0.004, respectively). Pain relief at 2 hours was 46% for K-diclofenac oral solution, 41.6% for diclofenac tablets, and 24.1% for placebo ($P < 0.0001$, 0.0035, and 0.0001, respectively). Sustained pain-free efficacy at 24 hours was statistically superior for both formulations of diclofenac over placebo and the oral solution over tablet formulations of diclofenac ($P < 0.0001$, 0.0005, 0.0077, respectively). In both studies, the oral K-diclofenac was statistically superior to placebo for relief of associated migraine symptoms, and was well tolerated with few adverse events. The most common adverse event was nausea, 4.6%, with oral suspension of diclofenac vs 3.5% with placebo. As with other NSAIDs, there is a black box warning for increased risk of cardiovascular events, including myocardial infarction and death. Those with frequent use and the elderly are likely at greatest risk.

NSAIDs may have some value in reducing the risk or progression of migraine chronification relative to other acute migraine medications. In epidemiological studies, NSAID doses up to 12 times per month appeared to have a protective role in lowering the risk of developing CM.[47] In another study where naproxen 500 mg was provided daily and compared with daily dosing of sumatriptan, 85 mg, and naproxen, 500 mg, it was observed that the naproxen group experienced a reduction of migraine frequency over 3 months whereas the sumatriptan/naproxen group had no reduction in migraine frequency. The authors suggested there may be a disease-modifying benefit for naproxen, and that the addition of naproxen to sumatriptan may have prevented further increases in migraine frequency that might be assumed to occur with sumatriptan alone when used at that frequency as a solo agent. In neither group was there clear evidence of subjects developing MOH.[112]

While NSAIDs are beneficial in migraine, they do have rare but potentially serious side effects associated with their use, especially when migraine attacks require frequent treatment. Gastrointestinal bleeding, kidney and liver toxicity, and, rarely, cardiovascular events, including myocardial infarction and heart failure, have all been associated with NSAID use.

Dihydroergotamine

Dihydroergotamine (DHE) is widely used as acute treatment for migraine, particularly by headache specialists. It is often employed in a variety of repetitive dose protocols to manage medication overuse headache and intractable migraine.[113] However, these can be effective first-line interventions as well.

DHE, like triptans, is a 5-HT1B/D agonist, but DHE has additional receptor activity at the 5-HT1A and 5-HT2 receptors as well as alpha-1 and -2 adrenergic receptors.[114] DHE binding to the α_2-adrenergic receptor blocks ATP-sensitive trigeminal neurons by decreasing membrane expression of the P2X$_3$ receptor protein, which may have implications[115] in primary nociception, hyperalgesia, and CGRP release.[116,117] In addition, DHE may also be more penetrant of the blood—brain barrier than triptans.[118] The benefit of using a drug with multiple receptor activity compared with one with limited, more specific receptor activity in a complex biological event like migraine has not been adequately investigated. In a single comparator study, SC sumatriptan was slightly more efficacious and provided earlier headache relief than DHE, but DHE was associated with less recurrence of migraine.[119] Given the potential heterogeneity of the migraine population, mechanisms exploring these unique subsets could prove to be a line of further investigation.

Oral DHE is poorly and erratically absorbed, and consequently of limited value in acute treatment of migraine. As a nasal spray, the bioavailability of DHE is about 32% of that obtained with an injection. DHE nasal spray is marketed as a 4-mg dose, with 2 mg being sprayed into each nostril. Clinical trials of DHE nasal spray demonstrated 2-hour pain-relief rates ranging between 61% and 30% vs placebo rates between 23% and 33%.[120] Associated symptoms of nausea, photophobia, and phonophobia were reduced relative to placebo, and there were lower recurrence rates of migraine relative to triptans. Nasal DHE is generally well tolerated, with the most common adverse events being rhinitis, dizziness, altered taste, site reactions, nausea, and vomiting. Often an anti-emetic such as metoclopramide is combined with DHE to minimize nausea, especially when administered intravenously.

An orally inhaled formulation of DHE has recently been investigated in clinical trials. In Phase III studies, 28.4% vs 10.1% ($P < 0.0001$) of subjects were pain free at 2 hours and 58.7% vs 34.5% ($P < 0.0001$) had pain relief with orally inhaled DHE vs placebo, respectively. Associated migraine symptoms of nausea, photophobia, and phonophobia were also statistically improved vs placebo. Compared with IV DHE, orally inhaled DHE had a lower incidence of both nausea (25% vs 62%) and dizziness (25% vs 44%), which may be explained by its lower C_{max}.[121] Importantly, patients using inhaled DHE have reported low recurrence of migraine at both 24 and 48 hours (6.5% and 10.3%, respectively).[122] Unlike oral triptans, DHE appears effective when administered late in the attack of migraine.[123]

In addition, orally inhaled DHE has been studied in healthy volunteers, smokers, and patients with asthma. In these populations there were no differences between baseline and 4-hour post-dose measurements of FEV1 between orally inhaled DHE and placebo.[124]

DHE in any formulation carries a risk of cardiovascular and cerebrovascular events, and this should be considered in selecting patients for its use. There are no data comparing vascular risk of triptans and DHE, but in general the same safety concerns noted with triptans should be considered when using DHE. DHE, like all ergot compounds, is category X in pregnancy.

Ergotamine

Oral and rectal ergotamine were more commonly used prior to the introduction of triptans to treat acute migraine. Studies demonstrate efficacy, but oral absorption is frequently problematic. Ergotamine, like triptans and DHE, is a 5-HT1B/D receptor agonist and, like DHE, also has significant activity at the 5-HT1A and 5-HT2 receptors and the α_1- and α_2-adrenergic receptors. This difference in receptor activity may account for an increase in vasoconstriction relative to triptans or even DHE. Many clinicians compound ergotamine into rectal suppositories with an anti-emetic such as metoclopramide. Dosage can be individualized by cutting the suppository. Ergotamine has been associated with serious cardiovascular and cerebrovascular events, and is Category X in pregnancy.

Neuroleptics and Anti-emetics

Several phenothiazines, chlorpromazine, prochlorperazine, and promethazine are commonly used as acute treatment for migraine and/or adjunctive treatment in combination with other acute treatments to potentiate their efficacy or provide anti-emetic benefit. These medications likely work by blocking dopamine. They are commonly used in emergency departments as a rescue therapy. In a randomized double-blind clinical trial comparing prochlorperazine, 10 mg, intravenously with metoclopramide, 20 mg, intravenously combined with diphenhydramine, 25 mg, intravenously (to minimize the risk of extrapyramidal side effects), both medications were determined to be efficacious treatment for acute migraine.[125] In a meta-analysis of 13 trials, phenothiazines were compared with placebo in 5 trials and to another active agent in 10 trials. They were found to be more effective than other agents combined. Overall, 48% (95% CI: 43—54) of patients treated with phenothiazines had complete relief of headache, and 78% (95% CI: 74—82) reported clinically meaningful relief.[126] Phenothiazines have demonstrated efficacy in migraine independent of their anti-emetic benefit, but they reduce the disability of nausea and are sedative, which in certain patients may be beneficial.

Phenothiazines are associated with important adverse events such as hypotension, excessive sedation, and extrapyramidal reactions with short-term use, and suicide, metabolic syndrome, and tardive dyskinesia with long-term use.

Opioids and Butalbital

The use of opiates and butalbital-containing products for acute migraine is controversial. Neither opioids nor butalbital products are recommended as first-line therapies for migraine. They may be appropriate for selected patients who cannot tolerate, do not respond to, or have contraindications to standard migraine-abortive medications.

Opioids have a good analgesic effect, and historically have been commonly employed for treatment of refractory headache in emergency departments and inpatient settings. Butorphanol as a nasal spray is considered to have Class A evidence by the US Consortium Guidelines. However, its use is generally discouraged by headache specialists because it is considered to have a high potential for MOH.[127] The *American Academy of Neurology Practice Parameter: Evidence-based Guidelines for Migraine Headache* cautions clinicians to limit the use of opioids and butalbital combination products to "two headache days per week on a regular basis." The American Migraine

Prevalence and Prevention population-based study data showed that MOH development is linked to baseline frequency of 10 headache days per month for both opioids and butalbital, although many headache specialists believe the threshold for MOH with these compounds may be much lower.[128] The 2012 update of the guidelines states that opioid use is positively associated with chronic headache conditions.[129]

A combination drug that contains butalbital, aspirin/acetaminophen, caffeine, and possibly codeine is prescribed frequently for migraine yet is a non-specific analgesic that treats symptoms rather than the migraine process. One large double-blind placebo-controlled study compared sumatriptan/naproxen, butalbital/acetaminophen/caffeine/codeine, and placebo in a population of severely impacted migraine subjects who in the past had used a butalbital-containing product. Compared with the butalbital combination and placebo groups, the subjects who took sumatriptan/naproxen reported a greater incidence of being pain free or achieving pain relief. Both sumatriptan/naproxen and the butalbital combination product were well tolerated. However, butalbital compounds are considered to have a significant risk of MOH, headache chronification, and drug dependency.

Common opioid side effects included dizziness, sedation, nausea, gastrointestinal discomfort, constipation, and vomiting.[130] Butalbital combination products without codeine are well tolerated, and reduce anxiety. The most common adverse events associated with their use are drowsiness and side effects of each component (i.e., aspirin, caffeine, and codeine).

SPECIAL POPULATIONS

Pregnancy and Nursing

No acute therapies have been specifically studied in pregnancy. The largest pregnancy registry is with sumatriptan.[131] This registry is relatively small, given that migraine is a common disorder in women of reproductive age, but to date there is no compelling evidence of adverse consequences on pregnancy attributable to sumatriptan. Whether that is a class effect is unknown. Triptans are Class C in pregnancy, requiring clinicians and patients to weigh the risks and benefits of their use in pregnancy.

Triptans are excreted in small amounts in breast milk. However, the American Academy of Pediatrics has endorsed sumatriptan and zolmitriptan for use by women breastfeeding.[132] Many clinicians recommend "pump and dump" as a strategy for women requiring triptans while breastfeeding. Accordingly, women are recommended to discard their breast milk 2–4 hours following use of triptan, and to use a shorter half-life triptan such as sumatriptan and zolmitriptan.

NSAIDs are commonly used in pregnancy, but are contraindicated in the third trimester because of concerns over bleeding and, possibly, premature closure of the ductus arteriosus. In addition, recent studies have suggested a risk for pulmonary hypertension in the newborn with maternal exposure to NSAIDs.[133] There is also theoretical risk in early pregnancy regarding implantation of the fertilized ovum and premature birth.[134]

Children and Adolescents

Recently, rizatriptan has been approved at a 5-mg dose for children ages 6–17 years weighing less than 88 pounds, and at a 10-mg dose for children weighing over 88 pounds and in adolescents 12–18 years of age.[135] Almotriptan is approved for adolescents aged 12–17 years as well.[136] Clinical trials are ongoing for several other triptans for use in children and adolescents with migraine.

Elderly

In general, migraine becomes less frequent and virulent for mature adults, especially for women after menopause, but this is not to suggest migraine goes away. A substantial number of mature adults continue to experience migraine with significant impact and disability. Many of these patients have been using triptans successfully for years. However, given their contraindication in coronary heart disease (CAD), their continued use may pose excessive risk as adults mature. It is therefore paramount that clinicians carefully assess risk factors for CAD, and be attentive to adverse events suggestive of cardiac adverse events.

PART 2

PREVENTATIVE PHARMACOLOGICAL AND NON-PHARMACOLOGICAL TREATMENT OF MIGRAINE

While preventative treatment of migraine is deemed desirable by both patient and provider, it is significantly underutilized in clinical practice. It is estimated that only 8–13% of patients with migraine are using preventative pharmacology, despite clinical guidelines suggesting it could benefit nearly 40% of the migraine population seeking medical care for migraine.[137,138] There are numerous reasons for the underutilization of preventative pharmacology, including failure of patients and providers to accurately estimate the disease severity of migraine and to focus on the attack rather than the disease progression of migraine. In addition, utilizing daily medication for an episodic condition can appear counterintuitive to many patients. While clearly not all patients with migraine will develop chronic migraine, it is estimated that 2–3% of the general population with episodic migraine transform to CM in a given year.[139] Given that migraine often persists over decades, it is reasonable to assume that a much higher percentage of patients with migraine will in their lifetime experience chronic migraine. The first step in assessing the need for preventative intervention is to assess where in the progression of the disease state of migraine each migraine patient resides, on a spectrum from an infrequent episodic syndrome to the serious chronic disease of chronic migraine.

Episodic migraine implies 14 or less days of headache per month, while chronic is 15 or more days per month. This non-specific division of migraine lacks the clinical sensitivity to determine preventative care needs for migraine patients. A more sensitive and clinically meaningful method is to stage the migraine population into one of four patterns of frequency: infrequent episodic, frequent episodic, very frequent episodic or transforming, and chronic (see Table 8.3).

STAGING: PREVENTIVE TREATMENT NEEDS BASED ON THE EVOLUTION OF MIGRAINE

Defining migraine is challenging.[140] For a person with rare attacks migraine is generally considered a nuisance, much like the flu. Although capable of producing intense symptoms and disability of relatively short duration, termination is followed by a prolonged period of normal function. On the other hand, for people with very frequent near-daily attacks, migraine is an unremitting chronic disease. The transformation of infrequent episodic to chronic migraine is only recently becoming understood and accepted. In fact, in the 1988 International Headache Society diagnostic criteria, CM was not included in the taxonomy of primary headache disorders despite other primary headaches being recognized as having chronic subtypes (i.e., chronic tension-type headache and chronic cluster headache).[141] Today, the pathophysiology and diagnosis of CM remains in transition and incompletely understood.

Preventing the progression of episodic to chronic migraine is considered a priority in the management of migraine patients. Generally, the control of migraine frequency is considered the purview of preventative pharmacology. However, studies suggest that patients without effective acute therapy are at greater risk of developing CM.[142] Thus, it may be better to consider prevention of migraine chronification to be an important goal of both acute and chronic migraine management. This suggests that optimization of both acute and preventive treatment is important to the long-term successful management of migraine patients.[73,143,144]

The frequency of migraine attacks varies considerably from person to person, and even for the same person at different times throughout his or her life. For most migraineurs, migraine begins as an episodic condition but has the potential to evolve or transform into CM. In fact, in the ICHD-II classification, CM is considered a complication of episodic migraine (EM).[145]

In addition, as migraine frequency worsens it is increasingly associated with other comorbid conditions

TABLE 8.3 Efficacy of Preventive Interventions Based on the Stage of Migraine*

Interventional strategy	Stage 1: Infrequent episodic migraine	Stage 2: Frequent episodic migraine	Stage 3: Transforming migraine	Stage 4: Chronic migraine
Education	++	++++	++++	++++
Non-pharmacological	+	++++	++++	++++
Acute pharmacological	++++	++++	+++	+++
Preventive pharmacological	+	+++	++++	++++
Topiramate	−	+++	++++	+++
Divalproex	−	+++	+++	?
OnabotulinumtoxinA	−	−	−	++++
Tricyclics	−	+++	+++	+++
Beta-blockers	−	+++	+++	++

*The relevance of these lines of demarcation extends beyond headache frequency.
Level of efficacy: −, not effective; +, somewhat; ++, mild; +++, moderate; ++++, high.

TABLE 8.4 Comorbidities of Chronic Migraine and Episodic Migraine*[146]

Comorbidity	Chronic migraine	Episodic migraine
Psychiatric disorders	46.3%	28.5
• Depression • Anxiety	n = 231	n = 2347
Non-headache pain	41.7%	33.3%
• Fibromyalgia • Chronic fatigue syndrome	n = 208	n = 2739
Vascular disease events	8.2%	3.3%
• Hypertension • Stroke	n = 41	n = 275
Other conditions	49.9%	37.5%
	n = 249	n = 3089

*Survey of 8726 migraine sufferers: chronic migraine n = 499, episodic migraine n = 8227.

that substantially increase the disease burden of migraine. It is essential that patients be assessed as to where they reside on a spectrum of migraine disease severity. It is also valuable to evaluate and understand a patient's physiological function between attacks of migraine, and assess carefully for comorbidities, as this frequently changes over time (Table 8.4).[146]

Staging migraine underscores the obvious fact that with increasing migraine frequency there is less time for the nervous system to recover between attacks. At some point, the frequency of migraine is such that the nervous system remains in a permanent or protracted state of migraine. A paramount goal of migraine management, shared by both acute and preventive therapeutic interventions, is to increase neurological recovery time between attacks of migraine. Considering that an attack of migraine can last 5 or more days (premonitory through resolution), complete, prompt, and prolonged neurological recovery with acute and preventative interventions plays a significant role in reducing long-term attack-related disability and, very likely, the risk of disease progression.

As migraine attacks become more frequent, the utilization of acute medication needs to be monitored closely to prevent MOH. When acute medication is ineffective or marginally effective, the need to re-dose with acute medication becomes almost inevitable and, consequently, medication utilization often increases. This is also more likely to occur when the use of effective medications or formulations is delayed because of poor education, cost, or availability of effective acute treatment. It is essential that clinicians understand the patient factors determining the use of acute medications, and not simply define MO as the number of

acute medications a patient is utilizing. It is also important to realize that MOH does not occur in all patients, and is dependent on both medication and patient factors. Many preventive medications are less effective when a patient has MOH. One exception to this generalization may be onabotulinumtoxinA.[147]

Migraine Stages to Chronification

Stage 1 – Infrequent Episodic Migraine

This patient population typically has two or fewer migraine attacks per month. While individual migraine attacks can be severe and disabling, they are self-limited and occur infrequently. Often this population views migraine as they would the flu and self-treats attacks, often not choosing to consider migraine as a medical problem. Patients with infrequent episodic migraine may respond positively to NSAIDs, OTC combination medications, or oral triptans. They should be instructed on early intervention with a goal of being pain free within 2 hours. These individuals can often identify their migraine triggers, and only rarely do they require preventive treatment.

Stage 2 – Frequent Episodic Migraine

Frequent episodic migraine is characterized by attack frequencies of 3–8 migraine days per month. Attacks are discernible, and there is generally complete recovery between episodes of migraine. Acute abortive medications remain a cornerstone of treatment; however, preventative non-pharmacological and pharmacological interventions should be provided for a significant portion of this population of migraineurs, especially if there are signs or symptoms suggesting comorbidity, recent increases in migraine frequency, marginal response to acute treatment, or significant migraine-related disability.

Stage 3 – Transforming Migraine

In Stage 3, migraine begins to lose its episodic nature as the frequency increases to 9–14 headache days a month. Acute medication use is often increasing, and frequently it is less effective at producing a sustained abortive response. Patients' concerns over missing work or other activities is increasing. Recovery between attacks is often blurred by days of lower-grade headache or incomplete migraine recovery. Indicators of comorbidities are common, as evidenced by symptoms such as insomnia, fatigue, and anxiety. Acute treatments continue to be essential, but patients and providers need to be vigilant of the risk of medication overuse and MOH.

Preventive pharmacological and non-pharmacological treatment is paramount for almost all of this patient population.

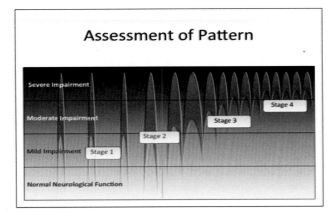

FIGURE 8.5 Pattern of chronic migraine. *Reproduced from Cady et al.,[148] Headache 2004; 44:426−35.*

Stage 4 − Chronic Migraine

Chronic migraine is defined as 15 or more headache days per month for at least 3 months. With CM, the nervous system is generating migraine at such frequency that rarely does the nervous system fully recover between attacks (Figure 8.5).[148] Preventative medications and non-pharmacological therapies are needed to reverse disease progression of chronic migraine. Acute treatment, while important, needs to be used judiciously and monitored closely. CM often requires interdisciplinary care and consultation, and referral is not uncommon. It is easier to prevent CM than to treat it.

PREVENTIVE MEDICATIONS FOR MIGRAINE (TABLE 8.5)

Only three drugs have been approved for migraine prophylaxis in the past 20 years: divalproex sodium (1996),[149] topiramate (2004),[150] and onabotulinumtoxinA (2011).[151] There are, in addition, two beta-blockers, propranolol and timolol, that also have FDA approval. Outside the US, the calcium channel blocker flunarizine is widely utilized. Numerous studies suggest that these and other migraine preventatives are significantly underutilized in clinical practice.[9,139,152]

Choosing preventive medication is based on the physician's experience and level of comfort with specific migraine prophylactic medications and with managing the range of comorbidities common to this patient population. The mechanisms by which prophylactic medications work are poorly understood, and choosing a specific medication is to a large extent based on efficacy data from clinical trials, patient

TABLE 8.5 Preventive Medications for Migraine

Generic treatment	Doses
Alpha₂-agonists	
Clonidine tablets	0.05 mg to 0.3 mg/day
Guanfacine tablets	1 mg
Anticonvulsants	
Divalproex sodium tablets*	500 mg to 1500 mg/day
Gabapentin tablets*	300 mg to 3000 mg
Levetiracetam tablets	1500 mg to 4500 mg
Topiramate tablets*	50 mg to 400 mg
Zonisamide capsules	100 mg to 400 mg
Antidepressants	
MAOIs	
Phenelzine tablets	30 mg to 90 mg/day
TCAs	
Amitriptyline tablets*	30 mg to 150 mg
Nortriptyline tablets	30 mg to 100 mg
SSRIs	
Fluoxetine tablets	10 mg to 40 mg
Sertraline tablets	25 mg to 100 mg
Paroxetine tablets	10 mg to 30 mg
Venlafaxine tablets	37.5 mg to 225 mg
Mirtazapine tablets	15 mg to 45 mg
Beta-blockers	
Atenolol tablets*	25 mg to 100 mg
Metoprolol tablets	50 mg to 200 mg
Nadolol tablets	20 mg to 200 mg
Propranolol tablets*	30 mg to 240 mg
Timolol tablets*	10 mg to 30 mg
Calcium channel antagonists	
Verapamil tablets*	120 mg to 720 mg
Nimodipine tablets	40 mg tid
Diltiazem tablets	30 mg to 60 mg tid
Nisoldipine tablets	10 mg to 40 mg qd
Amlodipine tablets	2.5 mg to 10 mg qd
NSAIDs for prevention	
Naproxen sodium tablets*	500 mg to 1100 mg/day
Ketoprofen tablets	150 mg/day
Mefenamic acid tablets	1500 mg/day
Flurbiprofen tablets	200 mg/day

(Continued)

TABLE 8.5 (Continued)

Generic treatment	Doses
Serotonergic agents	
Methysergide tablets*	2 mg to 12 mg
Cyproheptadine tablets	2 mg to 16 mg
Pizotifen tablets*	1.5 mg to 3 mg
Miscellaneous	
Montelukast sodium tablets	5 mg to 20 mg
Lisinopril tablets	10 mg to 40 mg
BotulinumtoxinA injection	25 units to 100 units (IM)
Feverfew tablets	50 mg to 82 mg/day
Magnesium gluconate tablets	400 mg to 600 mg/day
Riboflavin tablets	400 mg/day
Petasites 75 mg*	75 mg bid

*Evidence for moderate efficacy from at least two well-designed placebo-controlled trials.
IM, intramuscularly; NSAID, non-steroidal anti-inflammatory drug; SSRI, selective serotonin reuptake inhibitor; TCA, tricyclic antidepressant.

factors and comorbidities, and a clinician's comfort and experience with specific medications. In broad terms, the mechanisms of action of many of the preventive medications for migraine can be characterized as (1) serotonergics, (2) neuronal stabilizers (AEDs), (3) beta-blockers, and (4) 5-HT2 receptor antagonists (see Table 8.6).

Neuronal Stabilizers (Anti-epileptic Drugs) for Stages 2, 3, & 4

Topiramate (Evidence Level A)

Topiramate is arguably the best-studied and most efficacious migraine preventive for patients with frequent EM. Its mechanism of action is presumed to be reduction of neuronal hyperexcitability through several possible mechanisms: a state-dependent blocking of sodium channels, augmentation of GABA activity, antagonism of AMPA/kainate glutamate receptors, or inhibition of carbonic anhydrase.[153] It has strong Class A evidence supporting its efficacy in migraine prevention, and is used worldwide for

TABLE 8.6 Selecting Preventive Medications

Drug	Efficacy	Adverse events	Comorbid condition Relative contraindication	Relative indication
Beta-blockers	4+	2+	Asthma, depression, congestive heart failure, Raynaud's disease, diabetes	Hypertension, angina, performance anxiety
Antiserotonin				
Pizotifen	4+	2+	Obesity	
Methysergide	4+	4+	Angina, vascular disease	Orthostatic hypotension
Calcium channel blockers				
Verapamil	2+	1+	Constipation, hypotension	Aura, hypertension, angina, asthma
Flunarizine	4+	2+	Parkinson's, depression	Dizziness, vertigo
Antidepressants				
TCAs	4+	2+	Mania, urinal retention, heart block, constipation	Depression, anxiety, insomnia, pain
SSRIs	2+	1+		Depression, OCD
MAOIs	4+	4+	Mania, unreliable patient	Refractory depression
Anticonvulsants				
Divalproex/ Valproate	4+	2+	Liver disease, bleeding disorders	Mania, epilepsy, anxiety
Gabapentin	2+	2+	Hypersensitivity	Epilepsy, anxiety
Gabapentin topiramate	4+	2+	Liver disease, bleeding disorders, kidney stones	Mania, epilepsy, anxiety, obesity
NSAIDs	2+	2+	Ulcer disease, gastritis, CV events	Arthritis, other pain disorders, high risk of MOH with triptans

TABLE 8.7   Topiramate vs Placebo for Migraine Prevention Clinical Studies Review

	MIGR-001 2004 [159]	MIGR-002 2004 [150]	MIGR-003 2004 [156]	Diener 2007 [157]	Silberstein 2009 [158]	TPM/Frova 2012 [48]
	n = 487	*n* = 483	*n* = 176	*n* = 59	*n* = 306	*n* = 76
Efficacy TPM vs PL	54%/23%	49%/23%	37%/23%	22%/0%	37%/29%	80%/54% (after 8 weeks)
Safety withdrawal due to AEs	18%/10%	27%/12%	27%/10%		82%/70%	18%/4%
Drop-outs after randomized	35%/41%	47%/42%	32%/30%	12%/48%	44%/45%	n/a
ADVERSE EVENTS						
Anorexia	13%	13%	17%	6%	5%	n/a
Cognition	7%	10%	9%	6%	9%	n/a
Fatigue	11%	14%	19%	6%	12%	n/a
Nausea	16%	10%	13%	9%	9%	7
Paresthesia	47%	50%	55%	53%	29%	n/a
Weight loss	9%	11%	7%	n/a	n/a	n/a

TPM, topiramate; PL, placebo; Frova, frovatriptan; n/a, not available.

prevention of migraine.[154] The efficacy and safety of topiramate is supported by four large randomized controlled clinical studies (see Table 8.7).[48,150,155–158]

Topiramate is generally well tolerated, but significant adverse events do occur and drop-out rates in clinical trials and clinical practice are relatively high. It has few absolute contraindications, and is largely excreted unchanged through the kidneys. However, there are several important precautions, including renal lithiasis, suicidal ideation, low BMI, and, potentially, osteoporosis from metabolic acidosis.[159] It is Class D in pregnancy, and associated with cleft lip and palate defects. It is excreted in breast milk. At higher doses, it may reduce the efficacy of estrogen-based birth control pills.

Topiramate may be considered a first-line preventive in patients with EM who are overweight, binge eating, have a seizure disorder or neuropathic pain, or possibly are substance abusers. Clinically, the starting dose of topiramate is 25 mg daily, with doses being increased at weekly intervals until a dose of 100 mg is achieved. After evaluation for tolerability and adherence, dosage increases to 150–200 mg may be considered. Topiramate can often be dosed once a day to increase patient adherence.

Sodium Valproate (Evidence Level A)

Sodium valproate (divalproex sodium) was the first AED demonstrated to provide conclusive benefit as a migraine prophylactic. It enhances GABA, an inhibitory neurotransmitter, which may in part explain its efficacy in migraine. There are five randomized

placebo-controlled studies providing Class A evidence of efficacy of divalproex sodium in EM, and a single randomized placebo-controlled study of slow-release sodium valproate and extended-release divalproex sodium.

Divalproex sodium in slow-release and extended-release formulations was generally well tolerated in clinical trials. However, several adverse events, such as weight gain, hair loss, tremor, drowsiness, amenorrhea, and pancreatitis, are observed. In addition, it is Category X in pregnancy and associated with neural tube defects. Clinicians often prescribe concomitant folate for women in their reproductive years while on valproate. It is contraindicated in liver disease, pancreatitis, and bleeding disorders.

Generally, all formulations of divalproex sodium are started at a low dose and titrated over 3- to 5-day intervals. The target dose of divalproex sodium is between 500 and 1500 mg. The extended dose is targeted to 500–1000 mg and has the advantage of daily dosing, which may improve adherence.

In clinical practice, divalproex sodium may be considered the first-line intervention in a patient with concomitant epilepsy or bipolar disease, though the dosing for these diseases is typically higher than those used in migraine. Use of this drug requires frequent monitoring of liver function, pancreatitis, and pregnancy risk.

Other AEDs

Gabapentin (Evidence level U). Gabapentin has several small studies suggesting efficacy in migraine.[160,161] It is generally well tolerated in doses as high as 2400 mg per day. Common adverse events include fatigue, lethargy, and dry mouth. It is Category C in pregnancy, and is excreted in breast milk.

Lamotrigine (Evidence level: Not effective). Lamotrigine has been suggested as effective for preventing aura but not migraine, though the data supporting this are relatively weak.[162] A recent more rigorous comparator study did not demonstrate superiority against placebo, and it was inferior to topiramate.[163]

Oxcarbazepine (Evidence level: Possibly not effective). Oxcarbazepine in one Class II study did not demonstrate superiority to placebo.

Carisbamate (Evidence level: Not effective). Carisbamate in a large, randomized, placebo-controlled study failed to demonstrate benefit over placebo.[164]

The purpose of reporting these negative studies is to highlight that the migraine prophylactic benefit of AEDs should not be considered a class effect, as many have negative studies in migraine.

Beta-blockers

There are two beta-blockers approved by the FDA for prevention of migraine, and several others that have studies supporting their efficacy and are commonly used in clinical practice. Beta-blockers are presumed to reduce adrenergic drive, and may reduce thalamic signaling to the somatosensory cortex through a glutamate modulating mechanism attributable to beta-1 adrenergic activity.[165] Only beta-blockers with sympathomimetic activity are considered efficacious in migraine.

Beta-blockers are commonly used during pregnancy for various medical complications, but are Category C and have been associated with small-for-gestational-age infants (OR 1.97), preterm birth (OR 1.26), and increased perinatal mortality (OR 1.89).[166] Small amounts of some beta-blockers are excreted into breast milk, but the American Academy of Pediatrics generally considers their use to be safe during nursing.[167]

Propranolol (Evidence level A). Propranolol is approved by the FDA for prophylaxis of migraine, with several older clinical trials supporting its efficacy. Dosing is usually initiated at 20 mg daily and slowly increased as tolerated to a target dose of 80–160 mg daily.

Timolol (Evidence level A). Timolol is also approved by the FDA for migraine prophylaxis, but is used infrequently in clinical practice. Target dosing is between 10 and 12 mg qd.

Metoprolol (Evidence level A). Metoprolol is commonly used in clinical practice to prevent migraine, but it is not FDA approved. Dosing is initiated at 50 mg and slowly titrated to 200 mg as tolerated.

Atenolol (Evidence level B). Atenolol is used as a first-line intervention or when treatment is unsuccessful with other beta-blockers. It does not have FDA approval, but does have studies supporting its efficacy. Dosing is initiated at 25 mg and titrated up to 100 mg as tolerated. It is FDA pregnancy risk Category D.

Nadolol (Evidence level B). Nadolol has been found to be effective for CM and reducing the frequency and severity of migraine attacks. Dose ranges are generally between 20 and 160 mg qd. It is ideal for daily dosing.

Antidepressants

Tricyclic Antidepressants

Amitriptyline (Evidence Level B). Amitriptyline is a time-honored preventive treatment for migraine, with good efficacy but rather high dropout rates in clinical trials (see Table 8.5). Amitriptyline is also indicated in depression and used commonly in treatment of neuropathic pain, irritable bowel syndrome, insomnia, and anxiety. It is a serotonin reuptake inhibitor, increasing serotonergic activity, and is an agonist at numerous serotonin, α-adrenergic, and histamine receptors, which is at least in part its mechanism of action in both migraine and depression. It may also increase receptor sensitivity.

Adverse events are frequent at higher doses, and include anticholinergic effects (dry mouth, constipation, and orthostatic hypotension), tinnitus, weight gain, drowsiness, and cardiac arrhythmias. It is Category C in pregnancy, and excreted in small amounts in breast milk. It is contraindicated in certain cardiovascular diseases (especially after a myocardial infarction), severe liver disease, and mania. Amitriptyline can be considered a first-line preventive for migraine in patients with non-specific sleep disturbances, IBS, anxiety, depression, or concomitant chronic pain.

Nortriptyline (Evidence level C). Nortriptyline is an active metabolite of amitriptyline that has studies demonstrating efficacy in depression, EM, and chronic pain.[168,169] Its adverse event profile, contraindications, and pregnancy category are the same as amitriptyline. Nortriptyline may be used in the same populations as amitriptyline, and may be less sedating.

Protriptyline (Evidence level U). Protriptyline is a secondary amine with greater norepinephrine than serotonin reuptake inhibition. It is less sedating than amitriptyline and nortriptyline. It has less conclusive evidence for efficacy in migraine prevention.

Selective Serotonin/Norepinephrine Reuptake Inhibitors

Venlafaxine (Evidence level B). Venlafaxine is a selective serotonin/norepinephrine reuptake inhibitor (SSNRI) indicated for treatment of depression. It has one Class A study demonstrating efficacy in EM, and a Class 2 migraine prevention study demonstrating equal efficacy to amitriptyline but better tolerability.[170] It is often used to treat anxiety, panic, social phobias, and vasomotor symptoms.

It is generally well tolerated, though there is a withdrawal syndrome associated with sudden discontinuation. Adverse events include suicidal ideology, nausea, somnolence, dry mouth, dizziness, insomnia, nervousness, and sexual dysfunction. It is contraindicated in glaucoma, heart disease, hypertension, and pregnancy.

Venlafaxine may be useful in migraine patients with depression, anxiety, or perimenopausal vasomotor symptoms. Venlafaxine is considered a pregnancy Category C medication.

Selective Serotonin Reuptake Inhibitors

In general, selective serotonin reuptake inhibitors (SSRIs) have not demonstrated clear efficacy as migraine preventives. Fluoxetine has mixed results in clinical studies. Other SSRIs have not been thoroughly studied.[171]

Angiotensin Receptor Blockers and Angiotensin Converting-Enzyme Inhibitors

Lisinopril (Evidence level C). Lisinopril is an angiotensin-converting enzyme (ACE) inhibitor indicated for hypertension, with a single study demonstrating efficacy in migraine prevention (see Table 8.5). Its mechanism of action in migraine is unknown. Adverse events include cough, dizziness, and fainting. Concomitant use of NSAIDs with lisinopril, especially with diuretics, should be avoided due to the risk of acute renal failure. It is Category D in the first two trimesters of pregnancy, and Category C in the third trimester. It may have value as a preventive in a patient with hypertension and frequent migraine.

Candesartan (Evidence level C). Candesartan is an angiotensin receptor blocker (ARB) indicated for hypertension, with a single study that demonstrated efficacy in migraine. It was well tolerated. Adverse events for candesartan include dizziness, headache,

and fluid depletion. Candesartan should not be used concomitantly with NSAIDs or in pregnancy or lactation.

Calcium Channel Blockers (Evidence Level U)

There is contradictory evidence of the efficacy of calcium channel blockers in migraine prophylaxis, although there is substantial evidence that verapamil has efficacy in cluster headache.

Triptans (Table 8.8)

Frovatriptan (Evidence level A). Frovatriptan has a large Class 1 study showing efficacy for short-term prophylaxis of menstrual migraine when used at 2.5 mg bid or qd for 6 days beginning 2 days prior to menses. The adverse event profile was similar to placebo. Frovatriptan taken during prodrome to prevent the headache phase of migraine has been compared with daily topiramate and placebo in a single small study. Both active drugs were superior to placebo and similar in efficacy over a 2-month timeframe. There were far fewer dropouts in the frovatriptan arm than with topiramate.[48] All triptans are pregnancy Category C.

Naratriptan and zolmitriptan (Evidence level B). Naratriptan has Class 1 evidence as a short-term prophylactic for menstrually related migraine. Administered 1 mg bid for 5 days beginning 2 days before menses, it showed a significant reduction in anticipated menstrually related migraine.[172] Adverse events were similar to placebo. Similar results were observed with 2.5 mg of zolmitriptan taken bid or tid.[173]

No triptan has been studied as a short-term preventive therapy for migraine in a population of CM. However, triptans have been described as "bridge therapies." Daily doses of a triptan for a short period of time (ex. 5 days) have been shown to attenuate intractable migraine or assist in managing MOH on an outpatient basis.[174]

NSAIDs

There are few large clinical trials on the use of NSAIDs in the prevention of migraine. Ibuprofen, naproxen, and aspirin are the most commonly used treatment interventions by CM sufferers. In a 2009 study of 14,540 participants, assessing the medications used in acute treatment of episodic and chronic migraines, NSAIDs were used by 43.1% of the CM group, with ibuprofen (45.2% of NSAID use), naproxen (26.2%), and aspirin (23.6%) being the most common. Chronic migraineurs used NSAIDs an average of

TABLE 8.8   Triptan Formulations

Generic	Formulations	Doses	Maximum daily dose
Sumatriptan	Oral tablet	25 mg, 50 mg, 100 mg	200 mg
	Nasal spray	5 mg, 20 mg	40 mg
Sumatriptan	Oral tablet	6 mg	12 mg
	Nasal spray		
	Subcutaneous injection		
Zolmitriptan	Oral tablet	2.5 mg, 5 mg	10 mg
	Orally disintegrating tablet	2.5 mg, 5 mg	10 mg
Rizatriptan	Oral tablet	5 mg, 10 mg	30 mg
	Orally disintegrating tablet	5 mg, 10 mg	30 mg
Naratriptan	Tablet	1 mg, 2.5 mg	5 mg
Almotriptan	Tablet	12.5 mg	25 mg
Frovatriptan	Tablet	2.5 mg	7.5 mg
Eletriptan	Tablet	20 mg, 40 mg	80 mg
Sumatriptan/naproxen sodium	Tablet	85 mg/500 mg	170 mg/1000 mg

14.8 days per month. Understanding whether the driver for behavior is availability, cost, or efficacy is unknown. The only NSAID with an FDA approval for migraine is an oral suspension of potassium diclofenac.[113] Several OTC products have substantial evidence of efficacy in EM, including aspirin,[175] ibuprofen,[176] and naproxen sodium.[111] Evidence indicates that ibuprofen 200 and 400 mg are effective in reducing headache intensity and rendering patients pain free at 2 hours, with improvement in photophobia and phonophobia with 400 mg dosing.[110] The non-prescription combination of acetaminophen, aspirin, and caffeine was highly effective for the treatment of migraine headache pain as well as for alleviating the nausea, photophobia, phonophobia, and functional disability associated with migraine attacks.[177] Parenteral formulations of ketorolac are available as an intramuscular injection and nasal spray, and are commonly used as acute treatment and rescue therapies.[178,179]

PART 3

ATTACK-BASED CARE: CLINICAL APPROACH AND MEDICATIONS MANY PATIENTS FIND MOST USEFUL AND EFFECTIVE (FIGURE 8.6)

Step 1:

1. Stage migraine pattern and migraine phenotypes.
2. Provide appropriate patient-specific education and diaries.

3. Health hygiene – diet and lifestyle changes, exercise, and stress reduction.
4. Diet – a modified Mediterranean diet or gluten free diet; screen for gluten antibodies.
5. Biofeedback and cognitive behavioral therapy with a psychologist.

Step 2: Acute medications for initial therapy

1. Oral triptans are most effective when taken while the headache is mild.
2. Nasal formulations of triptans if nausea is an early or frequent symptom.
3. Oral liquid formulation of diclofenac (Cambia) and/or another NSAID to use as initial treatment or as an add-on medication when trying to control frequent use of acute medications.
4. OTC combinations of ASA/acetaminophen/caffeine if attack frequency is low and if patients have not failed with OTCs prior to evaluation.
5. Subcutaneous sumatriptan for disabling early morning migraine or rapid onset of headache and nausea.

Step 3: Back-up therapies

1. Nasal formulations of triptans to use for attacks with early nausea or for attacks treated beyond mild headache.
2. Subcutaneous sumatriptan if patients are prescribed a NSAID or an oral or nasal formulation of sumatriptan.
3. Dihydroergotamine as an alternative to triptan, especially in longer duration migraine.

Covering the bases of Acute Intervention with Attack-based Care

Initial Treatment	• **Oral triptan** • **NSAID** • **OTC**	*Used for attacks of migraine where intervention occurs during mild HA and before onset of meaningful GI disturbances*
Backup Treatment	• **Nasal triptans** • **Sc sumatriptan** • **DHE** • **NSAID-IN or IM**	*Used for advanced attack on when meaningful GI symptoms are present*
Rescue Treatment	• **Sc IM DHE** • **Sc sumatriptan** • **Neuroleptic such as chlorpromazine** • **Analgesics (rarely)**	*Used when initial and back up therapy has failed and to prevent treatment in ED or outpatient acute care centers*

FIGURE 8.6   Covering the bases of acute intervention with attack-based care. Patients frequently have a strategy for all three levels of acute care needs.

4. NSAIDs, especially diclofenac liquid, to use as an adjunct or back-up intervention.

Step 4: Rescue therapies

1. Subcutaneous sumatriptan if no recent exposure to triptans other than sumatriptan.
2. DHE if no recent exposure to triptans.
3. Phenothiazines if not dehydrated.
4. Analgesics (rarely).

Summary

Create acute strategies with patients that minimize frequent exposure to a single class of abortive medications. By dissecting the patient's migraine pattern and developing a strategy for different migraine phenotypes, efficacy and patient involvement improves.

PREVENTATIVE MEDICATIONS

First line:

1. Topiramate in patients with obesity, concern of weight gain, or personal or family history of diabetes; also when mood stabilization is indicated by patient history. Avoid with history of renal lithiasis and risk of pregnancy or inadequate birth control.
2. Sodium valproate with bipolar disease when low risk for pregnancy. Avoid in liver disease, polycystic ovarian disease.
3. Amitriptyline/nortriptyline for patients with mild to moderate sleep disturbance or mild symptoms suggesting depression or anxiety. Titrate for sleep restoration. For depression or perimenopausal vasomotor symptoms, use venlafaxine.
4. Propranolol, metoprolol, or atenolol are the beta-blockers helpful when the migraine trigger is performance anxiety or "let down" after a very stressful period of time, and also in patients with anxiety, premature ventricular contractions or palpitations, and hypertension.
5. Tizanidine for excessive muscle tension, especially with sleep difficulties. Generally rule out sleep apnea if risk factors present (day time sleepiness, snoring or observed apnea events).

Second line:

1. When first-line preventative is ineffective, add a second prophylactic medication before pushing the first drug to the point of significant adverse events. Tricyclics with AEDs or beta-blockers work well. Use venlafaxine instead of a beta-blocker whenever there is a suggestion of perimenopausal symptoms and vasomotor symptomatology; also if there is frank depression.
2. When goals are not met, refer patients to a headache specialist interdisciplinary clinic. An inpatient program is desirable when patients need separation from their current environment or have significant medical problems requiring inpatient care. Stress that upon return from the tertiary clinic, the patient

will receive support and long-term healthcare management needs from the referring clinician.

CONCLUSION

Managing patients with migraine is both intellectually satisfying and gratifying for healthcare professionals. As a collaborative partner, you truly can make a difference in the life of a patient living with migraine. In many ways, managing patients with migraine is a tutorial on medicine itself. Migraineurs are a population at risk because of migraine and the numerous comorbid diseases that are associated with the migraine population. Migraine patients desperately need a "good healthcare provider" who not only knows the "pharmacology of migraine" but also is willing to care for the patient with migraine. Managing migraine is a long-term commitment for both the patient and the healthcare professional.

References

1. Buse DC, Manack AN, Fanning KM, Serrano D, Reed ML, Turkel CC, et al. Chronic migraine prevalence, disability, and sociodemographic factors: results from the American Migraine Prevalence and Prevention Study. *Headache*. 2012;52:1456–1470.
2. Atlas of headache disorders and resources in the world 2011: a collaborative project of World Health Organization and Lifting the Burden. World Health Organization, Geneva 27, Switzerland, 2011.
3. Latinovic R, Gulliford M, Ridsdale L. Headache and migraine in primary care: consultation, prescription, and referral rates in a large population. *J Neurol Neurosurg Psychiatry*. 2006;77 (3):385–387.
4. Hirtz D, Thurman DJ, Gwinn-Hardy K, Mohamed M, Chaudhuri AR, Zalutsky R. How common are the "common" neurologic disorders? *Neurology*. 2007;68(5):326–337.
5. Stewart WF, Wood C, Reed ML, Roy J, Lipton RB; AMPP Advisory Group. Cumulative lifetime migraine incidence in women and men. *Cephalalgia*. 2008;11:1170–1178.
6. Aurora SK, Wilkinson F. The brain is hyperexcitable in migraine. *Cephalalgia*. 2007;27(12):1442–1453.
7. Lipton RB. Epilepsy and migraine as comorbid disorders: epidemiologic perspectives. *Adv Stud Med*. 2005;5(6E):S649–S657.
8. Silberstein S, Loder E, Diamond S, Reed ML, Bigal ME, Lipton RB, with the AMPP Advisory Group. Probable migraine in the United States: results of the American Migraine Prevalence and Prevention (AMPP) study. *Cephalalgia*. 2007;27(3):220–229.
9. Diamond S, Bigal ME, Silberstein S, Loder E, Reed M, Lipton RB. Patterns of diagnosis and acute and preventive treatment for migraine in the United States: results from the American Migraine Prevalence and Prevention Study. *Headache*. 2007;47:355–363.
10. Tepper S, Martin V, Burch S, Fu AZ, Kwong J, Downs KE. Acute headache treatments in patients with health care coverage: what prescriptions are doctors writing? *Headache & Pain*. 2006; 17:11–17.
11. Stovner L, Hagen K, Jensen R, Katsarava Z, Lipton R, Scher A, et al. The global burden of headache: a documentation of headache prevalence and disability worldwide. *Cephalalgia*. 2007;27:193–210.
12. Cady RK, Wendt JK, Kirchner JR, Sargent JD, Rothrock JF, Skaggs H. Treatment of acute migraine with subcutaneous sumatriptan. *JAMA*. 1991;265:2831–2835.
13. The Subcutaneous Sumatriptan International Study Group. Subcutaneous sumatriptan in the acute treatment of migraine. *N Engl J Med*. 1991;325:316–321.
14. Lipton RB, Pan J. Is migraine a progressive brain disease? *JAMA*. 2004;291:493–494.
15. Haut SR, Bigal ME, Lipton RB. Chronic disorders with episodic manifestations: focus on epilepsy and migraine. *Lancet Neurol*. 2006;5:148–157.
16. Bigal ME, Lipton RB. The prognosis of migraine. *Curr Opin Neurol*. 2008;21:301–308.
17. Geraud G, Keywood C, Senard JM. Migraine headache recurrence: relationship to clinical, pharmacological, and pharmacokinetic properties of triptans. *Headache*. 2003;43:376–388.
18. Cady RK, Sheftell F, Lipton RB, O'Quinn S, Jones M, Putnam DG, et al. Effect of early intervention with sumatriptan on migraine pain: retrospective analyses of data from three clinical trials. *Clin Ther*. 2000;22:1035–1048.
19. Goadsby PJ. The "Act when Mild" (AwM) study: a step forward in our understanding of early treatment in acute migraine. *Cephalalgia*. 2008;28(suppl 2):36–41.
20. Ng-Mak DS, Cady R, Chen Y-T, Ma L, Bell CF, Hu XH. Can migraineurs accurately identify their headaches as "migraine" at attack onset? *Headache*. 2007;47:645–653.
21. Dodick DW. Triptan nonresponder studies: implications for clinical practice. *Headache*. 2005;45(2):156–162.
22. Walling AD, Woolley DC, Molgaard C, Kallail KJ. Patient satisfaction with migraine management by family physicians. *J Am Board Fam Pract*. 2005;18(6):563–566.
23. Bigal M, Rapoport A, Aurora S, Sheftell F, Tepper S, Dahlof C. Satisfaction with current migraine therapy: experience from 3 centers in US and Sweden. *Headache*. 2007;47:475–479.
24. Sheftell FD, Cady RK, Borchert LD, Spalding W, Hart CC. Optimizing the diagnosis and treatment of migraine. *J Am Acad Nurse Pract*. 2005;17(8):309–317.
25. Cady R, Schreiber C, Farmer K, Sheftell F. Primary headaches: a convergence hypothesis. *Headache*. 2002;42:204–216.
26. Cady RK, Lipton RB, Rothrock JF. *Chronic Migraine: A Patient-Centered Guide to Effective Management*. Hamilton, ON, Canada: Baxter Publishing Inc. <www.chronicmigraine.org/diagnosisof-chronicmigraine/index#pg1>; 2013 Accessed 20.10.14.
27. Cady RK, Lipton RB, Hall C, Stewart WF, O'Quinn S, Gutterman D. Treatment of mild headache in disabled sufferers: results of the Spectrum Study. *Headache*. 2000;40:792–797.
28. Cady RK, Schreiber CP. Sinus headache or migraine? Considerations in making a differential diagnosis. *Neurology*. 2002;58(9 suppl 6):S10–S14.
29. Kaniecki RG. Migraine and tension-type headache: an assessment of challenges in diagnosis. *Neurology*. 2002;58(9 suppl 6):S15–S20.
30. Buse DC, Manack A, Serrano D, Turkel C, Lipton RB. Sociodemographic and comorbidity profiles of chronic migraine and episodic migraine sufferers. *J Neurol Neurosurg Psychiatry*. 2010;81(4):428–432.
31. Cady R, Farmer K, Dexter JK, Schreiber C. Cosensitization of pain and psychiatric comorbidity in chronic daily headache. *Curr Pain Headache Rep*. 2005;9(1):47–52.
32. Scher AI, Stewart WF, Lipton RB. The comorbidity of headache with other pain syndromes. *Headache*. 2006;46(9):1416–1423.
33. Negro A, D'Alonzo L, Martelletti P. Chronic migraine: comorbidities, risk factors, and rehabilitation. *Intern Emerg Med*. 2010;5 (suppl 1):S13–S19.
34. Silberstein SD, Lipton RB. Chronic daily headache, including transformed migraine, chronic tension-type headache, and

medication overuse. In: Silberstein SD, Lipton RB, Dalessio DJ, eds. *Wolff's Headache and Other Head Pain.* 7th ed. New York, NY: Oxford University Press; 2001:247–282.

35. Sheftell F, Tepper SJ. New paradigms in the recognition and acute treatment of migraine. *Headache.* 2002;42:58–69.

36. Cady R, Dodick DW. Diagnosis and treatment of migraine. *Mayo Clin Proc.* 2002;77:255–261.

37. Kelman L. The triggers or precipitants of the acute migraine attack. *Cephalalgia.* 2007;27(5):394–402.

38. Martin VT, Behbehani MM. Toward a rational understanding of migraine trigger factors. *Med Clin North Am.* 2001;85:911–941.

39. Buse DC, Andrasik F. Behavioral medicine for migraine. *Neurol Clin.* 2009;27(2):445–465.

40. Worthington I, Pringsheim T, Gawel MJ, Gladstone J, Cooper P, Dilli E, et al.; Canadian Headache Society Acute Migraine Treatment Guideline Development Group. Canadian headache society guideline: acute drug therapy for migraine headache. *Can J Neurol Sci.* 2013;40(5 suppl 3):S1–S80.

41. Lipton RB, Cady RK, Farmer K, Bigal ME. *Managing Migraine: A Healthcare Professional's Guide to Collaborative Migraine Care.* Hamilton, Ontario: Baxter Publishing Inc; 2008:95–96.

42. Brandes JL, Poole A, Kallela M, Schreiber CP, MacGregor EA, Silberstein SD, et al. Short-term frovatriptan for the prevention of difficult-to-treat menstrual migraine attacks. *Cephalalgia.* 2009;29(11):1133–1148.

43. Dowson AJ, Massiou H, Aurora SK. Managing migraine headaches experienced by patients who self-report with menstrually related migraine: a prospective, placebo-controlled study with oral sumatriptan. *J Headache Pain.* 2005;6(2):81–87.

44. Tuchman M, Hee A, Emeribe U, Silberstein S. Efficacy and tolerability of zolmitriptan oral tablet in the acute treatment of menstrual migraine. *CNS Drugs.* 2006;20(12):1019–1026.

45. Silberstein SD, Bradley K. DHE-45 in the prophylaxis of menstrually related migraine. *Cephalalgia.* 1996;16:371 [Abstract 39].

46. Silberstein SD, Holland S, Freitag F, Dodick DW, Argoff C, Ashman E; Quality Standards Subcommittee of the American Academy of Neurology and the American Headache Society. Evidence-based guideline update: pharmacologic treatment for episodic migraine prevention in adults: report of the Quality Standards Subcommittee of the American Academy of Neurology and the American Headache Society. *Neurology.* 2012;78:1337–1345.

47. Bigal ME, Serrano D, Buse D, Al Scher, Stewart WF, Lipton RB. Migraine medications and evolution from episodic to chronic migraine: a longitudinal population based study. *Headache.* 2008;48:1157–1168.

48. Cady RK, Voirin J, Farmer K, Browning R, Beach ME, Tarrasch J. Two center, randomized pilot study of migraine prophylaxis comparing paradigms using pre-emptive frovatriptan or daily topiramate: research and clinical implications. *Headache.* 2012;52(5):749–764.

49. Becker WJ. The premonitory phase of migraine and migraine management. *Cephalalgia.* 2013;33(13):1117–1121.

50. Giffin NJ, Ruggiero L, Lipton RB, Silberstein SD, Tvedskov JF, Olesen J, et al. Premonitory symptoms in migraine: an electronic diary study. *Neurology.* 2003;60:935–940.

51. Farmer K, Cady RK, Reeves D. The effect of prodrome on cognitive efficiency. *Headache.* 2003;43:518 [Abstract F1].

52. Farmer K, Cady R, Reeves D, Bleiberg J. Cognitive efficiency following migraine therapy. In: Olesen J, Steiner TJ, Lipton RB, eds. *Reducing the Burden of Headache: Frontiers in Headache Research.* New York, NY: Oxford University Press; 2003: 46–51.

53. Waelkens J. Domperidone in the prevention of complete classical migraine. *Br Med J (Clin Res Ed).* 1982;284(6320):944.

54. Luciani R, Carter D, Mannix L, Hemphill M, Diamond M, Cady R, et al. Prevention of migraine during prodrome with naratriptan. *Cephalalgia.* 2000;20(2):122–126.

55. Aurora SK, Barrodale PM, McDonald SA, Jakubowski M, Burstein R. Revisiting the efficacy of sumatriptan therapy during the aura phase of migraine. *Headache.* 2009;49(7): 1001–1004.

56. Edwards KR, Rosenthal RL, Farmer KU, Cady RK, Browning R. Evaluation of sumatriptan-naproxen in the treatment of acute migraine: a placebo-controlled, double-blind, crossover study assessing cognitive function. *Headache.* 2013;53:656–664.

57. Headache Classification Committee of the International Headache Society. The International Classification of Headache Disorders, 3rd edition (beta version). *Cephalalgia.* 2013;33:629–808.

58. Schürks M, Rist PM, Shapiro RE, Kurth T. Migraine and mortality: a systematic review and meta-analysis. *Cephalalgia.* 2011;31(12): 1301–1314.

59. Bates D, Ashford E, Dawson R, Ensink FB, Gilhus NE, Olesen J, et al. Subcutaneous sumatriptan during the migraine aura. Sumatriptan Aura Study Group. *Neurology.* 1994;44:1587–1592.

60. Foley KA, Cady R, Martin V, Adelman J, Diamond M, Bell CF, et al. Treating early versus treating mild: timing of migraine prescription medications among patients with diagnosed migraine. *Headache.* 2005;45:538–545.

61. Klapper J, Lucas C, Rosjo O, Charlesworth B. Benefits of treating highly disabled migraine patients with zolmitriptan while pain is mild. *Cephalalgia.* 2004;24:918–924.

62. Cady R, Elkind A, Goldstein J, Keywood C. Randomized, placebo-controlled comparison of early use of frovatriptan in a migraine attack versus dosing after the headache has become moderate or severe. *Curr Med Res Opin.* 2004;20(9):1465–1472.

63. Mathew NT. Early intervention with almotriptan improves sustained pain-free response in acute migraine. *Headache.* 2003;43:1075–1079.

64. Winner P, Mannix LK, Putnam DG, McNeal S, Kwong J, O'Quinn S, et al. Pain-free results with sumatriptan taken at the first sign of migraine pain: 2 randomized, double-blind, placebo-controlled studies. *Mayo Clin Proc.* 2003;78:1214–1222.

65. Burstein R, Jakubowski M, Rauch SD. The science of migraine. *J Vestib Res.* 2011;21(6):305–314.

66. Burstein R, Collins B, Jakubowski M. Defeating migraine pain with triptans: a race against the development of cutaneous allodynia. *Ann Neurol.* 2004;55:19–26.

67. Raskin NH. *Headache.* 2nd ed. New York, NY: Churchill Livingstone; 1988.

68. Diamond M, Cady R. Initiating and optimizing acute therapy for migraine: the role of patient-centered stratified care. *Am J Med.* 2005;118(suppl 1):18S–27S.

69. Cady RK, Schreiber CP, Farmer KU. Understanding migraine: advancing from clinical trials to clinical practice. *Headache Care.* 2004;1(3):183–190.

70. Burstein R, Cutrer MF, Yarnitsky D. The development of cutaneous allodynia during a migraine attack. Clinical evidence for the sequential recruitment of spinal and supraspinal nociceptive neurons in migraine. *Brain.* 2000;123:1703–1709.

71. Yarnitsky D, Goor-Aryeh I, Bajwa ZH, Ransil BI, Cutrer FM, Sottile A, et al. Wolff Award: possible parasympathetic contributions to peripheral and central sensitization during migraine. *Headache.* 2003;43(7):704–714.

72. Lipton RB, Buse DC, Fanning KM, Serrano D, Reed ML. Suboptimal treatment of episodic migraine may mean progression to chronic migraine. *Poster presented at the 2013 IHC/AHS Congress*; Boston, MA, June 26, 2013; Abstract LB02.

73. Lipton RB, Hamelsky SW, Dayno JM. What do patients with migraine want from acute migraine treatment? *Headache.* 2002;42(suppl 1):S3–S9.

74. Tfelt-Hansen P, Block G, Dahlof C, Diener HC, Ferrari MD, Goadsby PJ, et al.; the International Headache Society Clinical Trials Subcommittee. Guidelines for controlled trials of drugs in migraine: second edition. *Cephalalgia*. 2000;20:765−786.

75. Taylor FR, Kaniecki RG. Symptomatic treatment of migraine: when to use NSAIDs, triptans, or opiates. *Curr Treat Options Neurol*. 2011;13(1):15−27.

76. Dussor G. Serotonin, 5HT1 agonists, and migraine: new data, but old questions still not answered. *Curr Opin Support Palliat Care*. 2014;8(2):137−142.

77. Belvís R, Pagonabarraga J, Kulisevsky J. Individual triptan selection in migraine attack therapy. *Recent Pat CNS Drug Discov*. 2009;4(1):70−81.

78. Goldstein J, Silberstein SD, Saper JR, Elkind AH, Smith TR, Gallagher RM, et al. Acetaminophen, aspirin, and caffeine versus sumatriptan succinate in the early treatment of migraine: results from the ASSET trial. *Headache*. 2005;45(8):973−982.

79. The Diclofenac-K/Sumatriptan Migraine Study Group. Acute treatment of migraine attacks: efficacy and safety of a non-steroidal anti-inflammatory drug, diclofenac-potassium, in comparison to oral sumatriptan and placebo. *Cephalalgia*. 1999;19(4):232−240.

80. Durham PL. Inhibition of calcitonin gene-related peptide function: a promising strategy for treating migraine. *Headache*. 2008;48(8):1269−1275.

81. Durham PL, Russo AF. New insights into the molecular actions of serotonergic antimigraine drugs. *Pharmacol Ther*. 2002;94(1-2): 77−92.

82. Humphrey PPA, Feniuk W, Perren MJ, Connor HE, Oxford AW. The pharmacology of the novel 5HT1-like receptor agonist, GR43175. *Cephalalgia*. 1989;9(suppl 9):23−33.

83. Edvinsson L. Calcitonin gene-related peptide (CGRP) and the pathophysiology of headache. *CNS Drugs*. 2001;15(10):745−753.

84. Welch KM, Mathew NT, Stone P, Rosamond W, Saiers J, Gutterman D. Tolerability of sumatriptan: clinical trials and post-marketing experience. *Cephalalgia*. 2000;20(8):687−695.

85. Wammes-Van Der Heijden EA, Rahimtoola H, Leufkens HG, Tijssen CC, Egberts AC. Risk of ischemic complications related to the intensity of triptan and ergotamine use. *Neurology*. 2006;67(7):1128−1134.

86. Lugardon S, Roussel H, Sciortino V, Montastruc JL, Lapeyre-Mestre M. Triptan use and risk of cardiovascular events: a nested-case-control study from the French health system database. *Eur J Clin Pharmacol*. 2007;63(8):801−807.

87. Evans RW, Tepper SJ, Shapiro RE, Sun-Edelstein C, Tietjen GE. The FDA alert on serotonin syndrome with use of triptans combined with selective serotonin reuptake inhibitors or selective serotonin-norepinephrine reuptake inhibitors: American Headache Society position paper. *Headache*. 2010;50(6):1089−1099.

88. Gillman PK. Triptans, serotonin agonists, and serotonin syndrome (serotonin toxicity): a review. *Headache*. 2010;50 (2):264−272.

89. Sclar DA, Robison LM, Castillo LV, Schmidt JM, Bowen KA, Oganov AM, et al. Concomitant use of triptan, and SSRI or SNRI after the US Food and Drug Administration alert on serotonin syndrome. *Headache*. 2012;52(2):198−203.

90. Limmroth V, Katsarava Z, Fritsche G, Przywara S, Diener HC. Features of medication overuse headache following overuse of different acute headache drugs. *Neurology*. 2002;59(7):1011−1014.

91. Smith TR, Stoneman J. Medication overuse headache from antimigraine therapy: clinical features, pathogenesis and management. *Drugs*. 2004;64(22):2503−2514.

92. Aurora SK, Papapetropoulos S, Kori SH, Kedar A, Abell TL. Gastric stasis in migraineurs: etiology, characteristics, and clinical and therapeutic implications. *Cephalalgia*. 2013;33(6):408−415.

93. Lipton RB, Buse DC, Saiers J, Fanning KM, Serrano D, Reed ML. Frequency and burden of headache-related nausea: results from the American Migraine Prevalence and Prevention (AMPP) study. *Headache*. 2013;53(1):93−103.

94. Cady RK, Aurora SK, Brandes JL, Rothrock JF, Myers JA, Fox AW, et al. Satisfaction with and confidence in needle-free subcutaneous sumatriptan in patients currently treated with triptans. *Headache*. 2011;51(8):1202−1211.

95. Rothrock JF, Freitag FG, Farr SJ, Smith III EF. A review of needle-free sumatriptan injection for rapid control of migraine. *Headache*. 2013;53(suppl 2):21−33.

96. Rapoport A. The sumatriptan nasal spray: a review of clinical trials. *Cephalalgia*. 2001;21(suppl 1):13−15.

97. Charlesworth BR, Dowson AJ, Purdy A, Becker WJ, Boes-Hansen S, Farkkila M, on behalf of the ZINC I Study Group. Speed of onset and efficacy of zolmitriptan nasal spray in the acute treatment of migraine: a randomised, double-blind, placebo-controlled, dose-ranging study versus zolmitriptan tablet. *CNS Drugs*. 2003;17:653−667.

98. Djupesland PG, Skretting A. Nasal deposition and clearance in man: comparison of a bidirectional powder device and a traditional liquid spray pump. *J Aerosol Med Pulm Drug Deliv*. 2012;25(5):280−289.

99. Cady R, Messina J, Carothers J, Mahmoud R. Efficacy and safety of AVP-825, a novel breath powered™ powder sumatriptan intranasal treatment, for acute migraine. *Poster presented at the 66th Annual AAN Meeting*, Philadelphia, PA; April 26−May 3, 2014. #P7.203.

100. Goldstein J, Smith TR, Pugach N, Griesser J, Sebree T, Pierce M. A sumatriptan iontophoretic transdermal system for the acute treatment of migraine. *Headache*. 2012;52(9):1402−1410.

101. Smith TR, Goldstein J, Singer R, Pugach N, Silberstein S, Pierce MW. Twelve-month tolerability and efficacy study of NP101, the sumatriptan iontophoretic transdermal system. *Headache*. 2012;52(4):612−624.

102. Brandes JL, Kudrow D, Stark SR, O'Carroll CP, Adelman JU, O'Donnell FJ, et al. Sumatriptan-naproxen sodium for acute treatment of migraine: a randomized trial. *JAMA*. 2007;297:1443−1454.

103. Silberstein S, Tepper S, Brandes J, Diamond M, Goldstein J, Winner P, et al. Randomized, placebo-controlled trial of rofecoxib in the acute treatment of migraine. *Neurology*. 2004;62 (9):1552−1557.

104. Krymchantowski AV, Barbosa JS. Rizatriptan combined with rofecoxib vs. rizatriptan for the acute treatment of migraine: an open label pilot study. *Cephalalgia*. 2002;22(4):309−312.

105. Burstein R, Jakubowski M. Analgesic triptan action in an animal model of intracranial pain: a race against the development of central sensitization. *Ann Neurol*. 2004;55:27−36.

106. Tzeng SF, Hsiao HY, Mak OT. Prostaglandins and cyclooxygenases in glial cells during brain inflammation. *Curr Drug Targets Inflamm Allergy*. 2005;4(3):335−340.

107. Durham PL, Vause CV. Calcitonin gene-related peptide (CGRP) receptor antagonists in the treatment of migraine. *CNS Drugs*. 2010;24(7):539−548.

108. Suthisisang C, Poolsup N, Kittikulsuth W, Pudchakan P, Wiwatpanich P. Efficacy of low dose ibuprofen in acute migraine treatment: systemic review and meta-analysis. *Ann Pharmacother*. 2007;41(11):1782−1791.

109. Suthisisang CC, Poolsup N, Suksomboon N, Lertpipopmetha V, Tepwitukgid B. Meta-analysis of the efficacy and safety of naproxen sodium in the acute treatment of migraine. *Headache*. 2010;50(5):808−818.

110. Goldstein J, Silberstein SD, Saper JR, Ryan Jr. RE, Lipton RB. Acetaminophen, aspirin, and caffeine in combination versus ibuprofen for acute migraine: results from a multicenter,

double-blind, randomized, parallel-group, single-dose, placebo-controlled study. *Headache*. 2006;46(3):444–453.

111. Kahn K. Cambia® (Diclofenac Potassium for Oral Solution) in the Management of Acute Migraine. *US Neurology*. 2011;7 (2):139–143.

112. Cady R, O'Carroll P, Dexter K, Freitag F, Shade CL. SumaRT/Nap vs naproxen sodium in treatment and disease modification of migraine: a pilot study. *Headache*. 2014;54(1):67–79.

113. Raskin N. Repetitive intravenous dihydroergotamine as therapy for intractable migraine. *Neurology*. 1986;36(7):995–997.

114. Silberstein SD, McCrory DC. Ergotamine and dihydroergotamine: history, pharmacology, and efficacy. *Headache*. 2003;43 (2):144–166.

115. Masterson CG, Durham PL. DHE repression of ATP-mediated sensitization of trigeminal ganglion neurons. *Headache*. 2010;50 (9):1424–1439.

116. Kori S, Zhang J, Kellerman D, Armer T, Goadsby PJ. Sustained pain relief with dihydroergotamine in migraine is potentially due to persistent binding to 5-ht(1b) and 5-ht(1d) receptors. *Paper presented at AHS 54th Annual Meeting*; June 21-24, 2012; Los Angeles, CA.

117. Seybold VS. The role of peptides in central sensitization. *Handb Exp Pharmacol*. 2009;451–491.

118. Dahlöf C, Maassen Van Den Brink A. Dihydroergotamine, ergotamine, methysergide and sumatriptan – basic science in relation to migraine treatment. *Headache*. 2012;52(4):707–714.

119. Winner P, Ricalde O, Le Force B, Saper J, Margul B. A double-blind study of subcutaneous dihydroergotamine vs subcutaneous sumatriptan in the treatment of acute migraine. *Arch Neurol*. 1996;53(2):180–184.

120. Tepper SJ. Orally inhaled dihydroergotamine: a review. *Headache*. 2013;53(suppl 2):43–53.

121. Aurora SK, Silberstein SD, Kori SH, Tepper SJ, Borland BW, Wang M, et al. MAP0004, orally inhaled DHE: a randomized, controlled study in the acute treatment of migraine. *Headache*. 2011;51(4):507–517.

122. Rapoport AM. The therapeutic future in headache. *Neurol Sci*. 2012;33(suppl 1):119–125.

123. Silberstein SD, Young WB, Hopkins MM, Gebeline-Myers C, Bradley KC. Dihydroergotamine for early and late treatment of migraine with cutaneous allodynia: an open-label pilot trial. *Headache*. 2007;47(6):878–885.

124. Chang D, Kellermann S, Kori S, Meyer J, Zhou J, Armer T. Acute inhaled safety of MAP0004: studies in healthy volunteers, smokers and patients with asthma. *Poster presented at the 54th Annual Scientific Meeting of the AHS*. June 21–24, 2012.

125. Friedman B, Esses D, Solorzano C, Dua N, Greenwald P, Radulescu R, et al. A randomized controlled trial of prochlorperazine versus metoclopramide for treatment of acute migraine. *Ann Emerg Med*. 2008;52(4):399–406.

126. Kelly A, Walcynski T, Gunn B. The relative efficacy of phenothiazines for the treatment of acute migraine: a meta-analysis. *Headache*. 2009;49:1324–1332.

127. Loder E. Post-marketing experience with an opioid nasal spray for migraine: lessons for the future. *Cephalalgia*. 2006;26:89–97.

128. Tepper SJ. Medication-overuse headache. *Continuum (Minneap Minn)*. 2012;18:807–822.

129. Holland S, Silberstein SD, Freitag F, Dodick DW, Argoff C, Ashman E; Quality Standards Subcommittee of the American Academy of Neurology and the American Headache Society. Evidence-based guideline update: NSAIDs and other complementary treatments for episodic migraine prevention in adults: report of the Quality Standards Subcommittee of the American Academy of Neurology and the American Headache Society. *Neurology*. 2012;78(17):1346–1353.

130. Kelley NE, Tepper DE. Rescue therapy for acute migraine, Part 3: Opioids, NSAIDs, steroids, and post-discharge medications. *Headache*. 2012;52:467–482.

131. GlaxoSmithKline Pregnancy Registries. <http://pregnancyregistry.gsk.com/sumatriptan.html>.

132. Duong S, Bozzo P, Nordeng H, Einarson A. Safety of triptans for migraine headaches during pregnancy and breastfeeding. *Can Fam Physician*. 2010;56(6):53753–53759.

133. Bloor M, Paech M. Nonsteroidal anti-inflammatory drugs during pregnancy and the initiation of lactation. *Anesth Analg*. 2013;116(5):1063–1075.

134. Østensen ME, Skomsvoll JF. Anti-inflammatory pharmacotherapy during pregnancy. *Expert Opin Pharmacother*. 2004;5(3):571–580.

135. Maxalt prescribing information. <www.merck.com/product/usa/pi_circulars/m/maxalt/maxalt.html>.

136. Axert prescribing information. <www.axert.com/prescribing-information>.

137. Lipton RB, Bigal ME, Diamond M, Freitag F, Reed ML, Stewart WF; AMPP Advisory Group. Migraine prevalence, disease burden, and the need for preventive therapy. *Neurology*. 2007;68:343–349.

138. Bigal ME, Serrano D, Reed M, Lipton RB. Chronic migraine in the population: Burden, diagnosis and satisfaction with treatment. *Neurology*. 2008;71:559–566.

139. Scher AI, Stewart WF, Ricci JA, Lipton RB. Factors associated with the onset and remission of chronic daily headache in a population-based study. *Pain*. 2003;106:81–89.

140. Young WB, Kempner J, Loder EW. Naming migraine and those who have it. *Headache*. 2012;52(2):283–291.

141. Headache Classification Committee of the International Headache Society. Classification and diagnostic criteria for headache disorders, cranial neuralgias and facial pain. *Cephalalgia*. 1988;8(suppl 7):1–96.

142. Buse DC, Rupnow MF, Lipton RB. Assessing and managing all aspects of migraine: migraine attacks, migraine-related functional impairment, common comorbidities, and quality of life. *Mayo Clin Proc*. 2009;84(5):422–435.

143. Cady RK. The future of migraine: beyond just another pill. *Mayo Clin Proc*. 2009;84(5):397–399.

144. Lipton RB, Bigal ME. Migraine: epidemiology, impact, and risk factors for progression. *Headache*. 2005;45(suppl 1):S3–S13.

145. Headache Classification Committee of the International Headache Society. The International Classification of Headache Disorders, 2nd edition. *Cephalalgia*. 2004;24 (suppl 1):1–160.

146. Bagley CL, Rendas-Baum R, Maglinte GA, Yang M, Varon SF, Lee J, et al. Validating Migraine-Specific Quality of Life Questionnaire v2.1 in episodic and chronic migraine. *Headache*. 2012;52(3):409–421.

147. Silberstein SD, Blumenfeld AM, Cady RK, Turner IM, Lipton RB, Diener HC, et al. OnabotulinumtoxinA for treatment of chronic migraine: PREEMPT 24-week pooled subgroup analysis of patients who had acute headache medication overuse at baseline. *J Neurol Sci*. 2013;331(1-2):48–56.

148. Cady RK, Schreiber CP, Farmer KU. Understanding the patient with migraine: the evolution from episodic headache to chronic neurologic disease. A proposed classification of patients with headache. *Headache*. 2004;44:426–435.

149. Klapper J. Divalproex sodium in migraine prophylaxis: a dose-controlled study. *Cephalalgia*. 1997;17(2):103–108.

150. Brandes JL, Saper JR, Diamond M, Couch JR, Lewis DW, Schmitt J; MIGR-002 Study Group. Topiramate for migraine prevention: a randomized controlled trial. *JAMA*. 2004;291 (8):965–973.

151. FDA News Release. October 15, 2010; U.S. Food and Drug Administration today approved Botox injection (onabotulinumtoxinA) to prevent headaches in adult patients with chronic migraine. <www.fda.gov/NewsEvents/Newsroom/PressAnnouncements/ucm229782.htm>.

152. Vikelis M, Rapoport AM. Role of antiepileptic drugs as preventive agents for migraine. *CNS Drugs*. 2010;24(1):21−33.

153. White HS, Rho JM. *Mechanisms of Action of Antiepileptic Drugs*. West Islip, NY: Professional Communications, Inc; 2010.

154. Ramadan NM, Silberstein SD, Freitag FG, Gilbert TT, Frishberg BM. Evidenced-based guidelines for migraine headache in the primary care setting: pharmacological management for prevention of migraine. *American Academy of Neurology Practice Guideline*, April 2000. <www/aan.com/professional/practice>.

155. Silberstein SD, Neto W, Schmitt J, Jacobs D, for the MIGR-001 Study Group. Topiramate in migraine prevention: results of a large controlled trial. *Arch Neurol*. 2004;61:490−495.

156. Mirza N, Marson AG, Pirmohamed M. Effect of topiramate on acid-base balance: extent, mechanism and effects. *Br J Clin Pharmacol*. 2009;68(5):655−661.

157. Diener HC, Tfelt-Hansen P, Dahlöf C, Láinez MJ, Sandrini G, Wang SJ, et al. Topiramate in migraine prophylaxis: results from a placebo-controlled trial with propranolol as an active control. *J Neurol*. 2004;251:943−950.

158. Diener HC, Bussone G, Van Oene JC, Lahaye M, Schwalen S, Goadsby PJ; TOPMAT-MIG-201(TOP-CHROME) Study Group. Topiramate reduces headache days in chronic migraine: a randomized, double-blind, placebo-controlled study. *Cephalalgia*. 2007;27(7):814−823.

159. Silberstein S, Lipton R, Dodick D, Freitag F, Mathew N, Brandes J, et al. Topiramate treatment of chronic migraine: a randomized, placebo-controlled trial of quality of life and other efficacy measures. *Headache*. 2009;49(8):1153−1162.

160. Mathew NT, Rapoport A, Saper J, Magnus L, Klapper J, Ramadan N, et al. Efficacy of gabapentin in migraine prophylaxis. *Headache*. 2001;41:119−128.

161. Di Trapani G, Mei D, Marra C, Mazza S, Capuano A. Gabapentin in the prophylaxis of migraine: a double-blind randomized placebo-controlled study. *Clin Ter*. 2000;151(3):145−148.

162. Pascual J, Caminero AB, Mateos V, Roig C, Leira R, García-Moncó C, et al. Preventing disturbing migraine aura with lamotrigine: an open study. *Headache*. 2004;44(10):1024−1028.

163. Gupta P, Singh S, Goyal V, Shukla G, Behari M. Low-dose topiramate versus lamotrigine in migraine prophylaxis (the Lotolamp study). *Headache*. 2007;47(3):402−412.

164. Cady RK, Mathew N, Diener H-C, Hu P, Haas M, Novak GP, on behalf of the Study Group. Evaluation of carisbamate for the treatment of migraine in a randomized, double-blind trial. *Headache*. 2009;49:216−226.

165. Shields KG, Goadsby PJ. Propranolol modulates trigeminovascular responses in thalamic ventroposteromedial nucleus: a role in migraine? *Brain*. 2005;128(Pt 1):86−97.

166. Meidahl Petersen K, Jimenez-Solem E, Anderson JT, Petersen M, Brødbæk K, Køber L, et al. B-blocker treatment during pregnancy and adverse outcomes: a nationwide population-based cohort study. *BMJ Open*. 2012;2:4.

167. Hutchinson S, Marmura MJ, Calhoun A, Lucas S, Silberstein S, Peterlin BL. Use of common migraine treatments in breast-feeding women: a summary of recommendations. *Headache*. 2013;53(4):614−627.

168. Silberstein SD. Practice parameter: evidence-based guidelines for migraine headache (an evidence-based review): report of the Quality Standards Subcommittee of the American Academy of Neurology. *Neurology*. 2000;55(6):754−762.<www.neurology.org/content/55/6/754.full.html> Accessed 20.10.14.

169. Demaagd G. The pharmacological management of migraine, Part 2: preventative therapy. *Pharm Ther*. 2008;33(8):480−487.

170. Bulut S, Berilgen MS, Baran A, Tekatas A, Atmaca M, Mungen B. Venlafaxine versus amitriptyline in the prophylactic treatment of migraine: randomized, double-blind, crossover study. *Clin Neurol Neurosurg*. 2004;107(1):44−48.

171. Saper JR, Silberstein SD, Lake III AE, Winters ME. Fluoxetine and migraine: comparison of double-blind trials. *Headache*. 1995;35(4):233.

172. Newman LC, Mannix LK, Landy S, Silberstein S, Lipton RB, Putnam DG, et al. Naratriptan as short-term prophylaxis of menstrually associated migraine: a randomized, double-blind, placebo-controlled study. *Headache*. 2001;41:248−256.

173. Tuchman MM, Hee A, Emeribe U, Silberstein S. Oral zolmitriptan in the short-term prevention of menstrual migraine: a randomized, placebo-controlled study. *CNS Drugs*. 2008;22(10):877−886.

174. Tepper SJ, Tepper DE. Breaking the cycle of medication overuse headache. *Cleve Clin J Med*. 2010;77:236−242.

175. Diener HC, Hartung E, Chrubasik J, Evers S, Schoenen J, Eikermann A, et al.; Study group. A comparative study of oral acetylsalicylic acid and metoprolol for the prophylactic treatment of migraine: a randomized, controlled, double-blind, parallel group phase III study. *Cephalalgia*. 2001;21:120−128.

176. Scher AI, Lipton RB, Stewart WF, Bigal M. Patterns of medication use by chronic and episodic headache sufferers in the general population: results from the frequent headache epidemiology study. *Cephalalgia*. 2010;30:321−328.

177. Lipton RB, Stewart WF, Ryan Jr RE, Saper J, Silberstein S, Sheftell F. Efficacy and safety of acetaminophen, aspirin, and caffeine in alleviating migraine headache pain: three double-blind, randomized, placebo-controlled trials. *Arch Neurol*. 1998;55(2):210−217.

178. Duarte C, Dunaway F, Turner L, Aldag J, Frederick R. Ketorolac versus meperidine and hydroxyzine in the treatment of acute migraine headache: a randomized, prospective, double-blind trial. *Ann Emerg Med*. 1992;21:1116−1121.

179. Pfaffenrath V, Fenzl E, Bregman D, Färkkila M. Intranasal ketorolac tromethamine (SPRIX®) containing 6% of lidocaine (ROX-828) for acute treatment of migraine: safety and efficacy data from a phase II clinical trial. *Cephalalgia*. 2012;32:766−767.

9

Chronic Migraine: Diagnosis and Management

Roger K. Cady[1] and Paul L. Durham[2]

[1]Headache Care Center, Springfield, Missouri, USA [2]Center for Biomedical & Life Sciences, Missouri State University, Springfield, Missouri, USA

INTRODUCTION

Chronic migraine (CM) can be one of the most debilitating and challenging diseases managed by healthcare professionals. A day with a severe migraine is considered by the World Health Organization to be in the same disability classification as a day with quadriplegia or severe psychosis.[1] However, this only partially explains the disability attributable to CM. For most living with migraine, the extreme disruptions of daily life are intermittent and hopefully infrequent. However, for those with CM these disruptions are omnipresent and can persist for decades. In CM, migraine has progressed from an episodic occurrence to a chronic disease capable of interfering with and disrupting virtually every domain of life. The accumulated impact of living with frequent migraine is what makes it one of the 10 leading causes of worldwide disability.[2] It is not simply the disability of a single severe attack but the attrition of repeated attacks of migraine occurring unpredictably, often for days on end, and being repeated over the decades of a person's life, that makes CM so disabling. Given its attack frequency, symptom severity, and duration, CM is arguably the most disabling of all neurological diseases.

The responsibility of caring for patients with migraine is shared by primary and specialty healthcare professionals at all levels. Most patients with CM are managed in primary care, but successful clinical management of CM often requires collaboration with many other medical specialists, especially neurologists, psychologists, and headache specialists.[3] In addition, the future of migraine requires greater involvement from researchers and health policy leaders. This chapter, and indeed this book, is dedicated to assisting providers at all levels to better prevent and manage migraine.

RECOGNITION OF CHRONIC MIGRAINE

Only recently has CM become an independent medical diagnosis. Its inclusion in the most recent International Classification of Headache Disorders-III beta (ICHD) criteria is a welcome addition and a critically important advancement that will undoubtedly help identify and lead to more appropriate treatment of millions of people.[4] The ICHD criteria will also encourage much needed research into the transformation of episodic to chronic migraine, and an understanding of the disease state itself. The inclusion of CM into ICHD criteria will expand efforts to advance medical understanding of the epidemiology, natural history, pathophysiology, and treatment needs of CM patients. Indeed, understood and used appropriately, these criteria will become a critically important platform for future advancements in CM.

CHALLENGES AND IMPLICATION OF DEFINING A DIAGNOSIS FOR CHRONIC MIGRAINE

In 1988, the International Headache Society (IHS) proposed diagnostic criteria for headache and facial pain syndromes.[5] This landmark publication rapidly became the accepted diagnostic standard for headache disorders by the medical community as well as the diagnostic standard accepted in clinical trials by regulatory agencies worldwide.

The IHS diagnostic schema categorized headaches as being primary or secondary, based on the presence or absence of definable pathology. Primary headaches were considered syndromes and defined by symptoms, while secondary headaches were defined based on the presence of pathology as being the etiology of

S. Diamond, R. K. Cady, M. L. Diamond & V. T. Martin (Eds): Headache and Migraine Biology and Management.

DOI: http://dx.doi.org/10.1016/B978-0-12-800901-7.00009-4

the headache. Migraine was subdivided into migraine with or without aura, while most of the other primary headaches were subdivided into episodic or chronic.

In 2004, the IHS revised the diagnostic criteria (ICHD-II) and for the first time included the diagnosis of CM.[6] Consistent initially with the 1988 taxonomy, CM was defined as 15 or more days of IHS defined migraines without aura (a single headache phenotype) per month occurring for 3 or more consecutive months. This definition proved untenable, especially for migraine research, and in 2006 a new appendix for CM diagnosis was proposed that provided a radical departure from the taxonomy principles used in the original 1988 IHS taxonomy.[7]

The 2006 appendix definition proposed that more than a single primary headache phenotype should be included in the diagnosis of a primary headache. As such, CM was defined as 15 or more "headache days" a month for 3 or more consecutive months, with the usual caveats of at least five lifetime attacks of IHS migraine and exclusion of secondary pathology. However, rather than insisting on stereotypic symptoms to define each headache day, the appendix criteria required that only 8 or more days a month of headache needed to fulfill IHS criteria for migraine (or respond to migraine-specific medication before all the symptoms of migraine had occurred). Other headache days could be defined by headache symptomatology consistent with other primary headaches, such as probable migraine or tension-type headache. In 2013, the appendix definition of CM was essentially maintained and advanced in the ICHD-III beta criteria.[4] ICHD-II criteria also stated that CM should be considered a complication of episodic migraine (EM). Thus, while EM and CM are two different headache diagnoses, they appear to have a common etiology and pathophysiological relationship.

The implications of these changes are profound, as they support a spectrum of different headache presentations or phenotypes as constituting a single primary headache diagnosis. Implied in the diagnosis of CM is that these unique migraine presentations share enough common pathophysiology to be considered as a single diagnosis, and that the range of headache presentations observed during different "headache days" is attributable to chronic migraine.

The spectrum of headache described for CM is not unlike the spectrum of headache reported to exist in patients with episodic migraine with or without aura. A spectrum of headaches occurring in patients with episodic migraine responding to sumatriptan is well documented.[8,9] This spectrum is also the basis for the treatment paradigm of early intervention, where a migraine attack is treated without all necessary diagnostic features of migraine being present at the time of treatment (see *Chapter 8*).[10,11] However, in patients with episodic

migraine with or without aura, different headache presentations are required to be diagnosed independent of each other and each considered a unique IHS diagnosis. Ironically, the spectrum of headaches recognized in the diagnosis of CM, though the same as that observed in EM, is not currently recognized as migraine until there is a headache frequency of greater than 15 days per month. However, many clinicians and headache experts support the notion that the spectrum of migraine existing in patients with migraine with or without aura also shares common pathophysiological mechanisms.[12,13] Truly, the implications and novelty of the diagnostic criteria for CM can pave the way to expand diagnostic understanding of migraine, and perhaps lay the foundation for a disease model of migraine beginning with infrequent episodic migraine and ending with CM. Ultimately, these changes could fundamentally advance migraine research and the clinical management of migraine patients.

EPIDEMIOLOGY AND NATURAL HISTORY OF CHRONIC MIGRAINE

Migraine with and without aura occurs in approximately 12% of the adult population of the United States.[14] Epidemiological studies from other occidental populations are relatively consistent with this prevalence marker.[15] Japanese populations have a somewhat lower prevalence of migraine, as do those in South-East Asia, Africa, and the Middle East.[16] However, primary headache disorders are by far the most common neurological condition in the world for which medical consultation is sought.[2] Despite the prevalence of primary headache disorders, they are also one of the most misdiagnosed and undermanaged medical conditions on the planet. This is likely due in part to rapidity of changes in the medical understanding (misunderstanding) of migraine, failure to define and develop newer migraine-specific therapies, and confusion over the relationship of migraine to other primary headache diagnoses.

Chronic migraine evolves from the pool of people with EM, and is estimated to affect 1–2% of the general population in the United States.[17] This translates to about 5 million people in the US living with CM in a given year. CM has a similar female preponderance (3:1, F:M) as seen in EM but, interestingly, the prevalence of CM in women is bimodal, reaching peak prevalence in the late teens and again in the fourth decade of life. For men, the prevalence steadily increases until it peaks in the fourth decade of life.[18] In both genders, CM persists well into mature adulthood. In any given year, approximately 3% of the US population with EM will develop CM.[19] A recent epidemiological survey showed that in a 1-year time span, 26% of CM patients reverted back to EM, another 40% oscillated between CM and high frequency EM migraine, and 31% persisted in CM.[20] This suggests that a significant percentage of the population once diagnosed with CM will persist and be impacted by frequent EM and/or CM for a protracted period of time.

EPIGENETIC CONSIDERATIONS IN THE PATHOPHYSIOLOGY OF CHRONIC MIGRAINE

Pathophysiological distinctions inherent to only CM are in the infancy of scientific study, but the pathophysiological foundation for migraine in general is built on a concept of a "hyperexcitable" brain.[21-23] Given the relationship of these two diagnoses, it seems likely that genetic hyperexcitability of the nervous system is the basis of CM as well as EM. For decades experts have suggested that this hyperexcitable brain has a genetic basis, supported by the observation that migraine is familial, with as many as 80% of migraineurs reporting a first-degree relative with migraine.[24] In addition, results from twin studies demonstrate the odds ratio for migraine is significantly greater for identical vs fraternal twins.[25]

Armed with these clinical observations and the decoding of the human genome in the 1990s, millions of dollars were invested by the pharmaceutical industry in search of mutations in the DNA nucleotide sequence of genes that might explain the molecular and genetic basis of migraine. A popular theme throughout this period of exploration was that monogenetic mutations in various ion channels, signaling genes, and gap junction proteins might alter ion transfer in neurons in such a manner as to explain the hyperexcitability in the nervous system of migraineurs. Obviously this approach was thought to be a viable strategy to identify novel molecular targets for drug development.

The consequences of this scientific exploration, however, were not conclusive, as only a few relevant gene mutations were discovered and then only in some patients with rare forms of familial migraine.[26] To date, none of these mutations has translated into new drug discoveries. The unfortunate reality is that there has been little substantive progress in discovering DNA mutations in genes that strongly promote the neuronal hyperexcitability of patients with migraine and the general population.

A plausible explanation for why prevalent genetic mutations have not been identified for EM or CM is the complex biological nature of these neurological conditions, which are likely the consequence of numerous genetic factors rather than a single gene mutation.[27] It may well be that there are multiple normally expressed genes in the human genome with the potential to increase or decrease neuronal excitability, based

on whether or not they are actively functioning. Entertaining the possibility that environmental factors may be able to modify gene expression, neurobiologists and geneticists have begun seeking explanations for the "genetic" nature of migraine and the predicted hyperexcitable nervous system by asking an age-old question: "To what extent does nurture vs nature play a role in predisposing a person to migraine?"

Historically, an attack of migraine is assumed to occur when a person with the genetic predisposition for migraine interacts with a specific environmental change that "triggers" the attack.[28] This hypothesis presupposes that migraine triggers exert direct influences on the nervous system that launch an attack of migraine. However, these precipitating events (migraine triggers) rarely provoke migraine even close to 100% of the time, and it is often observed that migraine triggers can be present in the environment of a migraineur for long periods of time before migraine attacks actually occur or increase in frequency (for example, stress or menstruation). An explanation that has been advanced to explain this observation is that many "migraine triggers" are in reality environmental influences that affect gene expression rather than exerting direct influence on neuronal function. Exposure to the migraine risk environments, especially during critical time-points in development, may, at a cellular level, modify gene expression in neurons and other cells to a heightened level of sensitization (vigilance) of the nervous system into a more hyperexcitable state, thus increasing its susceptibility to attacks of migraine when an imbalance occurs. The novel science underpinning this viewpoint is called "epigenetics".[29]

Epigenetics is the study of how gene expression is controlled and orchestrated through the activation or deactivation of specific genes in response to environmental influences.[30] In essence, it is how gene expression adapts to changes in diet, social and physical environment, and lifestyle choices (Figure 9.1).

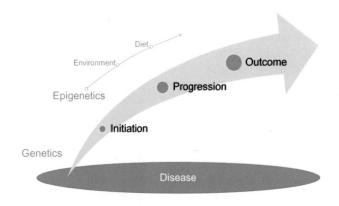

FIGURE 9.1 Impact of inherited genetic and epigenetic factors. © *Primary Care Network* 2014.

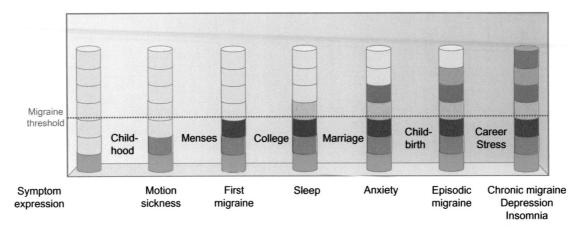

FIGURE 9.2   Gene expression and disease progression. © *Primary Care Network 2014.*

Accordingly, in a complex disease such as migraine, multiple gene expression rather than alteration of single gene mutation largely contributes to the hyperexcitable nervous system. Considering the possibility that multiple genes modulate neuronal excitability, it is plausible that the more genes a person inherits that are capable of transcribing proteins that enhance neuronal hyperexcitability, the more likely he or she is to have frequent migraines if and when his or her genome is exposed to specific migraine-provoking environments. In other words, individuals with high susceptibility to migraine may have multiple genes activated and capable of modulating neuronal hyperactivity, while those with low susceptibility to migraine have far fewer genes actively expressing proteins that increase neuronal excitability (Figure 9.2). This notion of differential gene expression may also better explain differences in migraine phenotypes, attack severity, and even pharmacological responsiveness. Further, and perhaps most importantly, epigenetics may explain how lifestyle, behavior, diet, and one's physical environment can ultimately determine the progression or remission of migraine without the presence of mutations to individual genes implicated in maintaining neuronal homeostasis. Epigenetics is a radical departure from the genetics used to understand diseases, such as phenylketonuria (PKU), that are caused by monogenetic single DNA nucleotide mutations.

Although we have approximately 20,000 genes capable of being expressed in each of our cells, typically roughly only around half (10,000–11,000) of the genes are expressed in any particular cell type.[31] The other 10,000 or so genes are turned off or deactivated by epigenetic influences through modification of DNA, RNA, and their associated proteins (histones). The two major molecular modifications of genes observed in human cells involve DNA methylation and histone deacetylation.[32] These molecular mechanisms coordinate to dynamically alter gene expression in response to environmental influences facing an organism. In general, hypermethylation of DNA in the promoter region of genes results in gene silencing, while hypomethylation in promoter regions results in gene activation and expression.[33] Methylation of DNA can be a permanent or a temporary response to specific environmental influences. Thus, changes in DNA methylation precipitated in response to environmental factors may be a major determinate of an individual's susceptibility to complex diseases such as migraine.

Another epigenetic mechanism involves histone modifications, which function to regulate how chromatin is packaged within the nucleus of the cell and controls the binding of transcription factors to DNA promoter sequences of genes.[34] Molecular modification of histone includes acetylation, phosphorylation, and/or methylation, and is mediated through enzymes that selectively add or remove functional groups critical for DNA transcription. The effect of histone modifications on gene expression is mediated through mRNA and can facilitate or prevent functional protein synthesis essential to gene expression.[35] In fact, a study with monozygotic twins showed that while patterns of DNA methylation were very similar in young twins, the patterns became quite distinct as twins' age and, from an epigenetic perspective, are each required to solve different environmental challenges.[36] Epigenetic mechanisms may help explain the importance of the psychosocial environment to the frequency and severity of chronic migraine and other chronic pain by providing a molecular link between inherited gene expression and environmental exposure to risk factors for initiation and progression of a specific pain disease.

At a molecular level, multiple signaling events are required to promote changes in the level of intracellular calcium and other ions that modify the voltage threshold for neuronal activation. This largely determines a cell's threshold for release of neurotransmitters. If the resting potential becomes more negative (hypopolarized) a greater stimulus is required to activate the neuron, while if the potential is more positive (hyperpolarized) a lesser stimulus is needed to initiate neuronal excitation and the cell is said to be in a sensitized state. This type of sensitization occurs in both peripheral and central nociceptive neurons. Expanding this cellular model to the system level, these functional changes can be objectively measured clinically to define peripheral and/or central sensitization. Both of these physiological events are implicated in the pathology of CM and are believed to be a major determinant of disease progression. Prolonged neuronal sensitization, as seen in CM, is proposed to result in nociceptive neurons reaching a "primed state" where each influencing epigenetic event(s) is moving the neuron to a more sensitized functional state.[37] Molecular messengers etiologically implicated in neuronal sensitization (cytokines, chemokines, ATP) in the primed neuronal state now function to activate pain pathways. At a cellular level, this transformation likely contributes to persistent hyperexcitability of both peripheral and central neurons that ultimately creates the neuronal networks associated with sustaining CM and other chronic pain states.

Given this background, it is highly probable that everyone inherits particular genes at the time of birth that predispose them to developing certain diseases. However, it is environmental interactions with these genetic predispositions that ultimately determine what, when, and if a specific disease actually occurs. Epigenetic influences have been proposed as an underlying mechanism for many complex diseases, such as chronic pain and many chronic inflammatory diseases.[38] Twin studies suggest that the heritability of a migraine phenotype ranges from 34% to 65%, thus supporting the critical role environmental factors play in the initiation and progression of the disease of migraine over time.[39,40] In essence, epigenetics provides a dynamic model whereby "permissive genes" under the influence of their environment ultimately determine which proteins are expressed and, over time, what disease(s) a person may experience.

As already suggested, the physiological foundation of migraine is the hyperexcitable nervous system. Changes in ion channels, receptors, gap junction proteins, and signaling molecules are likely mechanisms for the regulation of neuronal excitability. While people with migraine are almost certainly born with a genetic predilection for migraine, their exposure to specific environmental stimuli (external and/or internal) is equally important to gene expression and function. Ultimately, what precipitates migraine and whether or not it progresses into CM may be a consequence of the dynamic relationship of a person's genome with specific environmental influences. Hence, epigenetics may ultimately provide a dynamic blueprint for the clinical course of migraine throughout the lifetime of each individual migraineur.

Clinically, multiple factors might be considered as increasing the risk for CM through epigenetic interactions, and function to set the threshold for progression or severity of migraine. These factors might include stress, obesity, gastrointestinal absorption defects, medication overuse (MO) by people with addictive predilections, cigarette smoking, and disruptions or deprivation of sleep. It may also be that pharmacological interventions, especially those used chronically as prophylactics, may affect gene expression epigenetically rather than by directly acting on cell membrane receptors. In all instances, "environmental influences" may have the potential to alter gene expression. Interestingly, epigenetic factors are largely the same "risk factors" that have been recognized by clinicians as predisposing individuals to attacks of migraine, and recently several epidemiological studies have defined many of these same factors as posing risks for migraine chronification.[41,42]

The true implications of epigenetics for clinicians lies not in molecular pathways of DNA methylation, but instead in the ability to validate environmental, dietary, and lifestyle modification as congruent meaningful management strategies that prevent or reverse chronic migraine. While the triptan era provided effective acute treatment for millions of migraineurs, it also subtly suggested that targeted pharmacological "magic bullets" could be created to unravel the complexity of migraine. Thus, the hope was born that by defining mutations or errors in gene sequence, the "magic" receptor or ion channel causing migraine would be discovered and more selective pharmacological treatments would be created.

Epigenetics, on the other hand, ponders a plausible hypothesis that a meaningful ongoing dialog between patient and healthcare professional about the environmental (holistically defined) impact on a person's genetic predisposition to migraine is the critical component of effective and successful migraine treatment. This is perhaps why the most successful clinics specializing in managing CM patients use a comprehensive interdisciplinary care model that includes lifestyle, behavioral care, biofeedback, and other modalities in concert with pharmacological interventions. Perhaps, someday, migraine as a medical specialty will assist other therapeutic fields to integrate epigenetics into their understanding and management of chronic diseases.

TABLE 9.1　IHS Diagnostic Criteria for CM[4]

1. Chronic migraine	**A.** Headache (tension-type-like and /or migraine-like) on ≥15 days per month for >3 months and fulfilling criteria B and C
	B. Occurring in a patient who has had at least five attacks fulfilling criteria B–D for 1.1 *Migraine without aura* and/or criteria B and C for 1.2 *Migraine with aura*
	C. On ≥8 days per month for >3 months, fulfilling any of the following:
	1. Criteria C and D for 1.1 *Migraine without aura*
	2. Criteria B and C for 1.2 *Migraine with aura*
	3. Believed by the patient to be migraine at onset and relieved by a triptan or ergot derivative
	4. Aggravation or avoidance of physical activity
	D. Not better accounted for by another ICHD-III diagnosis

DIAGNOSIS OF CHRONIC MIGRAINE

Individuals receiving a diagnosis of CM should in most instances have a history, or at the very least a family history, of EM. The history of EM generally precedes CM by years, and there is a progression in attack frequency over time that eventually reaches the headache frequency to be diagnosed as CM. As discussed in *Chapter 8*, accompanying the increased frequency of migraine attacks is a lessening of the time the nervous system has to recover between attacks of migraine. In CM there is a lack of prolonged or complete neurological recovery between days with headache.

Ideally, a clinician should be able to trace the progression of EM from Stage 1 (infrequent episodic migraine) to Stage 2 (frequent episodic migraine) to Stage 3 (transforming migraine) to Stage 4 (chronic migraine). While this effort may be academically satisfying, it is often burdensome in a time-constrained outpatient practice. However, what is important is that the clinician establishes enough history to be confident that the current complaint of headache progression is not an occult secondary headache in a person with migraine.

At times, EM can rapidly transform into CM. This often occurs around a period of major stress, such as divorce, change in financial security, or death of a significant other, and even more frequently if there is a history of earlier life stress events.[43] As the current stressful events resolve, the frequency of migraine lessens — especially when proper treatment and support are provided. This group may be the 26% of the population of patients with CM observed to revert back to EM in epidemiological surveys.[20]

The diagnosis of CM needs to include an evaluation of the different headache phenotypes a person is experiencing. According to the ICHD criteria, there needs to be a threshold number of migraines fulfilling IHS criteria for migraine with or without aura. Ideally, this should occur on at least 8 days per month over a 3-month period (Table 9.1). However, frequently there are numerous other headache days where the phenotype included in the definition of CM does not fulfill

TABLE 9.2　Comorbid Conditions Associated with Chronic Migraine[42]

Comorbidity	Chronic migraine prevention
Allergies	60%
Sinusitis	45%
Depression	36%
High cholesterol	34%
High blood pressure	34%
Arthritis	34%
Chronic pain	31%
Anxiety	30%
Obesity	25%
Asthma	24%
Bronchitis	19%
Fibromyalgia	17–35%

IHS criteria for migraine without aura. These headache phenotypes may be consistent with probable migraine (missing one characteristic of IHS migraine without aura) or tension-type headache. In other words, patients with CM have a nervous system with the propensity to generate many primary headache phenotypes. While in all likelihood this tendency did not begin the day a person breached the threshold of 15 days of headache per month, it is fair to suggest that as migraine increases in frequency, so too does the tendency for the nervous system to express a spectrum of primary headache phenotypes.

Finally, in making a diagnosis of CM a healthcare professional needs to be attentive to a plethora of comorbid syndromes and diseases that are often associated with migraine chronification (Table 9.2). This is important because when comorbidity is not defined and appropriately treated it can be a cause of non-response to appropriate treatment and management of CM.[44]

Comorbidities can and perhaps should also be considered as epigenetic factors leading to chronification of migraine. Broad categorization of these factors may include psychological conditions such as anxiety, panic disorder, depression, and bipolar disease; gastrointestinal conditions such as irritable bowel syndrome, gluten sensitivity, and gastrointestinal reflux; metabolic conditions such as obesity and metabolic syndrome; and myofascial condition such as fibromyalgia and myofascial trauma. In addition, current and remote traumatic events need to be considered as having a profound and perhaps in some instances permanent epigenetic influence on migraine. This approach allows sincere and dedicated healthcare professionals to move from diagnostic understanding of a CM patient to providing effective management.

THE SUCCESSFUL MANAGEMENT OF CHRONIC MIGRAINE

The management of CM is complex, and requires investment of time, effort, and resources by both the patient and the healthcare professional. By the time most patients have evolved from episodic to chronic migraine, they have failed numerous treatment efforts and are plagued not only by near daily migraine but also by numerous comorbidities and significant disruption in many domains of their daily life. CM frequently impacts on work, social behavior and, sadly, personal and family dynamics. There is no simple formula to treat patients with CM. Each patient requires a unique blend of the best the science and art of medicine has to offer to break the bondage imposed by CM.

As migraine becomes chronic, it often transforms from an infrequent self-limited and self-managed condition to a syndrome that becomes a medical concern and eventually evolves into a destructive medical disease. Accompanying this transformation is often a sense of personal failure on the part of the patient for not being able to self-manage migraine. There is a tendency in patients for self-blame — that somehow they are "causing" migraine. It is essential to educate patients that migraine is a valid biological process.

CM requires individualized evaluation and management. It cannot be managed with a simple algorithm. Proper management requires attention to the multidimensional medical and psychological needs of the patients and, as is the case with most chronic diseases, it takes time and frequent adjustment and readjustment of the treatment plans to ultimately achieve success. Paraphrasing the quotation attributed to William Osler, "The good physician treats the disease; the great physician treats the patient who has the disease".[45]

Steps to the Management of Chronic Migraine

Step 1: Confidently Provide a Diagnosis of CM to the Patient

The diagnosis of CM must be understandable to the patient and serve as a template for constructing the management plan. CM is not a description of migraine duration or severity, and patients need to understand CM as a separate diagnostic entity. Tracing an individual's historical transformation of episodic to chronic migraine can become the basis for understanding the diagnosis and various risk factors that will need to be addressed in a management plan. Explaining the hyperexcitable nervous system can provide the basis for understanding the purpose and role of medications and non-pharmacological treatments important in controlling CM.

Step 2: Define Management Roles for the Patient and the Provider

It is often valuable to establish a collaborative dynamic where both provider and patient are recognized for the expertise each brings to solving the complex riddle of CM. Testing patients' health literacy by having them interpret medical explanations or instructions in their own words insures more congruent communication. Additionally, using interview strategies that increase the likelihood of patient participation can improve adherence. One such interview technique researched with primary care physicians and general neurologists, with positive results for both patient and clinician, is the "ask—tell—ask" method (see Chapter 19). In this model the patient is encouraged to assume greater responsibility for adhering to treatment, monitoring outcome, and communicating with the clinician. The clinician, on the other hand, has the responsibility to listen to and accept that the patient is communicating accurate information. In addition, the clinician is responsible for providing the management tools that are necessary to safely address the patient's therapeutic needs.

Step 3: Establish Agreed-upon Objective Goals and Boundaries

Treatment interventions should be considered as tools, not solutions or cures. Emphasis needs to be placed on preserving or enhancing quality of life through the medical management of CM. Therefore, it is essential to explain that managing migraine is a lifelong journey and success is best achieved through time, diligence, and perseverance. Goals need to be definable, objective, and agreed upon by both the patient and healthcare provider (HCP). Obtainable objective goals need to be established for behavioral changes, acute and preventative migraine pharmacology, and treatment of comorbidity. Adherence to a management plan and decision-making skills often become additional goals.

The use of a diary is an excellent method to monitor goals. Recording the number of headache days (or the "real" goal of headache-free days) or the number of times acute treatment has been utilized successfully, or recognizing new activities that are now being performed because of successful treatment, are all worthy uses of diaries. Diaries can also be a tool for focusing discussions between the patient and HCP on the success or shortcomings of management decisions. For example, asking patients which attacks they treated since their last visit that did not meet expectations or defined goals allows for a very different conversation than a question asking: "How are your migraines?" Armed with this information, the provider and patient can reassess and refine interventional strategies, medication formulations, or timing strategies of interventional efforts. When the frequency or pattern of headache worsens, the conversation can revolve around effectiveness of preventive medication, adherence with daily medication regiments, and/or lifestyle changes.

Step 4: Avoid Being Judgmental

It is bad enough to live with the impact of CM, but when guilt is confused with motivation the disease worsens. CM patients routinely report feeling guilty for behaviors that "cause" CM. It is far better to establish achievable, simple goals that nurture success rather than add to the trail of failures that often plague a person with CM. While a list of important changes a patient needs to make materializes in the clinician's mind, the patient may need to start low, go slow, but keep going forward. It is up to the patient to tackle the specifics of reversing CM.

Step 5: Establish Agreement on Management Decisions, Especially Medications

Unfilled prescriptions and lack of adherence are two of the unspoken realities accounting for failure of treatment of CM. It is estimated that over one-third of prescriptions are never filled, and a far greater percentage of patients are not adhering to instructions for using treatment medications.[46] This problem is amplified in the current healthcare system when pharmaceutical benefit managers, far removed from patient needs, influence medication use and misuse. Patients frequently turn to OTC products or other more accessible medications, such as opioids, for relief of migraine, and may not readily share this information with their HCP for fear of being dismissed as noncompliant. Discussions of medications by the HCP need to instill confidence, yet be explicit about important risks that both parties need to monitor.

NON-PHARMACOLOGICAL MANAGEMENT OF CHRONIC MIGRAINE

The non-pharmacological management of CM is often a misunderstood and neglected resource for CM management. Given the discussion on epigenetics, it is a critically important clinical resource for the patient. Non-pharmacological management can improve patient involvement and adherence with the overall care plan, and it provides options for care beyond pharmacology that can benefit a patient for a lifetime. While many types of non-pharmacological management exist, consider the interventions that a specific patient will accept and sustain over time. Psychologists, nutritionists, physical therapists, exercise coaches, acupuncturists, and alternative medical practitioners represent only a short list of medical specialists that can be useful in the management of CM. HCPs need to reinforce how lifestyle habits impact nervous system function as well as influence the effectiveness of medications, thereby improving patient outcomes. In this context, non-pharmacological management can be seen as a balance to various risk factors for migraine that are not always under a person's direct control, such as sudden weather changes or menses.

A diary is an effective way of integrating behavioral efforts, such as identifying risk factors, into a comprehensive management plan. Through monitoring success of balancing protective factors and avoiding risk factors, patients become involved in their own health care and recognize that lifestyle influences their sense of well-being. For example, an overweight patient is given nutritional and dietary advice and is instructed to use the diary as a tracking tool for changes in weight and physical activity. Knowing that the clinician will review the diary and the patient's progress motivates the patient to follow through. This is an opportunity to discuss the role of epigenetics and how the internal and external environment can influence the genetic sensitivity of the nervous system in both positive and negative ways.

Lifestyle Factors

Exercise

Exercise is considered a positive influence in reducing migraine frequency.[47] The challenge is that exercise may increase headache intensity and at times actually trigger migraine. Consequently, exercise needs to be prescribed based on the patient's level of physical fitness. Beginning with simple stretching exercises and graduating to low-impact water exercises permits the patient to succeed, opening a pathway to progress to a more active exercise regimen and lifestyle. Aerobic exercise has been documented to increase endorphin production.[48] It is also valuable to support types of exercise that are sustainable and deemed pleasurable.

Diet

Diet is often neglected, but is a critically important component in the management of a CM patient. Obesity is a known risk factor for migraine chronification, and many medications used as prophylactic medications in migraine are associated with weight gain.[49] For this reason, weight needs to be monitored. Choosing a preventive medication that is weight neutral, or, as with topiramate or bupropion, may assist weight loss, can be reassuring to the patient. Medical guidance is key for migraineurs to select a diet (for example, low carb, high protein, low fat, etc.). Adding commonsense diet messages about snacking, "empty" calories, and good nutrition can be valuable.

In addition, it is important to consider the gastrointestinal (GI) tract as the anatomic home of the enteric nervous system and acknowledge the rich communication occurring between the gut and the brain. Specific medical conditions, such as irritable bowel and metabolic syndrome, should be addressed and management of these conditions incorporated into a patient's care plan. In general, it is valuable to encourage good GI health.[50]

Sleep Hygiene

Sleep has long been recognized as nature's antidote in recovery from attacks of migraine. Ironically, too much or too little sleep is also recognized as a risk factor that can precipitate migraine. Thus, monitoring of sleep patterns and quality of sleep can be an important step in improving patient outcomes.

Basically, most patients with migraine are served well by maintaining scheduled sleep and wake patterns. Implementing successful sleep plans is initiated by evaluating sleep hygiene — i.e., behaviors that encourage or discourage sleep. The first rule is that the bedroom is used for sleep and intimate activity. Televisions, computers, and phones should be removed from the sleep environment. Pre-sleep activities should facilitate calming of the nervous system and a retreat from the chaotic activity of the day. There needs to be an established ritual for sleep that is consistently maintained as much as possible. Adjuncts for sleep preferred by the authors include melatonin and 5-hydroxytryptamine. Prescription medications for sleep may be helpful, but in general should be considered short-term adjunct therapies. Psychiatric causes of sleep disruption need to be diagnosed and appropriately treated.

Smoking Cessation

For a healthy lifestyle, smoking cessation should be encouraged for all migraine patients. Tobacco smoke is a recognized migraine trigger, and smoking increases serum levels of carbon monoxide and dioxide — both additional risk factors of migraine. Numerous aids are available to assist patients in this effort, but none is more important than persistent support by the clinician.

Behavioral Therapies for Chronic Migraine

Biofeedback Training

Biofeedback is the process of reversing the hyperexcitable nervous system by replacing the fight-or-flight reaction with the relaxation response. It is a process of bringing involuntary physiological functions under voluntary control.[51] The goal of thermal biofeedback is to train the nervous system to shut out excessive stimulation, perseverative thoughts, and emotional over-reactions. The efficacy of biofeedback training is largely dependent on the technique and skill of the therapist guiding the training. Rapport needs to be established between the patient and therapist. Often, the individual is connected to a machine that records the finger temperature and/or the microvoltage of a muscle. However, to be maximally effective, the patient needs to be guided through visualization to an imagined refuge where there is safety, serenity, and peace. There are three levels to biofeedback training. The first is relaxation, the second is behavioral retraining, and the third is physiological recalibration.

The goals of biofeedback are to alter the fight-or-flight responses that occur when confronted with stress, to change behavior patterns regarding stressful events in a positive manner, and to improve physiological resilience to present and past traumatic events. In essence, it is about discovering a healthier self.[52]

RELAXATION TRAINING

The person is taught diaphragmatic breathing at a very slow rate with the eyes closed. Usually this is accompanied by calming music, contracting and relaxing muscles, or the soothing sounds of nature. The goal of relaxation is to produce a physiological break from the fight-or-flight response for the nervous system.

BEHAVIORAL RETRAINING

A second level of biofeedback is the process of learning to stop thought patterns and behaviors with the potential to provoke migraine, and to focus on thoughts that are protective and ultimately elevate the migraine threshold. By using a biofeedback device, the physiological response to behavioral suggestions is measured and demonstrates to the individual that thoughts precipitate a physiological reaction. For example, a patient may find that a memory of an upsetting event lowers the finger temperature, indicating a stress response. During biofeedback, the patient visualizes this event and envisions a more confrontive response that empowers them to stand up to the threat rather than feel victimized. With practice and use of positive affirmations, the automatic response to these stimuli are erased, anxiety and negative talk are diminished, and finger temperature increases. This process uncovers negative thought processes that may defeat the person in developing protective behaviors for migraine. Once these negative thoughts are revealed, positive behaviors can be identified and practiced to improve management of physiological responses to changes.

PHYSIOLOGICAL RECALIBRATION

Studies have shown that profound psychological trauma produces enduring physiological changes in the nervous system.[53] For a person with a hyperexcitable brain and a low migraine threshold, trauma exaggerates the genetic hyperexcitability of the nervous system into one of chronic hypervigilance. Repeated traumatic assaults, including uncontrolled migraine or negative encounters with the healthcare system, can be central stimuli to chronification of migraine. Recalibration of nervous system function is accomplished by revisiting and reinterpreting major traumatic events in a detached state of deep relaxation. Skilled therapists guide this process and help migraineurs reframe catastrophic memory and reinterpret thoughts and emotional responses in a more physiologically protective manner. Instead of feeling victimized, the person may slowly realize that the trauma was not his or her fault, but an expression of another's pathology or other extenuating circumstances. Using deep states of relaxation, usually theta brain waves (4–8 Hz), a negative emotional experience is reexamined and reinterpreted without a fearful emotional charge.

Mindfulness

Mindfulness is a process of focusing on slow, diaphragmatic breathing and putting oneself in the "moment." Based on a Buddhist concept that dates back 2600 years, mindfulness has become a popular psychological movement against the current cultural theme that anything is possible by exerting the power of the mind over the body. Mindfulness in fact provides an alternate possibility. By appreciating the present, the individual is directed to listen to the body for signals of awareness in the moment and reunites body and mind connectedness. Unlike biofeedback, which relies on an exterior measure to gauge the level of tension in the body, mindfulness forces the person to concentrate on the internal functioning of the muscles, nerves, and blood vessels, which ultimately normalizes physiological response to the environment and reveals the importance of the moment. As a result, the body relaxes, perseverative ideas stop, working memory becomes more efficient, and fear decreases. There have been no studies that measure the brainwave that mindfulness generates, but the end result is much like biofeedback, where theta brain waves put the person in a state of heightened internal awareness without emotional reactivity.

Cognitive Behavioral Therapy

Cognitive behavioral therapy (CBT) is a systematic approach that addresses dysfunctional emotions, behaviors, and thought processes through goal-oriented psychotherapy. The initial effort of the therapist is to help the patient identify negative and often automatic thoughts and behaviors that are interfering with healthy function. Once identified, positive affirmations are consciously practiced by the patient until this more positive dialog replaces the maladaptive thinking or behavior. CBT is considered to be effective in conditions such as depression, anxiety, substance abuse, eating disorders, and schizophrenia.[54] Beck, considered to be the father of CBT, espoused the theory that negative thoughts can produce depression,[55] and CBT is widely accepted as an effective form of psychotherapy today, with Class A evidence in migraine.[56]

Goal-setting is a cornerstone of CBT. For most chronic migraineurs, their life revolves around debilitating headaches and medication. They fear that planning or engaging in activities will aggravate their migraine or disappoint others. By creating a goal, a person's focus changes from the headache to an attractive activity outside him- or herself. Goals may be simple, such as walking the dog daily, exercising in front of the TV, or visiting the library. As the range of activity increases, the goals likewise become wider and more far-reaching.

It is worthwhile considering defining specific goals, with all CM patients, that include measurable outcomes for acute and preventive treatment as well as lifestyle/behavior. These goals should be defined in a manner that is objective, obtainable, and dynamic.

Acupuncture

Several small studies support acupuncture and its use in CM. A small randomized study of 66 patients receiving either topiramate or acupuncture showed statistical superiority for the acupuncture group on reduction of headache days. Those receiving acupuncture went from 20.2 ± 1.5 days to 9.8 ± 2.8 days compared with 19.8 ± 1.7 days to 12.0 ± 4.1 days in the topiramate group ($P < 0.01$). Significant differences favoring acupuncture were also observed for all secondary efficacy variables. These significant differences still existed when the focus was on patients who were overusing acute medication. Adverse events occurred in 6% of the acupuncture group and 66% of the topiramate group.[57] In a recent Cochrane analysis of acupuncture for routine care, Linde and colleagues concluded that there is no evidence for an effect of "true" acupuncture over sham interventions. However, after reviewing 22 trials of 4419 participants with migraine, they concluded that migraine patients can benefit from acupuncture.[58]

Complementary and Alternative Medicine

The use of complementary and alternative medicine (CAM) by the CM population is common, although generally unstudied. Patients using CAM observe that pharmacotherapy is marginally effective in CM, expensive, and associated with numerous adverse events.[59] Several CAM therapies are in frequent use, such as yoga, osteopathic and chiropractic manipulation, and acupuncture. Integration of CAM can be effective, and invites patients to be proactive in their migraine care. These therapies, like all others, need to be monitored and have goals established for their use.

Osteopathic and Chiropractic Manipulative Therapy

The use of manipulative therapies for CM has been documented, although, as with many studies, the definition of CM is not always in agreement with ICHD criteria. Evidence-based guidelines for the chiropractic treatment of adults with headache were published in 2011 by the Canadian Chiropractic Association.[60] The guidelines were based on a review of 16 studies and five system reviews and support chiropractic care, including spinal manipulation, for the management of migraine and cervicogenic headaches. Spinal manipulation and multimodal multidisciplinary interventions, including massage, are recommended for management of patients with EM or CM. Review of nine studies also concluded that spinal manipulation therapy has an effect equivalent to first-line prophylactic prescription medications for tension-type headache and migraine headache.[61]

In complete contradiction to the previous studies, a systematic review of randomized clinical trials published in 2011 stated that there was no evidence to support the use of spinal manipulations for the treatment of migraine headaches.[62] There clearly is an opportunity for further clinical study in the efficacy of chiropractic manipulation for migraine therapy.

The osteopath approaches therapy from a holistic viewpoint, focusing on the entire body by using a variety of techniques to treat patients; these include diagnostic touch; high-velocity, low-amplitude thrust techniques; strain—counterstrain; and myofascial release. Osteopathic manipulative treatment (OMT) for migraine traditionally concentrates on relieving an ongoing headache, and reducing the severity or frequency of headache and accompanying symptoms. There are few clinical trials on the efficacy of osteopathic therapy for relief of migraine. A review article of current randomized controlled trials (RCTs) concluded that OMT, specifically spinal manipulation, was equally effective as propranolol and topiramate in the management of migraine.[63] There is a great need for large RCTs in the study of the effectiveness of OMT for headaches.

PHARMACOLOGICAL MANAGEMENT OF CHRONIC MIGRAINE

Few clinical trials have been specifically designed to study the efficacy and safety of preventive pharmacology in CM. Preventative pharmacology for EM migraine is discussed in *Chapter 8*, and its use in CM for the most part is essentially the same as in EM. It will not be repeated in this chapter except in circumstances unique to treating CM.

Prophylaxis of Chronic Migraine

OnabotulinumtoxinA (Evidence Level A)

OnabotulinumtoxinA is the only FDA approved medication for prophylactic treatment of CM. The basis for regulatory approval in North America, the UK, and Europe are two placebo-controlled double-blinded randomized parallel studies called PREEMPT 1 and 2.[64,65] Results of the pooled analyses from both studies included a population of 1384 adult subjects with a definition of CM that closely parallels the current ICHD definition of CM. Subjects were randomized to receive onabotulinumtoxinA or placebo 1:1. There was a mean decrease from baseline in frequency of headache days of 8.4 days vs 6.6 days for active treatment over placebo at 24 weeks, respectively ($P < 0.001$).[66] This was the primary endpoint. Secondary endpoints of cumulated headache hours, migraine episodes, moderate or severe headache days, headache episodes, and percent of patients with Headache Impact Test (HIT-6) score >60 (severe impact) were also significantly improved at the 24-week time-point. At 24 weeks, subjects were eligible to continue in an open-label phase of the study for an additional 28 weeks (Figures 9.3, 9.4).

Interestingly, in the PREEMPT studies there was no significant reduction in use of acute headache medications. In an *ad hoc* sub-analysis, it was demonstrated that populations overusing acute medications (as defined by ICHD-II criteria) during the active study period improved to the same degree as those not overusing acute medications.

Also noteworthy from the PREEMPT studies is that subjects randomized to placebo for the first 24 weeks of the study did not obtain the same degree of benefit as subjects receiving onabotulinumtoxinA at randomization. This difference persisted to the study endpoint at 52 weeks.[67] This may suggest that early intervention in CM may have important implications in disease modification. Paradoxically, in today's healthcare system, patients with longstanding CM are frequently

required to fail multiple oral preventives not FDA approved or studied in CM before being approved to receive onabotulinumtoxinA.[68] This likely represents the unfortunate fact that today short-term healthcare cost containment reigns supreme over long-term benefits and quality care of patients.

In a second study, onabotulinumtoxinA was compared with topiramate in 59 subjects with CM diagnosed by ICHD-II criteria.[69] Both drugs significantly reduced the number of headache days and neither was statistically superior to the other, which likely reflects the fact that the study was underpowered for this endpoint. However, onabotulinumtoxinA was associated with improvement in sleep, recreation, mood, presenteeism, and absenteeism relative to topiramate, and was much better tolerated. A total of 15 subjects discontinued the study prematurely: 8 topiramate subjects and 7 onabotulinumtoxinA subjects. For the topiramate group, half reported "adverse events" as the reason for dropping out.

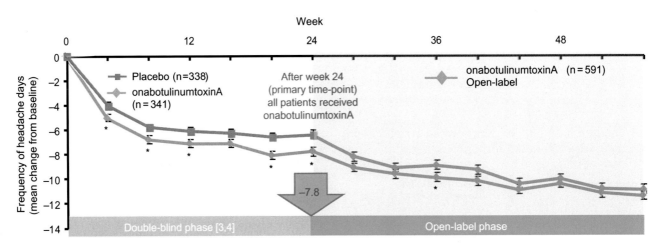

FIGURE 9.3   Mean change from baseline in headache days: Study 1. © *Primary Care Network 2014.*

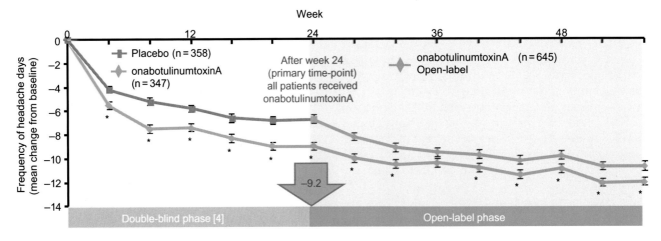

FIGURE 9.4   Mean change from baseline in headache days: Study 2. © *Primary Care Network 2014.*

A third, single-site analysis of 223 CM patients refractory to two or more oral preventative medications treated with two injection cycles of onabotulinumtoxinA 12 weeks apart noted mean reduction of 0.92 emergency department (ED) visits ($P < 0.001$), 0.39 urgent care visits ($P < 0.001$), and 0.11 hospitalizations ($P = 0.003$) following initiation of onabotulinumtoxinA treatment.[70] Analysis of treatment-related costs yielded a reduction of $1025 per patient. Compared with the 6 months predating initial treatment, patients had 55% fewer ED visits, 59% fewer urgent care visits, and 57% fewer hospitalizations for migraine during the 6-month treatment period ($P < 0.01$ for all).[68]

Interpreting the scope of data in these three studies into clinically meaningful constructs requires understanding the significant impact and disability imposed by CM. The benefit that patients with CM obtain with the return or increase of headache-free days, or at least reduction of baseline headache intensity, is immeasurable. OnabotulinumtoxinA is a valuable and important tool for improving the life of many patients with CM.

Topiramate

Clinical trials have explored the use of topiramate in treatment of CM. An additional related study was conducted to assess whether topiramate could prevent transformation of frequent EM to CM. In a large double-blinded placebo-controlled multicenter study, topiramate, while failing to meet its primary endpoints of a decrease in mean monthly total headache days and headache-free days, demonstrated efficacy on several secondary efficacy measures.[71] These included significant reduction of migraine/migrainous days, total headache days, and headache-free days, and improvement in quality of life scores. The mean reduction of monthly headache days was 5.8 for topiramate vs 4.7 days for placebo.[69] In another study, topiramate was not superior to placebo in preventing the transformation of EM to CM.[72] While topiramate may be useful both adjunctively and as a solo intervention in CM, its preferred role is in frequent EM.

Repetitive Dihydroergotamine

Repetitive dosing of dihydroergotamine (DHE) is widely used by headache specialists to treat status migraine, severe CM, and in patients with medication overuse headache (MOH) withdrawing from analgesics. DHE interacts with multiple serotonin receptors, including 5-HT1B/D and 5-HT1F, adrenergic and dopaminergic receptors (see *Chapter 8*). In CM and analgesic overuse headache, DHE is often administered repetitively over several days. Typical dosing is 0.5−1 mg every 8 hours for nine doses. Other protocols initiate tid dosing for 2 days, bid dosing for 2 days, and qd dosing for 1−2 days. Often an anti-emetic is administered with DHE. DHE does not cause dependency, and some suggest it has a lower risk for MOH than triptans. Repetitive DHE can be quite effective as both an acute intervention and a short-term prophylactic to break a cycle of intense daily headache in CM with or without MOH.

A recently approved orally inhaled delivery system for DHE appears to have fewer associated adverse events than IV DHE, particularly nausea, but there is no research available for administering repetitive DHE with the oral inhalation.[73]

Risks and contraindications for DHE are similar to those of triptans. Common adverse events include nausea, muscle cramps, abdominal pain, paresthesia, hypertension, and diarrhea. DHE is Category X in pregnancy.

Methysergide/Methergine

Methysergide is used for prevention of intractable migraine and cluster headache. It is no longer available in the US. It is an ergotamine, and its mechanism for migraine prevention is believed to be mediated through antagonism of the 5-HT2B receptor. It is known in rare instances to cause fibrosis, and consequently it is used infrequently.

Methylergonovine is an active metabolite of methysergide and has been used both as an acute and preventive treatment.[74] It has cholinergic side effects, such as nausea, vomiting, and diarrhea, as well as adverse events of dizziness and hypertension. It is contraindicated in pregnancy and pre-eclampsia. Typical doses are 0.2 to 0.4 mg tid. As with methysergide, a drug holiday is recommended on an annual basis and screening should be performed periodically to assess for development of fibrosis.

Phenelzine

Phenelzine is a non-selective and irreversible monoamine oxidase inhibitor (MAOI) which is used for refractory CM that is not responsive to other pharmacological preventive medications. By inhibiting amine metabolism, phenelzine significantly increases extracellular levels of serotonin, norepinephrine, dopamine, and melatonin. Consequently, tyramine-containing foods can cause a severe hypertensive crisis, and dietary restriction of tyrosine-containing foods is essential when this medication is used. A partial list of restricted foods includes smoked, aged, or pickled meats; fermented products, including alcohol; chocolate; and certain fruits and nuts.

Phenelzine has antidepressant and anti-anxiety effects, and is particularly useful in atypical depression.[75] Its use is limited by significant dietary restrictions, adverse events, and the fact that it cannot be used with triptans. Common adverse reactions include hypotension and weight gain, dry mouth, constipation, and insomnia. Liver function abnormalities can also occur. It can be used with close supervision in combination with amitriptyline or beta-blockers, but therapy should be initiated under inpatient observation. A 2007 review demonstrated elevated levels of neurotransmitters in the hypothalamus and amygdala of migraine patients.[76] Dosing is generally initiated at 15 mg per day and slowly increased to doses as high as 75 mg. The average dose in CM is typically 45 mg qd.

Naproxen vs Sumatriptan/Naproxen

A two-center double-blinded randomized parallel study compared 28 patients with CM using either naproxen 500 mg, or sumatriptan 85 mg plus naproxen 500 mg (randomized 1:1).[77] Subjects received daily study medication for 1 month, and then used the same study drug as acute treatment for the subsequent 2 months. At the end of the first month, there was a 60% reduction in headache days for naproxen compared with a 24% reduction for sumatriptan/naproxen. At the end of the third month, the reduction of headache days was maintained for the naproxen group while the sumatriptan/naproxen group had an 11% decrease in headache days. The duration of headache was also reduced to a greater degree for the naproxen group. However, the number of dropouts, especially due to lack of efficacy for naproxen as an acute treatment in CM, was considerably greater for naproxen than sumatriptan/naproxen. The authors hypothesized a protective role for naproxen and suggested that combining naproxen with sumatriptan was a better acute intervention, but did not reduce headache days possibly because naproxen offset the potential increase in headache frequency observed with frequent sumatriptan use. However, neither population demonstrated evidence of MOH throughout the study period.

Neurostimulation

Neurostimulation of various peripheral and cranial nerves has been hypothesized to benefit patients with CM but, while early pilot studies suggest promise, it is too early to make any definitive statement about their efficacy or safety in treatment of CM. Given that disclaimer, Silberstein and colleagues reported on a double-blind placebo-controlled randomized clinical trial of 157 subjects with CM using an implanted occipital nerve stimulation.[78] Subjects with CM were randomized (2:1) to receive active or sham stimulation devices. While the study failed to reach statistical significance for its primary endpoint (greater than 50% reduction in visual analog scale [VAS] score without increase in headache duration), it demonstrated significant reduction in pain and headache days, and

improved measures of quality of life. The authors suggested this modality may be beneficial, by reducing migraine-related disability. Additional studies are ongoing.

A small study has also been conducted with noninvasive vagal nerve stimulation (VNS) in CM. Silberstein and colleagues reported on a sham controlled blinded study of 59 subjects randomized 1:1 to receive either VNS or sham stimulation.[79] The self-administered treatment was delivered transcutaneously three times a day, with two 90-second stimulations given 5−10 minutes apart. Stimulation was provided to the area of the neck over the carotid pulse and, presumably on anatomical basis, the vagus nerve. All subjects had a 4-week run-in period followed by an 8-week sham-controlled period and a 6-month open-label period where all subjects received VNS. Results were suggestive that VNS could reduce the frequency of headache days (20.9 HA days at baseline; 19 HA days at 4 weeks; 12.1 at 6 months for VNS and 22 HA days at baseline; 22.2 at 4 weeks; 16.8 HA days at 6 months for sham stimulation). The authors point out that for a subset of subjects, this is accomplished without the inherent risks of additional pharmacology.

Repetitive Sphenopalatine Ganglia Blockade

A double-blind placebo-controlled study compared the efficacy of sphenopalatine ganglia (SPG) blockade with sham blockade in treatment of subjects meeting ICHD criteria for CM (unpublished data, 2014). Following a 1-month run-in period to establish the frequency of headache days, subjects were randomized 2:1 to receive SPG blockade with 0.5% bupivacaine or saline twice a week over a 6-week time period. Subjects with SPG blockade provided by bupivacaine and the Tx360® delivery device had a statistical reduction in headache intensity at 15 and 30 minutes, and 24 hours. There were progressive increments of improvement for each SPG blockade over the 6-week time period, and a statistical reduction in the number of headache days (3.9 days vs 0.58 days active vs sham, respectively). A statistical benefit was also observed at 1 month following the last SPG block for active, but not sham, blockade. Additionally, numerical benefit was maintained at 6 months for the active group. The authors explained that statistical benefit not achieved at 6 months was likely due to a higher dropout rate for those in the sham treatment group, given the 2:1 randomization. A more definitive study of this procedure is required.

Chronic Opioids

The use of long-term opioids for CM is a controversial subject. While most clinicians recognize that not all patients are sufficiently managed with current migraine pharmacology, they also understand that misappropriate use of opioids can worsen rather than improve the disease. Consequently, opioid medications are not recommended as first-line therapies for patients with migraine. They may, however, be appropriately prescribed for selected patients who cannot tolerate or do not respond to other acute medications, or who have contraindications to migraine-abortive medications. In these instances they need to be prescribed only after careful analysis of other treatment options, and patients must be monitored carefully in an ongoing management strategy. Opioids have a good analgesic effect and perhaps even better anxiolytic effect that may be of therapeutic value in select patients with CM. Opioids also are considered to have a high potential for MOH.[80]

Clinical studies on long-term use in CM have been mixed. In subjects selected after undergoing extensive screening, Robbins found that chronic opioid therapy was an effective strategy for treating CM. Positive response was defined as a 30% reduction of headache intensity in patients continuing long-acting opioids for at least 9 months, or improvement in headache frequency and/or severity over a 3-month baseline assessment.[81]

In a partially contradictory study, Saper and colleagues, reporting on 5-year follow-up of patients using long-acting opioids for treatment of intractable headache, suggested that less than 25% of patients actually benefited in a clear and measurable manner.[82] Over half of patients requested discontinuation of opioid medications soon after they were prescribed, and 25% continued medication with only marginal benefit. Another 23% demonstrated substantial benefit. The authors noted that in the successful patient population there was a reduction in hospitalization and emergency services. They suggested that a small group of patients, who have failed advanced care for CM, may benefit from chronic long-term opioid use. As in the Robbins' study, the Saper study underscored the critical importance of careful patient screening, selection, and monitoring.

The *American Academy of Neurology Practice Parameter: Evidence-based Guidelines for Migraine Headache* cautions clinicians to limit the use of opioids and other acute medications to "two headache days per week on a regular basis." The American Migraine Prevalence and Prevention population-based study data showed that MOH development is linked to baseline frequency of 10 headache days per month.[83] The 2012 update states that opioid use is positively associated with chronic headache conditions.[84] Common opioid side effects included dizziness, sedation, nausea, gastrointestinal discomfort, constipation, and vomiting.[85]

While no definitive statement can be made on the role of chronic opioids in the management of CM, available data suggest a small but significant population of patients benefit from chronic opioid therapy.

Co-Pharmacy

Co-pharmacy is combining two or more migraine prophylactic agents in an effort to enhance efficacy or reduce dosages of a single agent to limit adverse events. Even though there is little evidence evaluating specific combinations of agents, it is a widespread clinical practice. Commonly, co-pharmacy is utilized when migraine is associated with a comorbid disease. In these instances, an agent is selected for treatment of the comorbid condition and a second is a more standard migraine prophylactic agent. Common combinations include tricyclics and beta-blockers when a patient has frequent migraine and sleep disturbances or depression; topiramate and venlafaxine for obesity and depression; or onabotulinumtoxinA and divalproex sodium for a patient with bipolar mania. OnabotulinumtoxinA is particularly useful in a combination, as it has little potential for drug interactions.

Acute Medication for Management of Chronic Migraine

Dihydroergotamine

Dihydroergotamine (DHE) has been an abortive treatment for migraine for more than 60 years. Even throughout the triptan era, it was widely used by headache specialists. DHE is a 5-HT1B/D receptor agonist whose selective action causes cranial vasoconstriction and inhibits the release of calcitonin gene-related peptide (CGRP) from trigeminal afferents during migraine. However, DHE also has significant receptor agonist activity at the 5-HT1F receptor. The 5-HT1F receptors have been localized to the trigeminal ganglion and trigeminal nucleus caudalis, where agonists inhibit neuronal activation.[86] In addition, DHE has affinity for 5-HT1A and 5-HT2 receptors as well as alpha-1 and -2 receptors. Thus, the potential exists for DHE to benefit migraine through numerous receptor mechanisms. DHE has a relatively long terminal half-life of 9 hours. It also is noted to have prolonged receptor binding, and binds to the 5-HT1 receptor 8- to 14-times longer than sumatriptan.[87] This may in part explain why DHE is frequently used in headache specialty centers to treat intractable and prolonged migraine.

DHE is available as a nasal spray and as an injection (IV, IM, or SC). A newly formulated, orally inhaled formulation is pending approval by the FDA. Clinical trial data of orally inhaled DHE suggest significant efficacy at relieving headache and migraine-associated symptoms of nausea, photophobia, and phonophobia.

It was well tolerated, and appears safe and effective in smokers and patients with asthma.[88] Though DHE has not formally been studied in CM, there are several attributes that make it an attractive option for acute treatment in these patients.

Triptans

The pivotal trials of subcutaneous sumatriptan demonstrated that 70% of subjects with moderate to severe migraine had relief of headache and associated symptoms within 1 hour, and nearly 50% were pain free.[89,90] In the decades that followed, subcutaneous sumatriptan, six additional oral triptans, and two intranasal triptan products were brought to market. These drugs all have the same mechanism of action, and presumably treat very similar populations of migraine patients.

The efficacy of tablets decreased to approximately 40% for relief of headache and migraine-associated symptoms at 2 hours, and pain-free efficacy rates were approximately 29%.[91] Improvement in efficacy was obtained with a paradigm called "early intervention," where oral triptans were initiated early during the mild headache phase of an evolving attack.[10] Unfortunately, due to patients' behaviors, attack characteristics, or quantity limits on acute medications, it is estimated that only 50% of attacks are treated with early intervention. Further, there is debate as to whether early intervention in frequent migraine improves outcomes and reduces medication use, or encourages medication overuse by treating migraines that may not need treatment with triptans.

Although triptans are widely used in patients with CM, there are no data from studies specifically supporting their role in CM. In addition, there are no studies of rescue treatments for non-responders to acute treatment or treatment of headache recurrence in this population. Data would suggest that in the best of circumstances there are approximately 20–30% of migraine attacks for which triptans are ineffective, and consequently, in a population having 20 or more headache days per month, there is a significant potential for frequent non-response to medications and ensuing disability.

Whenever possible, triptans should be limited to fewer than 10 days of use per month when used to treat acute attacks in CM. If used more often, the clinician needs frequently to reassess the patient for evidence of MOH. Given the interplay between efficacy and frequency of use of acute interventions in CM, one can argue that when triptans are used, subcutaneous sumatriptan is preferred over oral formulations because of its higher efficacy and because, as an injection, there may be less potential for MOH.

ADVERSE EVENTS AND CONTRAINDICATIONS

Triptan-associated symptoms (adverse events) include nausea/vomiting, dizziness, warm/hot sensation, somnolence, paraesthesia, tingling, numbness, and tightness, heaviness, or pressure in the chest, neck, or throat.[92] Contraindications to triptans include arterial hypertension (untreated), coronary heart disease, cerebrovascular disease, Raynaud's disease, pregnancy and lactation, age under 18 (with the exception of almotriptan and rizatriptan), age above 65 years, and severe liver or kidney failure.[93] In the US, there is no contraindication based on age greater than 65. The only substantial pregnancy registry available for triptans is with sumatriptan.[94] To date, there are no known increases in birth defects associated with sumatriptan use during pregnancy, although the statistical power of the registry is far from being able to claim that sumatriptan is safe in pregnancy. All triptans are considered Category C in pregnancy. Sumatriptan is considered safe for nursing mothers by the American Academy of Pediatrics.[95] Despite the long list of potential adverse events, the fact is that triptans have proven to be a class of drugs with a low incidence of serious adverse events.

Several countries, including the UK, Germany, and Scandinavia, offer them to the public as an OTC product.[96]

Non-steroidal Anti-inflammatory Drugs

Ibuprofen, naproxen, and aspirin are the treatment interventions commonly used by CM sufferers. In a 2009 study of 14,540 participants assessing the medications used in acute treatment of EM and CM, non-steroidal anti-inflammatory drugs (NSAIDs) were used by 43.1% of the CM group, with ibuprofen (45.2% of NSAID use), naproxen (26.2%), and aspirin (23.6%) being the most common. Persons with CM used NSAIDs an average of 14.8 days per month.[97] It is unknown whether the motivation for this behavior is availability, cost, or efficacy.

The only NSAID with FDA approval for migraine is an oral suspension of potassium diclofenac.[98] Several OTC products, including aspirin,[99] ibuprofen,[100] and naproxen sodium,[101] have substantial evidence of efficacy in EM. There are no substantive studies demonstrating disease modification for any acute treatment medication. A single small pilot study suggested a reduction of migraine frequency over baseline at 3 months for naproxen when used daily for 1 month, and as the sole abortive for the subsequent 2 months.[102] Parenteral formulations of ketorolac are available as an intramuscular injection and nasal spray.[103,104] Ketorolac is widely used as an acute treatment for migraine, especially parenterally.

Intravenous Sodium Valproate

Intravenous (IV) sodium valproate has been studied as an abortive for prolonged attacks of migraine. Mathews and colleagues reported on a cohort of 61 patients treating 66 attacks of acute migraine with rapid infusion of 300 mg of IV sodium valproate.[105] A reduction of moderate to severe pain to mild or no pain was achieved in 37 of 66 attacks at 30 minutes. An additional 11 attacks had a greater than 50% reduction of pain intensity. The authors noted that 73% of attacks had significant improvement. Associated symptoms were also noted to improve, and there were no serious adverse events reported. Unfortunately, no large double-blind studies have been conducted to date. Dosage as high as 1 g as a rapid infusion has been recommended.

Phenothiazine/Metoclopramide

Phenothiazine and metoclopramide, both neuroleptics, are dopamine antagonists that have varying degrees of activity on the serotonergic, histaminic, adrenergic, and cholinergic neurotransmitter systems. They have been used as solo agents for acute treatment of migraine, and in combination with other medications such as triptans or DHE. As anti-emetics, their mechanism of action is thought to be as antagonists of 5-HT and dopamine receptors in the chemoreceptor trigger zone. In addition, they act as an alpha-adrenergic antagonist. Phenothiazines, including prochlorperazine, chlorpromazine, and promethazine, are commonly employed in the emergency department to treat severe nausea and vomiting accompanying migraine. Common side effects are sedation and drowsiness, akathisia, and dystonia.

According to a study on rescue therapy for acute migraine, Kelly and colleagues determined that phenothiazine and metoclopramide when used alone were superior to placebo.[106] Metoclopramide or prochlorperazine is frequently used in the emergency setting to effectively abort an acute migraine.

Addressing the 800-Pound Gorilla: Acute Medication Overuse and Misuse in Patients with Chronic Migraine

A common concern shared by patients with CM and their healthcare practitioner is the use of acute pharmacological interventions when a patient's headache frequency is high. There is little evidence-based information on acute treatment of patients with high treatment needs, as almost all clinical trials on acute treatment of migraine excluded populations that would today be classified with CM. On one side of this debate is the fact that the patients with high frequency headaches often have significant attack disability and hence would benefit from effective acute treatment. In addition, recent data suggest that poor optimization of acute therapy is a risk factor for development of CM.[107] The question left unanswered by this research is: "Does improving response to acute treatment reverse the progression of CM?"

On the other side of this debate are concerns that excessive use of acute medications may lead to MOH and a maintaining or worsening of CM. This is largely a debate of the chicken or the egg. Do patients frequently use acute treatments to maintain function, or does the medication pharmacologically maintain the headache pattern? Some studies demonstrate that discontinuing offending acute medications correlates with improvement in migraine frequency and quality of life[108]; however, other studies suggest the opposite.[109]

Medication Overuse and Medication Overuse Headache

One of the most perplexing challenges facing clinicians managing a patient with CM is MO and MOH. ICHD-II defines MO based on quantity of acute medication used. For opioids, combination analgesics, triptans, and ergotamine, MO is defined as >10 doses a month for greater than 3 months, and for NSAIDs and simple analgesics the threshold is ≥15 days per month for >3 months. Medication overuse is considered a risk factor for MOH (Box 9.1).

BOX 9.1

IHS DIAGNOSTIC CRITERIA FOR MEDICATION OVERUSE HEADACHE

IHS Diagnostic Criteria for Headache attributed to a substance or its withdrawal[4]:

8.2 Medication-overuse headache (MOH)
    A. Headache occurring on ≥ 15 days per month in a patient with a pre-existing headache disorder

    B. Regular overuse for >3 months of one or more drugs that can be taken for acute and/or symptomatic treatment of headache
    C. Not better accounted for by another ICHD-III diagnosis

---

These limits, while valuable in providing guidance for use of acute medications for patients with EM, often pose a hindrance to managing patients with CM, as their acute treatment needs frequently are greater than the consensus definition of MO. In fact, a majority of patients with a diagnosis of CM are in MO before the diagnosis of CM can be made.[110] Also it is challenging to control MO without patient participation, since many migraine medications are available without a prescription.

MOH is defined as an escalating chronic headache pattern associated with increasing use of acute treatment medication and is a clinical diagnosis, classified as a secondary headache disorder. However in the ICHD-III beta diagnostic criteria for CM, dual diagnosis of CM and MOH is encouraged. This is because a significant percentage of patients with greater than 15 days of headache per month will revert to an episodic pattern of migraine simply by withdrawal of overused acute medications. On the other hand, many patients do not improve following withdrawal of offending acute medications, and have CM. Recent studies have suggested that, at least for some preventive drugs, it is not necessary to withdraw acute medications before initiating prophylactic medication. Initiating topiramate while concomitantly withdrawing the offending acute medication has been shown to be effective, and likely more humane.[111]

Further, an *ad hoc* analysis of data from the PREEMPT study demonstrates equal efficacy for subjects with CM + MO and CM − MO.[112] Many questions remain unanswered about MOH. Does frequent use of abortive medications permanently alter pain mechanisms in migraine patients? Does the process of frequent migraine attacks independently sustain the headache frequency and intractability? Why is it assumed that medications with different mechanisms of action used at the same frequency over the same time span constitute medication overuse? In addition, there is the important clinical question as to whether patients should be subjected to months of treatment with preventive drugs that are not indicated in CM in order to receive medication that is indicated for CM. In the PREEMPT data, the group of subjects whose treatment with onabotulinumtoxinA was delayed by 24 weeks of placebo treatment failed to have the same outcome as those receiving onabotulinumtoxinA for the entire 52 weeks. This suggests there may be important therapeutic reasons to initiate effective treatment for CM early rather than later. The question of when to initiate preventive treatment in patients with presumed MOH has significant clinical relevance. Equally relevant is the issue of limiting effective acute care in populations of frequent migraine or CM. These and many other questions will need clarification in the future.

The use of assessment tools can expedite communication and management of MO and MOH. The patient diary is the most valuable tool for this purpose, as it provides a longitudinal record of changes in headache patterns and use of acute medications. Other assessment tools can enrich this assessment, such as the Migraine Disability Assessment Test (MIDAS), which can be used to measure changes in the headache impact score over a 3-month time period. An alternative tool is the HIT-6, which measures migraine impact over a 1-month period. Another useful tool for measuring optimization of acute treatment is the M-TOQ.[113] Screening for comorbidity can be accomplished with several tools, although perhaps the best for anxiety and depression is the PHQ-9 (Figure 9.5), simply because it measures both presence and the severity of these diseases. This test can also monitor change over time. It is worthwhile to consider that, at times, patients may be treating comorbidities with acute medications — i.e., treating anxiety with butalbital or insomnia with opioids. Recently, several electronic diaries have become available that can track headache and provide education to the patient. One particularly useful app for headache management is iHeadache® (www.iHeadacheApp.com), a free electronic diary.

Finally, it is important for clinicians to bear in mind that patients with CM can at times be demanding. Nursing and paraprofessional staff who are often at the frontline interfacing with these patients need support and encouragement, and it is advisable to create a team approach to care and manage this patient population.

Intravenous Magnesium

Magnesium is a drug/nutrient that has been successful in treating intractable headaches, menstrual migraines, and acute migraine, and has been suggested as a possible preventive migraine treatment. Magnesium reduces inflammation, relaxes muscles and blood vessels, and modulates calcium ion channels in cells, thus reducing the release of excitatory neurotransmitters. In a study of the efficacy of IV magnesium sulfate vs placebo with 30 patients, all patients responded to 1 g magnesium sulfate with 86.6% pain free and 13.4% with decreased pain intensity.[114] In another study of magnesium sulfate in the acute treatment of migraine without aura and migraine with aura, there was a statistical improvement of pain and all associated symptoms in the migraine with aura group. In the migraine without aura group, there was no significant difference in pain relief.[115]

PATIENT HEALTH QUESTIONNAIRE-9
(PHQ-9)

Over the last 2 weeks, how often have you been bothered by any of the following problems? (Use "✔" to indicate your answer)	Not at all	Several days	More than half the days	Nearly every day
1. Little interest or pleasure in doing things	0	1	2	3
2. Feeling down, depressed, or hopeless	0	1	2	3
3. Trouble falling or staying asleep, or sleeping too much	0	1	2	3
4. Feeling tired or having little energy	0	1	2	3
5. Poor appetite or overeating	0	1	2	3
6. Feeling bad about yourself—or that you are a failure or have let yourself or your family down	0	1	2	3
7. Trouble concentrating on things, such as reading the newspaper or watching television	0	1	2	3
8. Moving or speaking so slowly that other people could have noticed? Or the opposite—being so fidgety or restless that you have been moving around a lot more than usual	0	1	2	3
9. Thoughts that you would be better off dead or of hurting yourself in some way	0	1	2	3

FOR OFFICE CODING ___0___ + _____ + _____ + _____

=Total Score: _____

If you checked off any problems, how difficult have these problems made it for you to do your work, take care of things at home, or get along with other people?

Not difficult at all	Somewhat difficult	Very difficult	Extremely difficult
⑤	⑤	⑤	⑤

FIGURE 9.5   Patient Health Questionnaire-9 (PHQ-9). *Developed by Drs Robert L. Spitzer, Janet B.W. Williams, Kurt Kroenke and colleagues, with an educational grant from Pfizer Inc. No permission required to reproduce, translate, display or distribute.*

CONTINUITY OF CARE

A lesson learned from clinical trials of preventative treatment is that monitoring a patient, even with placebo intervention, has a positive effect on disease. Regular monitoring of patients with CM offers the clinician the opportunity to adjust therapy and motivate patients to adhere to treatment regimens and goals. It also provides an opportunity to identify changes in migraine and comorbidities early, and to provide timely proactive interventions.

CONSULTATION AND REFERRAL

Critical to the welfare of CM patients is the clinician's ability to match therapeutic need with appropriate resources. It is uncommon for a single clinician to be able to provide the array of tools required by the population of patients with CM. Consultation from other specialists may be useful for diagnostic clarity or to redirect therapeutic options of care. Frequently, headache specialists, neurologists, internists, gynecologists, and psychologists are consulted. Nurse practitioners and physician assistants also play an important role in the ongoing management of this patient population. More often, however, the need is to intensify management, and in this instance it is useful to consider interdisciplinary or multispecialty models of care. Typically, these models provide a combination of highly skilled professionals working collaboratively with the patient to deliver comprehensive and holistic treatment. At times, with severely impacted patients or patients with significant comorbid medical diseases, it is valuable to refer them to an inpatient program dedicated to the management of chronic headache.

PUTTING IT TOGETHER

Therapeutic needs of patients with CM are often numerous and complicated, and it is important to recognize that defining these needs often takes time and frequent reassessment. Even when patients are well controlled, their management needs will often change over time and circumstances. It is often useful to consider building a migraine tool box with the patient that includes education, acute and preventive pharmacology, and non-pharmacological tools. Given that patients will in most instances be selecting the "right" tool outside of direct medical supervision, it is critical that they receive the necessary education to make appropriate decisions. There are numerous resources to assist in that effort, including patient support groups, online support communities, and websites such as the website of the National Headache Foundation (www.headaches.org) or the American Headache Society (www.achenet.org). Regardless, regular patient follow-up by their HCP, generally spanning decades, is essential for most patients with CM.

SUMMARY

Prevention of disease progression is the most underappreciated yet most important contribution to care that a healthcare professional can provide to a person with migraine. Ideally, this begins when migraine is an episodic disorder. Providing education that validates the biological nature of migraine is an essential starting point. Concomitantly providing effective abortive medications with appropriate formulations can maintain control of migraine for years. For migraine that becomes more frequent and begins to transform, it is essential that patients and healthcare professionals recognize this warning as a time for re-evaluation and intensification of management. Implementation of both non-pharmacologic and pharmacologic prevention needs to be instituted, and the patient monitored closely. When CM is diagnosed, comprehensive efforts need to be initiated to include acute and preventive medications with known efficacy in CM. If such initiatives are not effective within a reasonable time period, the patient should be considered for referral for intensive interdisciplinary or multispecialty care.

CM is a serious complication of EM, and patients with CM have significant medical needs. However, treating CM patients with time, support, and understanding is frequently one of the most rewarding aspects of care a healthcare practitioner can experience.

APPENDIX

The Migraine Disability Assessment Test

The MIDAS (Migraine Disability Assessment) questionnaire was put together to help you measure the impact your headaches have on your life. The information on this questionnaire is also helpful for your primary care provider to determine the level of pain and disability caused by your headaches and to find the best treatment for you.

INSTRUCTIONS

Please answer the following questions about ALL of the headaches you have had over the last 3 months. Select your answer in the box next to each question. Select zero if you did not have the activity in the last 3 months.

_____ 1. On how many days in the last 3 months did you miss work or school because of your headaches?

_____ 2. How many days in the last 3 months was your productivity at work or school reduced by half or more because of your headaches? (Do not include days you counted in question 1 where you missed work or school.)

_____ 3. On how many days in the last 3 months did you not do household work (such as housework, home repairs and maintenance, shopping, caring for children and relatives) because of your headaches?

_____ 4. How many days in the last 3 months was your productivity in household work reduced by half of more because of your headaches? (Do not include days you counted in question 3 where you did not do household work.)

_____ 5. On how many days in the last 3 months did you miss family, social or leisure activities because of your headaches?

_____ Total (Questions 1–5)

_____ A. On how many days in the last 3 months did you have a headache? (If a headache lasted more than 1 day, count each day.)

_____ B. On a scale of 0–10, on average how painful were these headaches? (Where 0 = no pain at all, and 10 = pain as bad as it can be.)

Scoring: *After you have filled out this questionnaire,*

add the total number of days from questions 1–5 (ignore A and B)

MIDAS Grade	Definition	MIDAS Score
I	Little or no disability	0–5
II	Mild disability	6–10
III	Moderate disability	11–20
IV	Severe disability	21+

Please give the completed form to your clinician.

FIGURE 9A.1   **The Migraine Disability Assessment Test.** *This survey was developed by Richard B. Lipton, MD, Professor of Neurology, Albert Einstein College of Medicine, New York, NY, and Walter F. Stewart, MPH, PhD, Associate Professor of Epidemiology, Johns Hopkins University, Baltimore, MD.*

References

1. Menken M, Munsat TL, Toole JF. The global burden of disease study: implications for neurology. *Arch Neurol.* 2000;57 (3):418–420.

2. Stovner LJ, Hagen K, Jensen R, Katsarava Z, Lipton R, Scher A. The global burden of headache: a documentation of headache prevalence and disability worldwide. *Cephalalgia.* 2007;27: 193–210.

3. Bigal ME, Sheftell FD. Chronic daily headache and its subtypes. *CONTINUUM: Lifelong Learning in Neurology.* 2006;12 (6):133–152.

4. Headache Classification Committee of the International Headache Society. The international classification of headache disorders, 3rd edition (beta version). *Cephalalgia.* 2013;33: 629–808.

5. Headache Classification Committee of the International Headache Society. Classification and diagnostic criteria for headache disorders, cranial neuralgias and facial pain. *Cephalalgia.* 1988;8(Suppl 7):1–96.

6. Headache Classification Subcommittee of the International Headache Society. The international classification of headache disorders: 2nd edition. *Cephalalgia.* 2004;24(Suppl 1):1–160.

7. Headache Classification Committee, Olesen J, Bousser MG, Diener HC, et al. New appendix criteria open for a broader concept of chronic migraine. *Cephalalgia.* 2006;26:742–746.

8. Cady RK, Gutterman D, Saiers JA, Beach ME. Responsiveness of non-IHS migraine and tension-type headache to sumatriptan. *Cephalalgia.* 1997;17:588–590.

9. Lipton RB, Stewart WF, Cady R, Hall C, O'Quinn S, Kuhn T, et al. Wolff Award. Sumatriptan for the range of headaches in migraine sufferers: results of the Spectrum Study. *Headache.* 2000;40:783–791.

10. Goadsby PJ. The "Act when Mild" (AwM) study: a step forward in our understanding of early treatment in acute migraine. *Cephalagia.* 2008;28(Suppl 2):36–41.

11. Freitag FG, Finlayson G, Rapoport AM, Elkind AH, Diamond ML, Unger JR, et al. AIMS Investigators. Effect of pain intensity and time to administration on responsiveness to almotriptan: results from AXERT 12.5 mg Time Versus Intensity Migraine Study (AIMS). *Headache.* 2007;47(4):519–530.

12. Cady R, Schreiber C, Farmer K, Sheftell F. Primary headaches: a convergence hypothesis. *Headache.* 2002;42:204–216.

13. Goadsby PJ, Hargreaves RJ. Refractory migraine and chronic migraine: pathophysiological mechanisms. *Headache.* 2008;48: 1399–1405.

14. Lipton RB, Bigal ME, Diamond M, Freitag F, Reed ML, Stewart WF. Migraine prevalence, disease burden, and the need for preventive therapy. *Neurology.* 2007;68:343–349.

15. Lipton RB, Stewart WF, Diamond S, Diamond ML, Reed M. Prevalence and burden of migraine in the United States: data from the American Migraine Study II. *Headache.* 2001;41(7):646–657.

16. Takeshima T, Ishizaki K, Fukuhara Y, Ijiri T, Kusumi M, Wakutani Y, et al. Population-Based Door-to-Door Survey of Migraine in Japan: The Daisen Study. *Headache.* 2004;44:8–19.

17. Schwedt TJ. Chronic migraine. *BMJ.* 2014;348:1416.

18. Bigal ME, Lipton RB. Migraine at all ages. *Curr Pain Headache Rep.* 2006;10(3):207–213.

19. Bigal ME, Lipton RB. Concepts and mechanisms of migraine chronification. *Headache.* 2008;48:7–15.

20. Manack A, Buse DC, Serrano D, Turkel CC, Lipton RB. Rates, predictors, and consequences of remission from chronic migraine to episodic migraine. *Neurology.* 2011;76(8):711–718.

21. Chakravarty A. How triggers trigger acute migraine attacks: a hypothesis. *Med Hypotheses.* 2010;74(4):750–753.

22. Aurora SK, Wilkinson F. The brain is hyperexcitable in migraine. *Cephalalgia.* 2007;27(12):1442–1453.

23. Welch KM. Contemporary concepts of migraine pathogenesis. *Neurology.* 2003;61(8 Suppl 4):S2–S8.

24. Silberstein SD, Lipton RB, Goadsby PJ. *Headache in Clinical Practice.* 2nd ed. London, UK: Martin Dunitz, Ltd; 2002.

25. Graham JR. The natural history of migraine: some observations and a hypothesis. *Trans Am Clin Climatol Assoc.* 1953;64:61–74.

26. Terwindt GM, Ophoff RA, van Eijk R, Vergouwe MN, Haan J, Frants RR, et al. Involvement of the CACNA1A gene containing region on 19p13 in migraine with and without aura. *Neurology.* 2001;56:1028–1032.

27. Ghosh J, Pradhan S, Mittal B. Multilocus analysis of hormonal, neurotransmitter, inflammatory pathways and genome-wide associated variants in migraine susceptibility. *Eur J Neurol.* 2014;21(7):1011–1020.

28. Lance JW. Current concepts of migraine pathogenesis. *Neurology.* 1993;43(6 Suppl 3):S11–S15.

29. Berger SL, Kouzarides T, Shiekhattar R, Shilatifard A. An operational definition of epigenetics. *Genes Dev.* 2009;23(7):781–783.

30. Bird A. Perceptions of epigenetics. *Nature.* 2007;447(7143):396–398.

31. Alberini CM. Transcription factors in long-term memory and synaptic plasticity. *Physiol Rev.* 2009;89(1):121–145.

32. Strachan T, Read AP. *Human Molecular Genetics.* 2nd ed. New York, NY: Wiley-Liss; 1999:Chapter 8, Human gene expression. <www.ncbi.nlm.nih.gov/books/NBK7588/>.

33. Robertson KD, Jones PA. DNA methylation: past, present and future directions. *Carcinogenesis.* 2000;21(3):461–467.

34. Cooper GM. *The Cell: A Molecular Approach.* 2nd ed. Sunderland, MA: Sinauer Associates; 2000:Regulation of Transcription in Eukaryotes. <www.ncbi.nlm.nih.gov/books/NBK9904/>.

35. Rodríguez-Navarro S. Insights into SAGA function during gene expression. *EMBO Rep.* 2009;10(8):843–850.

36. Bell JT, Spector TD. A twin approach to unraveling epigenetics. *Trends Genet.* 2011;27(3):116–125.

37. Hucho T, Levine JD. Signaling pathways in sensitization: toward a nociceptor cell biology. *Neuron.* 2007;55(3):365–376.

38. Denk F, McMahon SB. Chronic pain: emerging evidence for the involvement of epigenetics. *Neuron.* 2012;73(3):435–444.

39. Gervil M, Ulrich V, Kaprio J, Olesen J, Russell MB. The relative role of genetic and environmental factors in migraine without aura. *Neurology.* 1999;53:995–999.

40. Mulder EJ, Van Baal C, Gaist D, Olesen J, Russell MB. Genetic and environmental influences on migraine: a twin study across six countries. *Twin Res.* 2003;6(5):422–431.

41. Scher AI, Lipton RB, Stewart W. Risk factors for chronic daily headache. *Curr Pain Headache Rep.* 2002;6:486–491.

42. Buse DC, Manack A, Serrano D, Turkel C, Lipton RB. Sociodemographic and comorbidity profiles of chronic migraine and episodic migraine sufferers. *J Neurol Neurosurg Psychiatry.* 2010;81(4):428–432.

43. Ashina S, Lipton RB, Bigal ME. Treatment of comorbidities of chronic daily headache. *Curr Treat Options Neurol.* 2008;10 (1):36–43.

44. Seng EK, Holroyd KA. Psychiatric comorbidity and response to preventative therapy in the treatment of severe migraine trial. *Cephalalgia.* 2012;32(5):390–400.

45. Xplore, Inc. BrainyQuote®. Hippocrates quotes. <www.brainyquote.com/quotes/authors/h/hippocrates.html>.

46. Tamblyn R, Eguale T, Huang A, Winslade N, Doran P. The incidence and determinants of primary nonadherence with prescribed medication in primary care: a cohort study. *Ann Intern Med.* 2014;160(7):441–450.

47. Ahn AH. Why does increased exercise decrease migraine? *Curr Pain Headache Rep.* 2013;17(12):379.

48. Köseoglu E, Akboyraz A, Soyuer A, Ersoy AO. Aerobic exercise and plasma beta endorphin levels in patients with migrainous headache without aura. *Cephalalgia*. 2003;23(10):972–976.

49. Jahromi SR, Abolhasani M, Meysamie A, Togha M. The effect of body fat mass and fat free mass on migraine headache. *Iran J Neurol*. 2013;12(1):23–27.

50. Volek JS, Phinney SD. *The Art and Science of Low Carbohydrate Living. An Expert Guide to Making the Life-Saving Benefits of Carbohydrate Restriction Sustainable and Enjoyable*. Miami, FL: Beyond Obesity LLC; 2011.

51. Sargent JD, Green EE, Walters ED. The use of autogenic feedback training in a pilot study of migraine and tension headaches. *Headache*. 1972;12:120–125.

52. Cady RK, Farmer K. *The adherence principle-empowering your healthy self*. Hamilton, Ontario: Baxter Publishing; 2011.

53. Kolb LC. A neuropsychological hypothesis explaining posttraumatic stress disorders. *Am J Psychiatry*. 1987;144(8):989–995.

54. Hofmann SG, Asnaani A, Vonk IJ, Sawyer AT, Fang A. The efficacy of cognitive behavioral therapy: a review of meta-analyses. *Cognit Ther Res*. 2012;36(5):427–440.

55. Beck AT. *Depression: Clinical, Experimental and Theoretical Aspects*. New York, NY: Harper & Row; 1967.

56. Buse DC, Andrasik F. Behavioral medicine for migraine. *Neurol Clin*. 2009;27(2):445–465.

57. Yang CP, Chang MH, Liu PE, Li TC, Hsieh CL, Hwang KL, et al. Acupuncture versus topiramate in chronic migraine prophylaxis: a randomized clinical trial. *Cephalalgia*. 2011; 31(15):1510–1521.

58. Linde K, Allais G, Brinkhaus B, Manheimer E, Vickers A, White AR. Acupuncture for migraine prophylaxis. *Cochrane Database Syst Rev*. 2009;(1):CD001218.

59. Wells RE, Bertisch SM, Buettner C, Phillips RS, McCarthy EP. Complementary and alternative medicine use among adults with migraines/severe headaches. *Headache*. 2011;51(7):1087–1097.

60. Bryans R, Descarreaux M, Duranleau M, Marcoux H, Potter B, Ruegg R, et al. Evidence-based guidelines for the chiropractic treatment of adults with headache. *J Manipulative Physiol Ther*. 2011;34(5):274–289.

61. Bronfort G, Assendelft WJ, Evans R, Haas M, Bouter L. Efficacy of spinal manipulation for chronic headache: a systematic review. *J Manipulative Physiol Ther*. 2001;24(7):457–466.

62. Posadzki P, Ernst E. Spinal manipulations for the treatment of migraine: a systematic review of randomized clinical trials. *Cephalalgia*. 2011;31(8):964–970.

63. Chaibi A, Tuchin PJ, Russell MB. Manual therapies for migraine: a systematic review. *J Headache Pain*. 2011;12(2):127–133.

64. Aurora S, Dodick D, Turkel C, DeGryse RE, Silberstein SD, Lipton RB, et al. OnabotulinumtoxinA for treatment of chronic migraine: results from the double blind, randomized, placebo-controlled phase of the PREEMPT 1 trial. *Cephalalgia*. 2010;30:793–803.

65. Diener H, Dodick D, Aurora S, Turkel CC, DeGryse RE, Lipton RB, et al. OnabotulinumtoxinA for treatment of chronic migraine: results from the double blind, randomized, placebo-controlled phase of the PREEMPT 2 trial. *Cephalalgia*. 2010;30 (7):804–814.

66. Dodick DW, Turkel CC, DeGryse RE, Aurora SK, Silberstein SD, Lipton RB, et al. OnabotulinumtoxinA for treatment of chronic migraine migraine: Pooled results from double blinded, randomized, placebo-controlled phases of the PREEMPT clinical program. *Headache*. 2010;50:921–936.

67. Royle P, Cummins E, Walker C, Chong S, Kandala N, Waugh N. Botulinum toxin type A for the prophylaxis of headaches in adults with chronic migraine: a single technology assessment. *Warwick Evidence*. 2011;<www2.warwick.ac.uk/fac/med/about/centres/warwickevidence/research/publications2011/>

68. Antonaci F, Dumitrache C, De Cillis I, Allena M. A review of current European guidelines for migraine. *J Headache Pain*. 2010;11:13–19.

69. Cady RK, Schreiber CP, Porter JA, Blumenfeld AM, Farmer KU. A multi-center double-blind pilot comparison of onabotulinumtoxinA and topiramate for the prophylactic treatment of chronic migraine. *Headache*. 2011;51(1):21–32.

70. Rothrock J, Bloudek L, Houle T, Andress-Rothrock D, Hanlon C, Varon S. Real-world economic impact of onabotulinumtoxinA in patients with chronic migraine [abstract]. *Neurology*. 2012;78 (Suppl 1):P03233.

71. Silberstein S, Lipton R, Dodick D, Freitag F, Mathew N, Brandes J, et al. Topiramate treatment of chronic migraine: a randomized, placebo-controlled trial of quality of life and other efficacy measures. *Headache*. 2009;49(8):1153–1162.

72. Lipton RB, Silberstein S, Dodick D, Cady R, Freitag F, Mathew N, et al. Topiramate intervention to prevent transformation of episodic migraine: the topiramate INTREPID study. *Cephalalgia*. 2011;31(1):18–30.

73. Aurora SK, Silberstein SD, Kori SH, Tepper SJ, Borland SW, Wang M, et al. MAP0004, orally inhaled DHE: a randomized, controlled study in the acute treatment of migraine. *Headache*. 2011;51(4):507–517.

74. Mueller L, Gallagher IM, Ciervo CA. Methylergonovine maleate as a cluster headache prophylactic: a study in review. *Headache*. 1997;37:437–442.

75. Jarrett RB, Schaffer M, McIntire D, Witt-Browder A, Kraft D, Risser RC. Treatment of atypical depression with cognitive therapy or phenelzine: a double-blind, placebo-controlled trial. *Arch Gen Psychiatry*. 1999;56(5):431–437.

76. D'Andrea G, Nordera GP, Perini F, Allais G, Granella F. Biochemistry of neuromodulation in primary headaches: focus on anomalies of tyrosine metabolism. *Neurol Sci*. 2007;28(Suppl 2):S97-S95

77. Cady R, Nett R, Dexter K, Freitag F, Beach ME, Manley HR. Treatment of chronic migraine: a 3-month comparator study of naproxen sodium vs SumaRT/Nap. *Headache*. 2014;54(1):80–93.

78. Silberstein SD, Dodick DW, Saper J, Huh B, Slavin KV, Sharan A, et al. Safety and efficacy of peripheral nerve stimulation of the occipital nerves for the management of chronic migraine: results from a randomized, multicenter, double-blinded, controlled study. *Cephalalgia*. 2012;32(16):1165–1179.

79. Silberstein SD, Da Silva AN, Calhoun AH, Grosberg BM, Lipton RB, Cady RK, et al. Poster presented at the 56th Annual Scientific Meeting of the American Headache Society. June 26–29, 2014 in Los Angeles, CA.

80. Katsarava Z, Buse DC, Manack AN, Lipton RB. Defining the differences between episodic migraine and chronic migraine. *Curr Pain Headache Rep*. 2012;16:86–92.

81. Robbins L. Long acting opioids for refractory chronic migraine. Pract Pain management, October 9, 2009.

82. Saper JR, Lake III AE, Hamel RL, Lutz TE, Branca B, Sims DB, et al. Daily scheduled opioids for intractable head pain: long-term observations of a treatment program. *Neurology*. 2004; 62(10):1687–1694.

83. Tepper SJ. Medication-overuse headache. *Continuum (Minneap Minn)*. 2012;18(4):807–822.

84. Holland S, Silberstein SD, Freitag F, Dodick DW, Argoff C, Ashman E. Quality standards subcommittee of the American Academy of Neurology and the American Headache Society. Evidence-based guideline update: NSAIDs and other complementary treatments for episodic migraine prevention in adults: report of the Quality Standards Subcommittee of the American Academy of Neurology and the American Headache Society. *Neurology*. 2012;78(17):1346–1353.

85. Kelley NE, Tepper DE. Rescue therapy for acute migraine, Part 3: Opioids, NSAIDs, steroids, and post-discharge medications. *Headache*. 2012;52(3):467–482.

86. Schaerlinger B, Hickel P, Etienne N, Guesnier L, Maroteaux L. Agonist actions of dihydroergotamine at 5-HT2B and 5-HT2C receptors and their possible relevance to antimigraine efficacy. *Br J Pharmacol*. 2003;140(2):277–284.

87. Kori S, Zhang J, Kellerman D, Armer T, Goadsby P. Sustained pain relief with dihydroergotamine in migraine is potentially due to persistent binding to 5-ht(1b) and 5-ht(1d) receptors. Paper presented at AHS 54th Annual Meeting; June 2012; Los Angeles, CA.

88. Chang D, Kellermann S, Kori S, Meyer J, Zhou J, Armer T. Acute inhaled safety of MAP0004: studies in healthy volunteers, smokers, and patients with asthma [poster]. 54th Annual Scientific Meeting of the American Headache Society; 2012 June 21–24; Los Angeles, CA.

89. Cady RK, Wendt JK, Kirchner JR, Sargent JD, Rothrock JF, Skaggs Jr. H. Treatment of acute migraine with subcutaneous sumatriptan. *JAMA*. 1991;5265(21):2831–2835.

90. The Subcutaneous Sumatriptan International Study Group. Treatment of migraine attacks with sumatriptan. *N Engl J Med*. 1991;325:316–321.

91. Ferrari MD, Roon KI, Lipton RB, Goadsby PJ. Oral triptans (serotonin 5-HT(1B/1D) agonists) in acute migraine treatment: a meta-analysis of 53 trials. *Lancet*. 2001;358(9294):1668–1675.

92. Olesen J, Goadsby PJ, Ramadan NM, Tfelt-Hansen P, Welch KMA, eds. *The Headaches*. 3rd ed. Philadelphia, PA: Lippincott, Williams & Wilkins; 2006.

93. Evers S, Afra J, Frese A, Goadsby PJ, Linde M, May A, et al. European Federation of Neurologic Societies. EFNS guideline on the drug treatment of migraine-revised report of an EFNS task force. *Eur J Neurol*. 2009;16:968–981.

94. GlaxoSmithKline. The sumatriptan/naratriptan/Treximet pregnancy registry. Interim report. Wilmington, NC: Kendle International Inc; 2009. <http://pregnancyregistry.gsk.com/sumatriptan.html>.

95. Ressel G. AAP updates statement for transfer of drugs and other chemicals into breast milk. American Academy of Pediatrics. *Am Fam Physician*. 2002;65(5):979–980.

96. Jonsson P, Hedenrud T, Linde M. Epidemiology of medication overuse headache in the general Swedish population. *Cephalalgia*. 2011;31(9):1015–1022.

97. Bigal ME, Borucho S, Serrano D, Lipton RB. The acute treatment of episodic and chronic migraine in the USA. *Cephalalgia*. 2009;29(8):891–897.

98. Kahn K. Cambia® (diclofenac potassium for oral solution) in the management of acute migraine. *US Neurology*. 2011;7(2):139–143.

99. Diener HC, Hartung E, Chrubasik J, Evers S, Schoenen J, Eikermann A, et al. Study group. A comparative study of oral acetylsalicylic acid and metoprolol for the prophylactic treatment of migraine: a randomized, controlled, double-blind, parallel group phase III study. *Cephalalgia*. 2001;21:120–128.

100. Scher AI, Lipton RB, Stewart WF, Bigal M. Patterns of medication use by chronic and episodic headache sufferers in the general population: results from the frequent headache epidemiology study. *Cephalalgia*. 2010;30:321–328.

101. Suthisisang CC, Poolsup N, Suksomboon N, Lertpipopmetha V, Tepwitukgid B. Meta-analysis of the efficacy and safety of naproxen sodium in the acute treatment of migraine. *Headache*. 2010;50(5):808–818.

102. Cady R, O'Carroll P, Dexter K, Freitag F, Shade CL. SumaRT/Nap vs naproxen sodium in treatment and disease modification of migraine: a pilot study. *Headache*. 2014;54(1):67–79.

103. Duarte C, Dunaway F, Turner L, Aldag J, Frederick R. Ketorolac versus meperidine and hydroxyzine in the treatment of acute migraine headache: a randomized, prospective, double-blind trial. *Ann Emerg Med*. 1992;21(9):1116–1121.

104. Pfaffenrath V, Fenzl E, Bregman D, Färkkila M. Intranasal ketorolac tromethamine (SPRIX®) containing 6% of lidocaine (ROX-828) for acute treatment of migraine: safety and efficacy data from a Phase II clinical trial. *Cephalalgia*. 2012;32(10):766–777.

105. Mathew NT, Kailasam J, Meadors L, Chernyschev O, Gentry P. Intravenous valproate sodium (depacon) aborts migraine rapidly: a preliminary report. *Headache*. 2000;40(9):720–723.

106. Kelley NE, Tepper DE. Rescue therapy for acute migraine, part 2: neuroleptics, antihistamines, and others. *Headache*. 2012;52(2):292–306.

107. Lipton RB, Buse DC, Fanning KM, Serrano D, Reed ML. Suboptimal treatment of episodic migraine may mean progression to chronic migraine. Poster presented at the 2013 IHC/AHS Congress; Boston, MA, June 26, 2013; Abstract LB02.

108. Farinelli I, Dionisi I, Martelletti P. Rehabilitating chronic migraine complicated by medication overuse headaches: how can we prevent migraine relapse? *Intern Emerg Med*. 2011;6(1):23–28.

109. Pini LA, Cicero AF, Sandrini M. Long-term follow up of patients treated for chronic headache with analgesic oveuse. *Cephalalgia*. 2001;21:878–883.

110. Krymchantowsky AV. Overuse of symptomatic medications among chronic (transformed) migraine patients. *Arq Neuropsiquiatr*. 2003;61:43–47.

111. Diener HC, Bussone G, Van Oene J, Lahaye M, Schwalen S, Goadsby P; TOPMAT-MIG-201(TOP-CHROME) Study Group. Topiramate reduces headache days in chronic migraine: a randomized, double-blind, placebo-controlled study. *Cephalalgia*. 2007;27:814–823.

112. Aurora S, Winner P, Freeman M, Spierings EL, Heiring JO, DeGryse RE, et al. OnabotulinumtoxinA for treatment of chronic migraine: pooled analyses of the 56-week PREEMPT clinical program. *Headache*. 2011;51(9):1358–1373.

113. Lipton RB, Kolodner K, Bigal ME, Valade D, Láinez MJ, Pascual J, et al. Validity and reliability of the Migraine-Treatment Optimization Questionnaire. *Cephalalgia*. 2009;29(7):751–759.

114. Demirkaya S, Vural O, Dora B, Topçuoğlu MA. Efficacy of intravenous magnesium sulfate in the treatment of acute migraine attacks. *Headache*. 2001;41(2):171–177.

115. Bigal ME, Bordini CA, Tepper SJ, Speciali JG. Intravenous magnesium sulphate in the acute treatment of migraine without aura and migraine with aura. A randomized, double-blind, placebo-controlled study. *Cephalalgia*. 2002;22(5):345–353.

10

Gender-Based Issues in Headache

Merle L. Diamond

Diamond Headache Clinic, Chicago, Illinois, USA

INTRODUCTION

This chapter examines the gender differences in migraine, and the role of hormones throughout the reproductive cycle in women. Previous studies have reported on the gender differences in migraine and other pain disorders (rheumatoid arthritis, fibromyalgia, and temporomandibular joint [TMJ] pain) (Figure 10.1).[1–4] In pre-adolescents, the prevalence is equal in males and females, but increases in girls about the time of menarche.[5] The lifetime prevalence of migraine in females has been reported to range between 16% and 32% in population-based studies.[6,7] In studies utilizing the International Classification of Headache Disorders (ICHD) diagnostic criteria, the estimated prevalence was 17% for females and 7.6% for males.[8] Studies have consistently shown that migraine occurs more frequently in females than males.[9] A recent report, utilizing data from the American Migraine Prevalence and Prevention (AMPP) study of over 160,000 subjects aged 12 years or older, again demonstrated higher prevalence of migraine in women at all ages.[10]

In female migraineurs, approximately 60% relate a menstrual relationship to their migraine attacks.[11] Migraine prevalence in females increases about the time of menarche, and peaks before menopause.[12,13] These headaches are impacted by changes in hormonal levels, whether in menses, pregnancy, contraception, or menopause. Martin and Behbehani have described the various events throughout a woman's life as a "different hormonal milieu" which greatly impacts those with migraine headaches.[14]

MENSTRUAL MIGRAINE

The earliest description of menstrual migraine was written by Johannes Van der Linden, in *De Hemicrania*

Menstrua.[15] In this treatise, Van der Linden describes the headaches experienced by his patient, the Marchioness of Brandenburg. The patient complained of unilateral headache associated with nausea and vomiting, which occurred monthly during her menstrual flow. The treatment of these headaches often focused on hormonal interventions. In 1919, the successful use of oral pituitary extract in menstrual headaches was reported.[16] During the 1930s, the use of a gonadotropin agent, emmenine, was reported as beneficial in migraine therapy.[17] About the same time, oestrin, an ovarian follicular hormone, was successfully injected as a treatment for migraine.[18] During the same decades, small doses of a gonadotropic substance extracted from pregnancy urine was reported as beneficial in menstrual migraine therapy.[19]

Somerville, during the 1970s, undertook a major analysis of the role of reproductive hormones in migraine[20] (Figure 10.2). Somerville noted that:

- Migraine occurrence was demonstrated during or after simultaneous decrease of estrogen and progesterone
- Migraine, not menstruation, was delayed by premenstrually administered estrogen[21]
- In contrast, progesterone delayed menstruation but not migraine
- Estrogen withdrawal did trigger menstrual migraine.

In another study, Somerville noted that patients needed to have several days of exposure to high estrogen levels before a drop in estrogen would precipitate migraine.[22]

Research has continued into the role of hormones in the development of menstrual migraine. In Britain, a study was undertaken to examine the prevalence of menstrual migraine.[23] In this investigation, 55 women aged between 17 and 50 years demonstrated a

S. Diamond, R. K. Cady, M. L. Diamond & V. T. Martin (Eds):
Headache and Migraine Biology and Management.

DOI: http://dx.doi.org/10.1016/B978-0-12-800901-7.00010-0

significant increase in migraine attacks around the first day of the menstrual period. No increase was noted at the time of ovulation. A restricted definition of "menstrual migraine" was used — migraine starting 2 days prior to and 3 days after onset of menstruation. Of the 55 women, only 4 were considered to be experiencing menstrual migraine. One-third of the group had headaches at menstruation but also at other times; their

headaches were described as menstrually related. Another one-third had headaches throughout the month with no increase at menstruation, and the remaining women had no menstrually related headaches. The authors concluded that hormonal treatments would be beneficial for those patients with menstrual migraine and those with menstrually related headaches, but not for other headache disorders.

Menses has been considered as the major factor in the risk and persistence of migraine.[24] Other studies have noted that the duration of menstrual migraine is longer than that of other headaches, and the response to therapy is less than for other forms of headache.[25]

The International Classification of Headache Disorders (ICHD-III beta) has defined pure menstrual migraine as attacks occurring over a 5-day period, starting 2 days before the menstrual period and continuing to the third day of the period, and at no other times of the cycle.[26] Diagnosis should be established for this pattern in two of three cycles. According to the criteria, pure menstrual migraine (A1.1.1) only occurs perimenstrually and not at other times of the month. This definition concurs with the findings of the 1990 British study by MacGregor and colleagues.[23] For those patients with headaches throughout the month

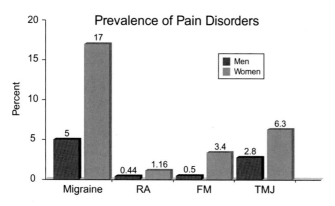

FIGURE 10.1 Prevalence of pain disorders.[1–4] *Reproduced with permission of Vincent Martin, MD, University of Cincinnati Medical Center.*

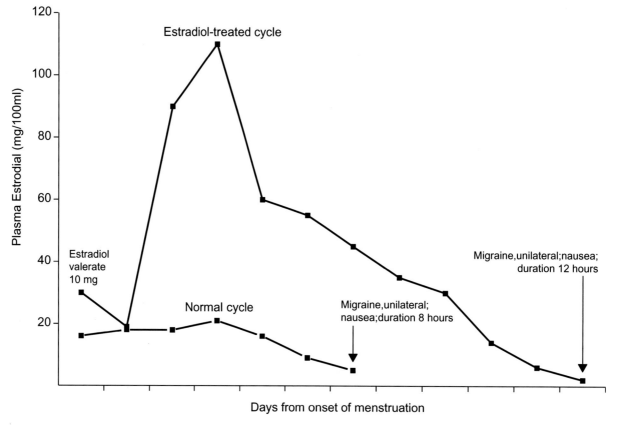

FIGURE 10.2 Estrogen-treated cycle (continuous line) and normal cycle (dashed line) in the same patient. The onset of migraine is postponed until the estrogen level falls. *Adapted from Somerville.[35]*

in addition to those occurring at menses, the diagnosis of menstrually related migraine (A1.1.2) is used. The criteria, from the ICHD-II edition, were studied by Marcus and coworkers.[27] These investigators concluded that in order to firmly establish the diagnosis, at least 2 months' of headache diaries should be evaluated. Inaccuracies were noted when relying on patient self-reporting, which often mislabeled their headaches as pure menstrual migraine.

Prevalence of menstrually associated migraine has been reported in population-based studies at between 20% and 60%.[28,29] Studies have established prevalence according to the definition used, and the study cohorts — whether population-based or clinic-based.[12,23,30] As noted in the study by Marcus et al.,[27] the results will also be influenced by the method of obtaining information — whether by interview (self-reporting) or diary analysis.[31]

Some patients will report a relationship between migraine attacks and ovulation. However, this link has not been confirmed by any epidemiological studies.[23,32,33] Studies have shown that the levels of ovarian hormones for the 7 days before menses and during one mid-cycle day are similar in subjects with menstrual migraine, those with non-menstrual migraine, and controls.[34]

Estrogen withdrawal is believed to be the trigger of menstrually associated migraine. This withdrawal occurs during the late luteal phase of the menstrual cycle. It also occurs during the placebo week for those patients using combined oral contraceptives.[21,22,31,35–37] Increased risk for perimenstrual migraine may be due to the impact of estrogen on pain-processing networks and vascular endothelium, which is implicated in the pathophysiology of migraine. The presence or absence of estrogen influences the serotonin-producing neurons, and we know that serotonin is involved in the pathophysiology of migraine.[38]

Later studies have supported the estrogen withdrawal theory by administering estrogen prior to menses. For example, Pradalier and colleagues treated migraine patients with transdermal estradiol patches, 100 mg, during the perimenstrual period and noted efficacy in preventing menstrual migraine.[39] The 50-mg patch was not effective. In a more recent study, Calhoun noted a 77% reduction in headache days during a cycle in a group of subjects who received conjugated estrogens, 0.9 mg, during the placebo week of their oral contraceptives.[40] It does appear that menstrual migraine was prevented by maintaining the serum estradiol levels above the 45 pg/mL range during the menstrual period.[14]

Prostaglandin release has also been implicated in menstrual migraine.[41] During perimenstruation, prostaglandins are released by a shedding endometrium due to progesterone withdrawal. In one investigation, headaches similar to migraine were induced in non-migraineurs by injecting prostaglandin E2 into these subjects.[42] Also, prostaglandin inhibitors, such as the non-steroidal anti-inflammatory agents (NSAIDs), have demonstrated efficacy in menstrually associated migraine.[43]

In the pathogenesis of menstrual migraine, progesterone has also been implicated. In a study which evaluated urinary progesterone metabolite levels, headache outcome measures were worse during the mid-luteal period when serum progesterone levels were moderately elevated.[44] These findings could suggest that low to moderate levels of progesterone may help prevent menstrually associated migraine, where higher levels of progesterone may trigger a migraine attack.

Decreased magnesium levels have also been implicated in menstrual migraine. In a study by Mauskop and colleagues, a deficiency in ionized magnesium was noted in 45% of menstrual migraine attacks, whereas this deficiency was only observed in 15% of non-menstrually related attacks.[45] This finding was confirmed by Facchinetti and coworkers, who administered oral magnesium during the latter 15 days of the menstrual cycle and were successful in preventing menstrual migraine.[46]

Treatment

Various studies of the treatments for menstrual and menstrually related migraine are listed in Table 10.1.[43,46–55] Because the clinical presentation of menstrual migraine is similar to other forms of migraine, acute interventions are similar. Dihydroergotamine (DHE), analgesics, and anti-emetics are all utilized for acute symptoms of menstrually associated migraine.

As a group, the triptans have been investigated in the treatment of menstrual migraine. In an early placebo-controlled study, sumatriptan 6 mg SC demonstrated significant benefit compared with placebo at 1 hour, and relief was maintained over 24 hours.[47] Oral sumatriptan in a randomized placebo-controlled study concluded that both the 50-mg and the 100-mg dose of the active drug was more effective than placebo.[48] For both menstrual and non-menstrual migraine, rizatriptan showed efficacy in two randomized trials.[49] At 2 hours, the subjects receiving oral rizatriptan, 10 mg, reported relief of moderate to severe headache. Zolmitriptan, in doses of 1.25–5 mg, was investigated in a placebo-controlled study of a single menstrual migraine attack.[50]

At a dose of 2.5 mg, oral naratriptan was evaluated in a randomized double-blind placebo-controlled study and it was found, in subjects receiving active

TABLE 10.1   Treatment of Menstrual and Menstrually Related Migraine

Medication	Dosage	Reference
Sumatriptan (SC)	6 mg	Solbach and Waymer [47]
Sumatriptan (oral)	50 mg or 100 mg	Nett et al. [48]
Rizatriptan (oral)	10 mg	Silberstein et al. [49]
Zolmitriptan (oral)	1.25–5 mg	Loder et al. [50]
Naratriptan (oral)	2.5 mg	Massiou et al. [51]
Frovatriptan (oral)	2.5 mg	MacGregor [52]
Eleptriptan (oral)	40–80 mg	Bhambri et al. [53]
Naproxen sodium (oral)	550 mg	Sances et al. [43]
Sumatriptan/naproxen	85 mg/500 mg	Martin et al. [54]
Dihydroergotamine	45 mg	Silberstein and Bradley [55]
Magnesium	360 mg	Facchinetti et al. [46]

drug, that a pain-free response was maintained from 4 to 24 hours.[51] Frovatriptan has been studied in several studies, and has demonstrated consistent results.[56] An optimal dose of frovatriptan, 2.5 mg, was cited for achieving and maintaining efficacy. At 24 and 48 hours after dosing, frovatriptan has also demonstrated significantly lower relapse rates and sustained pain-free response.[52] In comparison to other triptans, frovatriptan has a longer elimination half-life of 26 hours, which facilitates a longer duration of action.

Eletriptan has also been the focus of investigations into the treatment of menstrually associated migraine as well as headache outside of the menstrual period. In a recent report, data from five randomized controlled studies were reviewed.[53] The review included the responses of 2216 women. At 2 hours post-dose, headache recurrence was higher in the placebo group as opposed to those receiving 40 mg or 80 mg of frovatriptan. These authors observed that during the menstrual period recurrence of the migraine was more likely 2–24 hours post-dose, and sustained relief of nausea was less likely during that interval. This finding is representative of the observation that attacks of menstrual migraine are less responsive to treatment than is migraine occurring in non-menstrual intervals.

Because the occurrence of menstrually associated migraine is predictable, short-term preventive therapy can be undertaken.[57] The NSAIDs have been used in the prevention of migraine, with naproxen often used for menstrually associated migraine.[43] In that placebo-controlled study, Sances and colleagues investigated the use of naproxen sodium, 550 mg, versus placebo in 35 women. Treatment was started from 7 days premenstrual period through day 6 of the period, for 3 months. Headache severity and duration were improved, as well as the number of headache days and reduced use of analgesics.

Other NSAIDs used for prevention of these attacks include ketoprofen, mefenamic acid, flurbiprofen, and meclofenamate. In a recent study, a combined formulation of naproxen and sumatriptan versus placebo was investigated in 621 subjects with menstrual migraine and dysmenorrhea.[54] Non-headache symptoms associated with the menstrual period were also noted, including fatigue, bloating, and irritability, as well as menstrual pain symptoms (abdominal and back pain). The sumatriptan–naproxen sodium combination demonstrated superiority over placebo in relieving migraine symptoms, as well as tiredness, irritability, and abdominal and back pain. The authors concluded that this formulation could be beneficial for those experiencing menstrual migraine and dysmenorrhea.

The triptans have also been investigated for menstrual migraine prevention. A review of various triptans used in short-term prevention of menstrual migraine was presented in a recent article.[58] The authors concluded that the frequency, severity, and adverse events demonstrated that frovatriptan, 2.5 mg BID, and zolmitriptan, 2.5 mg TID are preferred for the treatment for menstrual migraine.

The efficacy of the ergotamines in migraine therapy led to an investigation of DHE, administered as a nasal spray.[55] The active drug was compared with placebo in 45 subjects with menstrual migraine. The treatment was started 2 days prior to the anticipated onset of the menstrual period, and continued for 6 days. DHE was beneficial in decreasing the severity of menstrual migraine.

The use of estrogens has also been investigated in menstrual migraine prevention. This treatment is discussed in the next section. Long-term prevention of menstrual migraine often involves the use of drugs used in migraine (other than menstrual migraine) prophylaxis. Agents that suppress ovulation and menses have been utilized, including danazol, tamoxifen, and gonadotropin-releasing hormone.[57] The number of subjects is small, and the long-term efficacy has not been proven.

Magnesium deficiency has been linked to the occurrence of menstrual migraine. A study involving a small cohort of subjects (20 women) was undertaken comparing placebo and magnesium pyrrolidone carboxylic acid 360 mg per day.[46] Treatment was started on day 15 of the month and continued through the start of the menses, for two periods. On active drug, the total pain index was decreased after 2 months of treatment.

Clinicians will continue to see patients who have undergone elective hysterectomy in order to prevent menstrually associated migraine. The impact of surgical menopause is recognized, as in an Italian study of the prevalence and characteristics of headaches in a large sample of postmenopausal women.[59] It was noted that in those patients who had undergone surgical ovariectomy, the natural course of migraine was worse than in the patients who experienced a physiological menopause. The concept of neuroprotective effects of estrogen in women remains controversial. An analysis of existing literature revealed that these effects depend on the age of the patient, the type of menopause (natural or surgical), and the stage of menopause (timing).[60] The effect on migraine has not been historically proven. Clinically, patients will probably continue their complaints of migraine.

CONTRACEPTION AND MIGRAINE

How does the use of contraceptive drugs affect migraine? Some form of contraception is being used by almost 62% of the 62 million of women in the US between the ages of 15 and 44 years[61]; of these, 64% will use a non-permanent form of contraception, including oral contraceptives, injectable contraceptives, hormonal patches, and the vaginal ring. If we apply data on female migraineurs, then 18% of those women will have experienced a migraine within the prior year, with the highest prevalence in women in their twenties and thirties.[62]

An increase in headaches or initial onset of headache, including migraine, has long been associated with the use of oral contraceptives, which are a combination of estrogen and progesterone. The placebo interval, in which the estrogen-containing agent is withheld, is known to cause an "estrogen-withdrawal"

headache.[37] Also, since the introduction of combined oral contraceptives (COCs) during the 1960s, the amount of estrogen in COCs has been reduced from 100 mg to 10 mg.[63] The use of COCs with high-dose estrogen has been associated with the risk of ischemic stroke. Any patient using oral contraceptives and who is a smoker needs to be counseled and encouraged to stop smoking.

Some healthcare practitioners will prescribe the lower-dose estrogen-containing COCs to their migraine patients in order to stabilize estrogen levels, independent of the contraceptive action.[64] However, data from the longitudinal Women's Health Study revealed that the risk of cardiovascular disease and stroke is greater in those patients with migraine with aura, and the frequency of the aura appears to correlate with an increase in the risk.[65] In that study of 27,798 women over the age of 45, there was a two-fold greater risk of stroke in those who experienced migraine with aura less than once a month. It increased to four-fold when the migraine aura was experienced once weekly.

In a large study involving centers in England and The Netherlands, the use of various combined oral contraceptives was reviewed in consideration of the risk of venous thrombosis.[66] In the review of 25 publications, which reported on 26 studies, the risk of venous thrombosis for COCs was similar in those agents containing 30−35 μg ethinylestradiol and gestodene, desogestrel, cyptoterone acetate, or drospirenone. These agents posed a 50−80% higher risk than COCs containing levonorgestrel. The authors observed that higher doses of the progestogen increased the risk of venous thrombosis.

The risk of thrombotic events did not decrease in the use of the contraceptive vaginal ring (NuvaRing™), which contains a combination of ethinyl estradiol and ethonogestrel.[67] In their review of multiple epidemiologic and database studies, the authors observed that the risk of thromboembolism was increased among users of the vaginal ring when compared with subjects using levonorgestrel-containing combined agents.

In a large population study, the occurrence of venous thrombosis was reviewed in subjects using hormonal contraceptives other than oral preparations.[68] This survey involved all Danish non-pregnant women (1,626,158) from 2001 through 2010. Venous thrombosis events were confirmed in 3434 subjects. In non-users of hormonal contraception, the incidence was 2.1 per 10,000 woman years. For women using transdermal combined contraceptive patches, the incidence of venous thrombosis was 9.7; for users of the vaginal ring, the incidence was 7.8.

The use of oral contraceptives in women with migraine with aura is controversial, and remains a subject of debate. In a very large study, representing

470,000 person-years, with an average follow-up of 26 years, migraine without aura was not associated with increased all-cause mortality risk.[69] The risk was present in those with migraine with aura.

In spite of the debate, COCs have been gaining acceptance as options for preventive treatment of menstrually associated migraine.[62] The newer, lower-dose COCs have reduced the risk of stroke and cardiovascular events, but their use in migraine with aura should be avoided. The use of oral contraceptives may be considered for those female migraineurs who do not have predictable periods and would not benefit from other therapies used during the menstrual period.[57] For some women who note an increase of their headaches during the placebo interval in the month on oral contraceptives, continuing estrogen during that week has been beneficial in preventing their menstrual migraine attacks.[40] In this study, the subjects received ethinylestradiol, 20 µg, for 21 days, followed by 0.9 mg of conjugated equine estrogen daily for 7 days. All 11 women diagnosed with menstrual migraine noted at least a 50% reduction in headache days per cycle.

Prior to initiating therapy with an oral contraceptive, the physician should screen for a history of thrombolic episodes or miscarriage, and a family history of stroke or sudden death. If there is concern, a hypercoagulability profile should be performed. The use of oral contraceptives to prevent menstrually associated migraine should never be undertaken in a patient who smokes.

A pilot study focused on the use of an extended-cycle vaginal ring contraceptive in patients with migraine with aura.[70] Twenty-three subjects completed the study, all diagnosed with menstrually related migraine, and received a transvaginal ring containing 0.120 mg etonogesterel/15 µg ethinyl estradiol. The results noted a decrease from baseline 3.23 migraine auras per month to 0.23 auras per month following treatment for 6 to 12 weeks. The authors also reported 91.3% elimination of menstrually related migraine attacks.

Another option is the use of transdermal estrogen. During the estrogen-free interval, the use of estradiol patches (0.05 mg) was consider suboptimal, and higher doses have shown higher efficacy.[37] In a 2012 report, the authors indicated that transdermal hormonal therapy can be used to improve menstrually related migraine.[71]

A recent study investigated the combined use of oral contraceptives and frovatriptan in menstrual migraine prophylaxis.[72] In this double-blind randomized placebo-controlled study, subjects were started on a 21/7 regimen of combined oral contraceptives that contained levonorgestrel and ethinyl estradiol. Frovatriptan or placebo was prescribed during the hormone-free intervals. Baseline values were compared with the 168-day extended regimen. Daily headache scores decreased during the extended COC treatment. For those subjects on frovatriptan, headache scores were lower during the hormone-free interval as compared with placebo. Withdrawing frovatriptan caused an increase in headache scores, despite the subjects remaining on the COC.

PREGNANCY AND MIGRAINE

Whether or not the patient has a history of menstrual migraine, pregnancy and lactation impact the frequency of migraine. Both of these events suppress ovulation and menses, resulting in changes in the hormonal milieu. For most women, the significantly high levels of estrogen during pregnancy decrease migraine frequency. In a study involving 47 subjects experiencing migraine without aura, improvement was noted by 47% during the first trimester of pregnancy,[73] 83% noticed improvement during the second trimester, and 87% during the third trimester (Figure 10.3). Total remission was reported by 11%, 53%, and 79% during the first, second, and third trimesters, respectively. Women with a prior history of menstrual migraine before the pregnancy had a higher incidence of no improvement in migraine attacks during the pregnancy. Regardless of prior history, migraine attacks occurred within 1 week of delivery in 34% of subjects, and 55% within the first month. Migraine recurrence was delayed in those subjects who were breastfeeding.

In a study of migraine during pregnancy by Somerville, which involved 200 subjects, 31 reported a prior history of migraine.[74] Headaches improved in 77% of the subjects during pregnancy. However, in 22.6% of these patients, headaches increased in either frequency and/or severity. Initial onset of headaches

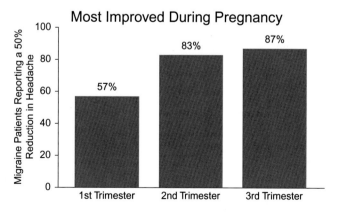

FIGURE 10.3 Migraine-without-aura subjects most improved during pregnancy. *Adapted from Sances et al.*[73]

TABLE 10.2   United States FDA Pharmaceutical Pregnancy Categories

Category A	Adequate and well-controlled human studies have failed to demonstrate a risk to the fetus in the first trimester of pregnancy (and there is no evidence of risk in later trimesters)
Category B	Animal reproduction studies have failed to demonstrate a risk to the fetus and there are no adequate and well-controlled studies in pregnant women OR Animal studies have shown an adverse effect, but adequate and well-controlled studies in pregnant women have failed to demonstrate a risk to the fetus in any trimester
Category C	Animal reproduction studies have shown an adverse effect on the fetus and there are no adequate and well-controlled studies in humans, but potential benefits may warrant use of the drug in pregnant women despite potential risks
Category D	There is positive evidence of human fetal risk based on adverse reaction data from investigational or marketing experience or studies in humans, but potential benefits may warrant use of the drug in pregnant women despite potential risks
Category X	Studies in animals or humans have demonstrated fetal abnormalities and/or there is positive evidence of human fetal risk based on adverse reaction data from investigational or marketing experience, and the risks involved in use of the drug in pregnant women clearly outweigh potential benefits
Category N	FDA has not yet classified the drug into a specified pregnancy category

occurred in 3.5% of subjects. Another study focused on headaches during the postpartum period in 71 subjects.[75] Postpartum headaches occurred in 39% of the patients, and 83% had a prior history of migraine.

For many years, the gold standard of headache treatment during pregnancy was no treatment. The oxytocic effects of the ergotamines and other agents with vasoconstrictor properties precludes their use in pregnancy. Due to the possibility of teratogenic effects, pregnant women are generally excluded in clinical drug trials. A classification of fetal risks due to pharmaceuticals was established in 1979 by the FDA; the categories are listed in Table 10.2.

For the acute treatment, Category B drugs include acetaminophen, caffeine, NSAIDs (after implantation and before 32 weeks), codeine (hydrocodone, oxycodone), butorphanol, and metoclopramide. The following are included in Category C: aspirin, butalbital, isometheptene mucate, phenothiazines, and the triptans. The ergots are classified as Category X.

Another rating system has been described: the TERIS rating system. It is designed to assess teratogenic risk to the fetus from drug exposure.[76] This rating is based on expert opinion and existing medical literature. The TERIS rating does not necessarily correlate with the FDA pregnancy categories. The TERIS risk ratings are:

- Undetermined
- None
- None–minimal
- Minimal
- Minimal–small
- High.

In order to determine the effects of triptan *in utero* exposure, an observational study was undertaken.[77] This study involved volunteer participation of healthcare practitioners throughout the world. The registry enrolled 680 evaluable exposed pregnant women, whose pregnancies resulted in 689 infants and fetuses (outcomes). The agent breakdown was sumatriptan (626), naratriptan (57), both sumatriptan and naratriptan (7), and sumatriptan/naproxen combination (6). In 528 subjects exposed to sumatriptan during the first trimester, 20 major birth defects were reported. One birth defect was reported in the subjects exposed to both sumatriptan and naratriptan. In the sumatriptan/naproxen exposure group, no major defects were reported.

This registry detected no signal of teratogenicity associated with major birth defects for sumatriptan. Enrollment for the naratriptan and sumatriptan/naproxen exposure groups was too small to provide definitive conclusions of the risks associated with exposure to these agents. The registry's scientific advisory committee concurred that due to low enrollment and high rates of loss to follow-up, the study would not be continued beyond its 16-year existence. They believed that prolonging the registry would not provide significant information to predict the risk of birth defects.

Preventive treatment of migraine during pregnancy may be undertaken, and usually involves agents used for migraine outside of pregnancy. The beta-blocker metoprolol and some SSRI antidepressants (fluoxetine, sertraline) are listed as Category B. Other beta-blockers and SSRIs are included in Category C, along with calcium channel blockers, gabapentin, and some tricyclic antidepressants (protriptyline, doxepin). Amitriptyline, nortriptyline, topiramate, and divalproex sodium are considered Category D. Topiramate has recently been added to this category.[78,79]

For those pregnant patients who are experiencing prolonged and disabling migraine attacks, emergency interventions may be considered. The patient may require fluid resuscitation. Pain control can be achieved via IV administration of metoclopramide, diphenhydramine, an opioid, and magnesium sulfate. Occipital nerve blocks are an option. If the headache episodes increase in frequency, prophylactic therapy may be considered as well as more aggressive management measures. The following medications should never be used during pregnancy: ergotamine, phenytoin, valproic acid, and lithium carbonate.

Acute non-systemic therapies include trigger-point injections, occipital nerve blocks, physical therapy, and intranasal or transdermal lidocaine. A recent study found that peripheral nerve blocks in pregnant women with headaches were effective, and could be performed safely during pregnancy. This intervention was indicated in patients who did not respond to standard regimens.[80] The study involved 11 pregnant women who received peripheral nerve blocks 24 times during their pregnancies. They received treatment for status migrainosus or as bridge therapy. Biofeedback and relaxation therapy may provide excellent non-drug options during pregnancy.

LACTATION

Soranus of Ephesus (150 AD) warned wet nurses to avoid drugs and alcohol lest it adversely affect the nursing infant. The same warning holds true today. However, we have previously discussed the occurrence of migraine during the postpartum period. It is known that drugs bond strongly to proteins in the milk, and the "pump and discard" procedure may be used during treatment for migraine.

Recently, a group of headache specialists reviewed the literature regarding migraine treatments in breast-feeding mothers.[81] Transfer of drugs into breast milk is typically described quantitatively, using plasma concentration rates. The recommended guide for safe use during lactation is a cut-off of 10%. The age of the infant should also be considered before starting treatment. Premature infants clear drugs less efficiently than their full-term counterparts. At approximately 7 months, the infant will be clearing drugs at a rate similar to adults.[82] When considering treatment options during lactation, it is efficacious to consult with the pediatrician.

When the patient returns for follow-up after delivery, the status of the migraine attacks should be explored. The patient should also be questioned about how long she intends to breastfeed. It should be noted that some drugs, such as valproate, which are contra-indicated during pregnancy, may be relatively safer during lactation.[74]

For the acute treatment of migraine in breastfeeding patients, sumatriptan and zolmitriptan have been approved by the American Academy of Pediatrics. In preventive therapy, it is advisable to start the drug at lower doses and titrate slowly. Amitriptyline or nortriptyline may be started at 10 mg rather than 25 mg. The initial dose of topiramate could be 15 mg instead of 25 mg, and the dose increased every 2–3 weeks rather than weekly. Initiating therapy at lower doses enables physicians to evaluate the infant for any adverse effects, including sedation. It should also be noted that infant metabolism increases rapidly during the first months of life.

MENOPAUSE

Menopause is defined as the absence of menses for at least 1 year. Typically, this hormonal milieu, with its varied symptoms, occurs over approximately 4 years.[83] During this time, the levels of estradiol and progesterone will have erratic fluctuations.[84] For those with migraine, different patterns of their headache occur.

It is essential that when a female patient of a certain age is evaluated for headaches, the healthcare practitioner questions her about other menopausal symptoms – including sleeping difficulties, hot flashes, night sweats, decreased libido, joint and muscle pain, memory loss, mood swings, anxiety/depression, fatigue, and skin changes. The use of a headache diary will enable the physician and the patient to recognize patterns and triggers of her headaches. For many patients, migraine headaches will improve or disappear. Some patients will experience an exacerbation of their headaches, or perhaps an initial onset of headaches, during menopause.

In addition to the challenge of changing hormone levels, some patients will be treated with hormone replacement therapy (HRT). As with oral contraceptives, many women will note an increase in their headaches with HRT. For others, HRT may help the headaches by maintaining stable levels of estrogen. In a British study involving 74 female patients at a London menopause clinic, 57% complained of headache and 29% reported migraine symptoms.[85] These headaches produced significant disability in the migraine group, with 80% reporting migraine attacks more frequently than once monthly.

Some studies have noted an improvement in migraine without aura during menopause.[86] In one study involving 1436 women at various stages of the menopause, 10.5% of postmenopausal patients reported a prevalence of migraine. This contrasts with 16.7% prevalence during the premenopausal and perimenopausal stages. A history of worsening of migraine during menopause has been correlated with a worsening of headaches in patients receiving HRT.[87]

If HRT is to be initiated, estrogen should be given continuously to avoid "estrogen withdrawal" migraine. Administration by non-oral routes is preferred, as it has a less negative effect on the headaches than oral formulations.[88] Continuous HRT with combinations of estrogen and progesterone seems to be better tolerated than cyclic combined therapy. However, combined agents have been associated with higher non-migraine risks.[89]

The risks of stroke and other complications with oral contraceptives used in migraine with aura were described earlier in the chapter. Few studies are available in investigating the risk of ischemic stroke in menopausal women with migraine who are receiving HRT. The Women's Health Study did not find a link between the use of postmenopausal HRT and ischemic stroke and myocardial infarction in women experiencing migraine, with or without aura.[90]

For migraine patients using HRT, a non-oral route is preferred. Transdermal estrogen patches provide for continuous dosing, preventing the "estrogen withdrawal" interval, and are better tolerated than oral hormones. Migraine frequency may initially increase with HRT, although the headaches usually improve with continued use.[74] If the patient presents with new onset headache after starting HRT, a careful evaluation is essential to rule out other, possibly morbid, causes. Thrombotic complications may occur with HRT. Retrospective analyses indicate that transdermal HRT does not present as high a risk as other forms of HRT.[91]

Alternative therapy is available, including the SSRIs (fluoxetine, paroxetine, venlafaxine) or gabapentin.[92] An adequate trial on the SSRIs is important, as there may be an initial exacerbation of the migraine attacks. The use of gabapentin has not been investigated in migraine prevention as much as the SSRIs.

CONCLUSION

The fluctuations of hormones throughout the reproductive years in women, as well as in the menopausal and postmenopausal stages, greatly impact the frequency, duration, and associated symptoms of migraine. Headaches described as menstrually associated migraine can be treated with therapies similar to those used in other migraine conditions. The hormones used in contraception and replacement therapy during menopause also complicate the management of migraine. As with treating any comorbid disorders during pregnancy and lactation, therapeutic choices must be selected judiciously and with the consultation of the OB-GYN and pediatrician. Being cognizant of the impact of these hormonal fluctuations on a female migraine patient will certainly facilitate management of the headaches.

References

1. Wolfe F, Ross K, Anderson J, Russell IJ, Hebert L. The prevalence and characteristics of fibromyalgia in the general population. *Arthritis Rheum.* 1995;38:19–28.
2. Lipton RB, Stewart WF, Diamond S, Diamond ML, Reed M. Prevalence and burden of migraine in the United States: data from the American Migraine Study II. *Headache.* 2001;41: 646–657.
3. Symmons D, Turner G, Webb R, Asten P, Barrett E, Lunt M, et al. The prevalence of rheumatoid arthritis in the United Kingdom: new estimates for a new century. *Rheumatology (Oxford).* 2002;41:793–800.
4. Isong U, Gansky SA, Plesh O. Temporomandibular joint and muscle disorder-type pain in US adults: The National Health Interview Survey. *J Orofac Pain.* 2008;22:317–322.
5. Stewart WF, Lipton RB, Celentano DD, Reed ML. Prevalence of migraine headache in the United States. Relation to age, income, race, and other sociodemographic factors. *JAMA.* 1992;267: 64–69.
6. Lipton RB, Bigal ME, Diamond M, Freitag F, Reed ML, Stewart WF; AMPP Advisory Group. Migraine prevalence, disease burden, and the need for preventive therapy. *Neurology.* 2007;68: 343–349.
7. Smitherman TA, Burch R, Sheikh H, Loder E. The prevalence, impact, and treatment of migraine and severe headaches in the United States: AS review of statistics from national surveillance studies. *Headache.* 2013;53:427–436.
8. Stewart WF, Simon D, Shechter A, Lipton RB. Population variation in migraine prevalence: a meta analysis. *J Clin Epidemiol.* 1995;48:269–280.
9. Stovner L, Hagen K, Jensen R, Katsarava Z, Lipton R, Scher AI, et al. The global burden of headache: a documentation of headache prevalence and disability worldwide. *Cephalalgia.* 2007;27: 193–210.
10. Buse DC, Loder EW, Gorman JA, Stewart WF, Reed ML, Fanning KM, et al. Sex differences in the prevalence, symptoms, and associated features of migraine, probable migraine and other severe headache: results of the American Migraine Prevalence and Prevention (AMPP) study. *Headache.* 2013;53:1278–1299.
11. Silberstein SD. The role of sex hormones in headache. *Neurology.* 1992;42(suppl 2):37–42.
12. Granella F, Sances G, Pucci E, Nappi RE, Ghiotto N, Nappi G. Migraine with aura and reproductive life events: a case–control study. *Cephalalgia.* 2000;20:701–707.
13. Karli N, Baykan B, Ertas M, Zarifoğlu M, Siva A, Saip S, et al. The Turkish Headache Prevalence Study Group. Impact of sex hormonal changes on tension-type headache and migraine: a cross-sectional population-based survey in 2,600 women. *J Headache Pain.* 2012;13:557–565.
14. Martin VT, Behbehani M. Ovarian hormones and migraine headache: understanding mechanisms and pathogenesis – Part 2. *Headache.* 2006;46:365–386.
15. Van der Linden JA. *De Hemicrania Menstrua.* London, UK: 1666.
16. Pardes IH. Pituitary headaches and their cure. *Arch Int Med.* 1919;23:174–184.
17. Blackie NH, Hossock JC. The treatment of migraine with emmenine. *Can Med Assoc J.* 1932;27:45–47.
18. Thompson AP. A contribution to the study of intermittent headache. *Lancet.* 1932;2:229–235.

19. Mofat WM. Treatment of menstrual migraine with small doses of gonadotropic extract of pregnancy urine. *JAMA*. 1937;108:612−615.

20. Somerville BW. The influence of progesterone and estradiol upon migraine. *Headache*. 1972;12:93−102.

21. Somerville BW. Estrogen-withdrawal migraine. I. Duration of exposure required and attempted prophylaxis by premenstrual estrogen administration. *Neurology*. 1975;25:239−244.

22. Somerville BW. Estrogen-withdrawal migraine. II. Attempted prophylaxis by continuous estradiol administration. *Neurology*. 1975;25:245−250.

23. MacGregor EA, Chis H, Vohrah RC, Wilkinson M. Migraine and menstruation: a pilot study. *Cephalalgia*. 1990;10:305−310.

24. Wöber C, Brannath W, Schmidt K, Kapitan M, Rudel E, Wessely P, et al. Prospective analysis of factors related to migraine attacks: The PAMINA study. *Cephalalgia*. 2007;27:304−314.

25. Granella F, Sances G, Allais G, Nappi RE, Tirelli A, Benedetto C, et al. Characteristics of menstrual and nonmenstrual attacks in women with menstrually related migraine referred to headache centres. *Cephalalgia*. 2004;24:707−716.

26. Headache Classification Committee of the International Headache Society. The international classification of headache disorders, 3rd edition (beta version). *Cephalalgia*. 2013;33:627−808.

27. Marcus DA, Bernstein CD, Sullivan EA, Rudy TE. A prospective comparison between ICHD-II and probability − Menstrual migraine diagnostic criteria. *Headache*. 2010;50:539−550.

28. Vetvik KG, MacGregor EA, Lundqvist C, Russell MB. Prevalence of menstrual migraine: a population-based study. *Cephalalgia*. 2014;34:280−288.

29. Couturier EG, Bomhof MA, Neven AK, van Duijn NP. Menstrual migraine in a representative Dutch population sample: prevalence, disability and treatment. *Cephalalgia*. 2003;23:302−308.

30. Granella F, Sances G, Zanferrari C, Costa A, Martignoni E, Manzoni GC. Migraine without aura and reproductive life events: a clinical epidemiological study in 1300 women. *Headache*. 1993;33:385−389.

31. Dowson AJ, Massiou H, Aurora SK. Managing migraine headaches experienced by patients who self-report with menstrually related migraine: a prospective, placebo-controlled study with oral sumatriptan. *J Headache Pain*. 2005;76:81−87.

32. MacGregor EA, Hackshaw A. Prevalence of migraine on each day of the natural menstrual cycle. *Neurology*. 2004;63:351−353.

33. MacGregor EA, Frith A, Ellis J, Aspinall L, Hackshaw A. Incidence of migraine relative to menstrual cycle phases of rising and falling estrogen. *Neurology*. 2006;67:2154−2158.

34. Epstein MT, Hockaday JM, Hockaday TD. Migraine and reproductive hormones throughout the menstrual cycle. *Lancet*. 1975;1:543−548.

35. Somerville BW. The role of estradiol withdrawal in the etiology of menstrual migraine. *Neurology*. 1972;22:355−365.

36. Lichten EM, Lichten JB, Whitty A, Pieper D. The confirmation of a biochemical marker for women's hormonal migraine: the depo-estradiol challenge test. *Headache*. 1996;37:367−371.

37. MacGregor EA, Hackshaw A. Prevention of migraine in the pill-free interval of combined oral contraceptives: a double-blind, placebo-controlled pilot study using natural oestrogen supplements. *J Fam Plann Reprod Health*. 2002;28:27−31.

38. MacEwan B. Estrogen actions throughout the brain. *Recent Prog Horm Res*. 2002;57:357−384.

39. Pradalier A, Vincent D, Beaulieu P, Baudesson G, Launay J. Correlation between oestradiol plasma level and therapeutic effect on menstrual migraine. In: Rose FC, ed. *New Advances in Headache Research*. Vol 4. London, UK: Smith-Gordon; 1994:129−132.

40. Calhoun AH. A novel specific prophylaxis for menstrual-associated migraine. *South Med J*. 2004;97:819−822.

41. Silberstein SD, Merriam GR. Estrogens, progestins, and headache. *Neurology*. 1991;41:786−793.

42. Carlson LA, Ekelund LG, Oro L. Clinical and metabolic effects of different doses of prostaglandin E1 in man. Prostaglandin and related factors. *Acta Med Scand*. 1968;183:423−430.

43. Sances G, Martignoni E, Fioroni L, Blandini F, Facchinetti F, Nappi G. Naproxen sodium in menstrual migraine prophylaxis: a double-blind placebo controlled study. *Headache*. 1990;30:705−709.

44. Martin VT, Wernke S, Mandell K, Ramadan N, Kao L, Bean J, et al. Defining the relationship between ovarian hormones and migraine headache. *Headache*. 2005;45:1190−1201.

45. Mauskop A, Altura BM. Role of magnesium in the pathogenesis and treatment of migraines. *Clin Neurosci*. 1998;5:24−27.

46. Facchinetti F, Sances G, Borella P, Genazzani AR, Nappi G. Magnesium prophylaxis of menstrual migraine: effects on intra-cellular magnesium. *Headache*. 1991;31:298−301.

47. Solbach MP, Waymer RS. Treatment of menstruation-associated migraine headache with subcutaneous sumatriptan. *Obstet Gynecol*. 1993;82:769−772.

48. Nett R, Landy S, Shackelford S, Richardson MS, Ames M, Lener M. Pain-free efficacy after treatment with sumatriptan in the mild pain phase of menstrually associated migraine. *Obstet Gynecol*. 2003;102:835−842.

49. Silberstein SD, Massiou H, Le Jeunne C, Johnson-Pratt L, McCarroll KA, Lines CR. Rizatriptan in the treatment of menstrual migraine. *Obstet Gynecol*. 2003;102:835−842.

50. Loder E, Silberstein SD, Abu-Shakra S, Mueller L, Smith T. Efficacy and tolerability of oral zolmitriptan in menstrually associated migraine: a randomized, prospective, parallel-group, double-blind, placebo-controlled study. *Headache*. 2004;44:120−130.

51. Massiou H, Jamin C, Hinzelin G, Bidaut-Mazel C. Efficacy or oral naratriptan in the treatment of menstrually-related migraine. *Eur J Neurol*. 2005;12:774−781.

52. MacGregor EA. A review of frovatriptan for the treatment of menstrual migraine. *Int J Womens Health*. 2014;6:523−535.

53. Bhambri R, Martin VT, Abdulsattar Y, Silberstein S, Almas M, Chatterjee A, et al. Comparing the efficacy of eletriptan for migraine in women during menstrual and non-menstrual time periods: a pooled analysis of randomized controlled trials. *Headache*. 2014;54:343−354.

54. Martin VT, Ballard J, Diamond ML, Mannix LK, Derosier FJ, Lener SE, et al. Relief of menstrual symptoms and migraine with a single-table formulation of sumatriptan and naproxen sodium. *J Womens Health*. 2014;23:389−396.

55. Silberstein S, Bradley K. DHE-45® in the prophylaxis of menstrually related migraine. *Cephalalgia*. 1996;16:371.

56. MacGregor EA, Pawsey JC, Hu X. Safety and tolerability of frovatriptan in the acute treatment of migraine and prevention of menstrual migraine: results of a new analysis of data from five previously published studies. *Gend Med*. 2010;7:88−108.

57. Newman L. Understanding the causes and prevention of menstrual migraine: the role of estrogen. *Headache*. 2007;47(suppl 2):S86−S94.

58. Yong H, Xiaofei G, Lin F, Lingjing J. Triptans in prevention of menstrual migraine: a systematic review with meta-analysis. *J Headache Pain*. 2013;14:7.

59. Neri I, Granella F, Nappi R, Manzoni GC, Facchinetti F, Genazzani AR. Characteristics of headache at menopause: a clinico-epidemiologic study. *Maturitas*. 1993;17:31−37.

60. Rocca WA, Grossardt BR, Shuster LT. Oophorectomy, menopause, estrogen, and cognitive aging: the timing hypothesis. *Neurodegener Dis*. 2010;7:163−166.

61. Guttmacher Institute. Contraceptive use in the United States. June, 2014. <www.guttmacherinstitute.org> Accessed 12.10.14.

62. Victor TW, Hu X, Campbell JC, Buse DC, Lipton R. Migraine prevalence by age and sex in the United States. A life span study. *Cephalalgia*. 2010;30:1065–1072.

63. MacGregor EA. Contraception and headache. *Headache*. 2013;53:247–276.

64. Calhoun A. Combined hormonal contraceptives: is it time to reassess their role in migraine? *Headache*. 2011;:648–660.

65. Kurth T, Slomke MA, Kase CS, Cook NR, Lee IM, Gaziano JM, et al. Migraine headache, and the risk of stroke in women: a prospective study. *Neurology*. 2005;64:1020–1026.

66. Stegeman BH, de Bastos M, Rosendaal FR, van Hylckama Vlieg A, Helmerhorst FM, Stijnen T, et al. Different combined oral contraceptives and the risk of venous thrombosis: systematic review and network meta-analysis. *BMJ*. 2013;347:5298.

67. Nguyen BT, Jensen JT. Evaluating the efficacy and safety of a progestin- and estrogen-releasing ethylene vinyl acetate copolymer contraceptive vaginal ring. *Expert Opin Drug Saf*. 2014;13: 1423–1430.

68. Lidegaard O, Hougaard Nielsen L, Skovlund CW, Løkkegaard E. Venous thrombosis in users of non-oral hormonal contraception: follow-up study, Denmark 2001–10. *BMJ*. 2012;344:32990.

69. Gudmundsson LS, Scher AI, Aspelund T, Eliasson JH, Johannsson M, Thorgeirsson G, et al. Migraine with aura and risk of cardiovascular and all cause mortality in men and women: prospective cohort study. *BMJ*. 2010;341.

70. Calhoun A, Ford S, Pruitt A. The impact of extended cycle vaginal ring contraception on migraine aura: a retrospective case series. *Headache*. 2012;52:1246–1253.

71. Tassorelli C, Greco R, Allena M, Terreno E, Nappi RE. Transdermal hormonal therapy in perimenstrual migraine: why, when and how? *Curr Pain Headache Rep*. 2012;16:467–473.

72. Coffee AL, Sulak PJ, Hill AJ, Hansen DJ, Kuehl TJ, Clark JW. Extended cycle combined oral contraceptives and prophylactic frovatriptan during the hormone-free interval in women with menstrual-related migraines. *J Womens Health (Larchmt)*. 2014;23: 310–317.

73. Sances G, Granella F, Nappi RE, Fignon A, Ghiotto N, Polatti F, et al. Course of migraine during pregnancy and postpartum: a prospective study. *Cephalalgia*. 2003;23:197–205.

74. Somerville BW. A study of migraine in pregnancy. *Neurology*. 1972;22:824–828.

75. Stein G, Morton J, Marsh A, Collins W, Branch C, Desaga U, et al. Headaches after childbirth. *Acta Neurol Scand*. 1984;69:74–79.

76. Silberstein S, Lipton RB, Goadsby PJ. *Headache in Clinical Practice*. 2nd ed. New York, NY: Martin Dunitz; 2002:257–267.

77. Ephross SA, Sinclair SM. Final results from the 16-year sumatriptan, naratriptan, and treximet pregnancy registry. *Headache*. 2014;54:1158–1172.

78. Marmura MJ. Safety of topiramate for treating migraines. *Expert Opin Drug Saf*. 2014;13:1241–1247.

79. Viana M, Terreno E, Goadsby PJ, Nappi RE. Topiramate for migraine prevention in fertile women: reproductive counseling is warranted. *Cephalalgia*. 2014;34:1097–1099.

80. Govindappagari S, Grossman TB, Dayal A, Grosberg BM, Vollbracht S, Robbins MS. Peripheral nerve blocks in pregnant patients with headache. *Obstet Gynecol*. 2014;123(suppl 1):147.

81. Hutchinson S, Marmura MJ, Calhoun A, Lucas S, Silberstein S, Peterlin BL. Use of common migraine treatments in breastfeeding women: a summary of recommendations. *Headache*. 2013;53:614–627.

82. Kearns GL, Abdel-Rahman SM, Alander SW, Blowey DL, Leeder JS, Kauffman RE. Developmental pharmacology – drug disposition, action, and therapy in infants and children. *N Engl J Med*. 2003;349:1157–1167.

83. McKinlay SM. The normal menopause transition: an overview. *Maturitas*. 1996;23:137–145.

84. MacGregor EA. Migraine, the menopause and hormone replacement therapy: a clinical review. *J Fam Plann Reprod Health Care*. 2007;245–249.

85. MacGregor EA, Barnes D. Migraine in a specialist menopause clinic. *Climacteric*. 1999;2:218–223.

86. Mattsson P. Hormonal factors in migraine: a population-based study of women aged 40 to 74 years. *Headache*. 2003;43:27–35.

87. Hodson J, Thompson J, al-Azzawi F. Headache at menopause and in hormone replacement therapy users. *Climacteric*. 2000;3:119–124.

88. MacGregor EA. Effects of oral and transdermal estrogen replacement on migraine. *Cephalalgia*. 1999;124–125.

89. Panay N, Studd J. Progestogen intolerance and compliance with hormone replacement therapy in menopausal women. *Hum Reprod Update*. 1997;3:159–171.

90. Kurth T, Gaziano JM, Cook NR, Logroscino G, Diener H-C, Buring JE. Migraine and risk of cardiovascular disease in women. *JAMA*. 2006;283–291.

91. Eisenberger A, Westhoff C. Hormone replacement therapy and venous thromboembolism. *J Steroid Biochem Mol Biol*. 2014;142: 76–82.

92. North American Menopause Society. Treatment of menopause-associated vasomotor symptoms: position statement of The North American Menopause Society. *Menopause*. 2004;11:11–33.

11

Cluster Headache

Frederick G. Freitag and Johnathon Florczak

Medical College of Wisconsin, Milwaukee, Wisconsin, USA

INTRODUCTION

Gerhard Van Swieten provided perhaps the first description of cluster headache.[1] In 1745, he wrote "A healthy robust man of middle age was suffering from troublesome pain, which came on every day at the same hour at the same spot above the orbit of the left eye." Over the centuries a variety of terms have been used for cluster headache, including migrainous neuralgia, red migraine, Sluder's syndrome, sphenopalatine neuralgia, histaminic cephalgia, and Horton's headache.

EPIDEMIOLOGY

Cluster headache is one of the three primary headache disorders. It is rarely compared to migraine headache. Its importance as a diagnostic entity lies in its character as a paroxysmal disorder with unpredictable severe facial and headache pain that is abrupt in onset and difficult to treat with conservative measures. The intractability of this disorder accounts for much of its morbidity without appropriate acute and preventative treatment, as well as adequate education and counseling of the patient and referring physicians. Among the primary headache disorders, it is uniquely debilitating due to the nature of its severe unrelenting pain that can be difficult to treat. The most intractable cases defy symptomatic control in spite of the standard of care in the form of multiple abortive and preventative agents. Although historically surgical intervention has been attempted, increasingly more options are available in the form of peripheral nerve and brain stimulation.

DIAGNOSTIC CLASSIFICATION AND CLINICAL DESCRIPTION

Cluster headache is classified, among the four types of primary headache disorders, in the trigeminal autonomic cephalalgias (TACs), which also include paroxysmal hemicrania, Short-lasting Unilateral Neuralgiform headache with Conjunctival injection and Tearing (SUNCT), Short-lasting Unilateral Neuralgiform headache with Autonomic features (SUNA), and hemicrania continua.[2] These conditions are included in Group 3 of the International Classification of Headache Disorders-III (ICHD-III), and are primarily differentiated from the other primary headaches of migraine or tension-type headaches in that these are "short-lasting" headaches. Cluster headaches are also differentiated from the fourth of the primary headache disorders, which includes entities such as primary stabbing headache, primary cough headache, primary exertional headache, primary headache associated with sex, and hypnic headache. Paroxysmal hemicrania is divided into the more common chronic paroxysmal hemicranias and the less common episodic paroxysmal hemicranias.

Before describing treatment options and other aspects of symptom management, we need to address the diagnostic criteria used in classifying these headaches. We must also perform a brief review of both imaging and experimental evidence that help to illustrate the pathophysiology of these disorders. According to the ICHD-III classification (Box 11.1), cluster headache is a condition of

attacks of severe strictly unilateral pain which is orbital, supraorbital, temporal or in any combination of these sites, lasting 15 to 180 minutes and occurring from once every other day to eight times per day and associated with one or more of various ipsilateral symptoms (conjunctival injection, lacrimation, nasal congestion, rhinorrhea, forehead and facial sweating, miosis, ptosis or eyelid edema.

Cluster headache can perhaps be considered as the prototype (as well as the most common) condition among the various headache disorders in the category of TACs. The other headache disorders in this

S. Diamond, R. K. Cady, M. L. Diamond & V. T. Martin (Eds):
Headache and Migraine Biology and Management.

DOI: http://dx.doi.org/10.1016/B978-0-12-800901-7.00011-2

BOX 11.1

INTERNATIONAL HEADACHE CLASSIFICATION III (BETA)[2]

3. Trigeminal Autonomic Cephalalgias (TACs)
   3.1 Cluster headache
       3.1.1 Episodic cluster headache
       3.1.2 Chronic cluster headache
   3.2. Paroxysmal hemicrania
       3.2.1 Episodic paroxysmal hemicrania
       3.2.2 Chronic paroxysmal hemicrania
   3.3 Short-lasting unilateral neuralgiform headache attacks
       3.3.1 Short-lasting unilateral neuralgiform headache attacks with conjunctival injection and tearing (SUNCT)
           3.3.1.1 Episodic SUNCT
           3.3.1.2 Chronic SUNCT
       3.3.2 Short-lasting unilateral neuralgiform attacks with cranial autonomic symptoms (SUNA)
           3.3.2.1 Episodic SUNA
           3.3.2.2 Chronic SUNA
   3.4 Hemicrania continua
   3.5 Probable trigeminal autonomic cephalalgia
       3.5.1 Probable cluster headache
       3.5.2 Probable paroxysmal hemicrania
       3.5.3 Probable short-lasting unilateral neuralgiform headache attacks
       3.5.4 Probable hemicrania continua

---

category, along with cluster, may best be known as "cluster headache variant".[3] As new cases have been reported in the literature, understanding of these headaches has evolved in that they appear to represent distinct headache types in themselves. These cluster headache variants differ in not only the duration of each headache type but also the typical frequency of attacks per bout. Other factors to consider are gender predilection, anatomical location, and response to various treatments. However, they all share core similarities in phenotype of being severe, unilateral, short-lasting headache with one or more prominent autonomic features that can be traced to hypothalamic involvement and/or excessive outflow from the trigemino-vascular system through the superior salivatory nucleus. The other TACs depart from the clinical behavior of cluster headaches.

The other TACs, or cluster headache variants, have a shorter duration than cluster headaches, but with cluster headache they share the typical unilateral location of the headache pain as well as the typical association with ipsilateral autonomic symptoms, such as lacrimation, rhinorrhea, conjunctival injection, ptosis, miosis, nasal congestion, eyelid edema, and ipsilateral face and forehead sweating. Not all, however, are associated with the typical restlessness and "pacing" behavior that is typically seen in cluster headache. Although patients with SUNCT are usually not restless, they may have features more attuned to trigeminal neuralgia. In SUNCT they may have a cutaneous trigger that can set off the pain, which may manifest by the avoidance of exertion or physical activity serving as a trigger, or to avoid exacerbating already severe pain. Between 40% and 60% of patients with cluster headache and paroxysmal hemicrania may have isolated migrainous features such as nausea, photophobia, phonophobia, or an aura.[4] Another feature that tends to characterize the cluster variants is the response specifically to the non-steroidal anti-inflammatory drug (NSAID) indomethacin, as both a diagnostic and therapeutic intervention.

The noted clinician John Graham described cluster headache patients with facial characteristics of *peau d'orange* skin, deep vertical facial creases, a ruddy complexion, and frequent facial telangectasia;[5] and as being heavy smokers,[6] the tallest individuals in their family, and having hazel-colored eyes. Patients were characterized as being tall, rugged males with a personality to match.[7] Trigger factors are common among patients with migraine headache. However, for patients with cluster headache these are unusual, except for consumption of alcoholic beverages. During a cycle, ingestion of alcoholic beverages may serve as a trigger for individual attacks of cluster headaches but does not elicit a cycle unto itself.

Diagnostic criteria in the IHC require that the pain must be unilateral, occurring during each of at least five attacks (Box 11.2).[2] There are both episodic and chronic forms of cluster headache. The headache must occur along with at least one of the following ipsilateral autonomic symptoms: unilateral ptosis/miosis; lacrimation; conjunctival injection; rhinorrhea/nasal congestion; forehead sweating; or eyelid edema. The typical duration will last from 15 minutes to 3 hours. Although patients may experience up to eight attacks in a day, they may typically have as few as two attacks per day or every other day.

Clinically, the pain of cluster headache is located in the unilateral retro- and periorbital areas with extension into the temple, and occasionally into the

---

BOX 11.2

CLINICAL FEATURES OF CLUSTER HEADACHE

- Males more commonly than females
- Peak occurrence in twenties
- Tobacco and alcohol use common
- Distinct facial characteristics
- Cycles begin near summer and winter solstices
- Individual attacks occur same time of day in given patient
- Attacks commonly occur during sleep

- Pain lasts 15 minutes to 2 hours
- Strictly unilateral
- Eye and temple region
- "Hot poker" pain description
- Rhinorrhea
- Lacrimation
- Partial Horner's syndrome

---

suboccipital area. An acute attack of cluster headache is brief when compared with other headaches, lasting 15 minutes to 2 hours. The diagnosis of cluster headache should be questioned if individual attacks of cluster headache are lasting over 4 hours. Although brief, the severe pain of cluster headaches is distinct from that of other headache disorders. The intensity of pain has led to it being called "suicide headache."

A hallmark of the diagnosis is its unilateral location, and in the majority of patients it always affects the same side. However, in about 14% of patients there may be a "side shift" with subsequent cycles. In only 3% of cluster headache patients is the headache pain, although still in a typical location, purported to occur without any clear autonomic symptoms. Also, with symptoms typical of an acute cluster headache cycle, 28% of affected patients may have "constant background pain" or some degree of persistent interictal pain that may be more dull and lingering, or pressure-like.

The pain is described as a boring, intense pain compared with the sore throbbing pain of migraine. A patient may use the description of a "hot poker being pushed into the eye." The migraineur will retreat to the dark quiet of the bedroom. However, patients with cluster headache may be found sitting and rocking, or pacing about the room clutching their head. These patients may also experience a less intense deep ache in the upper ipsilateral cervical paraspinal region during their attack. Also, they may report residual soreness in the temporal region between acute attacks of cluster headache.

During an acute attack, patients with cluster headache fit a certain phenotype regarding their overall clinical presentation or outward behavior. They are characteristically restless and cannot hold still, sometimes pounding their fists or applying firm pressure to their temples. Lying still may actually appear to exacerbate the pain.

Although there are clear diagnostic criteria, cluster headache and other TACs may be misdiagnosed either as "sinus type" headache or as a dental condition.

These headaches may also be confused, by less experienced clinicians, with either migraine headache or hemicrania continua. It is important to remember that in addition to meeting the above diagnostic criteria, one may also need to rule out other underlying secondary causes for headache, whether they are vascular, structural, or infectious in nature, particularly if there are other focal neurologic symptoms or signs from the history and neurologic exam. For example, the diagnosis may be established provided that the history and physical do not suggest any other secondary cause as the result of any systemic or constitutional symptoms, or no focal neurologic deficits are seen on a detailed neurologic exam. Also, in the context of any suspicion of a secondary headache cause, appropriate neuroimaging or laboratory studies should exclude any other cause for the headaches, particularly if presenting for the first time or presenting with a significant change in the headache pattern. This would not seem to be as essential if the headache is not occurring for the first time or the headaches do not seem to occur with any temporal association with any other separate underlying medical condition. Conditions that may fall within the differential diagnosis of cluster headache as symptomatic causes of secondary headache and that would need to be excluded include vertebral artery dissection, pseudo-aneurysm of the cavernous carotid, and aneurysms of the anterior cerebral, carotid, and vertebral arteries. Other structural lesions to be excluded involve other vascular processes such as arteriovenous malformations of the occipital lobe or middle cerebral artery, as well as giant cell arteritis. Neoplastic structural lesions may also mimic this condition, whether from cervical cord meningioma, pituitary adenoma, or sphenoid wing meningioma. Cluster headache symptoms may also overlap somewhat with facial trauma, Tolosa Hunt syndrome, and maxillary sinusitis.

The onset of cluster headache occurs between the ages of 11 and 80 years, but typically begins between the ages of 27 and 31, or later during middle age. Initial onset is noticeably later in life than migraine or tension-type headaches. It has been estimated that the mean time to diagnosis is 7 to 16 years, perhaps because it is a rare disorder. It is one of the rarer primary headaches, with historical estimates from Ekbom[8] determined to be 0.9% in a sample of Swedish Army recruits. It has been extrapolated by several studies to be in the range of 0.1% to 0.4% of the general population, although occurrence is much less than the incidence quoted for men and women with migraine of 6% and 18%, respectively. Prevalence was previously thought to be disproportionately higher in males, as it was once thought to be 6:1 male to female (M:F) ratio, but is now on average 2:1 to 4:1 M:F, perhaps depending on whether episodic or chronic cluster is being considered, according to various sources. It is thought that the gender discrepancy is even greater in chronic cluster headache. Torelli et al.,[9] in their review of a series of population studies, noted that Kudrow had found a 4.8:1 M:F ratio for episodic and 6.3:1 M:F ratio for chronic cluster headache.[9] In Torelli's own study, a population of 392 patients in an Italian sample displayed a ratio of M:F 3.6:1 for the episodic cluster group and 4.2:1 M:F ratio for the chronic cluster group. However, more recent studies place the M:F ratio in the range of 2.1:1.[10] The ratio in chronic cluster headache ranges from 3.5:1 to 7.5:1 M:F. Over the decades, this shift to greater male-to-female equity has occurred in the setting of increased employment of women and increased use of tobacco among women.[11] The causal relationship to the change in gender preponderance of these lifestyle changes is dubious, whereas improved disease recognition of cluster headache is not.

Although its transmission may not have been fully elucidated, the risk in first-degree relatives has been quoted by some as between approximately 14- and 39-fold.[12] It could be that the newer estimates of gender predilection are now higher due to previous bias coupled with improved surveillance, education, and awareness. Various series have found some genetic basis for a risk of inheritance of 4%.[13] Others[12] have noted an affected first-degree relative showing a 5- to 18-fold increased risk, and second-degree relatives with a 1- to 3-fold higher risk. Several possibilities have been raised for genes, including "HCRT, HCRTR1, and HCRTR2",[14] and that it would appear to have autosomal dominant inheritance.

The majority of cluster headache sufferers' headaches display the episodic pattern — that is, they have episodes lasting between 7 and 365 days with a pain-free remission of 1 month or more. About 15% of patients with cluster headache experience chronic cluster headache, in which one has to suffer more than a year without remission. If a remission occurs, it is brief and of less than 1 month duration between two distinct cluster bouts. Among those patients with episodic cluster, perhaps up to 27% of patients only experience a single lifetime episode.[15] Since these patients do not satisfy either episodic or chronic cluster criteria, they are classified as "cluster headache (not episodic or chronic)".[2]

Demographics

Among other demographic factors, those with cluster headache are believed to be more likely to smoke tobacco and to have a higher incidence of cardiac disease and peptic ulcer disease.[16] Manzoni found that over 75% of episodic cluster patients, and almost 90% of those with chronic cluster headache, were smokers. Traumatic brain injury has been associated with the occurrence of cluster headache,[17] with an incidence of previous head trauma among cluster headache patients ranging between 5% and 37% of the patients. Other specific traumatic events or procedures, such as enucleation of either eye[18] or dental extraction,[19] have also been reported.

Paroxysmal hemicrania, SUNCT, as well as the closely related variant SUNA and hemicrania continua differ between themselves and cluster headache in prevalence as well as the M:F gender ratio of each of these disorders. Paroxysmal hemicrania is a rare and largely benign indomethacin-sensitive condition that more commonly occurs in its chronic form (chronic paroxysmal hemicranias [CPHs]). CPH occurs in ages ranging from 1 to 81 years,[20] with a mean age of onset of 34 years. Roughly 21% of CPH cases reported a family history of migraine. Although there is a female preponderance of 2:1 in CPH, it occurs equally in the genders for the paroxysmal form. In the initial reports of SUNCT, it was believed that it affected males more commonly than females by almost 4:1.[21] More recent work suggests that the ratio is in the range of 1.3:1 to 1.5:1 M:F.[22] It is an exceedingly rare condition, with a mean age of onset of 51 years of age. All of these disorders are relatively rare, and little is available on the epidemiology of hemicrania continua or SUNA.[23]

CIRCADIAN AND CIRCANNUAL FEATURES

A visual mnemonic for cluster headache might be a picture of grapes, a calendar, and a clock, since the attacks occur in a group, like a bunch of grapes, and occur from once to multiple times per day over a period of several weeks to many months. These bouts of headache are interspersed by pain-free periods. A typical cycle of cluster headache lasts from 2 weeks to 3 months. The cycle is rarely shorter, but it can be longer and persist for as long as a year without resolution. During a bout or cycle of cluster headache, patients will have one or more attacks per day. The acute attacks have a circadian element, with the attacks of cluster headache occurring at night and near the end of a sleep cycle. Patients commonly can predict the time of onset of their cluster headache attacks to within several minutes, producing bizarre changes in lifestyle to try to avoid the attacks. A circannual pattern is also common for patients. The series of headaches may occur multiple times a year, or there may be remissions of up to 20 years. The onset of a cycle typically begins within several weeks of the summer or winter solstices.[24] Other patients experience their attacks during the spring or fall.[25] Each patient with cluster headache tends to have his or her own unique pattern of both circadian and circannual patterns.[26,27]

OTHER TRIGEMINAL AUTONOMIC CEPHALALGIAS

The cluster headache variants constitute a group of disorders that, similar to cluster headache, are relatively brief headaches associated with autonomic features (Table 11.1). The paroxysmal hemicranias are perhaps those most closely linked with this cluster headache variant terminology.

Paroxysmal hemicrania is characterized by attacks of severe unilateral pain, typically around the orbital, supraorbital, and temporal areas. It can occur in any or all of these areas in a given individual during a single attack or over their history of attacks. An acute attack lasts anywhere from 2 to 30 minutes. The attacks occur multiple times per day. A typical attack is associated with ipsilateral conjunctival injection, lacrimation, nasal congestion, rhinorrhea, forehead and facial sweating, miosis, and ptosis. These associated symptoms are similar to those of typical cluster headache other than for the brevity of the attacks. These are some of the few disorders which can be characterized by their absolute responsiveness to indomethacin.

In rare cases, an episodic form of this disorder has been described. These patients have "cycles" of their paroxysms occurring over periods of a week or longer. Similar to episodic cluster headache, a remission occurs lasting at least 1 month. Unlike cluster headache, the episodic form is the rarer of the two forms. Chronic paroxysmal hemicrania is the typical form of the disorder.

Similar to paroxysmal hemicrania is a disorder classified as short-lasting, unilateral, neuralgiform headache attacks. These attacks are similar to the paroxysmal hemicrania attacks in that they are located in the same unilateral location, although the pain occurs over any of the geographic reach of all branches of the trigeminal nerve. The attacks differ from those of paroxysmal hemicrania in several ways. The attacks may be less intense and briefer than in paroxysmal hemicrania, being as short as 1 second or maybe lasting as long as 10 minutes. These attacks may also occur as isolated stabbing sensations or as a series of stabs in a given attack. The attacks occur on at least 50% of the days. The short-lasting unilateral neuralgiform headaches may be divided into two subgroups: SUNCT, in which the autonomic symptoms are relegated to being associated with lacrimation (or tearing) and conjunctival injection; and SUNA, in which the autonomic symptoms are potentially more

TABLE 11.1   Cluster Variants

Feature	Cluster	Hemicrania continua	Chronic paroxysmal hemicrania	SUNCT	SUNA
Duration of attacks	15 minutes to 3 hours	Persistent and unilateral	2 to 30 minutes	1 second to 10 minutes	1 second to 10 minutes
Number of attacks per day	1–8	Constant	Multiple	Single or multiple 50% of days	Single or multiple 50% of days
Associated symptoms	Conjunctival injection, lacrimation, nasal congestion, rhinorrhea, forehead and facial sweating, miosis, and ptosis; photophobia and phonophobia	Conjunctival injection, lacrimation, nasal congestion, rhinorrhea, forehead and facial sweating, miosis, and ptosis; photophobia and phonophobia	Conjunctival injection, lacrimation, nasal congestion, rhinorrhea, forehead and facial sweating, miosis, and ptosis	Conjunctival injection and lacrimation	Conjunctival injection, lacrimation, nasal congestion, rhinorrhea, forehead and facial sweating
First-line treatment	Corticosteroids, verapamil	Indomethacin	Indomethacin	Lamotrogine	Gabapentin

widespread, as in paroxysmal hemicrania or cluster headache. Episodic and chronic forms of these disorders have been described.

It is important to recognize that both SUNCT and SUNA headache attacks may be triggered into occurring. Once one acute attack has abated, it can be retriggered immediately. By comparison, trigeminal neuralgia attacks cannot be immediately retriggered as there is a refractory period after each attack. During the evaluation of patients with SUNCT, neuroradiologic examination of the posterior fossa should be entertained since lesions in that area can mimic it.

Hemicrania continua may have been the other part of the originally described cluster headache variants as represented by the group with the continuous "background" headache. Based on the IHC criteria, the pain is persistent and unilateral.[2] Similar to the other variants, it is characterized by the occurrence of autonomic symptoms as seen in cluster headache. These symptoms include ipsilateral conjunctival injection, lacrimation, nasal congestion, rhinorrhea, forehead and facial sweating, miosis, and ptosis. Unlike the other TACs, migraine-like symptoms, including photophobia and phonophobia, are common in these patients. Similar to cluster headache, but not to the other variants, patients with hemicrania continua experience restlessness and agitation during the attacks. As in paroxysmal hemicrania, hemicrania continua is also sensitive to indomethacin. The pain is described as being of moderate intensity, and occurs steadily without a respite for periods of at least 3 months. Pain-free days may, however, occur on occasion.

PATHOPHYSIOLOGY

It is now thought, from animal data for experimental models as well as functional imaging studies, that cluster headache is in part due to hypothalamic dysfunction. This could account for its link to circadian rhythm, as well as explain both the autonomic features that are a hallmark of the disorder and account for the nature of the severe pain. The pathogenesis of cluster headache has become more established with the aid of functional neuroimaging. Both positron emission tomography (PET) and functional magnetic resonance imaging (fMRI) have shown activation of the inferior posterior hypothalamus in cluster headache, as well as the ipsilateral thalamus in a case of SUNCT.[22] Activation of this area also lends itself well to understanding the timing of the onset of cluster headache attacks, since evidence suggests that this activation follows a diurnal variation.

The hypothalamic theory of pathogenesis of cluster headache supports the findings that timing of cluster attacks are typically during the first REM phase of the night (onset occurring within 60 to 90 minutes of onset of sleep).[28] Melatonin has been linked to the natural sleep cycle; however, melatonin's effects may be far-ranging beyond its putative role in sleep.[29] Among the mechanisms influenced by melatonin that have implications for cluster headache is its anti-inflammatory effect, its role as a toxic free radical scavenger, and its ability to reduce proinflammatory cytokine production. Other effects that are of interest in cluster headache include the impact on nitric oxide synthase activity, and inhibitory effects on dopaminergic neurons. Melatonin also modulates analgesia via positive effects on the GABA-mediated pain mechanism as well as on opioid-induced analgesia.

Glutamate and serotonin levels are both modulated by melatonin, which has demonstrated neuroprotective effects. Involvement of the opioid-mediated pain process in cluster headache[30] revealed a marked decline in methionine enkephalin levels during a cluster cycle, coupled with a marked alteration in the metabolism of methionine enkephalin. A PET scan of the pineal gland shows decreased binding of opioids, as well as declines in the opioid receptors in both the hypothalamus and cingulated cortex, with long-standing cluster headache.[31]

Hemicrania continua also correlated with imaging findings, with particular anatomic locations showing activation in the hypothalamus during an acute cluster headache attack.[14] However, the activation is seen in the contralateral hypothalamus, as opposed to the ipsilateral location seen with cluster headache. It is thought that there is activation seen in the dorsal pons as well as the ventral midbrain and the pontomedullary junction in hemicrania continua.[32,33] It has also been noted that this activation in the contralateral posterior hypothalamus was suppressed by administration of indomethacin. It is thought that this connection of the potential hypothalamic dysfunction can also explain the periodicity of cluster headache, and both the circadian rhythms and the circannual patterns noted above.[26]

Increased levels of plasma cortisol and ACTH occur in patients with cluster headache, especially during the morning and evening, suggesting an alteration of the feedback circuit involving the hypothalamus, the pituitary, and the adrenal gland, indicating an alteration of the circadian rhythm. These significant relationships between biochemical parameters and the clinical patterns suggest a complex interplay between the hypothalamus, neuroendocrinological parameters, activity of the autonomic nervous system, and the pain of cluster headache.

Several mechanisms have been postulated whereby cluster headache may be mediated at a molecular level, including nitric oxide, histamine, and serotonin. Nitric oxide pathways may be involved with the mast cell regulation of the inflammatory response modulating refractoriness of cluster headache. A genetic link could not be established in cluster headache with either nitric oxide synthase[34] or histamine metabolism.[35]

The pain of cluster headache is modulated via the trigeminal vascular system. Previous theories had suggested other mechanisms. Kudrow suggested that there may be dysregulation of autoregulatory chemoreceptors for oxygen in the carotid circulation, producing vascular changes.[36,37] Others suggested either vascular dilatation of the ophthalmic artery, or an inflammatory process in the cavernous sinus or tributaries.[38]

The range of clinical symptoms in cluster headache is believed to be due in part to parasympathetic activation and/or a deficit of sympathetic outflow. Holland[14] noted that the suprachiasmatic nucleus has a role to play in both pain and autonomic processing. It has been established that stimulation of the anterior hypothalamus in general can modulate responses in the dorsal horn to noxious stimuli. Neuropeptides in the pathways OX1R and OX2R, or the "orexins," have been postulated to be possible mediators in cluster headache pathogenesis.[14] They indicated that the orexins may regulate many of these functions of the hypothalamus, whether circadian sleep/wake cycles, autonomic discharges, or processing of painful stimuli. These concepts are far from established, since there is contradictory evidence for this hypothesis.[14] They present evidence that during the period of time when cluster headache attacks are most likely to begin, there is an alteration in orexin secretion, which is felt to be downregulated to aid sleep onset. This alteration correlates again with onset of REM sleep, and it is notable that one of the genes (HCRT2) thought possibly to be responsible for cluster headache is for the orexin receptor OX2R. The HRCT2 gene is thought to increase risk for cluster headache, although no human data are available to support these associations from animal models of experimental head pain. Orexins may also have indirect effects on other inflammatory neuropeptides and neurotransmitters, such as dopaminergic, histaminergic, as well as noradrenergic and serotonergic systems. Somatostatin, another hypothalamic peptide hormone, and nitric oxide are thought to affect cluster headache pathogenesis.

Additional biochemical evidence points to a specific active locus in the posterior hypothalamus that has been identified.[39] This evidence suggests that changes in calcitonin gene-related peptide (CGRP) release and vasoactive intestinal peptide (VIP) appear to be involved in a trigeminal–vascular mechanism, accounting for the pain. Clues to the role of the hypothalamic involvement in pain modulation came from observations of circadian biological changes and neuroendocrine disturbances.[40]

Functional imaging studies have reinforced the model of cluster headache pain as involving dysfunction in the hypothalamus. May[41] reported that PET studies have shown activation in the ipsilateral posterior hypothalamus during acute and chronic pain. There are anatomical differences in the inferior posterior hypothalamus between those with cluster headache and normal healthy controls, identified by means of voxel-based morphometric analysis. Sanchez del Rio,[42] in 2004, noted that PET studies performed during headaches that were triggered by nitroglycerin administration showed activation in other areas during periods of acute headache pain, including the ipsilateral inferior hypothalamus as well as the contralateral ventral posterior thalamus, anterior cingulate cortex, and bilateral insula, associated with an increase in regional cerebral blood flow.

The cardiovascular changes seen in cluster headache may be related to this altered hypothalamic function. Changes in sympathetic activity[43] may contribute to the clinical symptoms through inhibition of sympathetic activity. Other changes in sympathetic and parasympathetic function are suggested by its clinical features.[44] Parasympathetic activation is suggested by the symptoms of lacrimation, conjunctival injection, and rhinorrhea. Sympathetic impairment is suggested by the findings of miosis and ptosis. Additional parasympathetic-mediated effects may occur in the form of nasal congestion, eyelid edema, and forehead sweating. Autonomic system involvement is a hallmark of the other TACs, such as chronic paroxysmal hemicrania.[45] In chronic paroxysmal hemicrania, changes in CGRP and VIP are similar to those that occur in cluster headache.

Sympathetic inhibition occurs in the periphery and in the CNS.[46] Decreases in plasma norepinephrine, when compared with controls, correlate with the duration, intensity, and frequency of the cluster headaches. Cerebrospinal fluid levels of norepinephrine and its metabolites are consistent with central changes in autonomic processing. Drummond suggests that there are central and peripheral autonomic disturbances occurring in cluster headache.[47] He suggests that lacrimation and nasal secretion during attacks are related to trigeminal-parasympathetic discharge. Drummond also notes that the ocular-sympathetic deficit and loss of thermoregulatory sweating and flushing on the symptomatic side of the forehead stem from the cervical sympathetic pathway to the face. Further, he argues that a peripheral rather than a central lesion produces signs of cervical sympathetic inhibition, resulting from compression of the sympathetic fibers traversing the internal carotid artery and arising from the cervical-sympathetic chain. The main trigger for vasodilation during attacks appears to be trigeminal mediated with the parasympathetic, which may underlie the increased response to vasoactive intestinal polypeptide. Although neither trigeminal-mediated parasympathetic discharge nor cervical sympathetic deficit is the primary trigger for attacks, these changes could contribute to the rapidity of the pain process.

A pericarotid inflammatory process stimulates trigeminal nociceptors. Activation of local nociceptors initiates a neurogenic inflammation and trigeminally mediated parasympathetic vasodilation. These culminate with the release of mast cell products, which intensify the inflammation and trigeminal discharge. There is evidence to support interictal changes in parasympathetic activity from altered clinical parameters of parasympathetic function between cycles of headaches.[48]

TREATMENT

Due to its abrupt severe onset, cluster headache, even more than other headache types, requires a treatment regimen that may involve co-pharmacy and rely upon daily maintenance therapy initiated at a low dose and increased to therapeutic effect as it is titrated upwards. It is also of importance to provide timely and short-term relief from the severe intractable pain of individual attacks of cluster headache. It may require time for a preventative medication to start working, and it may be necessary to initiate a transitional treatment as perhaps "mini-prophylaxis" until such time as the preventative regimen has had adequate time to impact the long-term clinical course of the headaches.

We will first focus on symptomatic or acute treatment for acute cluster headache pain. Due to the acute abrupt nature of pain quickly escalating to its maximum severity, parenteral and intranasal routes are obviously preferred to oral administration whenever possible.

Acute Treatment

Francis[49] reviewed the available evidence-based treatments for cluster headache. Criteria from this review included both headache response and pain-free response at 15 or 30 minutes from symptomatic trials. Acute treatments that had Level I evidence include: sumatriptan 6 mg subcutaneous; sumatriptan 20 mg nasal spray (NS); zolmitriptan 5 or 10 mg NS; zolmitriptan 10 mg tablet; and oxygen 100% administered at rates of either 6 L/min or 12 L/min (Table 11.2). Level II evidence included intranasal cocaine and lidocaine. Although the evidence was less significant, there may also be a role for treatment with dihydroergotamine (DHE).

TABLE 11.2   Acute Treatment of Cluster Headache

Drug	Dose
Oxygen	100% at 8–10 L/min for up to 15 minutes
Dihydroergotamine	1 mg IV, IM or SC at onset. Repeat q 1 hour up to 3 mg per day and 5 mg per week
	or
	1 to 2 mg NS at onset; repeat in 1 hour if needed; limit 6 mg per week
Sumatriptan	4 to 6 mg SC at onset; repeat in 1 hour if needed; limit to two doses per day, six doses per week
	Or
	20 mg NS, repeat in 1 hour if needed; limit to two sprays per day, six sprays per week
Zolmitriptan	5 to 10 mg at onset by oral or NS; limit to 10 mg per day, 30 mg per week

The severity and brevity of cluster headache attacks effectively precludes the use of simple or combination analgesics such as NSAIDs, barbiturates, opiates, or other oral agents. The only exception may be zolmitriptan, 10 mg, orally.

Triptans are effective in control of acute cluster headache pain. Since the typical cluster headache attack can last as little as 10 to 20 minutes, the onset of action and the time to peak concentration for oral triptans are too long to provide effective relief. The evidence supports injectable sumatriptan as the best choice for acute cluster headache treatment.[28]

Ample studies support the use of sumatriptan injection as an effective abortive agent, including two well-known double-blind placebo-controlled trials. The first, involving 49 subjects from the Sumatriptan Cluster Headache Study Group,[50] showed that 75% of patients injected with the drug were pain free at 20 minutes as compared with 26% of patients treated with placebo. Ekbom[51] showed positive results in a larger study ($n = 157$), in which a two-point change on a four-point scale in pain relief was demonstrated for 80% and 75% for 12 mg and 6 mg sumatriptan, respectively, as compared with 35% with placebo. A slightly less robust response was shown with sumatriptan injection in chronic cluster headache, with about 8% reduced efficacy compared with that found in use with episodic cluster headache. A number of double-blind placebo-controlled trials have also yielded similar results.[50,52–54]

Nasal administration of sumatriptan NS and zolmitriptan NS are other options. Sumatriptan NS has proven effective in the acute treatment of cluster headache.[55,56] In a double-blind placebo-controlled trial by van Vliet[56] of 118 subjects, 75% of whom had episodic cluster headache (ECH), he found that 57% of patients (both ECH and chronic cluster headache [CCH]) responded as compared with 26% of those receiving placebo. The results were statistically significant.

Zolmitriptan NS was shown to be beneficial for the acute treatment for ECH in two large double-blind placebo-controlled trials. The first, by Cittadini,[57] with 92 ECH and CCH subjects, reported headache relief in 62% of patients treated with zolmitriptan NS, 10 mg, and 42% with a 5-mg NS dose as compared with 23% receiving placebo. Response rates for both doses reached statistical significance. The second study, by Rapoport,[58] reported on results with 83 subjects. A two-point decrease in pain on a four-point scale was demonstrated in 63% and 50% with zolmitriptan NS 10 mg and 5 mg, respectively, versus 20% in patients receiving placebo.

Oral zolmitriptan, in a double-blind placebo-controlled trial of 91 subjects with ECH and 33 with CCH, proved effective in a 10-mg dose but not a 5-mg dose.[59] These results were only seen in ECH, with 47% of all treated patients responding as compared with 29% of those on placebo. The results were statistically significant.

Level I evidence exists for sumatriptan SC 6 and 12 mg. However, an open-label study demonstrated efficacy with the 4-mg dose.[60] These results, in effect, increase the number of doses per day or week that patients can utilize.

A number of patients are not able to tolerate triptans, and in some patients their use is contraindicated. Attack frequency may not be too high, thus precluding safe use for all of their attacks. The most commonly used alternative treatment for acute symptomatic treatment of cluster headache pain is probably high-flow oxygen therapy. Pure oxygen, when administered via face mask at 100% FiO₂, at rates of 8 to 15 L/min for up to 15 minutes, is effective. Although it will typically work within 5 minutes in the majority of patients, oxygen therapy is effective in about 70% of patients.

Fogan[61] showed, in a double-blind placebo-controlled trial, that oxygen at a FiO₂ of 100% given at a rate of 6 L/min compared with compressed air as a placebo given in the same manner in 19 subjects was significantly more beneficial. However, this rate of oxygen flow was lower than that routinely used today. Kudrow[62] reported, in 1981, that about 75% of 52 patients treated with oxygen had at least 7 of 10 attacks successfully suppressed. Rozen demonstrated that higher flow rates also effectively treat cluster headache.[63] Another study compared high-flow oxygen via a facemask with hyperbaric oxygen.[64] Previously, it was postulated that oxygen may be effective in aborting cluster headache by reducing cerebral blood flow.[65] More recently, newer evidence postulated a mechanism leading to reducing neuronal activation of the trigeminal-cervical complex after suitable stimulation of the trigeminal autonomic reflex as another explanation of this therapy's efficacy.[66]

Dihydroergotamine (DHC) may be effective in the acute treatment of cluster headache, based on a single trial with Level III evidence.[67] Despite this dearth of evidence, DHE has gained acceptance, through expert consensus and clinical judgment, as both a parenteral and NS treatment.

Intranasal lidocaine and cocaine have been shown to be effective in a small randomized double-blind controlled trial.[68] Intranasal administration can be challenging because of the need to instill the solution in proximity to the sphenopalatine ganglia. Despite the range of agents and their efficacy in Class I clinical trials, only about 70–80% of patients showed an adequate response to these therapies.[69,70]

Preventative Treatment

Preventative treatment of cluster headache follows a more diverse course than many of the other primary headache disorders. The comparative rarity of cluster headache, combined with the often brief duration of the cycle and the intensity of the pain, results in a relatively sparse literature of evidence-based trials of treatment.

Preventive treatment is similar in ECH and CCH. The choice of treatments should be evidence-based, with consideration for tolerability. It is challenging to compare amongst the treatments because of differences in methodology, outcome measures, and numbers of patients treated.

Short-term Treatments or Bridges in Therapy

Corticosteroids remain a primary choice for the treatment of cluster headache (Table 11.3). With this therapy, improvement in cluster headaches occurs rapidly and at doses that are generally well tolerated. The mechanism of action is unknown, but may include

TABLE 11.3 Preventive Therapy of Cluster Headache

Drug	Dose
Prednisone	40 mg per day to start for 3 to 5 days; taper by 5-mg decrements over no more than 3 weeks
Dihydroergotamine	1 mg IV q 8 hours for up to 15 doses
Naratriptan	2.5 mg BID for up to 7 days at a treatment
Lithium	300 mg BID to QID; titrate dose on tolerability and efficacy; monitor blood levels.
Verapamil	80 mg TID to 240 mg TID

anti-inflammatory effects and hypothalamic modulation. Horton[71] introduced corticosteroids in the preventive treatment of cluster headache. Both Kudrow and Couch[72,73] described using doses of 40 mg per day of prednisone. The duration of treatment with corticosteroids is typically brief, with the effective dose being used for up to a week before rapidly tapering it.

Another bridge therapy is intravenous DHE,[74] using a modification of the original Raskin protocol for migraine. It is well tolerated, and can be given in an office-based infusion suite.

Although methysergide was previously a mainstay of treatment, it is no longer available. Methysergide was believed to act via serotonin receptors. The effect in migraine treatment was deemed to be modulated by its antagonism at the 5-HT2B and 5-HT2C receptors. It was active at the 5-HT1A receptor, where it was a partial agonist. One of its metabolites is methylergometrine, which in humans may be responsible for its psychedelic effects. It was one of a family of compounds, synthesized by Sandoz Pharmaceuticals, that also included ergotamine and dihydroergotamine as well as the hallucinogenic agent, lysergic acid diethylamide. Its use in cluster headache was a carryover from its benefits in migraine preventive treatment. Methylergonovine, a metabolite of methysergide, has been used as a replacement. Its primary indication for treatment is dysfunctional uterine bleeding. Similar to methysergide, a drug-free hiatus every 6 months, and lasting for 1 month, should be followed with methylergonovine.[75]

Another bridge is the use of oral triptans, including both sumatriptan and naratriptan. Sumatriptan, 100 mg, three times a day may be given as a preventive therapy, for up to 7 days of use. However, a study failed to find significant benefits with this therapy.[76] Naratriptan has also been used as a preventive agent.[77] The results of the latter case study were encouraging.

Lithium

Lithium's use dates to the late 1970s. The mechanism of action is unknown,[78] but may involve hypothalamic modulation via serotonin. Two Class II randomized controlled trials of lithium therapy in cluster headaches have been reported. One study compared lithium 800 mg daily to placebo, in patients with ECH.[79] The other study compared lithium 900 mg daily with verapamil 360 mg per day,[80] in patients with CCH. Although no differences existed for the primary outcome measure in the blinded trial of lithium, in the verapamil comparator trial there was statistical superiority for verapamil when compared with baseline values. Drug-level monitoring and thyroid function testing should be conducted regularly. Side effects are dose related, including tremor and nausea.

Verapamil

Similar to lithium, two well-controlled trials were conducted on the use of verapamil, including the comparator study with lithium.[80] The second[81] trial compared verapamil 360 mg daily with placebo in a brief 2-week trial. Verapamil was statistically superior in both trials. Adverse events typical of verapamil include constipation, decreased blood pressure, and reduced heart rate. The mechanism of action is unknown, but may involve vascular smooth muscle and modulation of serotonin. In verapamil therapy, doses of 240 mg/day up to 720 mg/day may be required. At higher doses, EKG (electrocardiogram) monitoring should be performed.

Anti-epileptic Drugs

There is a paucity of data on the use of anti-epileptic drugs (AEDs), with almost all based on open-label studies of valproic acid and topiramate.[82–85] One double-blind placebo-controlled trial[86] of divalproex failed to show benefit.

Miscellaneous Therapies

Melatonin is believed to regulate circadian processes, and prompted a study of its use in cluster headache. In a single placebo-controlled trial[87] of melatonin 10 mg a day for 2 weeks, the results were statistically significant. Capsaicin, which acts to deplete Substance P from neuronal sites, has been investigated for use in cluster headache. Capsaicin applied topically as an intranasal treatment for cluster headache[88] suggested efficacy.

Many other therapies have been advocated over the years without evidence for efficacy, including testosterone, H1 antagonists, H2 antagonists, tricyclic antidepressants, misoprostol, hyperbaric oxygen, and stimulants. Lacking evidence from controlled clinical trials, the paucity of agents with evidence demonstrating efficacy may necessitate these lesser agents being used in refractory patients.

Intractable Cluster Headache Treatment

Occipital Nerve Blocks

Several placebo-controlled trials of occipital nerve block have been reported in the prevention of cluster headache. In the first,[89] subjects received a combination of short- and long-acting corticosteroids with a local anesthetic, compared to placebo. Of the active-treatment subjects, 85% responded, as compared with none of those receiving placebo. Onset of remission of cluster headaches occurred by the third day. A 2011 study had a 2-to-1 active-to-placebo response with the injection of a corticosteroid versus placebo.[90] As in the previous study, response time was brisk for the corticosteroid group. A more recent open-label study[91] showed that the duration of response was 3 weeks.

Surgery

Hypothalamic Stimulation

The first hypothalamic implantation and stimulation for CCH was reported in 2000.[92] The positive results from this study prompted other patients to be implanted.[93–96] About 50% of the patients achieved a 50% improvement in their cluster headaches. Results ranged from pain free to no response. Some maintained a pain-free remission for several years following discontinuation of treatment based on the stimulation method. Adverse events are not uncommon, but in general are not severe.

Jannetta Procedure

The most complicated of the procedures used for cluster headaches is microvascular decompression of the trigeminal nerve (the Jannetta procedure). Up to 75% of patients obtain at least a 50% reduction in the frequency of cluster attacks. The patients were followed for approximately 5 years.[97] One-third of the patients who initially responded had loss of effect over the course of time, and repeating the procedure produced no restoration benefit in most cases.

Occipital Nerve Stimulation

The neuroanatomical findings of convergence of cervical peripheral nerves with dural input of the trigeminal nerve onto second-order nociceptors in the trigeminocervical complex[98] has formed the foundation of exploration of this procedure in headache disorders. Multiple randomized controlled trials for occipital nerve stimulation (ONS) in migraine have been published,[99–101] of which the results overall have been mixed with broad ranges of improvement based on patient selection and endpoints. Also, a substantial issue with complications arose, including infections and lead migration, which may be in part related to the type of lead and technical issues of the surgery itself. The results in cluster headache appear to be somewhat more promising, although large-scale active versus sham trials are lacking. The procedure does at least appear to be safer than deep brain stimulation for refractory cluster headache.[102,103]

Other Procedures

Percutaneous retrogasserian injunction[104] with glycerol of the second division of the trigeminal nerve (while avoiding the first division of the nerve) has been performed in cluster headache with some efficacy. Radiofrequency rhizotomy of the sphenopalatine ganglion[105] demonstrates response in about 60% of patients with EC and in about 30% with CCH. These procedures, along with the results of occipital nerve stimulation, have led to stimulation of the sphenopalatine ganglia.

A randomized double-blind active versus sham-controlled clinical trial in refractory CCH was undertaken.[106] This study examined only acute treatment of individual attacks of cluster headache. Pain relief occurred rapidly in most patients. Adverse events are common in early treatment, but tend to abate with time and repeated use.

Gamma Knife Irradiation of the Trigeminal Root Outlet

Gamma knife irradiation of the trigeminal root outlet, a procedure more commonly used in the treatment of trigeminal neuralgia, has been suggested[107] for use in cluster headache, where five of six patients treated with this technique improved. Significant risks may be associated with the procedure, which can include corneal anesthesia and anesthesia dolorosa.

Histamine Desensitization

Intravenous histamine desensitization in the treatment of cluster headache has been employed since the 1930s, when it was developed by Bayard Horton, MD, at the Mayo Clinic.[108] Its use has waxed and waned over the subsequent decades, with both supporters and detractors. More recent reports[109,110] of IV histamine desensitization as part of the treatment of intractable cluster headache have indicated significant benefit. Subcutaneous administration of histamine has Level B evidence in the most recent migraine guidelines.[111] The mechanism of action is unknown, but may relate to downregulation of the cGMP-NO-L-arginine pathway and subsequent effects on CGRP and other vasoactive peptides associated with cluster headache.[112,113]

Treatment of the Other Trigeminal Autonomic Cephalalgia

Indomethacin is the prototypical drug that has been associated with the various cluster variant syndromes. It is the drug of choice for hemicrania continua and chronic paroxysmal hemicrania.[114] Doses of 100 to 200 mg per day, in divided doses, are typically used for initiation of treatment, and response is usually seen within the first 4 weeks of therapy. Dose reductions based on response are made over time, to achieve the lowest dose that maintains remission. Precautions regarding gastrointestinal adverse events should be provided to the patient. Monitoring of laboratory parameters for long-term safety is important. Indomethacin appears to be relatively exclusive in its effectiveness, and this may relate to its differences from other NSAIDs.[115,116]

For SUNCT or SUNA, no significant studies with indomethacin treatment have demonstrated consistent response. In SUNCT, the preferred agent appears to be lamotrigine. Patients with SUNA may respond better to gabapentin.[117]

There has been a host of reports of diverse therapies for all of the cluster variant syndromes. These treatments include tricyclic antidepressants, botulinum toxin, surgical interventions as used for cluster headaches, intravenous corticosteroids, various AEDs, intravenous lidocaine, and celecoxib. All of these therapies have helped at least one patient.

CONCLUSION

Cluster headache is a relatively rare condition among the primary headache disorders. Recognition of this disorder is important to provide optimal therapy. The treatment of cluster headache is focused on preventive therapies because of the high frequency of attacks that patients may experience. Acute therapies such as oxygen are highly reliable and safe. Preventive therapies may be staged to provide early response, such as may occur with corticosteroid therapy and various long-term, better tolerated treatments. Patients with chronic cluster headache tend to become refractory to medical treatment, and specialized medical therapies or surgery may be required for intractable cases.

References

1. Isler H. Episodic cluster headache from a textbook of 1745: van Swieten's classic description. *Cephalalgia*. 1993;13:172–174.
2. Headache Classification Committee of the International Headache Society. The International Classification of Headache Disorders, 3rd edition (beta version). *Cephalalgia*. 2013;33:629–808.
3. Medina JL, Diamond S. Cluster headache variant. Spectrum of a new headache syndrome. *Arch Neurol*. 1981;38:705–709.
4. Silberstein SD, Niknam R, Rozen TD, Young WB. Cluster headache with aura. *Neurology*. 2000;54:219–221.
5. Graham JR. Cluster headache. *Headache*. 1972;11:175–185.
6. Manzoni GC. Cluster headache and lifestyle: remarks on a population of 374 male patients. *Cephalalgia*. 1999;19:88–94.
7. Rogardo AZ, Harrison RH, Graham JR. Personality profiles in cluster headache, migraine and normal controls. *Arch Neurol (Madr)*. 1974;37:227–241.
8. Ekbom K, Svensson DA, Traff H, Waldenlind E. Age at onset and sex ratio in cluster headache: observations over three decades. *Cephalalgia*. 2002;22:94–100.
9. Torelli P, Colgno D, Manzoni G. Gender ratio in cluster headache. In: Olesen J, Goadsby P, eds. *Cluster Headache and Related Conditions*. New York, NY: Oxford University Press; 1999:48–51.
10. Granella F, on behalf of the Italian Cooperative Study Group on the Epidemiology of Cluster Headache (ICECH). Case control study on the epidemiology of cluster headache. In: Olesen J, Goadsby P, eds. *Cluster Headache and Related Conditions*. New York, NY: Oxford University Press; 1999:37–41.
11. Manzoni GC. Gender ratio of cluster headache over the years: a possible role of changes in lifestyle. *Cephalalgia*. 1988;18:138–142.
12. Leone M, Russel MB, Rigamonti A, Attansaio A, Grazzi L, D'Amico D, et al. Increased familial risk of cluster headache. *Neurology*. 2001;56:1233–1236.
13. Russel MB. Epidemiology and genetics of cluster headache. *Lancet Neurol*. 2004;3:279–284.
14. Holland P, Goadsby P. Cluster headache, hypothalamus and orexin. *Curr Pain Headache Rep*. 2009;13:147–154.
15. Manzoni GC, Terzano MG, Bono G, Micieli G, Martucci N, Nappi G. Cluster headache – clinical findings in 180 patients. *Cephalalgia*. 1983;3:21–30.
16. Manzoni GC. Cluster headache and lifestyle: remarks on a population of 374 male patients. *Cephalalgia*. 1999;18:88–84.
17. Italian Cooperative Study Group on the Epidemiology of Cluster Headache (ICECH). Case–control study on the epidemiology of cluster headache. I: Etiological factors and associated conditions. *Neuroepidemiology*. 1995;14(3):123–127.
18. Sörös P, Vo O, Gerding H, Husstedt IW, Evers S. Enucleation and development of cluster headache: a retrospective study. *BMC Neurol*. 2005;5:6.
19. Sörös P, Frese A, Husstedt IW, Evers S. Cluster headache after dental extraction: implications for the pathogenesis of cluster headache? *Cephalalgia*. 2001;21:619–622.
20. Klasser GD, Balasubramaniam R. Trigeminal autonomic cephalalgias. Part 2: Paroxysmal hemicrania. *Oral Surg Oral Med Oral Pathol Oral Radiol Endod*. 2007;104:640–646.
21. Sjaastad O, Saunte C, Salvesen R, Fredriksen TA, Seim A, Røe OD, et al. Shortlasting unilateral neuralgiform headache attacks with conjunctival injection, tearing, sweating, and rhinorrhea. *Cephalalgia*. 1989;9:147–156.
22. Matharu MS, Cohen AS, Boes CJ, Goadsby PJ. Short-lasting unilateral neuralgiform headache with conjunctival injection and tearing syndrome: a review. *Curr Pain Headache Rep*. 2003;7:308–318.
23. Sjaastad O1, Bakketeig LS. The rare, unilateral headaches. Vågå study of headache epidemiology. *J Headache Pain*. 2007;8:19–27.
24. Kudrow L. The cyclic relationship of natural illumination to cluster period frequency. *Cephalalgia*. 1987;7(Suppl 7):76–78.
25. Graham J. Cluster headache. The relation to arousal, relaxation and autonomic tone. *Headache*. 1990;30:145–151.
26. Pringsheim T. Cluster headache: evidence for a disorder of circadian rhythm and hypothalamic function. *Can J Neurol Sci*. 2002;29:33–40.
27. Costa A, Leston JA, Cavallini A, Nappi G. Cluster headache and periodic affective illness: common chronobiologic features. *Funct Neurol*. 1998;13:263–272.
28. Rosen T. Trigeminal autonomic cephalalgias. *Neurologic Clinics*. 2009;27:537–556.
29. Peres MF. Melatonin, the pineal gland and their implications for headache disorders. *Cephalalgia*. 2005;25:403–411.
30. Mosnaim AD, Wolf ME, Lee G, Puente J, Garmnerdsiri S, Freitag FG, et al. Plasma degradation of methionine–enkephalin by cluster headache patients (*In vitro* studies). *Headache Q*. 1990;1:79–83.
31. Sprenger T, Willoch F, Miederer M, Schindler F, Valet M, Berthele A, et al. Opioidergic changes in the pineal gland and hypothalamus in cluster headache: a ligand PET study. *Neurology*. 2006;66:1108–1110.
32. Matharu MS, Cohen AS, McGonigle DJ, Ward N, Frackowiak RS, Goadsby PJ. Posterior hypothalamic and brainstem activation in hemicrania continua. *Headache*. 2004;44:747–761.
33. May A, Bahra A, Büchel C, Turner R, Goadsby PJ. Functional Magnetic Resonance Imaging in spontaneous attacks of SUNCT: Short-lasting neuralgiform headache with conjunctival injection and tearing. *Ann Neurol*. 1999;46:791–794.

34. Sjöstrand C, Modin H, Masterman T, Ekbom K, Waldenlind E, Hillert J. Analysis of nitric oxide synthase genes in cluster headache. *Cephalalgia*. 2002;22:758−764.

35. Freitag FG. Presented at the National Headache Foundation Research Summit, Genetics of Headache: Do the Genetics of Histamine Metabolism Predict Response to Treatment with Histamine in Cluster Headache?, Palm Springs, California, February 10, 2004.

36. Kudrow L. The pathogenesis of cluster headache. *Curr Opin Neurol*. 1994;7:278−282.

37. Kudrow L, Kudrow DB. The role of chemoreceptor activity and oxyhemoglobin desaturation in cluster headache. *Headache*. 1993;33:483−484.

38. Carter D. Cluster headache mimics. *Curr Pain Headache Rep*. 2004;8:133−139.

39. Edvinsson L, Uddman R. Neurobiology in primary headaches. *Brain Res Brain Res Rev*. 2005;48:438−456.

40. Goadsby PJ. Pathophysiology of cluster headache: a trigeminal autonomic cephalgia. *Lancet Neurol*. 2002;1:251−257.

41. May A, Leone M, Boecker H, Sprenger T, Juergens T, Bussone G, et al. Hypothalamic deep brain stimulation in positron emission tomography. *J Neurosci*. 2006;26:3589−3593.

42. Sanchez del Rio M. Functional neuroimaging of headaches. *Lancet Neurol*. 2004;3:645−651.

43. Cortelli P, Guaraldi P, Leone M, Pierangeli G, Barletta G, Grimaldi D, et al. Effect of deep brain stimulation of the posterior hypothalamic area on the cardiovascular system in chronic cluster headache patients. *Eur J Neurol*. 2007;14:1008−1015.

44. Gouveia RG, Parreira E, Pavão Martins I. Autonomic features in cluster headache. Exploratory factor analysis. *J Headache Pain*. 2005;6:20−23.

45. Goadsby PJ, Edvinsson L. Neuropeptide changes in a case of chronic paroxysmal hemicrania − evidence for trigemino-parasympathetic activation. *Cephalalgia*. 1996;16:448−450.

46. Strittmatter M, Hamann GF, Grauer M, Fischer C, Blaes F, Hoffmann KH, et al. Altered activity of the sympathetic nervous system and changes in the balance of hypophyseal, pituitary, and adrenal hormones in patients with cluster headache. *Neuroreport*. 1996;7:1229−1234.

47. Drummond PD. Mechanisms of autonomic disturbance in the face during and between attacks of cluster headache. *Cephalalgia*. 2006;26:633−641.

48. Meineri P, Pellegrino G, Rosso MG, Grasso E. Systemic autonomic involvement in episodic cluster headache: a comparison between active and remission periods. *J Headache Pain*. 2005;6:240−243.

49. Francis GJ, Becker WJ, Pringsheim TM. Acute and preventative pharmacologic treatment of cluster headache. *Neurology*. 2010;75:463−473.

50. The Sumatriptan Study Group. Treatment of acute cluster headache with sumatriptan. *N Engl J Med*. 1991;325:322−326.

51. Ekbom K, Monstad I, Prusinski A, Cole JA, Pilgrim AJ, Noronha D. Subcutaneous sumatriptan in the acute treatment of cluster headache: a dose comparison study. The Sumatriptan Cluster Headache Study Group. *Acta Neurol Scand*. 1993;88:63−69.

52. Göbel H, Lindner V, Heinze A, Ribbat M, Deuschl G. Acute therapy for cluster headache with sumatriptan: findings of a one year long-term study. *Neurology*. 1998;51:908−911.

53. Ekbom K, Waldenlind E, Cole J, Pilgrim A, Kirkham A. Sumatriptan in chronic cluster headache: results of continuous treatment of eleven months. *Cephalalgia*. 1992;12:254−256.

54. Ekbom K, Krabbe A, Micieli G, Prusinski A, Cole JA, Pilgrim AJ, et al. Cluster headache attacks treated for up to three months with subcutaneous sumatriptan (6 mg) (Sumatriptan Long-Term Study Group). *Cephalalgia*. 1995;15:230−236.

55. Hardebo JE, Dahlöf C. Sumatriptan nasal spray (20 mg/dose) in the acute treatment of cluster headache. *Cephalalgia*. 1998;18:487−489.

56. van Vliet JA, Bahra A, Martin V, Ramadan N, Aurora SK, Mathew NT, et al. Intranasal sumatriptan in cluster headache: Randomized placebo-controlled double-blind study. *Neurology*. 2003;60:630−633.

57. Cittadini E, May A, Straube A, Evers S, Bussone G, Goadsby PJ. Effectiveness of intranasal zolmitriptan in acute cluster headache. A randomized placebo-controlled, double-blind cross-over study. *Arch Neurol*. 2006;63:1537−1542.

58. Rapoport AM, Mathew NT, Silberstein SD, Dodick D, Tepper SJ, Sheftell FD, et al. Zolmitriptan nasal spray in the acute treatment of cluster headache: a double-blind study. *Neurology*. 2007;69:821−826.

59. Bahra A, Gawel MJ, Hardebo JE, Millson D, Breen SA, Goadsby PJ. Oral zolmitriptan is effective in the acute treatment of cluster headache. *Neurology*. 2000;54:1832−1839.

60. Diamond S, Robbins L, Freitag FG. Acute treatment of cluster headache. *Practical Pain Manage*. 2010;10:56−62.

61. Fogan L. Treatment of cluster headache. A double blind comparison of oxygen vs. air inhalation. *Arch Neurol*. 1985;42:362−363.

62. Kudrow L. Response of cluster headache attacks to oxygen inhalation. *Headache*. 1981;21:1−4.

63. Rozen TD. High oxygen flow rates for cluster headache. *Neurology*. 2004;63:593.

64. Di Sabato F, Fusco BM, Pelaia P, Giacovazzo M. Hyperbaric oxygen therapy in cluster headache. *Pain*. 1993;52:243−245.

65. Kudrow L, Kudrow DB. Association of sustained oxyhemoglobin desaturation and onset of cluster headache attacks. *Headache*. 1990;30:474−480.

66. Akerman S, Holland PR, Lasalandra MP, Goadsby PJ. Oxygen inhibits neuronal activation in the trigeminocervical complex after stimulation of trigeminal autonomic reflex, but not direct dural activation of trigeminal afferents. *Headache*. 2009;49:1131−1143.

67. Andersson PG, Jespersen LT. Dihydroergotamine nasal spray in the treatment of attacks of cluster headache. A double-blind trial versus placebo. *Cephalalgia*. 1986;6:51−54.

68. Costa A, Pucci E, Antonaci F, Sances G, Granella F, Broich G, et al. The effect of intranasal cocaine and lidocaine on nitroglycerin-induced attacks in cluster headache. *Cephalalgia*. 2000;20:85−91.

69. Schürks M, Rosskopf D, de Jesus J, Jonjic M, Diener HC, Kurth T. Predictors of acute treatment response among patients with cluster headache. *Headache*. 2007;47:1079−1084.

70. Tyagi A, Matharu M. Evidence base for the medical treatments used in cluster headache. *Curr Pain Headache Rep*. 2009;13:168−178.

71. Horton B. Histaminic cephalgia. *Lancet*. 1952;2:92−98.

72. Kudrow L. Comparative results of prednisone, methysergide, and lithium therapy in cluster headache. In: Greene R, ed. *Current Concepts in Migraine Research*. New York, NY: Raven Press; 1978:159−163.

73. Couch JR, Ziegler DK. Prednisone therapy for cluster headache. *Headache*. 1978;18:219−221.

74. Mather PJ, Silberstein SD, Shulman EA, Hopkins MM. The treatment of cluster headache with repetitive intravenous dihydroergotamine. *Headache*. 1991;31:525−532.

75. Mueller L, Gallagher RM, Ciervo CA. Methylergonovine maleate as a cluster headache prophylactic: a study and review. *Headache*. 1997;37:437−442.

76. Monstad I, Krabbe A, Micieli G, Prusinski A, Cole J, Pilgrim A, et al. Preemptive oral treatment with sumatriptan during a cluster period. *Headache*. 1995;35:607−613.

77. Loder E. Naratriptan in the prophylaxis of cluster headache. *Headache*. 2002;42:56−57.

78. Treiser SL, Cascio CS, O'Donohue TL, Thoa NB, Jacobowitz DM, Kellar KJ. Lithium increases serotonin release and decreases serotonin receptors in the hippocampus. *Science*. 1981;213:1529−1531.

79. Steiner TJ, Hering R, Couturier EGM, Davies PTG, Whitmarsh TE. Double-blind placebo-controlled trial of lithium in episodic cluster headache. *Cephalalgia*. 1997;17:673−675.

80. Bussone G, Leone M, Peccarisi C, Micieli G, Granella F, Magri M, et al. Double-blind comparison of lithium and verapamil in cluster headache prophylaxis. *Headache*. 1990;30: 411–417.

81. Leone M, D'Amico D, Frediani F, Moschiano F, Grazzi L, Attanasio A. Verapamil in the prophylaxis of episodic cluster headache: a double-blind study versus placebo. *Neurology*. 2000;54:1382–1385.

82. Hering R, Kuritsky A. Sodium valproate in the treatment of cluster headache: an open clinical trial. *Cephalalgia*. 1989;9:195–198.

83. Gallagher RM, Mueller LL, Freitag FG. Divalproex sodium in the treatment of migraine and cluster headaches. *J Am Osteopath Assoc*. 2002;102:92–94.

84. Wheeler SD, Carrazana EJ. Topiramate-treated cluster headache. *Neurology*. 1999;53:234–236.

85. Mathew NT, Kailasam J, Meadors L. Prophylaxis of migraine, transformed migraine, and cluster headache with topiramate. *Headache*. 2002;42:796–803.

86. El Amrani M, Massiou H, Bousser MG. A negative trial of sodium valproate in cluster headache: methodological issues. *Cephalalgia*. 2002;22:205–208.

87. Leone M, D'Amico D, Moschiano F, Fraschini F, Bussone G. Melatonin versus placebo in the prophylaxis of cluster headache: a double-blind pilot study with parallel groups. *Cephalalgia*. 1996;16:494–496.

88. Solomon GD, Kunkel RS, Frame JR. Intranasal capsaicin cream in cluster headache. In: Rose FC, ed. *New Advances in Headache Research 3*. London, UK: Smith-Gordon; 1993:239–244.

89. Ambrosini A, Vandenheede M, Rossi P, Aloj F, Sauli E, Pierelli F, et al. Suboccipital injection with a mixture of rapid- and long-acting steroids in cluster headache: a double-blind placebo-controlled study. *Pain*. 2005;118:92–96.

90. Leroux E, Valade D, Taifas I, et al. Suboccipital steroid injections for transitional treatment of patients with more than two cluster headache attacks per day: a randomised, double-blind, placebo-controlled trial. *Lancet Neurol*. 2011;10:891–897.

91. Lambru G, Abu Bakar N, Stahlhut L, McCulloch S, Miller S, Shanahan P, et al. Greater occipital nerve blocks in chronic cluster headache: a prospective open-label study. *Eur J Neurol*. 2014;21:338–343.

92. Proietti Cecchini A, Mea E, Tullo V, Peccarisi C, Bussone G, Leone M. Long-term experience of neuromodulation in TACs. *Neurol Sci*. 2008;29(Suppl 1):S62–S64.

93. Franzini A, Leone M, Messina G, Cordella R, Marras C, Bussone G, et al. Neuromodulation in treatment of refractory headaches. *Neurol Sci*. 2008;29(Suppl 1):S65–S68.

94. Rasche D, Klase D, Tronnier VM. Neuromodulation in cluster headache. Clinical follow-up after deep brain stimulation in the posterior hypothalamus for chronic cluster headache, case report – Part II. *Schmerz*. 2008;22(Suppl 1):37–40.

95. Magis D, Schoenen J. Neurostimulation in chronic cluster headache. *Curr Pain Headache Rep*. 2008;12:145–153.

96. Leone M, Franzini A, Cecchini AP, Broggi G, Bussone G. Hypothalamic stimulation for cluster headache. *J Clin Neurosci*. 2008;15:334–335.

97. Lovely TJ, Kotsiakis X, Jannetta PJ. The surgical management of chronic cluster headache. *Headache*. 1998;38:590–594.

98. Magis D, Bruno MA, Fumal A, Gérardy PY, Hustinx R, Laureys S, et al. Central modulation in cluster headache patients treated with occipital nerve stimulation: an FDG-PET study. *BMC Neurol*. 2011;11:25.

99. Palmisani S, Al-Kaisy A, Arcioni R, Smith T, Negro A, Lambru G, et al. A six year retrospective review of occipital nerve stimulation practice – controversies and challenges of an emerging technique for treating refractory headache syndromes. *J Headache Pain*. 2013;14:67.

100. Mueller O, Diener HC, Dammann P, Rabe K, Hagel V, Sure U, et al. Occipital nerve stimulation for intractable chronic cluster headache or migraine: a critical analysis of direct treatment costs and complications. *Cephalalgia*. 2013;33:1283–1291.

101. Brewer AC, Trentman TL, Ivancic MG, Vargas BB, Rebecca AM, Zimmerman RS, et al. Long-term outcome in occipital nerve stimulation patients with medically intractable primary headache disorders. *Neuromodulation*. 2013;16:557–562.

102. Magis D, Gerardy PY, Remacle JM, Schoenen J. Sustained effectiveness of occipital nerve stimulation in drug-resistant chronic cluster headache. *Headache*. 2011;51:1191–1201.

103. Fontaine D, Christophe Sol J, Raoul S, Fabre N, Geraud G, Magne C, et al. Treatment of refractory chronic cluster headache by chronic occipital nerve stimulation. *Cephalalgia*. 2011;31:1101–1105.

104. Pieper DR, Dickerson J, Hassenbusch SJ. Percutaneous retrogasserian glycerol rhizolysis for the treatment of chronic intractable cluster headaches: long-term results. *Neurosurgery*. 2000;46:363–368.

105. Taha JM, Tew Jr JM. Long-term results of radiofrequency rhizotomy in the treatment of cluster headache. *Headache*. 1995;35:193–196.

106. Ansarinia M, Rezai A, Tepper SJ, Steiner CP, Stump J, Stanton-Hicks M, et al. Electrical stimulation of sphenopalatine ganglion for acute treatment of cluster headaches. *Headache*. 2010;50:1164–1174.

107. Ford RG, Ford KT, Swaid S, Young P, Jennelle R. Gamma knife treatment of refractory cluster headache. *Headache*. 1998;38:3–9.

108. Horton BT, Maclean AR, Craig WMcK. A new syndrome of vascular headache: results of treatment with Histamine: Preliminary Report. *Proc Staff Meet Mayo Clin*. 1939;14:257–260.

109. Diamond S, Freitag FG, Prager J, Gandhi S. Treatment of intractable cluster. *Headache*. 1986;26:42–46.

110. Diamond S, Freitag FG, Diamond ML, Urban GJ. IV histamine desensitization therapy in intractable cluster headache. *Headache Quarterly*. 1998;9:55–59.

111. Holland S, Silberstein SD, Freitag F, Dodick DW, Argoff C, Ashman E. Evidence-based guideline update: NSAIDs and other complementary treatments for episodic migraine prevention in adults. *Neurology*. 2012;78:1346–1353.

112. Taylor JE, Richelson E. Desensitization of Histamine H1 receptor-mediated Cyclic GMP formation in mouse neuroblastoma cells. *Mol Pharmacol*. 1979;15:462–471.

113. Sarchielli P, Alberti A, Codini M, Floridi A, Gallai V. Nitric oxide metabolites, prostaglandins and trigeminal vasoactive peptides in internal jugular vein blood during spontaneous migraine attacks. *Cephalalgia*. 2000;20:907–918.

114. Sjaastad O, Vincent M. Indomethacin responsive headache syndromes: chronic paroxysmal hemicrania and hemicrania continua. How they were discovered and what we have learned since. *Funct Neurol*. 2010;25:49–55.

115. Summ O, Evers S. Mechanism of action of indomethacin in indomethacin-responsive headaches. *Curr Pain Headache Rep*. 2013;17:327.

116. Summ O, Andreou AP, Akerman S, Goadsby PJ. A potential nitrergic mechanism of action for indomethacin, but not of other COX inhibitors: relevance to indomethacin-sensitive headaches. *J Headache Pain*. 2010;11:477–483.

117. Pareja JA1, Alvarez M, Montojo T. SUNCT and SUNA: Recognition and Treatment. *Curr Treat Options Neurol*. 2013;15:28–39.

12

Tension-Type Headache

Robert G. Kaniecki

Department of Neurology, University of Pittsburgh Medical Center, Pittsburgh, Pennsylvania, USA

CLASSIFICATION

A headache is considered secondary when it develops in close temporal relation to any medical condition which is known to cause headache. Primary headache disorders arise from biological disorders of the brain without clear evidence of structural or systemic causation. Tension-type headache (TTH) is the most common, and yet perhaps the most poorly defined, primary headache syndrome. Over the years, it has been variably labeled as "muscle contraction," "stress," "psychogenic," "ordinary," or "essential" headache. The term "tension-type" was initially formally selected by the first Classification Committee of the International Headache Society in 1988, and subsequently maintained through updated classifications published in 2004 and 2013.[1] It was initially chosen to reflect possible contributions from either mental or muscular tension. For years, tension-type headache was considered to be psychological in origin. Numerous studies published over the past 25 years, since that first formal classification, suggest neurobiological causation. The International Classification of Headache Disorders, 3rd edition/beta version (ICHD-III beta) recognizes infrequent episodic (averaging <1 day/month), frequent episodic (averaging 1 to 14 days/month), and chronic (averaging >14 days/month) subtypes (Box 12.1).[1] Division of tension-type headache along these temporal lines has been found to be clinically relevant.

Infrequent tension-type headache is a benign disorder rarely requiring medical attention. Most patients successfully self-treat. Frequent tension-type headache may require medical attention and prescription medications, or other forms of management. Most of these patients are successfully treated by primary care physicians. Chronic tension-type headache is a serious and often disabling disorder which is frequently refractory to even aggressive treatment programs. Such patients should be referred to neurologists or headache specialists. Although the ICHD-III beta system then classifies all tension-type headache categories into those with or without associated pericranial tenderness, such additional classification has not been shown to be clinically helpful. Finally, similar to other primary headaches, tension-type headache may be termed "probable" when missing one of the features required for the diagnosis of a subtype of tension-type headache. Such a distinction also seems to possess little clinical importance.

CLINICAL PRESENTATION

The clinical features of tension-type headache are non-specific, and similar phenotypic pictures may be seen in those with secondary headaches or with migraine. For this reason, the formal diagnosis of tension-type headache provides a list of largely negative criteria excluding other headache disorders. The pain is typically described as lacking focal localization (typically bilateral or diffuse), severity (mild to moderate), pulsation, or aggravation by physical activity (often stable or improved). There is no definable aura or neurological symptoms. Sensitivities to light, noise, or odor are typically absent or minor. Nausea may be present in the chronic subtype, but vomiting is never seen. There is no prodrome or postdrome. Cranial autonomic features such as conjunctival injection, tearing, nasal congestion, or rhinorrhea are not noted. All of these elements tend to thus exclude other primary headache disorders such as migraine and the trigeminal autonomic cephalgias. The ICHD-III beta criteria also ultimately require that other secondary headache disorders be excluded.[1]

In its most prevalent episodic form, tension-type headache is characterized by attacks of non-disabling headache lasting hours to days. The episodes are

DOI: http://dx.doi.org/10.1016/B978-0-12-800901-7.00012-4

BOX 12.1

DIAGNOSTIC CRITERIA FOR TENSION-TYPE HEADACHE

International Classification of Headache Disorders, 3rd edition

2.1 Infrequent Episodic Tension-type Headache

A. At least 10 episodes occurring on <1 day/month on average (<12 days/year) and fulfilling criteria B−D listed below

B. Headache lasting from 30 minutes to 7 days

C. Headache has at least two of the following pain characteristics:
- Bilateral location
- Mild or moderate intensity (may inhibit, but does not prohibit, activity)
- Pressing/tightening (non-pulsating) quality
- No aggravation through climbing stairs or similar routine physical activity

D. Both of the following:
- No nausea or vomiting (anorexia may still occur)
- No more than one of photophobia or phonophobia

E. Not better accounted for by another ICHD-III diagnosis

2.2 Frequent Episodic Tension-type Headache

A. At least 10 episodes of headache occurring on 1−14 days per month on average for >3 months (>12 and <180 days per year) and fulfilling criteria B−E

B. Headache lasting from 30 minutes to 7 days

C. Headache has at least two of the following pain characteristics:
- Bilateral location
- Mild or moderate intensity (may inhibit, but does not prohibit, activity)
- Pressing/tightening (non-pulsating) quality
- No aggravation through climbing stairs or similar routine physical activity

D. Both of the following:
- No nausea or vomiting (anorexia may still occur)
- No more than one of photophobia or phonophobia

E. Not better accounted for by another ICHD-III diagnosis

2.3 Chronic Tension-type Headache

A. Headache occurring on ≥15 days/month on average for >3 months (≥180 days/year) and fulfilling criteria B−E

B. Lasting hours or days, or may be unremitting

C. Headache has at least two of the following pain characteristics:
- Bilateral location
- Mild or moderate intensity (may inhibit, but does not prohibit, activity)
- Pressing/tightening (non-pulsating) quality
- No aggravation through climbing stairs or similar routine physical activity

D. Both of the following:
- No more than one of photophobia, phonophobia, or mild nausea
- Neither moderate nor severe nausea nor vomiting

E. Not better accounted for by another ICHD-III diagnosis

---

essentially featureless and without symptoms aside from head pain. Dull and steady discomfort is typically reported. Most individuals will describe the quality of the pain as an ache or pressure sensation. The latter may be felt from inside the cranium as a sense of expansion or swelling, while more commonly an external pressure or "vice" feeling is noted. Although often global, patients report bilateral pain that may affect frontal, temporal, or occipital locations. The pain may also shift location during or between attacks, but is unilateral in only 10−20% of cases.[2] Physical exertion is tolerated and frequently results in headache improvement. The majority of episodes develop during daytime hours, but occasionally are present upon awakening. Headache arousing a person from sleep is rarely tension-type in origin and should raise concerns for other conditions. Those with chronic tension-type headache (CTTH) may report discomfort of a relatively continuous nature that persists or progresses throughout the day, and may persist or improve with sleep.

Most of the discomfort of tension-type headache is felt in the head, but sometimes in the cervical region and occasionally in the face or jaw. Tenderness in the muscles of the scalp, neck, or jaw is common, and may be seen during or between headaches.[3] There may be some correlation between the degree of muscle tenderness and the frequency and severity of headache attacks.[4] A non-muscular cutaneous sensitivity to touch or temperature, known as cutaneous allodynia, may also be seen in some with chronic tension-type headache.[5]

The most frequently reported triggers for tension-type headache are mental or physical stressors. The headache develops during the exposure to stress, and, unlike migraine, is only rarely provoked by the "let-down" phase of stress. Changes in systemic homeostasis such as hunger, dehydration, and sleep deprivation also may incite an episode. Excessive alcohol or caffeine consumption, or withdrawal from these substances, may also result in tension-type headaches. Female hormone and weather changes are more commonly correlated with migraine, but may induce tension-type headache in susceptible individuals.[2]

Many patients with migraine are misdiagnosed with tension-type headache. One population-based survey found 32% of misdiagnosed patients fulfilling ICHD criteria for migraine were mistakenly labeled as tension-type headache.[6] Another discovered 71% of patients initially diagnosed with episodic tension-type headache (ETTH) had their diagnosis altered to migraine after a detailed review of diary data.[7] A third population study demonstrated that 84% of those with self-diagnosed "stress" or "tension" headaches met criteria for migraine.[8] Since over 90% of patients consulting clinicians for recurrent episodic headache disorders will ultimately be found to have migraine, and only 3% tension-type headache, it is imperative to first exclude migraine as a possibility before arriving at a tension-type headache diagnosis.[9]

There are many potential explanations for this difficulty in clearly distinguishing these two common primary headache disorders.[10] Migraine may present with pain that at times is bilateral and steady rather than pulsatile, similar to tension-type headache. In contrast, migraineurs report severe discomfort and aggravation of pain with routine activity even when these other pain descriptors are noted. Migraine may be triggered by stress and may exhibit cervical pain, both overlapping with tension-type headache. Yet such headaches should be associated with nausea or significant photophobia and phonophobia — features absent from tension-type attacks. Although misdiagnosis of migraine is important to understand, potential co-diagnosis of both headache disorders in the same patient is also important to recognize. The 1-year prevalence of episodic tension-type headache is similar across those populations without migraine, those with migraine without aura, and those with migraine with aura.[11] Most patients with episodic migraine will describe instances of tension-type headache attacks in addition to typical migraine, or such headache attacks evolving into migraine. Most patients with chronic migraine will report lower-grade headaches that are absent of many of their migraine features and suggestive of tension-type headache. Whether these episodes should be termed "tension-type" or "mild migraine" remains a matter of ongoing debate, although most in the United States tend to favor the latter nomenclature.[12]

DIAGNOSTIC TESTING

Since the clinical features of tension-type headache are non-specific, they may be seen with an assortment of secondary headache conditions linked mechanistically to an identifiable structural or physiological disorder.[13] The underlying etiologies of such secondary headaches may range from benign to serious conditions. Intracranial mass lesions, such as brain tumors, quite commonly present with a picture similar to tension-type headache.[14] Other headaches resulting from elevated intracranial pressure, such as idiopathic intracranial hypertension, may also present in such a fashion. Headaches from giant cell arteritis are more likely to present with a tension-type picture than with "textbook" severe unilateral temporal pain and jaw claudication. Systemic conditions such as anemia, hypothyroidism, and hepatic and renal disease may have headaches with similar features. Viral syndromes, streptococcal pharyngitis, and Lyme disease may appear in this manner.

Given the non-specific nature of the symptoms, it is imperative that the clinician evaluating a patient presenting with tension-type headache remains vigilant for the "red flags" which might indicate the presence of organic underpinnings (Box 12.2).[15] Delineation of the time course of headache development and completion of a thorough neurological examination are critical components in the assessment process. Headaches described as "new" or "different," as well as those which progress in frequency or intensity, should raise particular concern. Most secondary headaches declare themselves within a period of several weeks to several months. Those presenting with a stable pattern of headache for over 6 months' duration, and a normal neurological examination, are unlikely to have serious secondary pathology. For those with secondary headache possibilities, diagnostic work-up would begin with brain neuroimaging. Since those with tension-type headache are unlikely to present emergently, brain magnetic resonance imaging is preferred over computed tomography.[16,17] Occasionally other investigations may be necessary, including cervical spine or sinus imaging, cerebrospinal fluid analysis, or serum tests such as erythrocyte sedimentation rate, C-reactive protein, Lyme or Epstein-Barr virus titers, complete blood count, or thyroid function studies.[18] In the

---

BOX 12.2

RED FLAGS FOR SECONDARY HEADACHE DISORDERS

1. First/worst headache
2. Abrupt onset headache
3. Progression or fundamental change in pattern of headache
4. New headache in those less than 5 years old, greater than 50 years old

5. New headache with cancer, immunosuppression, or pregnancy
6. Headache with syncope or seizure
7. Headache triggered by exertion/Valsalva/sex
8. Neurologic symptoms greater than 1 hour in duration
9. Abnormal general or neurological examination

---

absence of alteration in consciousness, there is no role for electroencephalography in the evaluation of patients presenting with tension-type headache.[19] Although data from over 20 electromyographic studies are available, inconsistent and contradictory findings eliminate such testing from the diagnostic algorithm.[20] There is also no role for evaluating cortical evoked responses (visual, auditory, somatosensory evoked potentials) since these are typically normal in tension-type headache.[21]

EPIDEMIOLOGY AND IMPACT

Tension-type headache is the most common primary headache, with lifetime prevalence of between 30% and 78%.[22] Prevalence values vary based on population, age, and sex. Some variation may also be explained by case definitions and sampling methods. A Danish population study found the 1-year prevalence of ETTH to be 63%, with 71% of women and 56% of men affected. The 1-year prevalence of CTTH was 2% (5% in women, 2% in men).[23] One survey from the general population in the United States established an annual prevalence of 38.3% for episodic tension-type headache and 2.2% for chronic tension-type headache.[24] Pooled results from five population-based studies determined a mean lifetime prevalence of episodic tension-type headache of 46% (range 12—78%), while the population prevalence for chronic tension-type headache across populations is fairly consistent at 2—3%.[25]

Both genetics and environment are felt to influence the prevalence of tension-type headache. Population-based genetic epidemiologic studies and evaluations of twin pairs report an increased genetic risk for chronic but not episodic tension-type headache.[26] First-degree relatives appear to have a two- to four-fold increased risk of chronic tension-type headache compared with the general population. No genetic defect has yet been discovered.[27] Interestingly, no differences have been noted between identical and fraternal twin pairs.

Tension-type headache prevalence rates also vary by gender and age.[28] Gender ratios favor women in the range of 1.04 to 1.8. Most develop tension-type headache prior to age 30, with peak prevalence in the fifth decade and subsequent declines with age in both sexes.[29] Despite such declines, rates of ETTH remain above 25% in the seventh decade of life for both men and women, while 1.5% of men and 2.7% of women continue to report CTTH beyond age 60.[30] Correlation between prevalence of episodic tension-type headache and higher educational level has been shown, while increased prevalence among higher income groups has been more difficult to establish.[31]

Both episodic and chronic tension-type headaches are commonly seen in children and adolescents, with prevalence values similar to migraine in these age groups.[32] Prevalence rates for episodic tension-type headache among children and adolescents are reported to range from 10% to 25%, while those for chronic tension-type headache fall into the 0.1% to 5.9% range.[33] Headaches develop at a mean age of 7, with an increasing prevalence between ages 7 and 15. Headache duration, frequency, intensity, and medication use associated with attacks of episodic tension-type headache, are typically lower when compared to those with episodic migraine. Environmental influences in children are significant in the development of tension-type headache. Affected individuals are more likely to report depression, divorced parents, and fewer peer relationships when compared with migraine and headache-free controls.[34]

Due to high population prevalence, the socioeconomic impact of tension-type headache is greater than that of migraine or other primary headache disorders.[25] Activities of daily living are altered in 44% and discontinued in 18% of those experiencing such headache occurrences. Both direct costs, such as use of medical services, and indirect costs, such as those associated with decreased work-related productivity, are substantial. One Danish study determined that 12% of employed tension-type headache sufferers missed work each year, with an estimated total loss of work days each year approximating 820 per 1000 employees — three times the number seen with migraine.[35] This accounts for approximately 10% of all disease-related absenteeism in Denmark. Although societal impact is considerable, individual impact and disability is also substantial. Physical suffering, impairment of social and personal functioning, economic impact, and overall reduced quality of life are factors which are more difficult to quantify. Instruments, including the Migraine Disability Assessment Scale (MIDAS) and the General Health Questionnaire, indicate reduced function, and those with chronic tension-type headache are seven times more likely than controls to be classified as impaired on such quality-of-life instruments.[36] Frequency, rather than severity, of headache appears to deliver a greater impact on disability and quality of life.

COMORBID CONDITIONS

A number of medical and psychiatric disorders are seen with increased prevalence among those with tension-type headache. Insomnia may be one of the most common complaints among this patient population. Sleep deprivation appears to negatively impact pain thresholds.[37] Headache-free individuals with insomnia are more likely to develop tension-type headache over the following decade.[38] Those with tension-type headache are more likely to report sleep disturbances.[39] A blinded controlled polysomnographic study detected significantly greater rates of insomnia, daytime tiredness, anxiety, and reduced subjective sleep quality in tension-type headache patients when compared with controls. These patients, interestingly, had no total sleep-time differences and more slow-wave sleep, a marker of increased sleep quality, suggesting perhaps that tension-type headache patients may have increased sleep requirements.[40] Sleep disturbance is also a risk factor for poor prognosis in those with tension-type headache.[41]

Temporomandibular disorders are listed as a secondary cause of headache disorders, with the anatomic source of pain arising from either the joint or, more commonly, the masticatory musculature. These disorders also have been noted to be more prevalent among those with migraine and with tension-type headache.[42] The relationship may be bidirectional, with tension-type headache aggravating temporomandibular joint pain or *vice versa*.[43] Temporomandibular disorders have also been associated with cutaneous allodynia as a marker for central sensitization — a finding also common to those with chronic tension-type headache.[44]

Patients with tension-type headache should be screened for mood and anxiety disorders. Although the evidence is more robust for those with migraine, a number of studies have established a relationship between tension-type headache and major depression, generalized anxiety disorder, and panic disorder.[45] One meta-analysis demonstrated that children and adolescents with tension-type headache showed more psychopathological symptoms than healthy controls, and at a degree similar to that with migraine.[46] There is excellent evidence to suggest that relaxation and cognitive behavioral therapies are effective in reducing the frequency and severity of chronic pain, including chronic tension-type headache, in children and adolescents.[47]

PATHOPHYSIOLOGY OF TENSION-TYPE HEADACHE

Although the precise pathophysiologic mechanisms underlying tension-type headache are unknown, a neurobiological basis is now widely accepted.[48] Given the heterogeneity in pathophysiological abnormalities discovered, and the strong component of environment in this condition, it is plausible to suggest multiple potential mechanisms. Whether the pain originates from myofascial tissue or central nervous system dysfunction remains a matter of debate and a source of ongoing research. A number of insights on pathophysiologic aspects have been gained over the past decade, implicating a multifactorial process involving both peripheral myofascial factors and central nervous system components.[49] An element of excessive contraction of pericranial and cervical muscles led to one of the original labels applied to this disorder, "muscle contraction headache." Most believe that infrequent episodic tension-type headache likely arises from peripheral pain mechanisms, and chronic tension-type headache from central pain mechanisms, with frequent episodic tension-type headache somewhere in the middle. Environmental influences appear to exert greater influence than genetic factors in the development of tension-type headache.[29]

Stress is the most commonly reported contributing factor to tension-type headache.[50] Other triggering influences include hunger, dehydration, sleep deprivation, and other disorders of body homeostasis. Both physical and psychophysiological stressors may be implicated, and sometimes both may be present. Patients often report a link between stressful days and headache occurrences. Such a connection may be direct through effect on pathophysiological mechanisms, or indirect through impact on headache-related comorbidities such as depression, anxiety, or insomnia. Stress has been shown to increase attention and vigilance to pain, which may impact pain levels or reporting behaviors.[51] Negative influences on muscle activation and pain processing have been noted with stress, both likely contributing to headache.[52] Pain sensitivity is increased by stress more in tension-type headache when compared with controls, and this can be correlated in these patients with the development of a stress-induced headache.[53] One potential model proposes that stress may enhance activation and sensitization of myofascial tissue peripherally while decreasing pain thresholds centrally.[54]

Research has implicated myofascial activity as the most likely source of episodic tension-type headache.[55] A number of studies have demonstrated increased levels of pericranial muscle tenderness in tension-type headache patients when compared with controls. This tenderness has also been correlated directly with both frequency and intensity measures of headache. Since this abnormality has been detected on headache-free days as well as headache days, it is speculated that this represents a cause rather than an effect of headache occurrences. Muscle tenderness may be diffuse, associated with tightness or spasm, or linked with hyperirritable areas within tight muscle known as trigger points. Studies have shown a greater number of pericranial or cervical active and latent trigger points in patients with tension-type headache when compared with controls.[56] These may be found in cases of both episodic and chronic tension-type headache. Lower pressure-pain thresholds have also been shown in the trapezius, frontalis, and temporalis muscles in those with episodic and chronic tension-type headache when compared with controls.[57] Impaired cervical spine range of motion has been identified in this population, and more severe forward head posture in sitting and standing positions may be shown.[58] Positive correlations between these findings and the frequency and duration of headache have been found, and the cervical range of motion impairment may be directly correlated with the degree of forward head posture.

Electromyographic (EMG) studies have been used to enhance our understanding of myofascial dysfunction in tension-type headache. Surface recordings of pericranial and cervical muscle groups have yielded contradictory results, with only half reporting increased EMG activity.[59] Variations in methodological conditions or the use of non-standardized recording techniques may be partially to blame, or perhaps clinical variability in patient populations is responsible. A more consistent finding is the presence of focal areas of increased EMG activity within trigger points. It has been proposed that continuous activity within muscle motor units over a sustained period of time might excite peripheral nociceptors, perhaps through direct mechanical means, local ischemia, or through release of inflammatory mediators.[60] To date, microdialysis studies have not demonstrated abnormally elevated levels of lactate or inflammatory mediators in these muscles.[61]

Central mechanisms, specifically central sensitization of neurons, appear to be more relevant to the pathogenesis of chronic tension-type headache.[62] Allodynia, as a marker of central sensitization, can be established through pressure, thermal, or electrical means. Pressure pain thresholds have been shown to be normal in infrequent episodic tension-type headache but decreased in frequent episodic and chronic tension-type headache. Patients with chronic tension-type headache have been found also to be hypersensitive to electrical and thermal stimuli in a number of tissues (muscle, tendon, nerve) at cephalic and extracephalic sites.[63] These allodynic measures can be found both during and between headache occurrences. Population-based studies have demonstrated a relationship between increased pain sensitivity and both the prevalence and frequency of tension-type headache.[64]

Electrophysiological and neuroimaging studies have also indicated abnormalities in central nervous system function and structure associated with tension-type headache, particularly the chronic subtype. High-density EEG brain mapping reveals abnormal processing of somatosensory-evoked potentials and impaired inhibition of nociceptive input in patients with chronic tension-type headache.[65] Suppression of the R2 component of the blink reflex, reflecting nociceptive-specific trigeminal pathway dysfunction, was seen in chronic but not episodic tension-type headache patients.[66] Brain MRI scans of patients with chronic tension-type headache have displayed significantly reduced density of gray matter structures along the pain matrix, including cingulate and insular and orbitofrontal cortex, right posterior temporal lobe, bilateral parahippocampus, dorsal rostral and ventral pons, and right cerebellum.[30] This decrease in gray matter correlated positively with increasing headache duration in years.

Taking these research data into account, it is possible to develop a model for the pathophysiology of tension-type headache. Enhanced nociceptive input from pericranial and cervical muscle tissue, arising from an assortment of reasons, elicits changes in second-order neurons in the central nervous system. These neurons in the spinal trigeminal nucleus (brainstem) and dorsal horn (cervical spinal cord) undergo central sensitization and plastic changes altering their sensitivity to ongoing stimuli. The nociceptive input to higher-order sensory neurons will then be enhanced, potentially negatively impacting descending antinociceptive pathways and motor pathways. The former might further escalate incoming nociceptive input, while the latter activation of motor neurons might increase muscle activity in the periphery.[22]

MANAGEMENT OF TENSION-TYPE HEADACHE

Non-Pharmacological Treatments

Management of tension-type headache should include a combination of non-pharmacological and pharmacological measures.[31] Evidence of benefit from non-pharmacological measures is unfortunately limited. Avoidance of headache triggers, previously identified by the patient, is certainly of utmost importance, when possible. Most clinicians recommend regulation of sleep schedules, since both sleep deficiency and excess may contribute to more frequent headaches. Spacing caloric intake throughout the day into four to six portions, and encouragement of adequate hydration, are also often suggested. Stress management techniques such as prayer, meditation, or yoga can help. Regular exercise may be invaluable in improving overall pain management. Passive physical manipulation and active cervical muscle stretching may help some, but a systematic review investigating physiotherapy management of headache revealed insufficient data to support its use in this setting.[67] Behavioral therapies are quite useful adjuncts. Relaxation therapy and EMG-guided biofeedback may be helpful for children and adolescents as well as adults. Cognitive behavioral therapy may provide additional benefit in cases of significant depression or anxiety. Studies on acupuncture have been inconclusive.[68] Guidelines for the use of non-pharmacological therapies in tension-type headache have been made available by the European Federation of Neurological Societies (Table 12.1).[69]

Acute Pharmacological Therapies

Episodic tension-type headache is pharmacologically managed through acute headache remedies, many of which do not require a prescription (Table 12.2).[69] Simple analgesics, non-steroidal anti-inflammatory drugs (NSAIDs), and caffeine-containing combination agents are most commonly recommended. Many controlled studies have documented the efficacy of simple analgesics, including acetaminophen, aspirin, and NSAIDs, as well as combination analgesics. Due to the potential for end-organ toxicity and transformation from episodic to chronic tension-type headache, the use of acute medication should be strictly limited to an average of 2 to 3 days per week.

The choice of analgesic for an episode of tension-type headache may be based on drug effectiveness, results from prior therapeutic trials, and patient preference. The presence of absolute or relative contraindications for the use of such medications also may merit serious consideration. Aspirin is more effective than placebo

TABLE 12.1 Non-Pharmacological Preventive Therapies for Tension-type Headache

Treatment	Level of recommendation
EMG Biofeedback	A
Cognitive behavioral therapy	C
Relaxation training	C
Physical therapy	C
Acupuncture	C

Source: Guidelines from the European Federation of Neurological Societies.[69] A level A rating (effective) required at least one convincing class I study or two consistent convincing class II studies. A level B rating (probably effective) required at least one convincing class II study or overwhelming class III evidence. A level C rating (possibly effective) required at least two convincing class III studies.

TABLE 12.2 Acute Therapies for Tension-type Headache

Agent	Dose	Level of recommendation
Acetaminophen	500–1000 mg	A
Aspirin	500–1000 mg	A
Ibuprofen	200–800 mg	A
Ketoprofen	25–50 mg	A
Naproxen	375–550 mg	A
Diclofenac	12.5–100 mg	A
Caffeine	65–200 mg	B

Source: Guidelines from the European Federation of Neurological Societies.[69]

and comparable to the efficacy of acetaminophen in the relief of acute tension-type headache. The use of aspirin may be limited by the presence of bleeding disorders, asthma with aspirin intolerance, gastrointestinal conditions, or the use of anticoagulants. Acetaminophen is typically avoided in the presence of hepatic dysfunction. Given superior efficacy versus both aspirin and acetaminophen in comparative trials, NSAIDs are generally considered the drugs of choice for acute tension-type headache. Ibuprofen and naproxen sodium may be better tolerated than other agents in the class, but all may cause gastrointestinal upset. Their use may be limited by gastrointestinal or bleeding disorders, renal dysfunction, or in the presence of anticoagulation. Two special populations require specific mention when discussing the use of analgesics in tension-type headache. In children and adolescents, aspirin is often avoided due to the possibility of Reye syndrome, while only acetaminophen is considered safe throughout all three trimesters of pregnancy.

Caffeine-containing compounds may be useful in those cases that fail to respond to simple or NSAID analgesics. Clinical trials have shown addition of caffeine (130–200 mg) significantly increases the efficacy of aspirin, acetaminophen, and ibuprofen.[70] There is no evidence that the use of muscle relaxants is beneficial in acute attacks of tension-type headache. Studies of triptans in episodic tension-type headache establish efficacy figures similar to those seen with acute migraine, but the difference between drug and placebo in tension-type trials was insignificant. However, tension-type attacks in individuals with coexisting migraine do respond to acute triptan use with statistical superiority over placebo.[71] Another combination analgesic includes the sedative butalbital with acetaminophen or aspirin and also, often, with caffeine. Some of these compounds also include codeine. A number of studies have demonstrated efficacy of these agents in the treatment of episodic tension-type headache.[72] Most of these studies would now be deemed inadequate according to present International Headache Society guidelines for the conduction of clinical trials. Since these publications predated the release of formal ICHD criteria for headache disorders, the populations may have been heterogeneous. Butalbital-containing compounds may be helpful in those patients who have contraindications to, or have previously failed, first-line analgesics. However, such drugs must be used with caution and their supplies limited strictly to use of 2 days per week due to an elevated risk of medication overuse headache.[73] Given similarly high rates of medication overuse headache and the possibility of dependence or addiction with chronic use, it is also recommended that opioid compounds be avoided in the management of tension-type headache.

Parenteral therapies may be helpful for those patients with tension-type headache presenting in acute care settings.[74] Data from controlled clinical trials have demonstrated safety and efficacy of intravenous forms of metoclopramide, a dopamine agonist, and chlorpromazine, a phenothiazine, in the management of acute tension-type headache.[75] The mechanism for headache relief in this setting is unknown. The combination of metoclopramide with diphenhydramine has been found to be more effective than ketorolac, while subcutaneous sumatriptan was no more effective than placebo in a single trial.[76]

Preventive Pharmacological Therapies

For those patients unable to limit analgesic use to 2 days per week (or 10 days per month) due to high headache frequency, preventive pharmacotherapy may be considered.[33] More frequent use of analgesics leads to the potential for tolerance and dose escalation, side effects and end-organ damage, and medication overuse headache. The latter may be a key contributing factor to the transformation from episodic to chronic tension-type headache in certain susceptible individuals. Declining effectiveness or development of adverse events with acute medications may provide additional indications for preventive medications even in those with fewer than 10 headache days per month. These agents should be initiated at low doses and gradually advanced, based on efficacy and tolerability. Once an effective dose is reached, treatment is typically continued for 6 to 12 months, at which point a trial of medication discontinuation should be undertaken.

Tricyclic antidepressants are the most widely used agents in the prevention of tension-type headache. Prospective controlled trials are few, and only some demonstrate superiority over placebo. The data for the tricyclic antidepressant, amitriptyline, are the most extensive and convincing, and many consider this the drug of choice for tension-type headache prophylaxis. The initial dose should be low, generally beginning with 10- or 25-mg tablets and gradually escalating the dose over several weeks as necessary and tolerated. Clinically effective doses are felt to occur in the 30 mg to 150 mg range.[77] Since sedation is one of the most common side effects, this drug is particularly helpful for those with comorbid insomnia. Weight gain, dry mouth, tachycardia, orthostatic hypotension, and constipation are other common adverse events with amitriptyline. The mechanism of action remains to be determined, but effect on headache appears to be independent of the antidepressant effect (Table 12.3).[78]

Other tricyclic antidepressants may be helpful alternatives if amitriptyline is poorly tolerated. Nortriptyline, a metabolic product of amitriptyline, often produces fewer side effects. Clomipramine and doxepin both have clinical trial data suggesting numerical superiority to placebo, while the data for doxepin were also statistically significant.[79] Protriptyline has been reported to be helpful in chronic tension-type headache. Unlike other tricyclic antidepressants, weight gain and sedation are typically not problematic with this agent, but dry mouth, constipation, and tachycardia are common side effects.

If tricyclic antidepressants fail or are poorly tolerated, other medications are then tried. Studies of the selective serotonin reuptake inhibitors paroxetine and citalopram have shown no superiority over placebo in the management of tension-type headache. However, the noradrenergic-serotonin reuptake inhibitors mirtazapine and extended-release venlafaxine have been shown to be beneficial in single clinical trials.[80] The centrally-acting antispasticity agent, tizanidine, has shown efficacy in a cross-over clinical trial.[81] Although anti-epileptic drugs, such as topiramate and gabapentin, are sometimes used, there are no data demonstrating benefit in tension-type headache prophylaxis.

TABLE 12.3   Effects of Some Antidepressants

Drug	Serotonin inhibition	Norepinephrine inhibition	Dopamine inhibition	Sedative effects	Anticholinergic effects
Amitriptyline	Moderate	Weak	Inactive	Strong	Strong
Bupropion	Weak	Weak	Weak	None	None
Clomipramine	Potent	Inactive	Inactive	Mild	Mild
Desipramine	Weak	Potent	Inactive	Mild	Moderate
Doxepin	Moderate	Moderate	Inactive	Strong	Strong
Fluoxetine	Potent	Weak	Inactive	None	Mild to none
Fluvoxamine	Potent	Inactive	Inactive	Mild	Mild
Imipramine	Fairly potent	Moderate	Inactive	Moderate	Strong
Maprotiline	Weak	Moderate	Inactive	Moderate	Moderate
Mirtazapine	Potent	Weak	Inactive	Moderate	Mild
Nortriptyline	Weak	Fairly potent	Inactive	Mild	Moderate
Paroxetine	Potent	Weak	Weak	None	None
Protriptyline	Weak	Fairly potent	Inactive	None	Strong
Sertraline	Potent	Weak	Weak	None	None
Trazodone	Fairly potent	Weak	Inactive	Strong	Mild
Trimipramine	Weak	Weak	Inactive	Moderate	Moderate
Venlafaxine	Potent	Potent	Weak	Mild	Mild

Adapted from Diamond,[78] Diagnosing and Managing Headaches, 6th edn. Caddo, OK: Professional Communications 2008; pp. 140–141.

Overall, the data for benefit with different pharmacological agents in the prevention of tension-type headache are unimpressive. Two recent meta-analyses arrived at opposite conclusions, with one establishing benefit and the other a lack of benefit for prophylactic drugs.[25,35] Guidelines from the European Federation of Neurological Societies have listed a number of agents deemed effective by available data or expert consensus (Table 12.4).[69]

Given the potential for direct impact on the peripheral myofascial origins of tension-type headache, interest has turned towards muscular injections for those refractory to medical management. In one study of patients with frequent episodic tension-type headache, lidocaine injections into pericranial myofascial trigger points reduced both the frequency and severity of pain.[82] Onabotulinum toxin type A injections were found to be superior to saline injections in one small trial, but larger placebo-controlled trials failed to demonstrate therapeutic efficacy.[34]

Although neurostimulation has been used in the management of select patients with chronic migraine and chronic cluster, there is no evidence that such invasive procedures are beneficial to patients with chronic tension-type headache. Occipital nerve stimulation may benefit those with occipital neuralgia, a condition sometimes confused with tension-type

TABLE 12.4   Pharmacological Preventive Therapies for Tension-type Headache

Agent	Daily dose	Level of recommendation
Amitriptyline	30–75 mg	A
Mirtazapine	30 mg	B
Venlafaxine	150 mg	B
Clomipramine	75–150 mg	B

Source: Guidelines from the European Federation of Neurological Societies.[69]

headache. However, there is no role for this in the primary tension-type headache patient. Similarly, deep brain, vagal nerve, sphenopalatine ganglion, and supraorbital stimulators are not indicated.[83] A Cochrane review of transcutaneous electric nerve stimulation determined there was no conclusive evidence to support its use in the treatment of chronic or recurrent headaches, including tension-type headache.[84]

PROGNOSIS OF TENSION-TYPE HEADACHE

Although tension-type headache may recur for years or even decades, the long-term prognosis is generally considered to be favorable. Aside from the issues associated with transformation into chronic tension-type headache seen in a subset, there appear to be no complications or permanent ramifications to recurrent tension-type headache attacks. In a study of adult outpatients with tension-type headache followed over 10 years, 44% with chronic tension-type headache reported significant improvement or complete resolution, while 29% of those with episodic tension-type headache converted to the chronic subtype.[37] A Danish cross-sectional population study revealed a 45% 2-year remission rate among those with frequent episodic or chronic tension-type headache, while 39% continued to report episodic and 16% chronic tension-type headache.[38] Variables from this study associated with poor outcomes in tension-type headache include baseline chronic tension-type headache, coexisting migraine, sleep difficulties, younger age, and the unmarried state.

CONCLUSIONS

Tension-type headache is the most prevalent but least distinct headache in the general population. Due to this high prevalence, there is substantial societal impact. Diagnosis is based on meeting a set of clinical criteria which are designed to first exclude the presence of secondary or migraine headache disorders. The frequency of attacks is used to subcategorize tension-type headache into infrequent episodic, frequent episodic, and chronic subtypes. Precise pathophysiologic mechanisms underlying tension-type headache are unknown. Pericranial myofascial mechanisms are likely of importance in episodic tension-type headache, while sensitization of central nociceptive pathways and inadequate endogenous anti-nociceptive circuitry appear to be more relevant in chronic tension-type headache. Simple analgesics such as acetaminophen, aspirin, and the NSAIDS are beneficial in the majority. Occasional patients will require use of combination analgesics to address acute attacks. Scientific evidence supporting the use of both non-pharmacological and pharmacological preventive therapies in tension-type headache is quite limited. Amitriptyline is considered to be the drug of choice. Although tension-type headache attacks may occur over years or decades, the overall prognosis is considered to be favorable.

References

1. Headache Classification Committee of the International Headache Society. The International Classification of Headache Disorders, 3rd edition (beta version). *Cephalalgia*. 2013;33 (9):629–808.
2. Rasmussen BK. Migraine and tension-type headache in a general population: psychosocial factors. *Int J Epidemiol*. 1992; 21:1138–1143.
3. Drummond P. Scalp tenderness and sensitivity to pain in migraine and tension headache. *Headache*. 1987;27:45–50.
4. Jensen R, Rasmussen B, Pedersen B, Olesen J. Muscle tension and pressure pain thresholds in headache. A population study. *Pain*. 1993;52:193–199.
5. Ashina S, Babenko L, Jensen R, Ashina M, Magerl W, Bendtsen L. Increased muscular and cutaneous pain sensitivity in cephalic region in patients with chronic tension-type headache. *Eur J Neurol*. 2005;12:543–549.
6. Lipton R, Diamond S, Reed M, Diamond M, Stewart W. Migraine diagnosis and treatment: results from the American Migraine Study II. *Headache*. 2001;41:638–645.
7. Lipton R, Stewart F, Cady R, Hall C, O'Quinn S, Kuhn T, et al. Sumatriptan for the range of headaches in migraine sufferers: results of the Spectrum Study. *Headache*. 2000;40:783–791.
8. Kaniecki R, Ruoff G, Smith T, Barrett P, Ames M, Byrd S, et al. Prevalence of migraine and response to sumatriptan in patients self-reporting tension/stress headache. *Curr Res Med Opin*. 2006;22:1535–1544.
9. Tepper S, Dahlof C, Dowson A. Prevalence and diagnosis of migraine in patients consulting their physician with a complaint of headache. Data from the Landmark Study. *Headache*. 2004;44:856–864.
10. Kaniecki RG. Migraine and tension-type headache: an assessment of challenges in diagnosis. *Neurology*. 2002;58(suppl 6): S15–S20.
11. Ulrich V, Russell M, Jensen R, Olesen J. A comparison of tension-type headache in migraineurs and in non-migraineurs: a population-based study. *Pain*. 1996;67:501–506.
12. Cady R. The convergence hypothesis. *Headache*. 2007;47(Suppl 1): S44–S51.
13. Sacco S, Ricci S, Carolci A. Tension-type headache and systemic medical disorders. *Curr Pain Headache Rep*. 2011;15:438–443.
14. Nelson S, Taylor L. Headaches in brain tumor patients: primary or secondary? *Headache*. 2014;54:776–785.
15. Kaniecki R. Headache assessment and management. *JAMA*. 2003;289:1430–1433.
16. Loder E, Weizenbaum E, Frishberg B, Silberstein S, American Choosing Wisely Task Force. Choosing wisely in headache medicine: the American Headache Society's list of five things physicians and patients should question. American Headache Society Choosing Wisely Task Force. *Headache*. 2013;53:1651–1659.
17. Frishberg B, Rosenberg J, Matchar D, McCrory DC, Pietrzak MP, Rozen, TD et al. Evidence-based guidelines in the primary care setting: neuroimaging in patients with nonacute headache. Available at: <http://protocols.xray.ufl.edu/live_protocols/documents/guidance/UF/headache_guide.pdf>; Accessed April 2014.
18. Evans R. Diagnostic testing for the evaluation of headaches. *Neurol Clin*. 1996;14:1–26.
19. American Academy of Neurology Quality Standards Subcommittee. Practice parameter: the electroencephalogram in the evaluation of headache (summary statement). Report of the Standards Subcommittee. *Neurology*. 1995;45:1411–1413.
20. Rossi P, Vollono C, Valeriani M, Sandrini G. The contribution of clinical neurophysiology to the comprehension of tension-type headache mechanisms. *Clin Neurphysiol*. 2011;122:1075–1085.
21. Magis D, Vigano A, Sava S, d'Elia TS, Schoenen J, Coppola G. Pearls and pitfalls: electrophysiology for primary headaches. *Cephalalgia*. 2013;33:526–539.
22. Bendtsen L, Jensen R. Tension-type headache: the most common, but also the most neglected, headache disorder. *Curr Opin Neurol*. 2006;19:305–309.

23. Rasmussen BK. Epidemiology of headache. *Cephalalgia.* 1995;15:45–68.

24. Schwartz BS, Stewart WF, Simon D, Lipton R. Epidemiology of tension-type headache. *JAMA.* 1998;279:381–383.

25. Stovner L, Hagen K, Jensen R, Katsarava Z, Lipton R, Scher A, et al. The global burden of headache: a documentation of headache prevalence and disability worldwide. *Cephalalgia.* 2007;27:193–210.

26. Ostergaard S, Russell M, Bendtsen L, Olesen J. Increased familial risk of chronic tension-type headache. *BMJ.* 1997;314:1092–1093.

27. Russell M, Iselius L, Ostergaard S, Olesen J. Inheritance of chronic tension-type headache investigated by complex segregation analysis. *Hum Genet.* 1998;102:138–140.

28. Crystal S, Robbins M. Epidemiology of tension-type headache. *Curr Pain Headache Rep.* 2010;14:449–454.

29. Ulrich V, Gervil M, Olesen J. The relative influence of environment and genes in episodic tension-type headache. *Neurology.* 2004;62:2065–2069.

30. Kaniecki R. Tension-type headache in the elderly. *Curr Treat Options Neurol.* 2007;9:31–37.

31. Pryse-Phillips W, Findlay H, Tugwell P, Edmeads J, Murray TJ, Nelson RF. A Canadian population survey on the clinical, epidemiologic, and societal impact of migraine and tension-type headache. *Can J Neurol Sci.* 1992;19:333–339.

32. Lewis D, Gozzo Y, Avner M. The "other" primary headaches in children and adolescents. *Pediatr Neurol.* 2005;33:303–313.

33. Anttila P. Tension-type headache in childhood and adolescence. *Lancet Neurol.* 2006;5:268–274.

34. Monteith T, Sprenger T. Tension-type headache in adolescence and childhood: where are we now? *Curr Pain Headache Rep.* 2010;14:424–430.

35. Rasmussen B, Jensen R, Olesen J. Impact of headache on sickness absence and utilization of medical services: a Danish population study. *J Epidemiol Comm Health.* 1992;46:443–446.

36. Holroyd K, Stensland M, Lipchik G, Hill K, O'Donnell F, Cordingley G. Psychosocial correlates and impact of chronic tension-type headache. *Headache.* 2000;40:3–16.

37. Roehrs T, Hydae M, Blaisdell B, Greenwald M, Roth T. Sleep loss and REM sleep loss are hyperalgesic. *Sleep.* 2006;29:144–151.

38. Odegard S, Sand T, Engstrom M, Stovner L, Zwart J, Hagen K. The long-term effect of insomnia on primary headaches: a prospective, population-based cohort study (HUNT-2 and HUNT-3). *Headache.* 2011;51:570–580.

39. Odegard S, Engstrom M, Sand T, Stovner L, Zwart J, Hagen K. Associations between sleep disturbance and primary headaches: the third Nord-Trondelag Health Study. *J Headache Pain.* 2010;11:197–206.

40. Engstrom M, Hagen K, Bjork M, Stovner L, Stjern M, Sand T. Sleep quality, arousal, and pain thresholds in tension-type headache: a blinded controlled polysomnographic study. *Cephalagia.* 2014;34:455–463.

41. Lyngberg A, Rasmussen B, Jorgensen T, Jensen R. Incidence of primary headache: a Danish epidemiologic follow-up study. *Am J Epidemiol.* 2005;161:1066–1073.

42. Bellegaard V, Thede-Schmidt-Hansen P, Svensson P, Jensen R. Are headache and temporomandibular disorders related? A blinded study. *Cephalagia.* 2008;28:832–841.

43. Glaros A, Urban D, Locke J. Headache and temporomandibular disorders: evidence for diagnostic and behavioral overlap. *Cephalalgia.* 2007;27:542–549.

44. Fernandes-de-las-Penas C, Galan-del-Rio F, Fernandez-Carnero J, Pesquera J, Arendt-Nielsen L, Svensson P. Bilateral widespread mechanical pain sensitivity in women with myofascial temporomandibular joint disorder: evidence of impairment of central nociceptive processing. *J Pain.* 2009;10:1170–1178.

45. Jensen R, Stovner L. Epidemiology and comorbidity of headache. *Lancet Neurol.* 2008;7:354–361.

46. Balottin U, Fusar Poli P, Termine C, Molteni S, Galli F. Psychopathological symptoms in child and adolescent migraine and tension-type headache: a meta-analysis. *Cephalalgia.* 2012;33:112–122.

47. Eccleston C, Palermo T, Williams A, Lewandowski Holley A, Morley S, Fisher E, et al. Psychological therapies for the management of chronic and recurrent pain in children and adolescents. *Cochran Database Syst Rev.* 2014 May 5;5:CD003968.

48. Ashina M. Neurobiology of chronic tension-type headache. *Cephalalgia.* 2004;24:161–172.

49. Fumal A, Schoenen J. Tension-type headache: current research and clinical management. *Lancet Neurol.* 2008;7:70–83.

50. Nash J, Thebarge R. Understanding psychological stress, its biological processes, and impact on primary headache. *Headache.* 2006;46:1377–1386.

51. Janssen A, Arntz A. Anxiety and pain: attentional and endorphinergic influences. *Pain.* 1996;66:145–150.

52. Leistad R, Sand T, Westgaard R, Nilsen KB, Stovner LJ. Stress-induced pain and muscle activity in patients with migraine and tension-type headache. *Cephalalgia.* 2006;26:64–73.

53. Cathcart S, Petkov J, Pritchard D. Effects of induced stress on experimental pain sensitivity in chronic tension-type headache sufferers. *Eur J Neurol.* 2008;15:552–558.

54. Cathcart S, Winefield A, Lushington K, Rolan P. Stress and tension-type headache mechanisms. *Cephalalgia.* 2010;30:1250–1267.

55. Jensen R, Bendtsen L, Olesen J. Muscular factors are of importance in tension-type headache. *Headache.* 1998;38:10–17.

56. Fernandez-de-la-Penas C, Cuadrado M, Pareja J. Myofascial trigger points, neck mobility, and forward head posture in episodic tension-type headache. *Headache.* 2007;47:662–672.

57. Abboud J, Marchand A, Sorra K, Descarreaux M. Musculoskeletal physical outcome measures in individuals with tension-type headache: a scoping review. *Cephalalgia.* 2013;33:1319–1336.

58. Fernandez-de-Las-Penas C, Alonso-Blanco C, Cuadrado M, Pareja J. Forward head posture and neck mobility in chronic tension-type headache: a blinded, controlled study. *Cephalalgia.* 2006;26:314–319.

59. Rossi P, Vollono C, Valeriani M. The contribution of clinical neurophysiology to the comprehension of the tension-type headache mechanisms. *Clin Neurophysiol.* 2011;122:1075–1085.

60. Ashina M, Stallknecht B, Bendtsen L, Pedersen JF, Galbo H, Dalgaard P, et al. In vivo evidence of altered skeletal muscle blood flow in chronic tension-type headache. *Brain.* 2002;125:320–326.

61. Ashina M, Stallknecht B, Bendtsen L, Pedersen JF, Schifter S, Galbo H, et al. Tender points are not sites of ongoing inflammation — in-vivo evidence in patients with chronic tension-type headache. *Cephalalgia.* 2003;23:109–116.

62. Bendtsen L. Central sensitization in tension-type headache — possible pathophysiological mechanisms. *Cephalalgia.* 2000;20:486–508.

63. Lindelof K, Ellrich J, Jensen R, Bendtsen L. Central pain processing in chronic tension-type headache. *Clin Neurophysiol.* 2009;120:1364–1370.

64. Buchgreitz L, Lyngberg A, Bendtsen L, Jensen R. Increased prevalence of tension-type headache over a 12-year period is related to increased pain sensitivity: a population study. *Cephalalgia.* 2007;27:145–152.

65. Buchgreitz L, Egsgaard LL, Jensen R, Arendt-Nielsen L, Bendtsen L. Abnormal pain processing in chronic tension-type headache: a high-density EEG brain mapping study. *Brain.* 2008;131:3232–3238.

66. Sohn J, Choi H, Kim C. Differences between episodic and chronic tension-type headache in nociceptive-specific trigeminal pathways. *Cephalalgia*. 2013;33:330–339.

67. Lenssinck M, Damena L, Verhagena A. The effectiveness of physiotherapy and manipulation in patients with tension-type headache: a systematic review. *Pain*. 2004;112:381–388.

68. Davis M, Kononowech R, Rolin S, Spierings EL. Acupuncture for tension-type headache: a meta-analysis of randomized controlled trials. *J Pain*. 2008;9:667–677.

69. Bendtsen L, Evers S, Linde M, Mitsikostas DD, Sandrini G, Schoenen J. EFNS guideline on the treatment of tension-type headache – report of an EFNS task force. *Fur J Neurol*. 2010;17:1318–1325.

70. Migliardi J, Armellino J, Friedman M, Gillings DB, Beaver WT. Caffeine as an analgesic adjuvant in tension headache. *Clin Pharmacol Ther*. 1994;56:576–586.

71. Cady R, Gutterman D, Saiers J, Beach ME. Responsiveness of non-IHS migraine and tension-type headache to sumatriptan. *Cephalalgia*. 1997;17:588–590.

72. Friedman A, Boyles W, Elkind A, Fillingim J, Ford RG, Gallagher RM, et al. Fiorina with codeine in the treatment of tension headache – the contribution of components to the combination drug. *Clin Ther*. 1988;10:303–315.

73. Da Silva A, Lake A. Clinical aspects of medication overuse headaches. *Headache*. 2014;54:211–217.

74. Friedman B, Serrano D, Reed M, Diamond M, Lipton RB. Use of the emergency department for severe headache. A population-based study. *Headache*. 2009;49:21–30.

75. Weinman D, Nicastro O, Akala O, Friedman BW. Parenteral treatment of episodic tension-type headache: a systematic review. *Headache*. 2014;54:260–268.

76. Friedman BW, Adewunmi V, Campbell C, Solorzano C, Esses D, Bijur PE, et al. A randomized trial of intravenous ketorolac versus intravenous metoclopramide plus diphenhydramine for tension-type and all nonmigraine, noncluster recurrent headaches. *Ann Emerg Med*. 2013;62:311–318.

77. Diamond S, Baltes B. Chronic tension headache treated with amitriptyline: a double blind study. *Headache*. 1971;11:110–116.

78. Diamond S. *Diagnosing and Managing Headaches*. 6th ed. Caddo, OK: Professional Communications; 2008. [140–141].

79. Morland T, Storli S, Mogstad T. Doxepin in the prophylactic treatment of mixed "vascular" and tension headache. *Headache*. 1979;19:382–383.

80. Zissis N, Harmoussi S, Vlaikidis N, Mitsikostas D, Thomaidis T, Georgiadis G, et al. A randomized, double-blind, placebo-contrlled study of venlafaxine XR in outpatients with tension-type headache. *Cephalalgia*. 2007;27:315–324.

81. Fogelholm R, Murros K. Tizanidine in chronic tension-type headache: a placebo-controlled double-blind crossover study. *Headache*. 1992;42:509–513.

82. Karadas O, Gul H, Inan L. Lidocaine injection of pericranial myofascial trigger points in the treatment of frequent episodic tension-type headache. *J Headache Pain*. 2013 May 22;14:44.

83. Jurgens T, Leone M. Pearls and pitfalls: neurostimulation in headache. *Cephalalgia*. 2013;33:512–525.

84. Bronfort G, Nilsson N, Haas M, Evans R, Goldsmith CH, Assendelft WJ, et al. Non-invasive physical treatments for chronic/recurrent headache. *Cochran Database Syst rev*. 2004: CD001878.

13

Post-Traumatic Headache

Sylvia Lucas

Departments of Neurology and Neurological Surgery, University of Washington Medical Center,
Seattle, Washington, USA

INTRODUCTION

Headache is one of the most common symptoms after traumatic brain injury (TBI) and the most frequent chronic pain problem following TBI.[1–3] Although TBI has received wide media attention recently, primarily due to the high numbers of military and veteran TBIs and to the high-profile sports-related complications of TBI, this has been a worldwide public health issue for decades. It has kindled increased attention to post-traumatic headache (PTH), the most physical symptom after TBI. The most recent CDC numbers report 2.5 million TBIs per year in the US alone, occurring either in isolation or with other injuries. Annually, 53,000 people die from TBI and 275,000 people are hospitalized for non-fatal TBI,[4] with an estimated 3.2 million people in the US now living with TBI-related disability.[5]

Except at the very extremes of age, there are gender differences in rates of TBI. Males are at higher risk of sustaining a TBI, with the highest M:F ratio occurring in adolescence and young adulthood. In a CDC report of national emergency department (ED) visits, hospitalizations, and deaths, the M:F ratio was 1.5:1 in 2001, decreasing to 1.3:1 by 2010. The risk reverses in geriatric populations.[6]

The incidence of TBI rate, in general, has decreased over the past several decades. Using the US National Health Interview Survey (NHIS), the incidence of TBI between 1977 and 1981 (excluding death) was approximately 825/100,000 persons.[7] By 1991, estimates were 618/100,000 persons, using the same NHIS survey.[8] A retrospective population-based review in Olmstead County, Minnesota, found the incidence of TBI in those requiring medical attention was 558/100,000 person-years.[9] Since the majority of TBI etiology in the civilian population is motorized vehicle-related accidents, it may be expected that cultures depending primarily on cars or

other motorized vehicles would have higher rates of TBI. Similarly, in some countries assaults could play a greater role in etiology. A randomized door-to-door survey in six urban cities in the People's Republic of China found the incidence of TBI in the early 1980s was 56/100,000 persons.[10] The huge disparity between the US and Chinese cohorts may reflect the low use of automobiles in China at the time of the study, as well as fewer assaults. The general decreasing trend in the US may be due in part to the use of seat belts in motor vehicles, and increased use and design improvements in protective helmets during sports activities and during vehicle use. In New Zealand, 32% of individuals had a TBI requiring medical attention by age 25,[5] with a total incidence of 790/100,000 person-years. In the US, 75% of TBI is mild TBI (mTBI); in the New Zealand cohort, 95% of brain injuries were mTBI.[5] Falls are the leading cause of TBI among children ages 0 to 4 years, and in adults over the age of 65.[11] In US EDs between 2006 and 2010, for the age group 15 to 44 years old, TBI-related visits resulted from virtually equal rates of motor vehicle traffic accidents, assaults, and falls.[4]

Studies of sports-related concussion, resulting almost always from mTBI, suggest that the numbers in the US, based on medical reporting or chart review, underestimate the problem. Many who participate in sports either do not recognize or minimize their injury, or deliberately fail to seek attention because of a strong desire to remain in the game. The incidence of sports-related concussion is now an astonishing 3.8 million per year among the 44 million children and 170 million adults who participate in organized sport activities.[12]

These statistics also fail to incorporate the burden of TBI in military personnel. Blast exposure has been the most common cause of injury to US troops in the recent Middle East conflicts (Operation Iraqi Freedom [OIF] and Operation Enduring Freedom [OEF]).[13,14] The Congressional Research Service (as of January 10,

S. Diamond, R. K. Cady, M. L. Diamond & V. T. Martin (Eds):
Headache and Migraine Biology and Management.

DOI: http://dx.doi.org/10.1016/B978-0-12-800901-7.00013-6

2014) reports that US military TBI totaled 287,911 between 2000 and 2013. Of those, 237,360 were mild, 23,319 were moderate, and 7224 were severe or penetrating; 20,008 were not classifiable.[15]

TRAUMATIC BRAIN INJURY, CONCUSSION, AND POST-CONCUSSIVE SYNDROME

By definition, concussion is an mTBI; however, the term "concussion," as commonly used, is not strictly interchangeable with mTBI. The American Congress of Rehabilitation Medicine (ACRM)[16] and the Concussion in Sport consensus panel[17] have defined these terms. The ACRM definition of mTBI is

> a traumatically induced physiologic disruption of brain function including: 1) any period of loss of consciousness; 2) any loss of memory for events immediately before or after the accident; 3) any alteration in mental state at the time of the accident such as feeling dazed, disoriented, or confused; and, 4) focal neurological deficits that may or may not be transient.

However, to meet criteria for mTBI, loss of consciousness must not exceed 30 minutes and the duration of post-traumatic amnesia must be less than 24 hours. In addition, the Glasgow Coma scale must range from 13 to 15. The 3rd International Concussion in Sport consensus panel defined concussion as "a complex pathophysiological process affecting the brain, induced by traumatic biomechanical forces." Both definitions emphasize that injury is a disruption in brain function rather than a predominantly structural lesion, but, essentially, all concussion results from mTBI but not all mTBI is concussion.

The post-concussive syndrome (PCS) is a collection of symptoms that should characterize the underlying condition. However, symptoms following a brain injury may be non-specific, with great variability between patients. Although headache is the most common physical symptom following TBI, it usually does not occur in isolation. PCS can include physical or somatic, psychological, and cognitive symptoms. The most recognized and cited definitions of PCS are from the Diagnostic and Statistical Manual of Mental Diseases, 4th edition (DSM-IV)[18] and the International Classification of Diseases, 10th revision (ICD-10)[19] (Table 13.1).

In one prospective longitudinal study of symptoms in over 700 subjects at 1 month after TBI, the most common symptoms reported were fatigue; headache; dizziness; difficulties with memory, sleeping, and concentrating; irritability; blurred vision; anxiety; and increased light and sound sensitivity. Severity of brain injury was correlated with a number of symptoms, and more severe injuries tended to be associated with a greater proportion of cognitive and psychological symptoms in addition to physical symptoms.[20]

TABLE 13.1   Symptoms of Post-concussive Disorder using Criteria from International Classification of Diseases, 10th edition (ICD-10) and Diagnostic and Statistical Manual of Mental Disorders, 4th edition (DSM-IV)

Post-concussional syndrome (ICD-10)	Post-concussional disorder (DSM-IV)
The syndrome occurs after head trauma (usually sufficient to result in loss of consciousness).	**A.** History of head trauma that has caused significant cerebral concussion.
At least three of the following features should be present. Objective evidence to substantiate the symptoms is often negative.	**B.** Difficulty with attention (concentrating, shifting of focus of attention, performing simultaneous tasks) or memory (learning or recalling information) based on neuropsychological testing.
1. Headache	**C.** Three or more of the following occur shortly after the trauma and last for at least 3 months:
2. Dizziness	1. Becoming fatigued easily
3. Fatigue	2. Disordered sleep
4. Irritability	3. Headache
5. Difficulty in concentrating and performing mental tasks	4. Vertigo or dizziness
6. Impairment of memory	5. Irritability or aggression on little or no provocation
7. Insomnia	6. Anxiety, depression or affective lability
8. Reduced tolerance to stress, emotional excitement or alcohol. Symptoms may be accompanied by feelings of depression or anxiety, resulting from loss of self-esteem or fear of permanent brain damage; such feelings enhance the original symptoms, and a vicious cycle results; some patients become hypochondriacal and may adapt a permanent sick role.	7. Changes in personality
	8. Apathy or lack of spontaneity
	D. The symptoms in B and C have their onset after head trauma or else represent a substantial worsening or pre-existing symptoms spontaneously.
	E. The disturbance causes significant impairment in social or occupational functioning and represents a significant decline from a previous level of functioning.
	F. The symptoms do not meet criteria for dementia due to head trauma and are not better accounted for by another mental disorder.

Data from DSM-IV[19] and ICD-10.[18]

When studying isolated sports concussion, females appear to be more susceptible to concussion and have more post-concussive cognitive changes.[12,21] Girls have nearly twice the rate of concussion when comparing similar sports – for example, boys playing baseball and girls playing softball.[22,23] The etiologies for this discrepancy are probably not only biomechanical (head to ball ratio, smaller head/neck mass ratio, weaker neck muscles, and greater angular velocity in girls compared with boys[21]), but possibly sociological as well.

EPIDEMIOLOGY OF POST-TRAUMATIC HEADACHE

Although many studies have been undertaken of symptoms following brain injury, variability in case ascertainment, TBI subgroup severity, subject selection bias, inclusion criteria, and different follow-up times have made the comparison and interpretation of these studies difficult. Also, because of litigation concerns and secondary gain issues,[24] the lack of objective findings and self-reporting of symptoms are often discounted. Despite the symptom prominence of headache after TBI,[1,25,26] some of these study design variables may explain the wide range of prevalence from 30% to 90% in retrospective studies.[1,27–29] Key epidemiologic studies from civilian and military populations are listed in Table 13.2.[20,25,26,30–39]

Although there are few large prospective studies, in the largest civilian study to date 452 subjects admitted to inpatient rehabilitation services following a moderate to severe TBI were followed for 1 year. The majority of subjects were men injured in vehicle-related accidents with an average age of 44 years. Initial evaluation was performed in the hospital, with telephone follow-up at 3, 6,

TABLE 13.2   Recent PTH Studies in Civilian and US Active Duty Service Members

Authors	Year/place	Number	Study design	Population	Key results
Blume et al.[30]	2012/US	462	Prospective cohort	Age 5–17 mTBI = 402 mod–severe TBI = 60 controls = 122 (arm injury)	Prev HA after mTBI. After 3 months = 43%;1 year = 41%. Prev HA after mod–severe TBI 3 months = 37%; 1 year = 35%; control 3 months = 26%; 1 year = 34%
Dikmen et al.[20]	2010/US	732	Prospective case–control	Age >16 Any TBI	HA the week prior to: 1 month post-TBI = 55% 1 year post-TBI = 26%
Erickson[31]	2011/US	100	Retrospective cohort	100 consecutive soldiers in headache clinic with chronic PTH	77% of patients had blast-related PTH; >95% met migraine criteria
Faux and Sheedy[32]	2008/Australia	100	Prospective case–control; ED	Age >16 Mod–severe TBI	Prev at injury = 100% At 1 month = 30% At 3 months = 15%
Hoffman et al.[25]	2011/US	452	Prospective in-person enroll Phone interview 3, 6, 12 months	Age >16 Mod–severe TBI	Cum incidence at 1 year = 71% Baseline prev = 47% At 1 year = 44%
Hoge et al.[33]	2008/US	2525	Cross-sectional survey	Soldiers returning from 1-year Iraq deployment	HA was the only symptom associated with concussion adjusting for mood disorders
Kuczynski et al.[34]	2013/Canada	670	Prospective ED cohort Retrospective chart review of treatment cohort from brain injury clinic; phone interview 7–10 days after injury and monthly until resolution	Age 0–18 mTBI (no extracranial injury)	Prev PTH = 11% at 16 days At 3 months = 8% ED cohort migraine = 54% Clinic cohort: migraine = 39% (mixed HA, MOH, mood disorders with HA excluded)

(Continued)

TABLE 13.2   (Continued)

Authors	Year/place	Number	Study design	Population	Key results
Lieba-Samal et al.[35]	2011/Austria	100	Prospective, phone interview	Age 18–65	Prev acute PTH at 7–10 days = 66%
				Excluded whiplash, medication overuse, pre-existing chronic PTH	All resolved by 3 months
					Migraine/probable migraine = 35%
Lucas et al.[26,36]	2012/US	452	Prospective in-person enroll; phone interview 3, 6, 12 months	Age >16	Mod–severe TBI baseline migraine/probable migraine = 52%
	2014/US	212		2012 mod–severe TBI	1 year = 54%
				2014 mTBI	mTBI cum incidence = 92%
					Migraine/probable migraine at 3 months = 49%; 1 year = 49%
Stovner et al.[37]	2009/ Lithuania	217	Prospective ED cohort, case–control; questionnaires at 3 months, 1 year	Age 18–60 LOC < 15 min	Prev HA 3 months = 66%
					1 year = 65%
					Migraine 3 months = 19%
					1 year = 21%
Theeler et al.[38,39]	2010/US 2012/US	1033	Cross-sectional, survey-based	Soldiers with concussion in post-deployment evaluation over 5 months in 2008	HA in 98% of soldiers; 37% met PTH criteria; migraine type in 89%; 20% CDH of which 55% met PTH criteria

CDH, chronic daily headache; ED, emergency department; HA, headache; MOH, medication overuse headache; PTH, post-traumatic headache; mTBI, mild traumatic brain injury; TBI, traumatic brain injury. mod; moderate; prev, prevalence; cum, cumulative.
Table adapted from Theeler et al.[72]

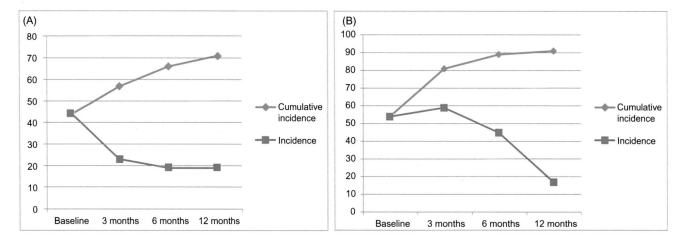

FIGURE 13.1   (A) Incidence and cumulative incidence of all headache after moderate to severe TBI (*n* = 452). Total incidence is the proportion of cases developing headache for the first time after TBI at each time-point (not all subjects are assessed at each time-point). (B) Incidence and cumulative incidence of headache after mTBI (*n* = 212). Total incidence is the proportion of cases with headache for the first time after TBI at each time-point (not all subjects are assessed at each time-point). *(A) Adapted from Hoffman et al.[25]; (B) Figures adapted from Lucas et al.[31]*

and 12 months.[25] Of the 452 subjects, 71% reported headache during the first year. Prevalence was 46% at the initial inpatient interview, and remained high with 44% reporting headache 1 year after TBI. About 28% of headaches developed after the initial evaluation, occurring at 6 and even 12 months after injury, although most PTH developed within 3 months of injury (Figure 13.1A).

Thus, an appreciable fraction of headaches developed after the 7-day window requirement in the ICHD-II criteria (used for this study) defining PTH.[40]

In a subsequent study of mTBI, the same research group using similar headache assessment tools followed a group of 212 subjects admitted to the hospital with mTBI, as well as other, primarily orthopedic, injuries.

Within 1 week after TBI, this group was evaluated in the hospital, with telephone follow-up at 3, 6, and 12 months after hospitalization. An intriguing difference between the moderate-to-severe TBI and the mTBI studies was the higher cumulative incidence of headache at 91% over 1 year for mTBI (Figure 13.1B). Prevalence of new or worse headache in this cohort was 54% within 1 week of injury, and remained high at 58% 12 months post-injury.[36] Others have also noted a higher prevalence of PTH following mTBI compared with more severe TBI.[3,27,28,41]

POTENTIAL RISK FACTORS FOR POST-TRAUMATIC HEADACHE

In addition to severity of injury, age and gender have been examined as possible risk factors for PTH. There is a consistent agreement that prior history of headache is a risk factor for development of PTH.[25,26,34,35,37]

Reports have been inconsistent that female gender is a risk factor for PTH. In large prospective studies by the same group,[25,36] female gender was not a risk factor for new or worse PTH in those sustaining a moderate to severe TBI or an mTBI.[25,26] In a Danish study of concussion or suspected concussion, an adjusted odds ratio (OR) of women compared with men was 2.6 (1.2–5.8) for PTH, whereas OR for other post-concussive symptoms was 1.0.[42] In a study of children with mTBI, no difference was demonstrated between girls and boys with persistent PTH either in an ED cohort or a clinic treatment cohort.[34] In a historical cohort study using a national Veterans Administration 2011 database of over 470,000 Iraq and Afghanistan war veterans, although headache diagnoses were more prevalent in women (18%) over men (11%), most of the difference was due to a higher prevalence of migraine; PTH was found to be less prevalent in women than in men.[43]

In studies that have examined age as a risk factor for PTH, older age (>60) was related to a lower prevalence of PTH following mTBI[36] and a similar result was found in a study of all levels of TBI severity.[20]

THE PHENOTYPE OF POST-TRAUMATIC HEADACHE

There is no single phenotype of PTH. Although PTH is among the most prevalent of the secondary headache disorders as currently classified in the International Classification of the Headache Disorders, 3rd edition, beta version (ICHD-III),[44] there is an absence of defining clinical characteristics that differentiate this PTH from any other primary or secondary headache disorder. Most PTHs have phenotypes indistinguishable from the primary headache disorders, though some are not classifiable. The importance of a phenotypic classification is not only of epidemiological interest, but may also reveal potential diagnostic markers for research into the physiology of PTH as well as provide a useful clinical benefit in guiding treatment.

In civilian populations, some early reports suggested that the headaches meeting tension-type headache (TTH) criteria were the most prevalent type of PTH,[1] although in some instances the diagnostic criteria provided by ICHD were not utilized. More recent studies support migraine or probable migraine as the most common PTH type. In the largest prospective longitudinal study of PTH to date of headache following moderate to severe brain injury, migraine or probable migraine was the most common PTH phenotype for headaches that persist over 1 month.[25,26] Migraine or probable migraine was found in over 52% of those reporting headache at baseline evaluation, and in 54% of those with headache at 1 year.[25,26] In individuals without a prior history of headache, migraine/probable migraine was found in 62% of those reporting headache at baseline, and in 53% at 1 year. Migraine/probable migraine was also the most common headache phenotype in a large prospective study of headache after mTBI in civilians.[36] These two headache types were found in 49% of subjects, with headache at 3, 6, and 12 months after injury with TTH never exceeding 40% over that year (Table 13.3). Interestingly, this study population consisted of 76% males injured primarily in vehicle accidents. The migraine or probable migraine headache type seen after

TABLE 13.3   Post-traumatic Headache Phenotypes Following Mild TBI

	3 months *n* (%)	6 months *n* (%)	12 months *n* (%)
Total subjects who had headache	126	139	109
Migraine or probable migraine	62 (49%)	67 (48%)	53 (49%)
Tension-type	47 (37%)	55 (40%)	35 (32%)
Cervicogenic	5 (4%)	6 (4%)	4 (4%)
Not classifiable	12 (10%)	11 (8%)	17 (16%)

Percentage (of column) is of those with new or worse headache at each time-point.
Adapted from Lucas et al.[36]

TBI was much more frequent than would be expected for primary migraine or probable migraine in males in the general population.[45] Cervicogenic headaches made up 10% or fewer of classifiable headaches at any time-points in these studies.

In a smaller series of 41 patients in whom headache classification after mTBI was reported, migraine followed by TTH were the most common headache types found.[46] In a large ED cohort of children, migraine was found to be the most common headache type, reported in 55% of those children who had headache after mTBI.[34] Another prospective study of 100 patients presenting to a trauma surgery department with mTBI found migraine or probable migraine to be the most common headache type, followed by TTH or probable TTH using ICHD-II criteria[35] (see Table 13.2).

In prior studies[37,47,48] in which TTH was the most common PTH after mTBI, some potential study factors underlying differences might be that one study was based on select clinical samples with a long interval between injury to consultation,[47] and another study was a retrospective chart review.[48] Some studies will report more than one headache type, whereas in the two large prospective studies discussed above,[25,26,36] only the most severe headache type that a subject reported at each time period was determined. Thus, if both a migraine and tension-type headache were present over that year at the same time, migraine headaches would be the reported headache.

PTH frequency is higher in those who have more severe PTH. Comparing civilians who had PTHs that resembled migraine or probable migraine with those with PTH that resembled TTH or cervicogenic head-ache, those with migraine-like headaches were most likely to describe headaches occurring several days a week or daily.[26] In fact, 23% of individuals sustaining moderate to severe TBI reported headache frequency of >15 days of headache per month (criteria for chronic daily headache, [CDH]) at all time-points up to 1 year after TBI. Following mTBI, for those who experienced headache several times a week or daily, 62% of the headache types were migraine in this high-frequency group at the 1-year time-point.[36] In contrast, 4–5% of those with headache in the general population report CDH.[49–53] In fact, head and neck injury accounts for approximately 15% of CDH cases.[54]

Rare cases of PTH in the civilian population are phenotypically similar to cluster headache,[55] hemicrania continua,[56] chronic paroxysmal hemicrania,[57] and short-lasting, unilateral, neuralgiform headache attacks with conjunctival injection and tearing (SUNCT syndrome).[58] Some PTH patterns are associated with specific cranio-cerebral injuries.[51] Leakage of cerebrospinal fluid (CSF) can produce low CSF pressure headaches.

Post-craniotomy headaches can complicate post-surgical treatment of TBI. However, no significant correlation has been demonstrated between acute neuroimaging abnor-malities and presence or absence of PTH in a study of moderate to severe TBI over 1 year.[59]

POST-TRAUMATIC HEADACHE IN MILITARY SETTINGS

TBI in deployed military personnel is high and distinguished from civilian TBI by extreme occupational risk, both physically and psychologically. In the Millenium Cohort Study of over 77,000 US Army soldiers screened before and after deployment, soldiers with com-bat exposure had significantly increased odds of a new onset headache disorder after deployment compared with those who did not deploy.[60] Soldiers who deployed but did not experience combat did not have an increase in new onset headaches after deployment compared with non-deployers, emphasizing the importance of combat exposure.[60] Most recent US military PTHs develop fol-lowing blast exposure.[38,61–63] In one study of 978 US Army soldiers, blast exposure occurred in over 80% reporting headaches after return from deployment. On average, these soldiers were exposed to five or more blasts occurring within 60 feet.[64] Of soldiers with chronic PTH, 77% had blast-induced TBI.[31]

Similar to findings in civilian studies, PTH occurring in military cohorts may not meet ICHD criteria for onset within 1 week of injury. In recently deployed US Army soldiers, nearly 40% of PTH occurred within 1 week of mTBI, 20% occurred within 1 month, and just over 40% occurred after 1 month following head injury.[38]

In a study of 2525 US Army infantry soldiers, head-ache was the only symptom significantly associated with concussion after adjusting for depression and anx-iety disorder.[33] Also similar to civilian TBI cohorts, migraine appears to be the most common headache phenotype. The prevalence of migraine is markedly higher in US Army soldiers upon return from deploy-ment, with 36% of US Army soldiers in a single combat brigade reporting headaches with features of migraine.[65] Military PTH meet criteria for migraine in the majority of cases, ranging from 60% to 97%, depending on the study population and methodology.[31,38,66,67] Although migraine is clearly the predominant headache pheno-type, in clinical studies in which additional headache types were assessed many soldiers with PTHs have additional headache types, including tension-type, cra-niofacial neuralgiform pain disorders, and rarely, tri-geminal autonomic cephalalgias.[67,68]

A striking feature of these studies of PTH in the mili-tary is the high prevalence of CDHs in this population.

Among 978 US Army soldiers with deployment-related concussion, 20% reported headaches on 15 or more days per month in the preceding 3 months, with a median of 27 headache days per month. Of the soldiers with CDHs, 55% had their headaches start within 1 week of a concussion.[69] Similarly, among 100 US Army soldiers with chronic PTHs seen in a headache clinic, the average headache frequency was 17 days per month.[31] Migraine features are present in 70% or more of the chronic PTH disorders in the military studies.

Although a full discussion of comorbid conditions is beyond the scope of this chapter, PTSD criteria are found in 40–50% of soldiers with chronic PTH.[31,39] The post-concussive syndrome in service members is arguably more complex than in the civilian populations, with a highly individual interplay of chronic PTH, PTSD, and the PCS, which differentiates itself from civilian injury by blast exposure being the predominant injury mechanism.[51,70,71] Direct attribution of cognitive, psychological, and pain symptoms in the chronic post-traumatic syndrome is difficult when CDH and psychiatric disorders coexist in someone who has sustained multiple concussions in combat.[72]

PHYSIOLOGY OF POST-TRAUMATIC HEADACHE

The physiology of TBI is an area of intense research, using a varied array of model systems.

As the primary symptom following TBI, the initiation as well as the persistence of PTH may be linked to the physiologic changes associated with TBI. Clinical similarities between PTH and primary headache disorders may indicate shared physiologic changes that develop following TBI and those seen in the primary headache disorders. The initial injury may result from a mechanical or pressure-wave-induced stimulus causing a neuronal depolarization, likely to be very rapid and relatively extensive, particularly given frequent loss of consciousness of variable duration. This could result in a propagating effect, depending on extent and type of injury, but also the injury may be complicated by structural damage such as hemorrhage or depressed skull fracture, in turn initiating other series of events caused by the injury.

Numerous factors are likely to contribute to development of PTH, as extensive response to injury is seen after TBI. Inflammation and cytokine changes, alterations in cerebral metabolism and hemodynamics, axonal injury, glial cell activation, and abnormality in neuropeptide and neurotransmitter activity are some changes seen in closed-head TBI. Cytokine concentration in the central nervous system (CNS) can increase markedly, primarily driven by activated glia following injury. Post-mortem studies have shown rapid changes after injury, with elevated messenger RNA levels of interleukin 1, 6, and 10 as well as tumor necrosis factor-alpha (TNF-α).[73] The function of the elevated cytokines is only partially understood and they may have both pro- and anti-inflammatory functions in the injury response, affect permeability of the blood–brain barrier (BBB),[74] and be involved in the initiation of pain.[75]

Microglia can become activated within minutes of a brain injury, as seen by changes in their structure and function, aggregation to area of injury, and the release of cytokines, complement factors, chemokines, proteolytic enzymes, and reactive oxygen species.[76] Activated microglia can also have either pro- or anti-inflammatory effects, depending on the local environment, with some microglial modulators such as PARP-1, PPAR agonists, and mGLU5 increasing neuronal survival in animal models of TBI.[73] Prolonged microglial activation, which may occur with repetitive head trauma, as well as continual release of cytotoxic proinflammatory substances, could perpetuate activation contributing to progressive symptoms.[77,78]

Blood–brain barrier function can be altered by injury to the brain or by a transient change in response to the injury, allowing inflammatory and other components of peripheral blood into the CSF compartment. The consequent inflammatory response, although mixed with both pro- and anti-inflammatory activity, can result in neuronal and glial changes beyond the site of injury. Trafficking of chemokines, matrix metalloproteinases, toll-like receptors, and P2X7 receptors can perpetuate the immune changes, neuron-to-glia signaling, and sensitization processes that are involved in maintenance of pain.[79–82]

It is likely that the linkage between TBI and PTH results from alterations in meningeal, cerebrovascular, neuronal, and glial structural and functional changes, possibly including or resulting in cortical spreading depression (CSD)[83] and activation of the trigeminovascular system.[84] CSD may be the link between TBI-induced and migraine-induced trigeminovascular changes.[83] Another possible link may be through activation of calcitonin gene-related peptide (CGRP) which can be released during CSD[85] or trigeminovascular activation and may be a reasonable biomarker for PTH, although it could be involved in many functions resulting in vasodilation. TBI stimulates the transient receptor potential (TRP) V1 channel which increases release of CGRP from trigeminal ganglia,[86] possibly explaining migraine type PTH through the vasodilator effect involved in neuroinflammation and pain.[87]

Diagnostic markers for concussion and TBI are currently the "holy grail" of research efforts and an active area of research. In 2010, the Biospecimens and

Biomarkers Working Group[88] defined common data elements for TBI and reported that the APOE-e4 allele could be useful as a biomarker based on an association with cognitive decline and poor functional outcome after TBI.

A recent study using diffusion tensor imaging (DTI) found decreases in fractional anisotropy in white-matter tracts of those developing chronic pain after TBI.[89] Functional MRI (fMRI) has shown that the default-mode network (DMN), a network of brain areas active in a resting, non-focused state, is disrupted after TBI.[90] Imaging studies have an advantage over animal models for study of PTH in that headache can be more directly targeted as a symptom. However, to date, no definitive relationships have been shown between structural or functional imaging and PTH. Imaging is likely to be an extremely useful biomarker method in the future.[91]

MANAGEMENT OF POST-TRAUMATIC HEADACHE

Following TBI, those who sustain a moderate to severe injury, or who have coexisting other severe physical injuries, are likely to have those injuries addressed prior to evaluating PTH. Additionally, the severe "other" injuries may preclude attention to headache. After TBI, those with PTH may seek medical attention if the headache causes significant pain, disability, or inability to function. Difficulty with concentration or attention to work or school may be secondary to a constant headache, or part of the PCS, or both, leading to significant work-loss and social functioning. Fear of the headache pain and its unpredictability, or fear that the headache reflects significant underlying brain damage from the injury, as well as other reasons, may bring people to practitioners. Many people will initially seek care from primary care providers, or from sports medicine specialists if the injury is sports-related. If prolonged treatment and rehabilitation is necessary for associated musculoskeletal injuries, a physiotherapist may be the specialist providing care. Many layers of physicians may be seen before a headache specialist or practitioner who is knowledgeable regarding PTH is found.

The treatment of PTH is empiric. To date, no strong evidence from clinical trials is available to direct the treatment of PTH. In a review of interventions for PTH,[92] there were no Class I studies and only one Class II study in 23 subjects for management of PTH using manual spine therapy versus cold packs, finding that the group receiving spine therapy had a decrease in headache intensity 5 weeks after treatment but the treatment effect was lost after 8 weeks.[93] In a retrospective case series of 28 children and adolescents with a mean age of 15 years, peripheral nerve blocks demonstrated a good therapeutic effect in 93%, with 71% reporting immediate relief from PTH with a mean decrease in headache intensity of 94%.[94]

No pharmacological study above a Class III study has been published. In a retrospective study of 100 patients after mTBI with CDH, headache frequency and severity were measured after at least 1 month of treatment with valproic acid. Analgesic, NSAID, chiropractic, and physical therapy treatments were allowed during this treatment intervention. After treatment, 44% of subjects had a 24% to 50% improvement, and another 16% had more than 50% improvement.[95] A subgroup analysis of subjects in a larger prospective study of amitriptyline for depression[96] compared 10 patients with depression and headache after mTBI versus trauma without mTBI. Those who did not have mTBI ($n = 12$) did better on headache severity and depression scores than those with mTBI ($n = 10$). In 20 subjects with at least a 3-month history of headaches, a topical compounded ketoprofen was used. Although results showed headache index decreased over 1 month, there was no subgroup analysis in the three PTHs in this study.[97] Anecdotal reports of successful treatment of acute PTH with sumatriptan or dihydroergotamine after concussion, or PTH in a civilian population, support the use of migraine-specific therapies.[98,99]

In military settings, the Defense and Brain Injury Center (DVBIC) has provided PTH treatment recommendations for deployed and non-deployed members based on Class IV study evidence.[64,100,101]

In one retrospective analysis of active duty service members with mTBI and PTH, triptans were effective as acute agents treating a predominantly migraine phenotype. Topiramate was an effective prophylactic agent, associated with a significant reduction in headache frequency, but overall the number of patients using different medications was small, which prevented treatment recommendations.[31] A relatively high non-responder rate was noted, in that only 35% of soldiers were considered responders to preventive medication.[31]

Headaches after mTBI are not well managed. Many of those with headache are likely to self-treat and use over-the-counter pain medication. In one prospective natural history study of 212 subjects with mTBI,[36] a high headache burden was found at 3, 6, and 12 months after injury. Prevalence of headache was never less than 58% of subjects at any time-point. Despite this high prevalence and the fact that the primary PTH type was migraine, more than 70% of those with headache at any time-period used acetaminophen or a non-steroidal anti-inflammatory agent (NSAID) for headache treatment.

Less than 5% of this group used triptans, and in the group with a migraine headache type only 19% endorsed complete relief after medication use.[102]

Expert opinion has suggested treating PTH according to its clinical characteristics using the primary headache disorder classification criteria.[24,103,104] Based on the close phenotypic similarity of PTH to primary headaches, one approach to treatment decisions is to use primary headache characterization of the PTH as migraine, probable migraine, tension-type headache, or other primary disorder. Because the treatment of migraine and probable migraine is the same, these migraines will be discussed together under the term "migraine." Acute treatment of a PTH is crucial if the headache is severe or disabling. The goal of acute headache treatment is to educate those with headache to treat early with effective therapy. Many patients will wait to see if their headache worsens, or choose over-the-counter (OTC) products because of cost or lack of access to medical care. If their usual pattern is that of migraine, and if simple analgesic agents or other OTC products do not effectively treat their PTH, then migraine-specific therapy is recommended. Two treatment approaches are used for migraine headache. Acute or abortive therapy treats an episodic headache as it occurs, and preventive therapy is used on a daily basis when attack frequency is high or when there is a suboptimal response to acute therapy.[105] Acute and preventive therapy is discussed in detail in (Chapter 8).

Overuse of abortive medication may contribute to persistence of headache after injury through the development of medication overuse headache (MOH), also known as rebound headache. MOH is reviewed in Chapter 20. Few epidemiologic data are available on the extent of this problem. Although it is extremely important to recognize the possibility of MOH, especially given the high rate of use of OTC products by those who have sustained a TBI,[102] it is important to examine the development of persistent headache and associated medication use and consider changing or withdrawing suspected medication before concluding that PTH is worsened or prolonged because of medication overuse. Education regarding medication overuse is key, and the assistance of a diary or other record of medication use is important. However, it is also important to consider possible cognitive limitations of those who have sustained a TBI, and whether assistance from caregivers or family members may be needed when PTH management is discussed.

Prevention of TBI and recurrent concussion remains the single most important management concern. Targeted efforts by government agencies (both civilian and military), sports leagues, coaches and trainers, and brain injury foundations, physicians and parents are needed to prevent TBI.

THE POST-TRAUMATIC HEADACHE ICHD-III CRITERIA

Post-traumatic headache is one of the most prevalent secondary headache disorders. As currently classified by the International Classification of Headache Disorders (ICHD-III),[44] PTH is defined by latency from injury, severity of injury, and transition from an acute to a chronic form of PTH at 3 months following the injury (Box 13.1). In the revised ICHD-III criteria, one very useful change to the classification criteria was the substitution of "persistent" for "chronic" when the PTH was longer than 3 months in duration; this avoids confusion with the definition of chronic headache as occurring more than 15 days a month.

As a secondary headache disorder, a temporal relationship exists between the time of injury and onset of headache. The ICHD criteria specify that the PTH must occur within 7 days of injury, or after regaining consciousness in the case of moderate to severe injury. This arbitrary latency may underestimate the prevalence of PTH. Clinically, in several civilian and military studies of PTH, latency is beyond the 7-day requirement. Civilian moderate to severe TBI-induced PTH occurred beyond 7 days in 28% of those who developed PTH.[25] In a military population study, up to 60% of PTH occurred after the first week following an mTBI,[38] and in another study of chronic headache seen in an army neurology clinic only 36% of those with head or neck injury reported headache within 1 week after trauma.[67] A pediatric ED cohort of 670 mTBI subjects found 11% were symptomatic with acute PTH at 15.8 days (mean), and 7.8% had increased headache at 3 months after injury.[34] A pediatric case series report of PTH 3 months after injury[106] also suggests a delayed onset of PTH.

As recent tensor imaging data demonstrate brain changes in volume occurring years after TBI,[107] and as adaptive or maladaptive changes in inflammatory, cytokine, vascular, and BBB occur, it is likely that TBI may be more fluid in development than a static time-of-injury event, accounting for a possible progressive development of symptoms. These phenomena would preclude a specific cut-off point, and warrant relaxing the 7-day cut-off period. An immediate versus a delayed onset also may ascertain whether underlying physiologic changes differ after brain injury. It is notable that in the beta form of the ICHD-III criteria a comment is made regarding a place for clinical decision-making being part of the latency argument (ICHD-III). The 7-day interval attempts to ensure a higher specificity for PTH caused by TBI, but has a corresponding loss of sensitivity. The PTH criteria discussed in the Appendix of the beta version of ICHD-III

BOX 13.1

INTERNATIONAL CLASSIFICATION OF HEADACHE DISORDERS CRITERIA FOR POST-TRAUMATIC HEADACHE

5.2.1 Acute headache attributed to moderate or severe traumatic injury to the head

Diagnostic criteria:

A. Any headache fulfilling criteria C and D
B. Injury to the head associated with at least one of the following:
  1. Loss of consciousness for >30 minutes
  2. Glasgow Coma Scale (GCS) score <13
  3. Post-traumatic amnesia lasting >24 hours
  4. Altered level of awareness for >24 hours
  5. Imaging evidence of traumatic head injury such as intracranial hemorrhage and/or brain contusion.
C. Headache is reported to have developed within 7 days after one of the following:
  1. The injury to the head
  2. Regaining consciousness following the injury to the head
  3. Discontinuation of medication(s) that impair ability to sense or report headache following the injury to the head
D. Either of the following:
  1. Headache has resolved within 3 months after the injury to the head
  2. Headache has not yet resolved but 3 months have not yet passed since the injury to the head
E. Not better accounted for by another ICHD-III diagnosis

5.2.1.1 Persistent headache attributed to moderate or severe traumatic injury to the head

Diagnostic criteria: as for 5.2.1 except for D.

D. Headache persists for >3 months after the injury to the head

5.2.1.2 Acute post-traumatic headache attributed to mild traumatic injury to the head

Diagnostic criteria:

A. Any headache fulfilling criteria C and D
B. Injury to the head fulfilling both of the following:
  1. Associated with none of the following:
    a) Loss of consciousness for >30 minutes
    b) Glasgow Coma Scale (GCS) score <13

c) Post-traumatic amnesia lasting >24 hours
d) Altered level of awareness for >24 hours
e) Imaging evidence of traumatic head injury such as intracranial hemorrhage and/or brain contusion
  2. Associated immediately following the head injury with one or more of the following symptoms and/or signs:
    a) Transient confusion, disorientation or impaired consciousness
    b) Loss of memory for events immediately before or after the head injury
    c) Two or more other symptoms suggestive of mild traumatic brain injury: nausea, vomiting, visual disturbances, dizziness and/or vertigo, impaired memory and/or concentration
C. Headache is reported to have developed within 7 days after one of the following:
  1. The injury to the head
  2. Regaining consciousness following the injury to the head
  3. Discontinuation of medication(s) that impair ability to sense or report headache following the injury to the head
D. Either of the following:
  1. Headache has resolved within 3 months after the injury to the head
  2. Headache has not yet resolved but 3 months have not yet passed since the injury to the head
E. Not better accounted for by another ICHD-III diagnosis

5.2.2 Persistent headache attributed to mild traumatic injury to the head

Diagnostic criteria: as for 5.2.1.2 except for D.

D. Headache persists for >3 months after the injury to the head

Adapted from The International Classification of Headache Disorders (ICHD), 3rd edition, beta version.[44]

includes a delayed onset headache diagnostic criteria attributed to traumatic brain injury (see, for example, A5.2.1, p. 688).[44]

Despite concerns regarding potential secondary gain or medico-legal issues, the occurrence of PTH beyond the 1-week latency, in an otherwise low-risk population, cannot possibly account for the significant proportion of subjects developing migraine-type PTH within 1 to 3 months following head injury, as incidence of migraine in young adult males is about 0.021% per month.[108]

CONCLUSIONS

Post-traumatic headache is the most common physical symptom reported after TBI. High-profile professional athletes and the large numbers of returning soldiers and veterans of war who sustain TBI, PTH, and other sequelae of head injury have driven a heightened interest in the recognition, treatment, and, hopefully, prevention of TBI-related symptoms. PTH is frequent, with recent data from civilian adult and pediatric as well as military populations reporting that PTH may be more of a chronic and persistent problem than previously thought.[109] In both the civilian and military populations, most PTHs could be classified using primary headache disorder criteria and most frequently met criteria for migraine or probable migraine.

Data from both civilian and military populations may support changes in the ICHD classification criteria for PTH. Although the requirement that a PTH occurs within 1 week of injury may underestimate PTH frequency, there is insufficient evidence to recommend a specific cut-off point. The occurrence of an immediate versus a delayed onset headache may be important in understanding the dynamic nature of an acute brain injury, and the documentation of PTH onset following brain injury will itself be important in future studies. Currently, many providers who treat PTH use treatments based on a clinical similarity to a primary headache. Whether or not similar phenotypes in the primary and secondary headache disorders will respond similarly to treatment is unknown. Controlled blinded clinical trials are needed to determine effective treatments for PTH. Before effective trials can be launched, clinically meaningful subgroups are necessary in order to study a presumptive similar biological response to injury. Headache phenotype, delayed versus immediate onset of headache, even type of injury or accompanying features, such as aura, may be part of the subgroup classification. Eventually, structural and functional imaging, genetic susceptibility to injury, and biomarkers reflecting the robustness of inflammatory response systems could be important in determining subtypes of PTH and the most effective treatment.

Currently, no PTH-specific recommendations for treatment exist, and the high incidence, prevalence, and frequency of PTH requires an evidence-based approach to treatment for those who sustain brain injury. Clearly, more research funding is necessary to further the study of individuals with TBI and PTH to improve quality of life and reduce associated disability.

References

1. Lew HL, Lin PH, Fuh JL, Wang SJ, Clark DJ, Walker WC. Characteristics and treatment of headache after traumatic brain injury: a focused review. *Am J Phys Med Rehabil.* 2006;85(7):619−627.
2. Nampiaparampil D. Prevalence of chronic pain after traumatic brain injury: a systematic review. *JAMA.* 2008;300(6):711−719.
3. Uomoto JM, Esselman PC. Traumatic brain injury and chronic pain: differential types and rates by head injury severity. *Arch Phys Med Rehabil.* 1993;74(1):61−64.
4. Centers for Disease Control and Prevention (CDC). Traumatic brain injury statistics 2012. <www.cdc.gov/traumaticbraininjury/statistics.html2012>.
5. Corrigan JD, Selassie AW, Orman JA. The epidemiology of traumatic brain injury. *J Head Trauma Rehabil.* 2010;25(2):72−80.
6. Annegers JF, Grabow JD, Kurland LT, Laws ER. The incidence, causes, and secular trends of head trauma in Olmsted County, Minnesota, 1935−1974. *Neurology.* 1980;30(9):912−919.
7. Fife D. Head injury with and without hospital admission: comparisons of incidence and short-term disability. *Am J Public Health.* 1987;77(7):810−812.
8. Sosin DM, Sniezek JE, Thurman DJ. Incidence of mild and moderate brain injury in the United States, 1991. *Brain Inj.* 1996;10(1):47−54.
9. Leibson CL, Brown AW, Ransom JE, Diehl NN, Perkins PK, Mandrekar J, et al. Incidence of traumatic brain injury across the full disease spectrum: a population-based medical record review study. *Epidemiology.* 2011;22(6):836−844.
10. Wang CC, Schoenberg BS, Li SC, Yang YC, Cheng XM, Bolis CL. Brain injury due to head trauma. Epidemiology in urban areas of the People's Republic of China. *Arch Neurol.* 1986;43(6):570−572.
11. Coronado VG, Xu L, Basavaraju SV, McGuire LC, Wald MM, Faul MD, et al. Surveillance for traumatic brain injury-related deaths − United States, 1997−2007. *MMWR Surveill Summ.* 2011;60(5):1−32.
12. Daneshvar DH, Nowinski CJ, McKee AC, Cantu RC. The epidemiology of sport-related concussion. *Clin Sports Med.* 2011;30(1):1−17:vii.
13. Warden D. Military TBI during the Iraq and Afghanistan wars. *J Head Trauma Rehabil.* 2006;21(5):398−402.
14. Eskridge SL, Macera CA, Galarneau MR, Holbrook TL, Woodruff SI, MacGregor AJ, et al. Injuries from combat explosions in Iraq: injury type, location, and severity. *Injury.* 2012;43(10):1678−1682.
15. Fischer H. A Guide to US Military Casualty Statistics: Operation Inherent Resolve, Operation New Dawn, Operation Iraqi Freedom, and Operation Enduring Freedom. Report ID#-RS22452. Congressional Research Service. 02/05/2013. <www.fas.org/sgp/crs/natsec/RS22452.pdf> Accessed 25.04.14.
16. American College of Rehabilitation Medicine. Definition of mild traumatic brain injury. *J Head Trauma Rehabil.* 1993;8:86−87.
17. McCrory P, Meeuwisse W, Johnston K, Dvorak J, Aubry M, Molloy M, et al. Consensus statement on concussion in sport − The 3rd international conference on concussion in sport held in Zurich, 2008. *PMR.* 2009;1(5):406−420.
18. American Psychiatric Association. *Diagnostic and Statistical Manual of Mental Disorders, 4th ed. text rev.* Washington, DC: American Psychiatric Association; 2000.

19. World Health Organization. *International Classification of Diseases.* 10th ed. Geneva, Switzerland: WHO; 1990.

20. Dikmen S, Machamer J, Fann JR, Temkin NR. Rates of symptom reporting following traumatic brain injury. *J Int Neuropsychol Soc.* 2010;16(3):401−411.

21. Dick RW. Is there a gender difference in concussion incidence and outcomes? *Br J Sports Med.* 2009;43(suppl 1):i46−50.

22. Lincoln AE, Caswell SV, Almquist JL, Dunn RE, Norris JB, Hinton RY. Trends in concussion incidence in high school sports: a prospective 11-year study. *Am J Sports Med.* 2011;39 (5):958−963.

23. Gessel LM, Fields SK, Collins CL, Dick RW, Comstock RD. Concussions among United States high school and collegiate athletes. *J Athl Train.* 2007;42(4):495−503.

24. Seifert TD, Evans RW. Post-traumatic headache: a review. *Curr Pain Headache Rep.* 2010;14(4):292−298.

25. Hoffman J, Lucas S, Dikmen S, Braden CA, Brown AW, Brunner R. Natural history of headache following traumatic brain Injury. *J Neurotrauma.* 2011;28:1−8.

26. Lucas S, Hoffman JM, Bell KR, Walker W, Dikmen S. Characterization of headache after traumatic brain injury. *Cephalalgia.* 2012;32(8):600−606.

27. Keidel M, Diener HC. [Post-traumatic headache]. *Nervenarzt.* 1997;68(10):769−777.

28. Evans RW. Post-traumatic headaches. *Neurol Clin.* 2004;22 (1):237−249:viii.

29. Linder S. Post-traumatic headache. *Curr Pain Headache Rep.* 2007;11(5):396−400.

30. Blume HK, Vavilala MS, Jaffe KM, Koepsell TD, Wang J, Temkin N, et al. Headache after pediatric traumatic brain injury: a cohort study. *Pediatrics.* 2012;129:e31−39.

31. Erickson JC. Treatment outcomes of chronic post-traumatic headaches after mild head trauma in US soldiers: an observational study. *Headache.* 2011;51(6):932−944.

32. Faux S, Sheedy J. A prospective controlled study in the prevalence of post-traumatic headache following mild traumatic brain injury. *Pain Med.* 2008;9:1001−1011.

33. Hoge C, McGurk D, Thomas J, Cox A, Engel C, Castro C. Mild traumatic brain injury in US soldiers returning from Iraq. *N Engl J Med.* 2008;358(5):453−463.

34. Kuczynski A, Crawford S, Bodell L, Dewey D, Barlow KM. Characteristics of post-traumatic headaches in children following mild traumatic brain injury and their response to treatment: a prospective cohort. *Dev Med Child Neurol.* 2013;55:636−641.

35. Lieba-Samal D, Platzer P, Seidel S, Klaschterka P, Knopf A, Wöber C. Characteristics of acute post-traumatic headache following mild head injury. *Cephalalgia.* 2011;31(16):1618−1626.

36. Lucas S, Hoffman JM, Bell KR, Dikmen S. A prospective study of prevalence and characterization of headache following mild traumatic brain injury. *Cephalalgia.* 2014;34(2):93−102.

37. Stovner LJ, Schrader H, Mickeviciene D, Surkiene D, Sand T. Headache after concussion. *Eur J Neurol.* 2009;16(1):112−120.

38. Theeler BJ, Flynn FG, Erickson JC. Headaches after concussion in US soldiers returning from Iraq or Afghanistan. *Headache.* 2010;50(8):1262−1272.

39. Theeler BJ, Flynn FG, Erickson JC. Chronic daily headache in US soldiers after concussion. *Headache.* 2012;52(5):732−738.

40. Headache Classification Subcommittee of the International Headache Society. The international classification of headache disorders: 2nd edition. *Cephalalgia.* 2004;24(suppl 1):9−160.

41. Couch JR, Bearss C. Chronic daily headache in the posttrauma syndrome: relation to extent of head injury. *Headache.* 2001;41 (6):559−564.

42. Jensen OK, Thulstrup AM. [Gender differences of post-traumatic headache and other post-commotio symptoms. A follow-up study after a period of 9−12 months]. *Ugeskr Laeger.* 2001;163 (37):5029−5033.

43. Carlson KF, Taylor BC, Hagel EM, Cutting A, Kerns R, Sayer NA. Headache diagnoses among Iraq and Afghanistan war veterans enrolled in VA: a gender comparison. *Headache.* 2013;53 (10):1573−1582.

44. Headache Classification Subcommittee of the Internaitonal Headache Society. International Classification of Headache Disorders, 3rd edition (beta version). *Cephalalgia.* 2013;33(9): 629−808.

45. Victor TW, Hu X, Campbell JC, Buse DC, Lipton RB. Migraine prevalence by age and sex in the United States: a life-span study. *Cephalalgia.* 2010;30(9):1065−1072.

46. Martins H, Ribas V, Martins B, Ribas R, Valenca M. Post-traumatic headache. *Arq Neuropsiquiatr.* 2009;2009:43−45.

47. Haas DC. Chronic post-traumatic headaches classified and compared with natural headaches. *Cephalalgia.* 1996;16(7):486−493.

48. Baandrup L, Jensen R. Chronic post-traumatic headache − a clinical analysis in relation to the International Headache Classification 2nd Edition. *Cephalalgia.* 2004;25(2):132−138.

49. Castillo J, Muñoz P, Guitera V, Pascual J. Kaplan Award 1998. Epidemiology of chronic daily headache in the general population. *Headache.* 1999;39(3):190−196.

50. Scher AI, Stewart WF, Liberman J, Lipton RB. Prevalence of frequent headache in a population sample. *Headache.* 1998;38(7):497−506.

51. Gironda RJ, Clark ME, Ruff RL, Chait S, Craine M, Walker R, et al. Traumatic brain injury, polytrauma, and pain: challenges and treatment strategies for the polytrauma rehabilitation. *Rehabil Psychol.* 2009;54(3):247−258.

52. Silberstein SD. Practice parameter: evidence-based guidelines for migraine headache (an evidence-based review): report of the Quality Standards Subcommittee of the American Academy of Neurology. *Neurology.* 2000;55(6):754−762.

53. Silberstein SD. Clinical practice guidelines. *Cephalalgia.* 2005;25 (10):765−766.

54. Couch J, Lipton R, Stewart W, Scher A. Head or neck injury increases the risk of chronic daily headache: a population-based study. *Neurology.* 2007;69(11):1169−1177.

55. Clark ME, Bair MJ, Buckenmaier CC, Gironda RJ, Walker RL. Pain and combat injuries in soldiers returning from Operations Enduring Freedom and Iraqi Freedom: implications for research and practice. *J Rehabil Res Dev.* 2007;44(2):179−194.

56. Lay CL, Newman LC. Post-traumatic hemicrania continua. *Headache.* 1999;39(4):275−279.

57. Matharu MJ, Goadsby PJ. Post-traumatic chronic paroxysmal hemicrania (CPH) with aura. *Neurology.* 2001;56(2):273−275.

58. Piovesan EJ, Kowacs PA, Werneck LC. [S.U.N.C.T. syndrome: report of a case preceded by ocular trauma]. *Arq Neuropsiquiatr.* 1996;54(3):494−497.

59. Lucas S, Devine J, Bell K, Hoffman J, Dickmen S. Acute neuroimaging abnormalities associated with post-traumatic headache following traumatic brain injury (P04.020). *Neurology.* 2013;80 (American Academy of Neurology meeting abstracts).

60. Jankosky CJ, Hooper TI, Granado NS, Scher A, Gackstetter GD, Boyko EJ, et al. Headache disorders in the millennium cohort: epidemiology and relations with combat deployment. *Headache.* 2011;51(7):1098−1111.

61. Ruff RL, Riechers RG, Wang XF, Piero T, Ruff SS. For veterans with mild traumatic brain injury, improved post-traumatic stress disorder severity and sleep correlated with symptomatic improvement. *J Rehabil Res Dev.* 2012;49(9):1305−1320.

62. Terrio H, Brenner L, Ivins B, Cho JM, Helmick K, Schwab K, et al. Traumatic brain injury screening: preliminary findings in a US Army Brigade Combat Team. *J Head Trauma Rehabil.* 2009;24 (1):14–23.

63. Taber KH, Warden DL, Hurley RA. Blast-related traumatic brain injury: what is known? *J Neuropsychiatry Clin Neurosci.* 2006;18(2):141–145.

64. Theeler BJ, Erickson JC. Post-traumatic headache in military personnel and veterans of the Iraq and Afghanistan conflicts. *Curr Treat Options Neurol.* 2012;14(1):36–49.

65. Theeler B, Mercer R, Erickson J. Prevalence and impact of migraine among US Army soldiers deployed in support of Operation Iraqi Freedom. *Headache.* 2008;48(6):876–882.

66. Ruff RL, Ruff SS, Wang XF. Headaches among Operation Iraqi Freedom/Operation Enduring Freedom veterans with mild traumatic brain injury associated with exposures to explosions. *J Rehabil Res Dev.* 2008;45(7):941–952.

67. Theeler BJ, Erickson JC. Mild head trauma and chronic headaches in returning US soldiers. *Headache.* 2009;49 (4):529–534.

68. Finkel AG, Yerry J, Scher A, Choi YS. Headaches in soldiers with mild traumatic brain injury: findings and phenomenologic descriptions. *Headache.* 2012;52(6):957–965.

69. Theeler BJ, Erickson JC. Post-traumatic headaches: time for a revised classification? *Cephalalgia.* 2012;32(8):589–591.

70. Ruff RL, Riechers RG, Ruff SS. Relationships between mild traumatic brain injury sustained in combat and post-traumatic stress disorder. *F1000 Med Rep.* 2010;2:64.

71. Lew H, Otis J, Tun C, Kerns R, Clark M, Cifu D. Prevalence of chronic pain, post-traumatic stress disorder, and persistent post-concussive symptoms in OIF/OEF veterans: polytrauma clinical triad. *J Rehabil Res Dev.* 2009;46(6):697–702.

72. Theeler B, Lucas S, Riechers RG, Ruff RL. Post-traumatic headaches in civilians and military personnel: a comparative, clinical review. *Headache.* 2013;53(6):881–900.

73. Frugier T, Morganti-Kossmann MC, O'Reilly D, McLean CA. In situ detection of inflammatory mediators in post mortem human brain tissue after traumatic injury. *J Neurotrauma.* 2010;27(3):497–507.

74. de Vries HE, Blom-Roosemalen MC, van Oosten M, de Boer AG, van Berkel TJ, Breimer DD, et al. The influence of cytokines on the integrity of the blood–brain barrier in vitro. *J Neuroimmunol.* 1996;64(1):37–43.

75. Marchand F, Perretti M, McMahon SB. Role of the immune system in chronic pain. *Nat Rev Neurosci.* 2005;6(7):521–532.

76. Loane DJ, Byrnes KR. Role of microglia in neurotrauma. *Neurotherapeutics.* 2010;7(4):366–377.

77. Block ML, Zecca L, Hong JS. Microglia-mediated neurotoxicity: uncovering the molecular mechanisms. *Nat Rev Neurosci.* 2007;8 (1):57–69.

78. Gao HM, Hong JS. Why neurodegenerative diseases are progressive: uncontrolled inflammation drives disease progression. *Trends Immunol.* 2008;29(8):357–365.

79. Chodobski A, Zink BJ, Szmydynger-Chodobska J. Blood–brain barrier pathophysiology in traumatic brain injury. *Transl Stroke Res.* 2011;2(4):492–516.

80. Korn A, Golan H, Melamed I, Pascual-Marqui R, Friedman A. Focal cortical dysfunction and blood–brain barrier disruption in patients with post-concussion syndrome. *J Clin Neurophysiol.* 2005;22(1):1–9.

81. Milligan ED, Watkins LR. Pathological and protective roles of glia in chronic pain. *Nat Rev Neurosci.* 2009;10(1):23–36.

82. Mayer CL, Huber BR, Peskind E. Traumatic brain injury, neuroinflammation, and post-traumatic headaches. *Headache.* 2013;53(9):1523–1530.

83. Lauritzen M, Dreier JP, Fabricius M, Hartings JA, Graf R, Strong AJ. Clinical relevance of cortical spreading depression in neurological disorders: migraine, malignant stroke, subarachnoid and intracranial hemorrhage, and traumatic brain injury. *J Cereb Blood Flow Metab.* 2011;31(1):17–35.

84. Packard RC. Epidemiology and pathogenesis of post-traumatic headache. *J Head Trauma Rehabil.* 1999;14(1):9–21.

85. Geppetti P, Rossi E, Chiarugi A, Benemei S. Antidromic vasodilatation and the migraine mechanism. *J Headache Pain.* 2012;13(2):103–111.

86. Loyd DR, Weiss G, Henry MA, Hargreaves KM. Serotonin increases the functional activity of capsaicin-sensitive rat trigeminal nociceptors via peripheral serotonin receptors. *Pain.* 2011;152(10):2267–2276.

87. Raddant AC, Russo AF. Calcitonin gene-related peptide in migraine: intersection of peripheral inflammation and central modulation. *Expert Rev Mol Med.* 2011;13:e36.

88. Manley GT, Diaz-Arrastia R, Brophy M, Engel D, Goodman C, Gwinn K, et al. Common data elements for traumatic brain injury: recommendations from the biospecimens and biomarkers working group. *Arch Phys Med Rehabil.* 2010;91(11): 1667–1672.

89. Mansour AR, Baliki MN, Huang L, Torbey S, Hermann KM, Schnitzer TJ, et al. Brain white matter structural properties predict transition to chronic pain. *Pain.* 2013;154(10):2160–2168.

90. Zhou Y, Milham MP, Lui YW, Miles L, Reaume J, Sodickson DK, et al. Default-mode network disruption in mild traumatic brain injury. *Radiology.* 2012;265(3):882–892.

91. Huang J, Cao Y. Functional MRI as a biomarker for evaluation and prediction of effectiveness of migraine prophylaxis. *Biomark Med.* 2012;6(4):517–527.

92. Watanabe T, Bell K, Walker W, Schomer K. Systematic review of interventions for post-traumatic headache. *PMR.* 2012;4: 129–140.

93. Jensen OK, Nielsen FF, Vosmar L. An open study comparing manual therapy with the use of cold packs in the treatment of post-traumatic headache. *Cephalalgia.* 1990;10(5): 241–250.

94. Dubrovsky AS, Friedman D, Kocilowicz H. Pediatric post-traumatic headaches and peripheral nerve blocks of the scalp: a case series and patient satisfaction survey. *Headache.* 2014;54(5): 878–887.

95. Packard RC. Treatment of chronic daily post-traumatic headache with divalproex sodium. *Headache.* 2000;40(9):736–739.

96. Saran A. Antidepressants not effective in headache associated with minor closed head injury. *Int J Psychiatry Med.* 1988;18(1): 75–83.

97. Friedman MH, Peterson SJ, Frishman WH, Behar CF. Intraoral topical nonsteroidal antiinflammatory drug application for headache prevention. *Heart Dis.* 2002;4(4):212–215.

98. Gawel MJ, Rothbart P, Jacobs H. Subcutaneous sumatriptan in the treatment of acute episodes of post-traumatic headache. *Headache.* 1993;33(2):96–97.

99. McBeath JG, Nanda A. Use of dihydroergotamine in patients with post-concussion syndrome. *Headache.* 1994;34(3):148–151.

100. Defense and Veterans Brain Injury Study Group. Clinical guidance for evaluation and management of concussion/ mTBI-acute/subacute. Deployed and CONUS Setting Versons. 2012. <www.dvbic.org/> Accessed 02.11.12.

101. Schultz BA, Cifu DX, McNamee S, Nichols M, Carne W. Assessment and treatment of common persistent sequelae following blast-induced mild traumatic brain injury. *NeuroRehabilitation*. 2011;28(4):309–320.

102. DiTommaso C, Hoffman JM, Lucas S, Dikmen S, Temkin N, Bell KR. Medication usage patterns for headache treatment after mild traumatic brain injury. *Headache*. 2014;54(3):511–519.

103. Lucas S. Headache management in concussion and mild traumatic brain injury. *PMR*. 2011;3(10 suppl 2):S406–412.

104. Evans RW. Expert opinion: post-traumatic headaches among United States soldiers injured in Afghanistan and Iraq. *Headache*. 2008;48(8):1216–1225.

105. Silberstein SD. Preventive treatment of migraine. *Rev Neurol Dis*. 2005;2(4):167–175.

106. Mishra D, Kaur S. Post-traumatic headache: an uncommon but treatable entity. *J Clin Diagn Res*. 2014;8(2):169–170.

107. Farbota KD, Sodhi A, Bendlin BB, McLaren DG, Xu G, Rowley HA, et al. Longitudinal volumetric changes following traumatic brain injury: a tensor-based morphometry study. *J Int Neuropsychol Soc*. 2012;18(6):1006–1018.

108. Stewart WF, Linet MS, Celentano DD, Van Natta M, Ziegler D. Age- and sex-specific incidence rates of migraine with and without visual aura. *Am J Epidemiol*. 1991;134(10):1111–1120.

109. Scher AI, Monteith TS. Epidemiology and classification of post-traumatic headache: what do we know and how do we move forward? Comment on Lucas et al., "Prevalence and characterization of headache following mild TBI". *Cephalalgia*. 2014;34 (2):83–85.

14

Headache and the Eye

Benjamin Frishberg

UCSD School of Medicine, The Headache Center of Southern California, Encinitas, California, USA

INTRODUCTION

"The eye is the window to the soul," and may also be a window into the etiology of a variety of headache disorders. The primary headache disorders may be associated with transient visual disturbances, eye pain, photophobia, visual loss, and abnormalities of the appearance of the eye. Secondary headache disorders may be associated with loss of vision, double vision, and pupillary abnormalities, as well as a red eye. The signs and symptoms of the patient typically provide enough information to allow the examiner to create a comprehensive differential diagnosis, develop a diagnostic testing algorithm, and create a reasonable treatment plan. In this chapter we will take a symptom-based approach to help develop a differential diagnosis, and discuss the various disorders that are commonly and not so commonly seen in the medical practitioner's office.

BASICS OF THE BEDSIDE EYE EXAMINATION

When evaluating a headache patient, especially when there are any visual issues, it is important to do an adequate examination of the visual system.[1]

Visual Acuity

Check distance visual acuity with a Snellen chart in the office, making sure the patient is wearing his or her distance correction. If there is no correction available, or there is a significant reduction in acuity, try asking the patient to look through a pinhole, which will typically improve acuity if the issue is refractive. Create a pinhole with a safety pin and a piece of thin cardboard. At the bedside, a hand-held acuity card is

adequate, provided patients can wear their reading glasses. Record your findings.

The Pupil

Assess the size and reactivity of the pupil and note any anisocoria. If a patient has a complaint of visual loss, or vision is not normal on examination, check for an afferent pupillary defect (Marcus-Gunn Pupil) using the swinging flashlight test. If a pupil shows no apparent reactivity, use your ophthalmoscope to obtain a magnified view of the pupil and observe for constriction.

Visual Fields

Standing 24–30 inches (60–75 cm) from the patient's face, ask that he or she covers each eye sequentially, while you instruct the patient to count fingers in each of the four quadrants. One can increase sensitivity by asking the patient to view both hands held at the same height in both the right and left visual fields, and asking if one hand is seen more clearly than the other. One can be more sensitive by using a red target in each field, comparing the intensity of the color. This procedure is especially useful for bitemporal hemianopia in patients with pituitary tumors. Peripheral visual field constriction may be checked by asking the patient to identify a wiggling finger entering the visual field from the periphery and comparing it with your own field if the finger is held midway between the patient and the examiner.

Eye Movements

Check gaze in all directions, specifically looking for loss of normal eye movements or listening for a complaint of diplopia from the patient. At the far ranges of

DOI: http://dx.doi.org/10.1016/B978-0-12-800901-7.00014-8

right and left gaze, many individuals will complain of blurring or slight doubling, which is usually normal. If the patient has a complaint of diplopia, determine if it is horizontal, vertical, or oblique. Cover each eye separately to determine if the diplopia is truly binocular. If it is monocular, it is not a neurologic issue but an ophthalmologic issue related to astigmatism or other refractive, retinal, or lenticular processes that can be identified by an eye specialist.

Funduscopic Examination

The fundus examination may be the most important, but often overlooked, part of the headache examination. As increased intracranial pressure typically is associated with headache, and those entities that increase intracranial pressure, such as tumor, infection, and idiopathic intracranial hypertension, should not be missed, it is imperative to obtain an adequate look at the optic nerve. The examination is ideally performed in dim illumination. If the pupil is too small to afford a good view, there is rarely a contraindication to dilation in the eye.

General Examination

Observe and document any redness of the eye. Other features that may be significant include proptosis, significant globe tenderness, periorbital edema, as well as ptosis or lid retraction. Always palpate the temporal arteries for tenderness and thrombosis.

Recalling the anatomy of the trigeminal nerve,[2] and considering that the ophthalmic branch of the trigeminal nerve (V1) innervates the structures in and around the eye (in particular the meninges), it is important to assess trigeminal function carefully[1]. This assessment will allow accurate localization and guide the differential diagnosis and further testing. Examination can be easily performed with a light brushing of a fingertip or cotton swab, a fresh safety pin, or a cold stimulus, and comparing sensation between the two sides in each of the three divisions of the trigeminal nerve. The corneal reflex can be a slightly more objective test comparing the subjective sensation between the corneal sensation of both eyes, and observing the patient response to corneal stimulation. This is best accomplished using a rolled corner of a tissue or rolled wisp of cotton. This procedure will not work if the patient is wearing contacts, as the corneal sensation is muted in contact wearers. Nasal tickle, using the same rolled tissue, is also useful for V2 sensation. If patients note any gingival sensory changes, this examination can be performed using a broken tongue depressor or a clean pin.

MIGRAINE-RELATED VISUAL AND EYE SYMPTOMS

Migraine-Related Visual Aura

The visual aura of migraine is a fascinating phenomenon that occurs in about 20% of migraine sufferers. "Migraine with visual aura" is the official term, but other names include migraine equivalents, migraine accompaniments, ocular migraine, optical migraine, eye migraines, and, ophthalmic migraine. The typical migraine visual aura is a binocular visual disturbance. Most patients describe the disturbance as monocular as it tends to occur in either the right or left visual field, which they interpret as occurring in the eye on the side of the disturbance. It is present with eyes open or closed. Careful questioning almost always reveals that it occurs as a binocular phenomenon. If the patient reports scotomas with both eyes open, this strongly suggests the disturbance is indeed binocular. Most patients do not alternately cover each eye to determine binocularity. Although there are many descriptions of the visual phenomena that occur with migraine, the typical disturbance starts as a small, often bright, flashing semicircle just to the side of fixation in one hemi-field, which then gradually builds up and moves across the visual field. The visual disturbance increases over several minutes, moves across the field, and gradually dissipates. The typical duration is 15–30 minutes, but may be as little as 5 minutes or as long as 60 minutes and still be within the usual definition of typical aura. Classic descriptions of the visual disturbance include brightness, movement, zigzags, or kaleidoscopes. The term "teichopsia" was coined by Sir Hubert Airy, himself a migraine sufferer, who wrote extensively and insightfully about his migraine aura.[3] It is a term exclusively used to describe the angled fortification-like "walls" that are so typical of the migraine visual aura. Other common patient descriptions include: flashing lights (photopsias), shattered glass, rain on a windshield, holes in the vision, heat rising, "blinded like after a flashbulb," missing half of everything, tunnel vision, and a sense of brightness. Figure 14.1 shows the build of up a migraine aura.

The visual disturbance of migraine may occur without associated headache. The visual disturbance may also be followed by other transient neurologic symptoms, including paresthesias in the face and arm, as well as a language disturbance (aphasia) with difficulty finding words and comprehending what people are saying. When this march of symptoms occurs with one symptom following the next symptom, it is virtually pathognomonic of migraine. Transient ischemic attacks (TIAs) are typically much shorter in duration, consist of negative visual symptoms, occur suddenly,

FIGURE 14.1 The migraine aura represented as a clock. *Top left:* Initial small paracentral scotoma; *top right:* Enlarging scotoma 7 minutes later; *bottom left:* Scotoma obscuring much of central vision 15 minutes later; *bottom right:* Break-up of scotoma at 20 minutes. *Reproduced from Hupp et al.[4], with permission.*

and have no build-up. Migraine visual aura without headache may develop later in life, often after age 60 in patients with or without a previous history of migraine. New onset of migraine visual aura without headache after age 60 is common, and usually quite benign. Although occipital lobe seizures and TIAs are in the differential, it is usually quite simple to distinguish migraine from these other more serious disorders. Occipital seizures are usually brief, lasting up to 3 minutes, and are associated with elementary visual hallucinations that do not have the zig-zag pattern seen with migraine.

Transient ischemic attacks (TIAs) are typically brief, lasting a few minutes, and associated with a negative visual disturbance with loss of visual field. These patients do not have gradual build-up of the visual disturbance, nor do they see motion, brightness, or flashing. It is not unusual for patients with migraine visual aura without headache to have clusters of events followed by long periods lasting months or years without spells. Triggers for this disorder vary from those for migraine without aura, with most patients noting no clear triggers except for certain patterns of light or seeing a glint of light or reflection that seems to precipitate the episodes.[5]

Persistent visual phenomena may be seen in several headache-related conditions. Prolonged visual aura is defined as any visual aura lasting more than 60 minutes. Although it may be seen occasionally as part of a typical migraine with aura, one has to pay particularly close attention to this symptom as it may indicate a stroke or intracranial process. If a patient has a persistent aura in association with homonymous visual field loss, this may indicate an occipital lobe infarction.

Migrainous infarction is a well-described syndrome occurring in patients with a history of migraine with visual aura. They experience a typical migraine visual aura, but the aura and visual field loss do not dissipate in the usual timeframe and often persist for days or weeks.[6] This condition is a neurologic emergency and requires immediate Emergency Department referral for possible thrombolysis, if seen within 4.5 hours.

Some patients with migraine may have persistent migraine aura. These patients experience aura symptoms that lasts days to years. The persistent visual aura typically consists of positive visual phenomena usually described as formed shapes or figures or unformed lights or sparkles.[7] Other ocular and neurologic conditions may cause similar visual symptoms. Thus, a diagnosis of persistent migraine visual aura is generally one of exclusion.

Another condition, persistent positive visual phenomena or "visual snow," is seen most commonly in migraineurs, and consists of a continuous pixilation of vision. It is described by patients as similar to snow on a cathode ray tube TV, or continuous small flashing dots throughout the visual field of both eyes. Typically, this condition is continuous, lasts for years, does not interfere with vision, may begin in childhood, and does not respond to treatment. Although it is generally considered benign, the condition may be disabling for some patients.[8,9]

Retinal Migraine

Transient monocular loss of vision lasting several minutes may be seen as a manifestation of migraine. The official definition requires the loss of vision be associated with, or is followed by, a headache with migraine features. However, many documented cases of recurrent transient unilateral loss of vision are associated with vasospasm of the retinal circulation without associated headache.[10] This entity is typically recurrent and benign. Differentiating it from embolic amaurosis fugax is clearly imperative. This condition is typically a diagnosis of exclusion. The use of calcium channel blockers has been shown to significantly reduce the frequency of these events.

PHOTOPHOBIA AND EYE PAIN

Isolated eye pain in the absence of findings on a comprehensive examination of the eye and cranial nerves is the rule rather than the exception in headache medicine. In evaluating patients with primary headache disorders, most will have a normal eye examination, with the exception of the trigeminal autonomic cephalalgias (TACs) in which redness, anisocoria, and periorbital edema will be present.[11] A number of headache syndromes, including migraine, are centered in and around the eye. Patients often consult their eye-care professional for initial evaluation, as they assume the headache or eye pain emanates from the eye. Migraine patients complain of eye pressure in the eye itself, or more commonly the pain feels like it is located behind the eye. The first division of the trigeminal nerve (V1) innervates the eye and orbital structures, so any process that stimulates the V1 fibers may result in eye pain, and these include temporomandibular joint dysfunction, sinusitis, and myofascial pain. Patients with unilateral eye discomfort, especially when it is continuous, may be found on examination to display occipital or suboccipital trigger points on the ipsilateral side, with the underlying eye pain having a cervicogenic origin. These patients often respond to trigger point injections with lidocaine or an occipital nerve block.[12]

Trochlear Headache

Trochlear headache is a recently described headache disorder characterized by pain emanating from the medial orbit. The inflammatory form is called "trochleitis," and a non-inflammatory form is termed "primary trochlear headache." Trochleitis is characterized by swelling and tenderness of the trochlea, and aggravation by gazing up or down. The trochlea is easily palpated with the finger pad under the superior medial orbital rim. Swelling may be observed clinically, or documented on orbital imaging studies. Tenderness is easy to elicit.[13] These patients most often present with new onset of chronic daily headache. The quality is typically achy and may be associated with photophobia and aggravation by eye movement, especially reading. This disorder usually responds to steroid injections into the trochlear region.

Photophobia

Photophobia is defined as pain with normal or dim light.[14] Unilateral photophobia suggests an inflammatory process in the affected eye, but may be seen with the trigeminal autonomic cephalalgias. Bilateral photophobia is a consistent complaint with migraine, and migraine patients tend to be more sensitive to light in general, including some patients who report light as a trigger to migraine headache.[15] Tension-type headache may have mild photophobia, but it is not typical. Chronic photophobia may be part of chronic migraine, but may also be seen in traumatic brain injury, benign essential blepharospasm, and agoraphobia.

HEADACHE AND THE RED EYE

Redness of the eye is related to vascular dilation. It usually involves the conjunctival vessels, but may affect the scleral vessels as well. The trigeminal autonomic cephalalgias, such as cluster headache or SUNCT (Short-lasting Unilateral Neuralgiform headache attacks with Conjunctival injection and Tearing), typically are associated with autonomic features that include injection of the conjunctiva as well as tearing and ptosis. Some patients with migraine will have conjunctival injection, typically on the side of the headache, which dissipates when the headache is gone, but it is less prominent than with the TACs.[16]

Glaucoma

Acute angle closure glaucoma is associated with very high intraocular pressure (>40 mmHg), with redness of the involved eye due to dilation of the larger scleral vessels (Figure 14.2). Eye pain usually is present, which may be severe and associated with nausea and vomiting. These patients may also have a mid-position fixed pupil due to iris ischemia.[17] Although it would be unusual for a patient with acute angle closure glaucoma to present to a primary care or neurology office, failure to recognize and treat this entity may lead to permanent vision loss. Some patients have intermittent angle closure with recurrent episodes of eye pain with redness that may masquerade as migraine or even cluster headache. An eye-care professional can evaluate for narrow angles with use of a gonioscope lens which allows for visualization of the trabecular meshwork that comprises the drainage pathway of the anterior chamber of the eye. Narrowed angles may be treated with laser iridotomy, which creates an alternative drainage pathway for the aqueous humor. Chronic open angle glaucoma, which is a much more common form of glaucoma, is not associated with pain or headache, but is associated with insidious loss of visual field with progressive optic nerve cupping. There is no contraindication to pupillary dilation in chronic open angle glaucoma, only in those untreated patients with narrow angles as dilation may lead to acute angle closure. Those who are treating migraine patients with topiramate need to consider the rare side effect of acute glaucoma associated with induced myopia that typically occurs in the first few weeks of treatment and presents with unilateral or bilateral visual blurring and eye pain. These patients need immediate ophthalmologic evaluation.[18]

There are multiple medications that are relatively contraindicated in patients with narrow angles, including the tricyclic antidepressants such as amitriptyline and nortriptyline. The classes of medications that have the potential to induce angle closure are topical anticholinergic or sympathomimetic dilating drops, tricyclic antidepressants, monoamine oxidase inhibitors, antihistamines, anti-parkinsonian drugs, and antipsychotic medications. This is based purely on their ability to dilate the pupil, which may precipitate acute angle closure. There are no contraindications to the use of these medications in patients with open angle glaucoma, which make up the vast majority of glaucoma patients.

Cavernous Sinus Fistula

A high-flow fistula representing a direct artery to vein communication within the cavernous sinus may occur as a result of head trauma or, more rarely, after aneurysmal rupture of an intracavernous aneurysm. A low-flow fistula may occur related to a dural arteriovenous malformation. These conditions are usually associated with chronic redness of one eye, but may involve both eyes if there is vascular communication between the right and left cavernous sinus. The redness is chronic, secondary to venous engorgement, and may be associated with arterialization of the scleral vessels (Figure 14.3). Fistulas tend to produce dull, aching pain, which may be unilateral or bilateral and

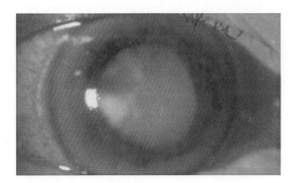

FIGURE 14.2 Acute angle closure glaucoma with corneal haze and vascular congestion.

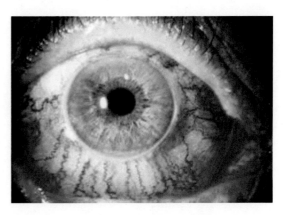

FIGURE 14.3 Scleral vessel engorgement in a cavernous sinus fistula.

frontal. The high-flow fistula is usually quite dramatic with prominent proptosis, loss of full eye movements, and a loud cranial bruit that may be heard by the patient and may be detected by the examiner using a stethoscope bell over the orbit. The more common low-flow dural fistula causes scleral injection and may be associated with diplopia and subjective pulsatile tinnitus.[19] Both of these conditions may be treated by occluding the involved vessels using interventional neuroradiologic techniques.

Inflammation

The combination of visual blurring, redness, and photophobia suggests an inflammatory process of the eye. Iritis is anterior, and involves the iris and anterior chamber. If the condition involves the posterior portion of the eye, it is termed "uveitis." Uveitis may be associated with optic nerve head swelling, macular edema, and significant loss of vision. Iritis tends to be more painful with prominent photophobia, redness, and tearing. Keratitis — corneal inflammation — is extremely painful, and patients have difficulty keeping their eye open due to the pain.[20] All of the inflammatory entities may be associated with periorbital pain radiating into one or both sides of the head. The eye symptoms are typically prominent enough that a primary ophthalmologic etiology is apparent. This may not be true for posterior scleritis, which tends to cause a deep dull aching pain in and behind the eye without much in the way of other symptoms or findings on exam.[21]

The diagnosis of posterior scleritis is usually confirmed by a retina specialist or with imaging, which shows thickening and enhancement of the sclera. It is especially important to consider the uveomeningitic syndromes, specifically Vogt-Koyanagi-Harada syndrome, Behçet's disease, sarcoidosis, and Wegener's granulomatosis (now termed "granulomatosis with polyangitis").[22,23] Behçet's is a small vessel vasculitis associated with mouth ulcers, genital ulcers, and uveitis, which commonly involves the central nervous system. This disorder may be associated with venous sinus thrombosis with secondary increase in intracranial pressure, resulting in headaches, and papilledema (Figure 14.4). It is most commonly seen in individuals from the Mediterranean region, but may be seen in all populations. The uveomeningitic syndromes are associated with meningeal inflammation with cells in the spinal fluid, headaches, and uveitis.

Idiopathic Orbital Inflammatory Syndrome

One of the most important conditions that presents with eye pain and headache is idiopathic orbital inflammatory syndrome (IOIS), previously called "orbital

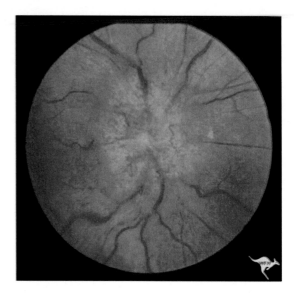

FIGURE 14.4 Typical papilledema. *Reproduced from NOVEL (Neuro-ophthalmology Virtual Education Library, http://novel.utah.edu), with permission.*

pseudotumor".[24] This condition is idiopathic, and typically presents with eye pain — often with acute onset. One usually observes signs of eye inflammation, deep aching pain, and often diplopia and proptosis. Neuroimaging typically shows enhancement and inflammation of orbital structures. This condition may be mistaken for orbital cellulitis, and is exquisitely sensitive to treatment with steroids, with resolution of pain within hours after treatment. Recently, there has been a subclass of IOIS termed "IgG4 disease" due to deposits of IgG4 in the inflammatory tissue. This condition may be associated with systemic involvement with sclerosing inflammation, including pachymeningitis, and has been found to respond to immunosuppression.[25] IOIS is usually a diagnosis of exclusion, as many systemic disorders, including lymphoma, fungal infection, Wegener's, Behçet's, and Sjögren's, have been reported to cause similar symptoms and findings.

Other causes of a painful red eye include infection (especially herpes simplex and herpes zoster [Figure 14.5]), severe dryness, and corneal abrasion. It is usually not difficult to rule out a primary ophthalmic problem from one of the primary or secondary headache disorders due to the objective eye findings. It is important to involve your ophthalmologist early when the eye is involved and the diagnosis is not clear. These specialists have the tools, experience, and training to manage the primary eye issues that cause pain.

HEADACHE AND VISUAL LOSS

Loss of vision may be separated into several categories, depending on etiology. Vascular causes include

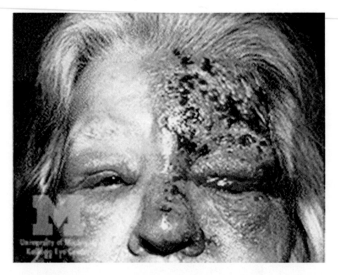

FIGURE 14.5 Herpes zoster ophthalmicus. Trigeminal herpes with uveitis and keratitis. *Reproduced courtesy of Jonathan Trobe, MD, University of Michigan Kellogg Eye Center.*

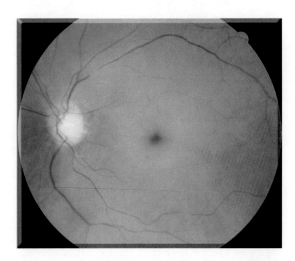

FIGURE 14.6 Central retinal artery occlusion with attenuation of the arterioles and macular cherry red spot. *Reproduced from Life in the Fast Lane (http://lifeinthefastlane.com/ophthalmology-befuddler-026-2), with permission.*

retinal artery occlusion (Figure 14.6), retinal ischemia, aneurysm, hemorrhagic and ischemic stroke, and cerebral venous sinus disease. Inflammatory diseases, such as optic neuritis, sarcoidosis, giant cell arteritis, and the many causes of uveitis, have been discussed previously. Most of these conditions may be associated with eye pain and headache. Acute loss of vision is typically an emergency situation (Table 14.1).

Vascular

Stroke can be ischemic or hemorrhagic. Ischemic stroke is associated with a 25% incidence of headache.

TABLE 14.1 Neuro-Ophthalmic Emergencies

Acute loss of vision and headache	Diplopia and headache
Giant cell arteritis	Giant cell arteritis
Pituitary apoplexy	Pituitary apoplexy
Aneurysmal rupture	Aneurysmal rupture
Meningitis	Third nerve palsy
Acute angle closure glaucoma	Meningitis
Carotid dissection	Orbital infection
Orbital infection	

BOX 14.1

CAUSES OF THUNDERCLAP HEADACHE WITH VISION LOSS

- Subarachnoid hemorrhage
- Cervical arterial dissection
- Cerebral venous sinus thrombosis
- Pituitary apoplexy
- Intracranial hemorrhage
- Reversible cerebral vasospasm syndrome

Embolic stroke and strokes of the posterior circulation are more likely to cause headache.[26,27] Cortical stroke may cause a homonymous visual field defect, typically a hemianopia or quadrantanopsia. The field defect may occur in isolation with no other neurologic findings in patients with infarction of the occipital lobe. The headache is nearly always on the side of the infarct. The older patient with acute unrelenting headache should always be evaluated for visual field defects, as the patient may be unaware of visual impairment, especially with a right parietal-occipital process with associated neglect.

Intracranial hemorrhage (ICH) presents in a manner similar to an ischemic infarct but with a higher incidence of headache, and associated alteration of consciousness and nausea. However, the overall incidence of headache in ICH is only about 33%, and does not help differentiate hemorrhage from infarct.[28]

Ruptured cerebral artery aneurysm presents in a dramatic fashion with thunderclap headache. Box 14.1 provides a differential diagnosis for other causes of acute onset of thunderclap headache. The headache usually occurs over seconds to peak intensity. The sudden increase in intracranial pressure may cause an acute hemorrhage into the retina and vitreous of the eye with sudden loss of vision, which is referred to as Terson's syndrome.[29] The hemorrhages may be demonstrated on CT, but are easily seen on funduscopic exam. Vision loss may also occur from rupture of an ophthalmic artery aneurysm or from vasospasm-related infarction that may occur after aneurysmal rupture.

Arterial Dissection

Arterial dissection may occur in the carotid or vertebral arteries. Cervical arterial dissection tends to occur in younger patients, often in their thirties and forties, and is due to a tear in the intimal wall of the vessel. The dissection can be spontaneous, due to significant trauma, or associated with trivial activity such as getting one's hair shampooed or a vigorous workout at the gym. Vertebral artery dissection causes pain in the occiput and neck, but may radiate anteriorly to the orbits. Vertebral dissection may cause unilateral or bilateral occipital lobe infarction or involvement of the brainstem with subsequent diplopia, nystagmus, and other eye movement disorders.

Carotid artery dissection usually causes pain in the neck radiating into the head, face, and eye in at least 60% of cases. A pathognomonic feature is an ipsilateral Horner's syndrome, which occurs in up to 40% of cases. These dissections may cause emboli with subsequent branch or central retinal artery occlusion as well as cortical infarcts or transient ischemic attacks.[30]

Cerebral Venous Sinus Thrombosis

Cerebral venous sinus thrombosis is a relatively rare condition that may cause increased intracranial pressure and subsequent papilledema due to inadequate venous drainage. This condition is in the differential diagnosis of idiopathic intracranial hypertension, and is most commonly seen post-partum and in patients with an underlying coagulopathy. It is also associated with dehydration, malignancy, oral contraceptives, and Behçet's disease.[31] Headache is the most common associated symptom, occurring in over 90% of recognized patients.[32] There are no identifying characteristics to the headache, but it can be severe. The acute thrombosis may be associated with a thunderclap headache. This condition may be devastating, as it can cause venous infarcts, uncontrolled intracranial hypertension, and death. The diagnosis is made with use of MR venography.

Optic Neuritis

Optic neuritis refers to an acute unilateral inflammation of the optic nerve, typically associated with an immune-mediated demyelinating condition — specifically, multiple sclerosis or neuromyelitis optica. Optic neuritis may be idiopathic, or associated with connective tissue disorders and infection. Prior to loss of vision, patients typically develop pain in and around the eye for several hours to several days. Vision loss takes place gradually over hours to days, and may be mild or severe. Of those patients with demyelinating optic neuritis, 90% have pain. The pain is very specific, as it occurs or increases with eye movements.[33] Vision typically improves in 3–6 weeks to 20/30 or better, and the pain usually resolves within a week. The pain is steroid-responsive. Some patients with optic neuritis experience mild ongoing orbital and retro-orbital pain that may persist for months or years.

Giant Cell Arteritis

In a patient over the age of 50 who develops acute loss of vision in association with headache, one must always consider a diagnosis of giant cell arteritis (GCA). This condition is seen more commonly in women than men, and has the highest incidence in people of Northern European background.[34] The incidence increases with advancing age, and the average age of onset is in the eighth decade. When visual loss is present in one eye, there is a high risk of visual loss in the contralateral eye in the ensuing days if treatment is not provided. GCA involves the midsize and large arteries that branch from the aorta, with a predilection for involvement of the extracranial carotid circulation. Visual loss results from involvement of the ciliary artery, resulting in anterior or posterior ischemic optic neuropathy, or from involvement of the ophthalmic artery, resulting in central retinal artery occlusion. Involvement of other extracranial arteries may result in jaw claudication, sore throat, orbital ischemia with diplopia, and possibly ischemic skin ulcers. The presence of significant proximal muscle soreness, especially in the morning, suggests the diagnosis of polymyalgia rheumatica (PMR), which may be a prodrome to GCA. If an elderly patient displays the symptoms of PMR and in addition has new headache, GCA must be ruled out on an urgent basis. Some patients will experience a prodrome of recurrent transient loss of vision (amaurosis fugax) in one eye prior to permanent loss, offering an opportunity to intervene and save vision. The sedimentation rate is elevated in >90% of cases but may be normal, especially initially. C-reactive protein is another marker of systemic inflammation and may be markedly elevated prior to elevation of the sedimentation rate; it is another useful marker for this disease. Temporal artery biopsy is the gold standard for diagnosis.[35] When there is any suspicion for GCA, it is always wise to start prednisone, 60 mg or more per day, while awaiting the biopsy. If there is visual involvement, higher-dose steroids or hospitalization may be necessary. Vision loss in GCA can be devastating and irreversible. When considering the possibility of GCA it is prudent to obtain a biopsy to establish the diagnosis, due to the necessity of long-term steroid therapy with its expected complications in the elderly population.

The Orbital Apex Syndrome

The posterior orbit is a crowded space containing the optic nerve, the vascular supply of the eye, the first division of the trigeminal nerve, and the eye muscles. Any inflammation or mass in that location may cause headache and vision loss, as well as ophthalmoplegia.[36] The pain is usually in the distribution of V1, but if there is more posterior involvement of the cavernous sinus, V2 may also be involved. Sudden enlargement of a vascular lesion (such as an orbital varix or hemangioma), or the presence of a malignant lesion such as lymphoma, head and neck tumors, or metastatic disease, may all be associated with pain and vision loss. Infections and inflammation are other potential causes of an acute orbital apex syndrome. Invasive fungal infections, such as aspergillus or mucormycosis, are well known to cause sudden and devastating loss of vision. These infections typically have a moderate to severe amount of accompanying orbital pain or headache. Sinus disease, specifically sphenoid sinusitis or a sphenoid sinus mucocele, may invade the posterior orbit, leading to vision loss. Inflammatory disorders such as idiopathic orbital inflammatory disease and granulomatosis with polyangitis (Wegener's granulomatosis) may also be associated with the orbital apex syndrome. Orbital congestion with crowding of the apex may also be seen with thyroid eye disease, and with carotid–cavernous fistulas. Early neuroimaging is imperative, and is often diagnostic. MRI, with and without contrast, is recommended, and there should be thin cuts through the orbits. An MRI of the brain alone is not sufficient.

HEADACHE AND THE ABNORMAL PUPIL

Nothing seems to trouble a healthcare provider more than an enlarged pupil. Bolstered by the media, there is an assumption that a dilated pupil always indicates some serious problem. Abnormal pupils, either too big or too small, may be seen as part of several primary headache disorders, but also may portend a serious underlying process (Table 14.2).

Headache and the Small Pupil

When dealing with anisocoria, it is always important to determine which pupil is abnormal – the larger or the smaller pupil. In addition, up to 40% of

TABLE 14.2   Causes of an Enlarged Pupil

Etiology	Findings
Pharmacologic dilation	Very large pupil, unreactive, unilateral or bilateral
Acute glaucoma	Mid-position, sluggish or no response, red eye, unilateral, pain
Third nerve palsy	Moderately dilated, no response, diplopia and ptosis
Adie's pupil	Markedly dilated, unreactive, painless
Benign pupillary mydriasis	Mild to moderate dilation, normally reactive
Essential anisocoria	Very mildly enlarged, <1 mm, normally reactive, changeable

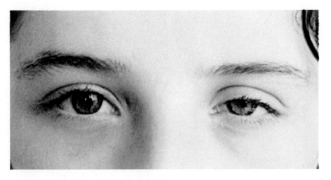

FIGURE 14.7   Horner's syndrome with ptosis, reverse ptosis, and meiosis.

individuals have mild anisocoria, termed "essential anisocoria," which is a normal variant but is typically no more than 0.6 mm.[37] If the pupils are normally reactive and there is no history of eye trauma or the use of eye drops, a smaller pupil may be indicative of a sympathetic defect, in which case we term it a "Horner's pupil." If associated ptosis and anhydrosis occur, then we term it "Horner's syndrome," which infers an oculosympathetic defect (Figure 14.7). In the absence of sympathetic input the pupil is smaller than the contralateral pupil, and the degree of anisocoria increases in a less illuminated environment. A Horner's pupil may be confirmed by the use of cocaine drops, which inhibit reuptake of norepinephrine at the myoneural junction of the iris muscle. Since there is no sympathetic input, there is no tonic release of norepinephrine and, therefore, no dilation will occur in a Horner's pupil, although the normal eye will dilate. Isolated Horner's syndrome may be seen with trigeminal autonomic cephalalgias, including cluster headache, chronic paroxysmal hemicrania, and SUNCT (Short-lasting Unilateral Neuralgiform headache with Conjunctival injection and Tearing), but Horner's syndrome is not seen with migraine.

Painful Horner's Syndrome

Several potential causes of a painful Horner's syndrome have been identified, including the primary headache disorders. The most important and relatively frequent cause is dissection of the carotid artery.[38] Any patient who presents with sudden anisocoria associated with ptosis (patients usually notice the ptosis and not the pupil), with new ipsilateral neck, head, or face pain, is considered to have a carotid dissection until proven otherwise. An MRI of the skull base will usually show an abnormality in the wall of the carotid artery, and CT angiography or MR angiography will usually confirm the dissection. Carotid dissection is the most common cause of acute stroke in patients under age 40. Other potential causes of painful Horner's syndrome include a lesion of the orbit or cavernous sinus, but these patients almost always have other localizing neurologic deficits.

Headache and the Large Pupil

An enlarged pupil may be reactive or unreactive, and the differential diagnosis and implications are quite different. An enlarged, normally reactive pupil is usually not pathologic. It calls into question whether the contralateral smaller pupil is not the pathologic pupil.

Third Nerve Palsy

The pupil constrictors are mediated through the third nerve. Disorders of the third nerve may be associated with an enlarged and poorly reactive or unreactive pupil. Typically, with a third nerve palsy some involvement of other structures innervated by the third nerve will be noted, including the lid, and the superior rectus, inferior rectus, medial rectus, and the inferior oblique muscles. Patients with third nerve palsy often have ptosis, diplopia, and an enlarged and poorly reactive pupil (Figure 14.8). Sudden onset of a headache associated with a partial or complete third nerve palsy is the *sine qua non* of an enlarging posterior communicating artery aneurysm, and constitutes a true medical emergency. In these cases, a CT or MR angiogram needs to be performed on an emergent basis with neurosurgical evaluation to follow. These aneurysms may suddenly expand, leak, or rupture, with subsequent high morbidity and mortality.

In older patients or those with vascular risk factors, especially diabetes, one can see an acute third nerve palsy on a microvascular basis. This condition is far more common than other causes of third nerve palsy. These patients often present with several days of periorbital aching pain followed by diplopia and ptosis. The degree of pain can be substantial, and may last for several weeks. Complete recovery of eye muscle function typically occurs between 6 and 12 weeks after onset. With microvascular third nerve palsies, there is the so-called "rule of the pupil." This states that a patient with a complete third nerve palsy (complete ptosis, and total loss of the superior rectus, inferior rectus, and medial rectus function) who has an entirely normal pupil does not have a compressive lesion. Although this may hold true >95% of the time, it is

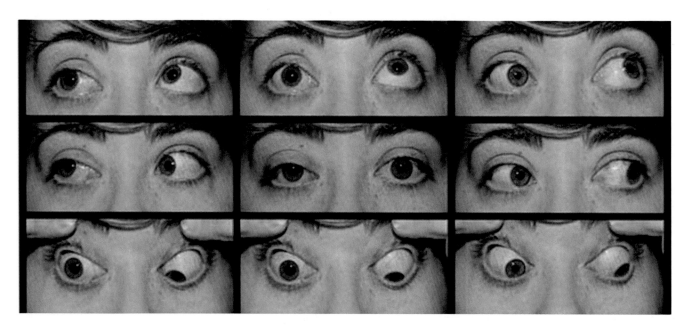

FIGURE 14.8 Partial third nerve palsy with decreased supraduction, infraduction, and adduction. *Reproduced courtesy of John Chen, MD, PhD, from Ophthalmic Atlas Images (http://webeye.ophth.uiowa.edu/eyeforum/atlas/pages/aberrant-regeneration-3rd-nerve-palsy.htm), Eyerounds, The University of Iowa, with permission.*

not 100%; therefore, the pupil-sparing third nerve palsy should always be imaged as an emergency procedure. The best imaging study for intracranial aneurysm is the CT angiogram.[39]

Benign Pupillary Mydriasis

Benign pupillary mydriasis is an entity associated with intermittent and typically painless dilation of one pupil. It is typically the same eye, but has been reported to occur in either eye. It is usually seen in women under the age of 40 who carry a diagnosis of migraine.[40] The episodes may last from minutes up to several days. The pupil itself reacts normally, but patients may note some mild visual blurring. The dilation may occur with a migraine or may be painless. It is important to recognize this entity to avoid unnecessary or invasive testing.

Acute Glaucoma

With acute angle closure glaucoma, iris ischemia due to very high intraocular pressure will cause the pupil to be mid-position and sluggish or unreactive. Typically, the eye is red and there is associated eye pain or frontal headache that may be severe.

Adie's Pupil

Adie's pupil, a painless unilateral and permanent dilation of the pupil, is seen in otherwise healthy young people and is due to involvement of the ciliary ganglion, postulated to be related to a viral infection. There are unconfirmed reports of an increase incidence of Adie's pupil in migraine patients. There is acute and painless dilation of the pupil, often with marked dilation and what appears to be complete loss of constriction to light or a near stimulus. Acutely, there is loss of accommodation with blurring of near vision. Under the magnification of the slit lamp, one can observe areas of the iris that do show some limited iris movement causing sector contraction and vermiform movements. The Adie's pupil will constrict to dilute pilocarpine (0.1%) due to denervation hypersensitivity, which is diagnostic. Over months and years the pupil slowly gets smaller, but never recovers its reactivity. Adie's pupil is typically benign and not usually associated with headache or eye pain.

Pharmacologic Pupil

Rarely, one may see a patient with a markedly dilated pupil with no other findings, but who may complain of headache and light sensitivity. Pharmacologic dilation typically causes a very enlarged pupil that does not react to light. This condition may occur secondary to accidental exposure or factitious disease (Munchausen syndrome), whereby a patient purposely uses a pharmacologic agent and then seeks medical help. A common accidental exposure occurs with scopolamine patches, with inadvertent rubbing of the eye after touching the patch. Always inquire about a recent sea voyage and possible use of the motion sickness patch. It can also be seen in patients who have been working outside pulling weeds, as some, like Jimson weed, have anticholinergic activity. Nurses who work with atropine and even with sympathomimetic agents used in respiratory therapy may also have a transiently dilated pupil. Factitious patients typically use dilating medical drops. One can test for a pharmacologic pupil by instilling 1% pilocarpine, which normally will constrict an abnormally dilated pupil related to a third nerve palsy or Adie's pupil. This procedure will have no effect on a pharmacologically blockaded pupil. If pharmacologic blockade can be demonstrated, reassurance should be given and unnecessary testing can be avoided.

HEADACHE AND DOUBLE VISION

Diplopia may be monocular or binocular. When a patient presents with a complaint of double vision, it is imperative to determine if the image becomes single with closing either eye. Essentially, monocular diplopia is always a primary ophthalmic problem, usually caused by astigmatism. Other less common causes include a dislocated lens, corneal irregularity, cataract, or macular disease.

When binocular diplopia is present, it indicates abnormal position of the eyes related to eye muscle imbalance. This condition can be related to a cranial nerve problem, neuromuscular junction disorder, primary muscle problem, or mechanical restriction from an orbital process. Diplopia can also be related to a nonparalytic strabismus with loss of compensation, which is entirely benign. Convergence spasm (spasm of the near reflex) is a functional disorder when patients unconsciously cross their eyes causing blurring, which is associated with pupillary constriction, and is usually stress related. When diplopia is present, examination may show obvious weakness in one or more of the extraocular muscles; however, this may be more subtle and only identified by an ophthalmologist or neuro-ophthalmologist. When headache is centered around one eye and diplopia is present, it is suggestive of a local inflammatory or infectious process, or of a mass lesion in the orbit, cavernous sinus, parasellar region, or posterior fossa.[41] When headaches are more diffuse, one must consider the possibility of increased intracranial pressure or intracranial hypotension, both of which may be associated with sixth nerve palsy (Figure 14.9).

Diplopia associated with headache often indicates a potentially serious disorder and usually requires neuroimaging. MRI of the brain and orbital structures, with

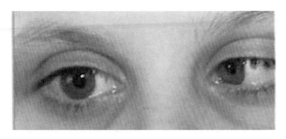

FIGURE 14.9   Right sixth nerve palsy. *Reproduced from NOVEL (Neuro-ophthalmology Virtual Education Library, http://novel.utah.edu), with permission.*

and without contrast, is the ideal method to view all of the involved structures (Box 14.2).

Increased Intracranial Pressure

Sixth nerve palsy may be a false localizing sign in patients with increased intracranial pressure, and may be unilateral or bilateral. Increased intracranial pressure may be associated with tumor, brain abscess, meningo-encephalitis, subdural hematoma, intracranial hemorrhage, ruptured aneurysm, and hydrocephalus. The causes should be obvious on neuroimaging, or with spinal fluid analysis. These serious and potentially life-threatening disorders may be acute or indolent. The most common cause of increased intracranial pressure in otherwise healthy-appearing patients with headache is idiopathic intracranial hypertension (IIH, previously called "pseudotumor cerebri"). The key finding on physical examination is papilledema. When performing a headache examination, it is imperative to obtain a good view of the fundus and optic nerve in order to observe for papilledema. It may take several days to several weeks for papilledema to become apparent with an acute increase in intracranial pressure. The classic ophthalmologic symptom associated with papilledema is the presence of transient visual obscurations. These may be unilateral or bilateral, and typically occur with changes in position. Patients note brief loss of vision, described as either dimming or complete blackout lasting a few seconds, with position changes, especially standing from a lying or seated position, or bending over and then standing erect. Transient visual obscurations may occur dozens of times a day, and have no prognostic implications. Another common symptom is pulsatile tinnitus, which is usually heard by the patient as a "whooshing" noise that is pulse synchronous. It may also be heard by the examiner with the bell of a stethoscope placed over the orbit or mastoid region, in a quiet environment.

Pseudotumor cerebri syndrome can be divided into the idiopathic form: idiopathic intracranial hypertension typically seen in obese young women, and that caused by venous sinus disease or as a toxic effect of medication. Of IHH patients, 90% present with headache associated with papilledema. It may be seen as a

---

BOX 14.2

DIFFERENTIAL DIAGNOSIS OF DIPLOPIA WITH HEADACHE OR EYE PAIN

1. Idiopathic orbital inflammatory syndrome (orbital pseudotumor)
2. Idiopathic orbital apex-cavernous sinus syndrome (Tolosa-Hunt syndrome)
3. Other specific orbital inflammatory syndromes
   a. Granulomatosis with polyangitis (Wegener's granulomatosis)
   b. Sarcoid
   c. Lupus
4. Trochleitis
5. Giant cell arteritis
6. Neoplastic
   a. Orbital metastases
   b. Cavernous sinus or skull base metastasis
   c. Orbital lymphoma
   d. Meningeal carcinomatosis or lymphomatosis
   e. Pituitary adenoma with or without apoplexy
   f. Craniopharyngioma

   g. Meningioma
   h. Cranial nerve schwannoma
7. Microvascular third, fourth, sixth nerve palsy
8. Infectious
   a. Herpes zoster ophthalmicus
   b. Fungal infection (aspergillus and mucor)
   c. Orbital cellulitis
   d. Orbital abscess
   e. Cerebral abscess
   f. Cavernous sinus thrombosis
   g. Meningitis
9. Thyroid eye disease (Grave's Ophthalmopathy)
   a. Aneurysm
10. Trauma
   a. Orbital fracture with muscle entrapment
   b. Muscle trauma
   c. Post-traumatic fourth nerve palsy
   d. Fourth and sixth nerve palsies

---

consequence of using acne medications, specifically isotretinoin, as well as doxycycline, minocycline, and tetracycline.[41] IIH is also associated with the use of, or withdrawal from, steroids. The headaches associated with IIH tend to be daily and have features of migraine, especially in patients with a migraine history. Non-specific imaging findings may include increased fluid in the optic nerve sheath, empty sella, and, possibly, small ventricles, although small ventricles are typical in young people. MR venography may show venous sinus narrowing or thrombosis. Venous sinus narrowing may be congenital, as the venous drainage of the brain is quite variable, or be seen as a consequence of increased intracranial pressure which may compress the venous structures. Marked venous sinus stenosis or cerebral venous sinus thrombosis may be the primary cause of elevated intracranial pressure, and can be treated with stenting and/or anticoagulation. Persistent elevation of intracranial pressure with papilledema may lead to permanent loss of vision and is best managed by a neuro-ophthalmologist or a neurologist experienced in caring for these patients, and coordinating management with an ophthalmologist. In the typical obese young woman with IIH, weight loss may be curative. Recent studies suggest as little as a 3% loss of weight may stop further loss of vision. The headaches may respond to migraine medications, including beta-blockers, topiramate, and the tricyclic antidepressants. Care must be taken to avoid further weight gain. Treatment typically involves diet, use of acetazolamide, and careful monitoring. For patients who do not respond to conservative therapy, optic nerve sheath decompression or spinal fluid shunting may be considered. There is no role for repeated lumbar puncture.[42]

Intracranial Hypotension

Low spinal fluid pressure is typically associated with a spinal fluid leak, and is one of the causes of a sixth nerve palsy, albeit rare. It is most commonly encountered after a lumbar puncture, in which case it is referred to as a "post-LP headache." It is commonly seen after an epidural block, when the needle passes through the dura and into the subarachnoid space, and can also be seen following head trauma with resultant skull fracture with leakage through the cribriform plate resulting in cerebrospinal fluid (CSF) rhinorrhea or even CSF otorrhea. Spontaneous intracranial hypotension nearly always results from CSF leaks from nerve root sleeves in the spine, and only rarely from the skull base. The triad of orthostatic headaches, diffuse patchy meningeal enhancement on MRI, and low CSF pressure is typically noted, but may be variable.[43] The headache usually disappears with lying down, and the headache is likely related to sinking of the brain in light of decreased spinal fluid. MRI may demonstrate meningeal enhancement, low-lying cerebellar tonsils, flattening of the brainstem, and abnormal pituitary enhancement. Recognizing the postural features helps identify this entity. Acutely, it may respond to bed rest, but most patients will require high-volume epidural blood patching. If not successful, then identification of the leak is attempted and can be surgically repaired if found.

Thyroid Eye Disease

Patients with a history of hyperthyroidism may develop inflammation of the eye muscles, which may result in proptosis, lid lag, lid retraction, and diplopia. As part of this syndrome, the patient may experience significant eye discomfort related to dryness and orbital congestion. Headache itself would be an unusual manifestation of thyroid eye disease (TED). Once there is more than minimal discomfort, the diagnosis becomes clear. TED may occur months or even years prior to overt hyperthyroidism, so the absence of a history of hyperthyroidism does not rule out TED.

Microvascular Cranial Neuropathy

One of the more common causes of acute diplopia with pain is microvascular disease involving the blood supply of cranial nerves III, IV, and VI. This is typically seen in a more senior population who have one or more risk factors for atherosclerosis, including diabetes, hypertension, hyperlipidemia, and smoking. Patients typically develop deep aching pain behind one eye followed 1–2 days later by acute diplopia. (There is an extensive discussion of third nerve palsy in the section on Headache and the Abnormal Pupil, above). While cranial nerve CN III is commonly microvascular, both sixth nerve palsy and fourth nerve palsy may occur on a microvascular basis. Although more serious disease needs to be ruled out with appropriate lab work and neuroimaging, microvascular involvement of cranial nerves III, IV, and VI remains the most common cause of painful ophthalmoplegia in patients over age 60.[44,45] The pain is always unilateral in the affected orbital region, may be quite severe, and typically dissipates within 2 weeks. The diplopia typically completely recovers within 6–12 weeks. Failure of recovery suggests another etiology.

Pituitary Apoplexy

Patients with known or undiscovered pituitary lesions are at risk for sudden hemorrhage within the pituitary gland. The typical presentation is sudden onset of headache with diplopia and/or vision loss. If the hemorrhage encroaches on the chiasm or intracranial optic nerves, visual loss may occur. The vision loss can be bilateral, complete, and devastating, or result in a unilateral optic neuropathy or chiasmal syndrome with bitemporal hemianopia. If the hemorrhage is more lateral and into the cavernous sinus structures, diplopia may result. The headache may be similar to that of a subarachnoid hemorrhage, and may present as a thunderclap headache or be more indolent. A CT without contrast is usually adequate to see the blood in the sellar region. These patients may have associated acute pituitary failure leading to adrenal insufficiency, which can be life-threatening. Thus, coverage with steroids is imperative. Pituitary apoplexy is one of the few neuro-ophthalmologic emergencies.

Ophthalmoplegic Migraine

Ophthalmoplegic migraine is a rare entity most often seen in children. It has been characterized as migraine for decades; however, with neuroimaging, it is apparent that the majority of patients have enhancement of the third nerve. It was recently reclassified from the migraine category to the cranial neuropathy category in the International Classification of Headache Disorders-III.[46] It was suggested that it might more appropriately be called "ophthalmoplegic cranial neuropathy".[47] Patients usually develop a third nerve palsy with significant headache, with symptoms lasting days to weeks, and it tends to be recurrent. The headache may have migraine characteristics. After multiple attacks, the diplopia and pupillary dilation may become permanent. The mean age of onset is age 8. Some response may be achieved with steroids.

CONCLUSION

When dealing with headache, one must be an interrogator, a detective, an analyst, and a therapist. The eye and visual system are intimately connected with the many signs and symptoms seen in the primary and secondary headache disorders. In this chapter, we have reviewed the clinically relevant eye examination as well as the commonly associated visual symptoms, and synthesized a differential diagnosis for the various clinical presentations. Diligence in history taking and in performing a thorough examination will allow the astute clinician to arrive at the diagnosis prior to additional testing. Recognizing the more serious and emergent conditions and making referrals to appropriate specialists or emergency care settings should insure better outcomes for our patients.

References

1. Donohoe CD. The role of the physical examination in the evaluation of headache. *Med Clin North Am.* 2013;97(2):197–216.
2. Frishberg BM. Clinical examination of the trigeminal nerve. In: Miller N, Newman N, eds. *Walsh and Hoyt's Clinical Neuro-Ophthalmology.* 5th ed. Baltimore, MD: Williams & Wilkins; 1997:1649–1661.
3. Lepore FE. Dr. Airy's "Morbid Affection of the Eyesight": lessons from Teichopsia Circa 1870. *J Neuroophthalmol.* 2014;34(3):311–314.
4. Hupp SL, Kline LB, Corbett JJ. Visual disturbances of migraine. *Surv Ophthalmol.* 1989;33:221–236.
5. Cutrer FM, Huerter K. Migraine aura. *Neurologist.* 2007;13(3):118–125.
6. Laurell K, Lundström E. Migrainous infarction: aspects on risk factors and therapy. *Curr Pain Headache Rep.* 2012;16(3):255–260.
7. San-Juan OD, Zermeño PF. Migraine with persistent aura in a Mexican patient: case report and review of the literature. *Cephalalgia.* 2007;27:456–460.
8. Schankin CJ, Maniyar FH, Sprenger T, Chou DE, Eller M, Goadsby PJ. The Relation Between Migraine, Typical Migraine Aura and "Visual Snow". *Headache.* 2014;May 9
9. Schankin CJ, Maniyar FH, Digre KB, Goadsby PJ. "Visual snow" — a disorder distinct from persistent migraine aura. *Brain.* 2014;137 (Pt 5):1419–1422.
10. Kosmorsky GS. Angiographically documented transient monocular blindness: retinal migraine? *Br J Ophthalmol.* 2013;97 (12):1604–1606.
11. May A. Diagnosis and clinical features of trigemino-autonomic headaches. *Headache.* 2013;53(9):1470–1478.
12. Armbrust KR, Kosmorsky GS, Lee MS, Friedman DI. A pain in the eye. *Surv Ophthalmol.* 2014;59(4):474–477.
13. Smith JH, Garrity JA, Boes CJ. Clinical features and long-term prognosis of trochlear headaches. *Eur J Neurol.* 2014;21(4):577–585.
14. Digre KB, Brennan KC. Shedding light on photophobia. *J Neuroophthalmol.* 2012;32(1):68–81.
15. Vanagaite J, Pareja JA, Storen O, White LR, Sand T, Stovner LJ. Light-induced discomfort and pain in migraine. *Cephalalgia.* 1997;17:733–741.
16. Lai TH, Fuh JL, Wang SJ. Cranial autonomic symptoms in migraine: characteristics and comparison with cluster headache. *J Neurol Neurosurg Psychiatry.* 2009;80(10):1116–1119.
17. Renton BJ, Bastowros A. Acute Angle Closure Glaucoma (AACG): an important differential diagnosis for acute severe headache. *Acute Med.* 2011;10(2):77–78.
18. Cole KL, Wang EE, Aronwald RM. Bilateral acute angleclosure glaucoma in a migraine patient receiving topiramate: a case report. *Emerg Med.* 2012;43(2):89–91.
19. Miller NR. Dural carotid-cavernous fistulas: epidemiology, clinical presentation, and management. *Neurosurg Clin N Am.* 2012;23(1):179–192.
20. Deibel JP, Cowling K. Ocular inflammation and infection. *Emerg Med Clin North Am.* 2013;31(2):387–397.
21. Maggioni F, Ruffatti S, Viaro F, Mainardi F, Lisotto C, Zanchin G. A case of posterior scleritis: differential diagnosis of ocular pain. *J Headache Pain.* 2007;8(2):123–126.
22. Brazis PW, Stewart M, Lee AG. The uveo-meningeal syndromes. *Neurologist.* 2004;10(4):171–184.
23. Ismail AM, Dubrey SW, Patel MC. Recurrent headaches: a case of neurological Behçet's disease. *Br J Hosp Med.* 2013;74(10):592–593.
24. Lutt JR, Lim LL, Phal PM, Rosenbaum JT. Orbital inflammatory disease. *Semin Arthritis Rheum.* 2008;37(4):207–222. [Review].

25. Wallace ZS, Deshpande V, Stone JH. Ophthalmic manifestations of IgG4-related disease: single-center experience and literature review. *Semin Arthritis Rheum*. 2014;43(6):806−817.

26. Rothrock JF. Headaches caused by vascular disorders. *Neurol Clin*. 2014;32(2):305−319.

27. Tentschert S, Wimmer R, Greisenegger S, Lang W, Lalouschek W. Headache at stroke onset in 2196 patients with ischemic stroke or transient ischemic attack. *Stroke*. 2005;3:e1−e3.

28. Runchey S, McGee S. Does this patient have a hemorrhagic stroke? Clinical findings distinguishing hemorrhagic stroke from ischemic stroke. *JAMA*. 2010;30:2280−2286.

29. Koskela E, Pekkola J, Kivisaari R, Kivelä T, Hernesniemi J, Setälä K, et al. Comparison of CT and clinical findings of Terson's syndrome in 121 patients: a 1-year prospective study. *J Neurosurg*. 2014;120(5):1172−1178.

30. Lyrer PA, Brandt T, Metso TM, Metso AJ, Kloss M, Debette S, et al.; For the Cervical Artery Dissection and Ischemic Stroke Patients (CADISP) Study Group Clinical import of Horner syndrome in internal carotid and vertebral artery dissection. *Neurology*. 2014;82(18):1653−1659.

31. Stam J. Thrombosis of the cerebral veins and sinuses. *N Engl J Med*. 2005;352:1791−1798.

32. Uzar E, Ekici F, Acar A, Yucel Y, Bakir S, Tekbas G, et al. Cerebral venous sinus thrombosis: an analysis of 47 patients. *Eur Rev Med Pharmacol Sci*. 2012;1:1499−1505.

33. Toosy AT, Mason DF, Miller DH. Optic neuritis. *Lancet Neurol*. 2014;13(1):83−99.

34. Waldman CW, Waldman SD, Waldman RA. Giant cell arteritis. *Med Clin North Am*. 2013;97(2):329−335.

35. Kale N, Eggenberger E. Diagnosis and management of giant cell arteritis: a review. *Curr Opin Ophthalmol*. 2010;21(6):417−422.

36. Yeh S, Foroozan R. Orbital apex syndrome. *Curr Opin Ophthalmol*. 2004;15(6):490−498.

37. Lam BL, Thompson HS, Corbett JJ. The prevalence of simple anisocoria. *Am J Ophthalmol*. 1987;104(1):69−73.

38. Biousse V, Touboul PJ, D'Anglejan-Chatillon J, Lévy C, Schaison M, Bousser MG. Ophthalmologic manifestations of internal carotid artery dissection. *Am J Ophthalmol*. 1998;126(4):565−577.

39. Yanovitch T, Buckley E. Diagnosis and management of third nerve palsy. *Curr Opin Ophthalmol*. 2007;18(5):373−378.

40. Skeik N, Jabr FI. Migraine with benign episodic unilateral mydriasis. *Int J Gen Med*. 2011;4:501−503.

41. Friedman DI. Idiopathic intracranial hypertension. *Curr Pain Headache Rep*. 2007;11:62−68.

42. Kosmorsky GS. Idiopathic intracranial hypertension: pseudotumor cerebri. *Headache*. 2014;54(2):389−393.

43. Mokri B. Spontaneous CSF leaks: low CSF volume syndromes. *Neurol Clin*. 2014;32(2):397−422.

44. Brazis PW. Isolated palsies of cranial nerves III, IV, and VI. *Semin Neurol*. 2009;29(1):14−28.

45. Tamhankar MA, Biousse V, Ying GS, Prasad S, Subramanian PS, Lee MS, et al. Isolated third, fourth, and sixth cranial nerve palsies from presumed microvascular versus other causes: a prospective study. *Ophthalmology*. 2013;120(11):2264−2269.

46. The International Headache Society (IHS). *IHS Classification ICHD-III*. London, UK: IHS. Available at: <www.ihs-classification.org/_downloads/mixed/International-Headache-Classification-III-ICHD-III-2013-Beta.pdf>.

47. Gelfand AA, Gelfand JM, Prabakhar P, Goadsby PJ. Ophthalmoplegic "migraine" or recurrent ophthalmoplegic cranial neuropathy: new cases and a systematic review. *J Child Neurol*. 2012;27(6):759−766.

15

Cranial Neuralgias, Sinus Headache, and Vestibular Migraine

Jan Lewis Brandes

Nashville Neuroscience Group, and Vanderbilt University School of Medicine, Nashville, Tennessee, USA

INTRODUCTION

The etiology of pain occurring in the face, eyes, teeth, or nasal regions is often challenging to diagnose. What may initially appear as intuitively obvious frequently is not, and intuition is invariably not as reliable for diagnosis as might be anticipated. While intuition in medicine may be of extraordinary benefit, the exception becomes the rule in terms of facial pain.

Various factors contribute to the failure of intuitive diagnosis: pain location, perceived triggering events, symptoms resulting from parasympathetic activation, response to acute medications, and, perhaps, denial. When recurring patterns of headache are approached intuitively, the diagnosis may well be incorrect and hence consistently effective strategies for treatment are either never offered or rejected. Perhaps, as in no other place in headache medicine, understanding the pathophysiology may curtail the inherent bias created by location, triggers, associated symptoms, and the desire for easily managed conditions.

The phrase "pain in the neck" originated in the early 1900s, and, along with "pain in the arse," surfaced about the same time many electrical items were invented. The strict definition is an annoyance, or a nuisance, which originated as euphemisms for less polite terms.

ANATOMY OF FACIAL PAIN

The neuroanatomy of the face and head is complex, with significant redundancy and overlap of sensory dermatomes, pain referral patterns, and autonomic reflexes. The first and second divisions of the trigeminal nerve provide sensory innervation to the nose and paranasal sinuses. These afferent fibers project via the trigeminal ganglion to the trigeminal sensory nucleus in brainstem.

Sympathetic and parasympathetic nerve fibers provide the autonomic innervation to the nose. Parasympathetic nervous system activation, via the superior salivatory nucleus of the seventh cranial nerve, may result in symptoms such as nasal congestion, rhinorrhea, or lacrimation. These symptoms are common to many primary headache disorders, such as migraine and cluster, as well as sinus pathologies such as infection and allergy. Confusion is amplified by the fact that the formal International Classification of Headache Disorders (ICHD) criteria for migraine does not include commonly occurring autonomic symptoms.[1] In addition, nasal engorgement potentially has been postulated by some authors to produce mucosal contact in areas with or without allodynia-related pain.[2,3] This, these investigators suggest, may support a role for contact point surgery even in patients with underlying migraine.[4,5]

Rhinogenic Headache

Rhinogenic headaches are those with primary pathophysiology in the nose with facial or structural head pain, and should be considered as secondary headaches. Acute rhinosinusitis and chronic or recurring rhinosinusitis both may cause headache. Both acute and chronic rhinosinusitus as underlying etiology for headache pain may include, by ICHD-III criteria, a temporal relationship with infection and visualization through imaging or endoscopy, or clinical evidence such as congestion/purulent drainage, with exacerbation of headache in response to pressure over the paranasal sinuses (Boxes 15.1,15.2).[1]

S. Diamond, R. K. Cady, M. L. Diamond & V. T. Martin (Eds):
Headache and Migraine Biology and Management.

DOI: http://dx.doi.org/10.1016/B978-0-12-800901-7.00015-X

BOX 15.1

11.5.1 Headache Attributed to Acute Rhinosinusitis

A. Any headache fulfilling criterion C.

B. Clinical, nasal endoscopic and/or imaging evidence of acute rhinosinusitis.

C. Evidence of causation demonstrated by at least two of the following:
   1. Headache has developed in temporal relation to the onset of the rhinosinusitis
   2. Either or both of the following:

   a. Headache has significantly worsened in parallel with worsening of the rhinosinusitis
   b. Headache has significantly improved or resolved in parallel with improvement in, or resolution of, the rhinosinusitis
   3. Headache is exacerbated by pressure applied over the paranasal sinuses
   4. In the case of a unilateral rhinosinusitis, headache is localized ipsilateral to it.

D. Not better accounted for by another ICHD-III diagnosis.

BOX 15.2

11.5.2 Headache Attributed to Chronic or Recurring Rhinosinusitis

A. Any headache fulfilling criterion C.

B. Clinical, nasal endoscopic, and/or imaging evidence of current or past infection or other inflammatory process within the paranasal sinuses.

C. Evidence of causation demonstrated by at least two of the following:
   1. Headache has developed in temporal relation to the onset of chronic rhinosinusitis

   2. Headache waxes and wanes in parallel with the degree of sinus congestion, drainage and other symptoms of chronic rhinosinusitis
   3. Headache is exacerbated by pressure applied over the paranasal sinuses
   4. In the case of a unilateral rhinosinusitis, headache is localized ipsilateral to it.

D. Not better accounted for by another ICHD-III diagnosis.

However, a majority of physician- and/or patient-diagnosed "sinus headache" has in numerous studies been demonstrated to meet diagnostic criteria for migraine and respond to migraine-specific medications.[6,7] In the Sinus, Allergy, and Migraine Study (SAMS), more than half (52%) of 100 consecutively consulting patients believed they had sinus headache, but the results determined they had migraine with or without aura, and 23% had probable migraine.[8] In another study of 2991 patients with a history of self-described or physician-diagnosed "sinus" headache and no previous diagnosis of migraine, 80% met International Headache Society (IHS) criteria for migraine without or with aura, and 8% met IHS criteria for migrainous headache.[9]

CRANIAL NEURALGIAS

Cranial neuralgias refer to pain associated with specific locations and distributions for cranial nerves and the overlapping areas supplied by the upper cervical nerve roots. Each pattern is typically named for the anatomical distribution of the individual cranial nerve. These neuralgias may be idiopathic or may have a secondary cause, related to infection, trauma, or inflammation.

Some consider all cranial neuralgias to be secondary, or caused by an underlying condition, but often no structural, infectious, or inflammatory cause can be determined. Course and duration may be self-limited, and response to treatment may vary considerably, suggesting that an identifiable underlying cause may be elusive in many cases.

The most common cranial neuralgias are trigeminal and glossopharyngeal. Occipital neuralgia is also frequent and has overlapping areas of pain with distributions supplied by the upper cervical nerve roots.

Location, description, pattern, and distribution of pain are the most useful tools for diagnosis. Imaging studies are indicated when cranial neuralgias are present, as the need to identify secondary causes is critical. Magnetic resonance imaging (MRI) and, often, lumbar puncture are required to eliminate secondary causes. Once diagnosis is achieved, secondary infectious and inflammatory causes may be treated and cases involving nerve root compression may require surgical intervention. Surgical therapies such as neural augmentative surgery are controversial, at best, and the placement of dorsal root electrodes for subsequent stimulation of the subcutaneous nerve to control pain has not been evaluated in well-controlled randomized placebo-controlled trials.

Types of Cranial Neuralgias

Occipital Neuralgia

Occipital neuralgia is defined as paroxysmal stabbing or shooting pain in the dermatomes of the nervus occipitalis major and/or nervus occipitalis minor. Onset of pain may occur suboccipitally and radiate into the occipital region and vertex. Short-term infiltration with local anesthetic is used to confirm the diagnosis. Dysesthesia and hypoesthesia may accompany the pain in the affected area.[10]

Location of the occipital nerves renders patients vulnerable to trauma, and damage or irritability of the nervus occipitalis major or minor is frequently the cause. For some patients, the pain may be described as aching rather than a sharp jabbing or stabbing pain. Tenderness to touch and palpation may be present where the nerve crosses between the mastoid and the occipital protuberance.

Migraine headache pain may originate in these same locations, usually without the dysesthesia, and with associated features of nausea and/or vomiting, and photo- and phonophobia. For patients with migrainous headache and associated posterior allodynia, location should not be considered for diagnosis.

If the pain in the occipital region is continuous and without sensory changes, referred pain should be considered from the C2–C3 facet joints, posterior fossa, or from the first division of the trigeminal nerve, of which the descending spinal tract converges with the C2–C3 afferent fibers on second-order neurons in the upper three segments of the spinal cord. The distinction in origin of pain could be assessed based on the response to infiltration of the tender area by blockade of the second cervical ganglion or local anesthetic.[11]

Occipital neuralgic pain may represent primary or secondary causes. Rheumatoid arthritis, cervical spondylosis, direct trauma to the occipital area, and whiplash or other injuries to the cervical spine may precipitate the pattern of pain. Migrainous features and duration, along with triggering factors and history of trauma, may help in distinguishing etiology and subsequent direction of treatment. Clinical confusion often results when ocular pain is also present along with dizziness, nausea, congested nasal passages, and tinnitus. These associated symptoms may be attributed to the overlap of the C2 dorsal root and the nucleus trigeminus pars caudalis.[12,13]

For non-migraine occipital pain, occipital nerve block or facet joint blocks may be warranted. Their role in migraine is currently being investigated in clinical trials. Pharmacological therapies include older drugs such as carbamazepine and tricyclic antidepressants, along with newer agents such as gabapentin, duloxetine, and pregabalin. No large studies have been performed with botulinum toxin. In one small uncontrolled series, six patients with occipital neuralgia pain had relief with botulinum injections for the sharp stabbing pain, but no improvement in the aching pain.[14] A single retrospective trial with botulinum toxin showed reduction in pain and improvement in pain disability index, with duration of relief averaging 16.3 ± 3.2 weeks.[15]

Other interventional management strategies, trials of pulsed radiofrequency (PRF) treatment of the nervi occipitales and the C2 ganglion spinal (dorsal root ganglion), and subcutaneous neurostimulation of the nervi occipitales, demonstrated mixed results. All trials and case series describing results have been limited to very small numbers of medically refractory patients without control populations. Adequately powered, randomized placebo-controlled trials are needed in this patient population.

Recent evidence-based guidelines published in 2012 by renowned pain specialists from Europe and the US currently suggest a single infiltration of the nervus occipitalis major with corticosteroids and local anesthetic for test block in occipital neuralgia.[16] If symptoms remain resistant to infiltration, PRF treatment of the nervus occipitalis may then be considered. If pain persists, PRF treatment of the dorsal root ganglion C2–C3 may be considered. However, the consensus group recommends that PRF treatment be performed in a clinical trial setting. Based on cost and invasiveness, subcutaneous nerve stimulation ranks last, and should only be performed in experienced centers per these recommendations.[16]

Glossopharyngeal Neuralgia

The onset of glossopharyngeal neuralgia manifests as deep stabbing pain in the throat radiating to the tongue, tonsil, or ear, and is unilateral in location. Attacks often last less than 1 minute and may occur in clusters from weeks to months. Pain may be triggered by swallowing, yawning, head turning, chewing, or drinking. Occasionally, vasovagal symptoms will be present. If presyncope, syncope, or cardiac arrhythmias occur, the condition may be termed "vagoglossopharyngeal neuralgia." Careful examination for structural lesions or focal infections is necessary, and may avoid misdiagnosis. Coexistence or overlap with other cranial nerve distributions should raise concern for an underlying CNS lesion.[17]

Trigeminal Neuralgia

Trigeminal neuralgia is the best understood of the cranial neuralgias. When an underlying vascular compression is identified as the etiology of the trigeminal nerve root irritation, the term "secondary trigeminal neuralgia" is correct. Since many patients have their trigeminal pain controlled by medication and do not undergo surgery to identify a vascular loop, the preferred diagnosis by the IHS classification is that of classical trigeminal neuralgia rather than primary trigeminal neuralgia.[1]

Consensus would also recommend avoiding the term "atypical" facial pain, or atypical trigeminal neuralgia, since many patients do share a pattern of pain. If no identifiable cause is elucidated the term "persistent idiopathic facial pain" has been suggested by the IHS, until these conditions are better understood.[1]

Trigeminal neuralgia symptoms are typically episodic, severe, shooting electrical jabs or jolts of pain, commonly located in the same trigeminal nerve root distribution. Attacks may occur spontaneously or may occur with external triggers such as touching the face, brushing teeth, chewing or talking, or wind hitting the face. Duration of pain varies from seconds to minutes, and attacks may occur repeatedly over several days to weeks with pain-free episodes between attacks.

Surgical options for trigeminal neuralgia are generally reserved for patients who do not respond to preventive therapies or for whom preventive therapies prove intolerable because of side effects, or are contraindicated for other medical reasons. Trigeminal nerve root decompression is not always effective, but if a vascular loop is identified as the cause of the neuralgia, outcomes are improved.

Trigeminal pain may progress to involve all three divisions of the trigeminal nerve over time, but the first division (ophthalmic) is involved in a minority of cases and about 35% of patients have division II and division III distribution pain simultaneously. In patients under the age of 50 years the diagnosis of multiple sclerosis should be considered, as 3% of cases are reported to occur in the setting of MS.[18]

Treatment of Cranial Neuralgias

Neuromodulators are the first line of treatment for cranial neuralgias. Both medical and surgical modalities are used to treat trigeminal neuralgia. Virtually all patients are initially treated with drug therapy. Important milestones in the treatment of trigeminal neuralgia were carbamazepine, clonazepam, and baclofen. Other drugs employed in the treatment of cranial neuralgias are valproic acid, lamotrigine, gabapentin, pregabulin, oxcarbazepine, topiramate, and zonisamide. All have significant side effects, and some patients may require polytherapy to achieve remission from pain.[19] As with migraine preventive strategies, recommended approaches are to start with a low dose and advance the dose on a weekly basis, allowing the patient to stop the dosage escalation if pain freedom is achieved or significant adverse medication reactions occur. Once pain freedom occurs, the difficult decision concerning how long to continue the drug or drugs has limited clinical trial data to use for guidelines. In general, if a patient can still be easily triggered by an external stimulus, or if a patient inadvertently misses one or two doses with recurrence of neuralgiform pain, the patient should be encouraged to continue on prophylaxis. Indeed, the patient may perhaps increase dosage until no triggers or breakthrough pain is experienced, even with a missed dose. Use dosing strategies to improve compliance, and try to obtain long-acting formulations where possible. Do not give up on a medication until an adequate trial is given, both for duration of dosing and dose itself. Once an adequate trial has been accomplished, if only partial benefit is seen, always consider polytherapy.

Surgical options should be investigated for patients who have failed medical therapy, those who initially responded but later became treatment resistant, or those for whom side effects of therapeutic medications are intolerable.

Surgical options include extracranial peripheral denervation by thermal, chemical, or traumatic means. Lidocaine and bupivacaine are frequently used for short-term relief. These interventions usually provide pain relief. However, even with neurectomy the pain may return within 20 to 30 months. Percutaneous denervation of the Gasserian ganglion and retrogasserian rootlets may be accomplished with radiofrequency ablation, glycerol, or balloon compression. Microvascular decompression is the treatment of choice for patients in whom a vascular loop is identified as compressing the trigeminal nerve at the root entry zone. For patients who fail microvascular decompression, repeat exploration is not well studied. Gamma knife radiosurgery is reported to be of benefit in medical and surgical non-responders;[20,21] however, the time to pain relief may be relatively slow, and long-term follow-up of gamma knife radiosurgery and its long-term complications is lacking.[20,21]

Persistent Idiopathic Facial Pain (Previously Atypical Facial Pain)

The term "atypical facial pain" is increasingly questioned as a diagnosis, as it implies a common causality underlying diagnosis. Since, by its nebulous nature, the term is most typically applied to any pain about the face or head that does not meet established or recognized diagnostic criteria, its use implies that the diagnosis and underlying cause are not established. Thus, the "atypical facial pain" diagnosis may be akin to that of other "wastebasket" diagnoses.

In 1988, Neil Raskin proposed the term "facial pain of unknown cause."[22] Consensus would follow to recommend avoiding the term "atypical facial pain," including those conditions described as "atypical trigeminal neuralgia," since many patients share a pattern of pain. If no identifiable cause is elucidated, the IHS currently recommends the term "persistent idiopathic facial pain" (PIFP).[1]

The term "atypical facial pain" was initially used to separate other stereotypical facial pain patterns from trigeminal neuralgia. Since the advent of sophisticated imaging techniques, secondary facial pain syndromes are usually easily identified. For the facial pain patterns for which the pathology is beyond the resolution of current imaging, the phrase "of unknown cause" may be helpful in supporting the search for ultimate underlying etiology and may serve to reassure both patient and clinicians that the search is not over. In consideration of this problem, the terms "PIFP" or "persistent facial pain of unknown cause" are considered appropriate replacement terms. These should only be applied after structural and pathophysiological causes are excluded and the neurological examinations are normal. For focal patterns, input from dentists, oral surgeons, and otolaryngologists may be indicated.

PIFP best applies to patients whose facial pain involves a pattern outside those of the major facial neuralgias and for whom the pain is steady, often unilateral, typically involves facial structures, and occasionally radiates to cervical areas. Paroxysms of pain and trigger zones are often absent and depression is often present. These three characteristics are not invariable, however, and the literature which continues to include them in the diagnostic characteristics of persistent facial pain tends to recall the approach argued against by the evolution of our understanding (or lack thereof) of these patterns.

Facial migraine remains in the initial differential diagnosis for these patients. When facial pain meets criteria for IHS migraine in terms of unilaterality, quality of pain, severity, and response to routine activity, along with associated features of nausea and/or vomiting, and photo- and phonophobia, the diagnosis becomes clearer. However, some patients with migrainous facial pain and fewer associated features may begin down the path of "sinus headache" if not closely evaluated.

SINUS HEADACHE

Patient and clinician perceptions of "sinus headache" have resulted in incorrect diagnosis. Various studies have shown that when patients self-report their headaches as "sinus" in origin, physicians are more likely to agree with the diagnosis.[6,7] The lay term of "sinus headache" continues to be perpetuated in the media. Primary care physicians and patients alike appear to continue to prefer the choice of "sinus headache" as the medical term when patients experience facial pain or pressure — symptoms traditionally associated with sinus disease. A more accurate term could be "rhinogenic headache," but "sinus headache" continues to be the more imprecise term that actually most commonly represents a misdiagnosis of migraine and leads to inadequate and often improper treatment.

Allergy triggers for migraine and the autonomic nervous system activation occur in approximately 45% of migraineurs. This link may contribute to the misperception of "sinus headache," and fosters continuation of the migraine as a misdiagnosed sinus headache by patients, physicians, and others — often in primary care settings. Over the past decades, specialists in otolaryngology, allergy, neurology, and headache have conducted multiple studies showing that the majority of patients presenting with "sinus headache" are actually experiencing migraine when appropriate diagnostic criteria are systematically applied. Undaunted by this research and its consensus findings, the label of "sinus headache" persists in primary care, in patient's minds, in the media, and in the over-the-counter pharmaceutical industry. Meanwhile, the aforementioned specialists often see patients with histories of "sinus headache" transformed into chronic migraine after years of inappropriate and ineffective care. These patients are only referred when the disorder becomes unmanageable with "sinus treatment."

Why does this mythology persist? Unilateral autonomic symptoms are often mistaken for typical sinus symptoms. Patel and colleagues performed a systematic literature review of >1400 abstracts that included adult patients with the diagnosis of "sinus headache".[6] Following a thorough neurological and otolaryngologic evaluation, they found that <5% of the cases described had organic causes for "sinus headache." They concluded that the majority of those presenting with "sinus headache" in the absence of significant acute inflammatory findings could be diagnosed with migraine.[6] Barbanti and colleagues found that 45.8% of 177 migraineurs evaluated had unilateral autonomic symptoms − nasal congestion, lacrimation, orbital edema, rhinorrhea, conjunctival infection − associated with their "sinus headaches".[23] Likewise, among 100 patients, all with self-diagnosed "sinus headache" evaluated at a tertiary rhinology clinic, none were found to have evidence of sinus disease of mucosal contact points on nasal endoscopy or computed tomography (CT) scan.[24]

Eross and colleagues, at the Mayo Clinic, applied IHS diagnostic criteria to 100 individuals who had "sinus headache" and found that 63% actually had migraine, 23% had probable migraine, and only 3% had headache secondary to rhinosinusitis.[8] Of the subjects, 76% reported pain in the second division of the trigeminal nerve and 62% experienced bilateral maxillary and forehead pain. The most commonly reported triggers for these patients were weather changes (83%) and allergen exposures (62%). Of note, seasonal variation was reported as a "sinus headache" trigger in 73% of patients, and yet only 1% of these patients met criteria for cluster, which is known to have seasonal variation. The most commonly associated symptoms were nasal congestion (56%), eyelid edema (37%), rhinorrhea (25%), and conjunctival injection (22%). Cleverly, the authors suggested that the diagnosis of "sinus headache" was based on "guilt by provocation, location, and association".[8]

In a smaller study of self-diagnosed "sinus headache" sufferers by Mehle and Kremer, all patients received sinus CT scans, which revealed that 74% had IHS-defined migraine. Only 5 of these 26 patients also had significant sinus disease.[25] This latter study does reiterate the importance of careful history in distinguishing migraine from rhinogenic headache, while serving to remind us that the diagnoses are not mutually exclusive. They may indeed occur in the same patient. Timing and pattern of attacks may be the most helpful approach in these patients.

Treating a "sinus headache" with antibiotics, steroids, and "sinus" medication, instead of treating the actual migraine headache, runs the inherent risk of diminished efficacy and reduced pain relief. Effectiveness may be further complicated by antihistamines, vasoconstrictors, and steroids, which may improve some underlying migraine symptoms. However, treatment response should never be allowed to confirm or support a headache diagnosis, with the notable exception of indomethacin response as required for a hemicrania continua diagnosis.

Schreiber and colleagues showed the likelihood of this reverse approach to treatment in a study of "sinus headache" patients, where, after noting this population met diagnostic criteria for IHS migraine, they utilized the migraine-specific acute treatment sumatriptan and found that 66% of these headaches were reduced to mild or no pain after 2 hours.[26] In 2008, Kari and DelGaudio also looked at triptan use in sinus headache patients. Of those patients who had no evidence of sinusitis, 82% had significant response to triptan use.[27] Of interest, a telling aspect of patient behavior with respect to study participation was observed when 34% of these "sinus headache" patients withdrew or failed to follow up, being "often reluctant to accept a diagnosis of migraine" despite normal nasal endoscopy and sinus CT. Therein lies a critical issue for clinicians − a diagnosis of migraine in lieu of more patient-acceptable and even primary care physician-acceptable "sinus headache diagnosis" requires time, patience, and a clear explanation regarding misleading triggers, response to acute medications, location of pain, and associated autonomic features (nasal congestion, lacrimation, eyelid edema, rhinorrhea).

For an individual patient, the migraine-specific triptans and FDA-approved valproate, based on multiple studies,[27−29] suggest that at least two-thirds of "sinus headaches" respond to treatment. However, in spite of the parallels with sinus medications, neither patients nor their clinicians were moved closer to accurate diagnosis.

Familiarity with the diagnostic criteria, with awareness of overlapping triggers and autonomic features, may help support clinicians striving for accuracy of diagnosis and treatment. Until then, one wonders whether revitalization of the older term "facial migraine" for these patients could help remove them from the diagnostic morass which continues.

Collaborative discussions and sharing of patient profiles between primary care, allergists, neurologists, otolaryngologists, ophthalmologists, and headache specialists may assist in dispelling myths and help patients as they too seek care and understanding of their recurrent patterns of headache. Likewise, inclusion of common autonomic symptoms in diagnostic criteria of the IHS may expand clinician awareness that what is considered "sinus headache" is indeed migraine.

For patients with dual diagnoses of migraine and "sinus" disease, the role of mucosal contact points remains an inadequately studied concern. Contact points are defined as a place within the nasal cavity where two opposing mucosal surfaces border each other. Sinus "abnormalities" may or may not include mucosal contact points. The role that mucosal contact points play in producing headache or facial pain was established by Harold G. Wolff in his text showing that stimulation of the meninges could produce pain in the face and sinus areas.[30] Wolff also found that stimulation in sinus areas could produce facial pain and headache, suggesting the referral of pain between the stimulated region and the location of pain.

Treatment of contact point-induced headache remains controversial; no randomized controlled trials have been conducted. Of concern among the prospective studies is the lack of long-term follow-up, which is critical for disorders such as migraine, with the waxing and waning attacks over long periods of time.

In an interesting prospective study of patients with frequent or medication-resistant migraine, Novak and Makek, who had promoted the use of nasal surgery as treatment for migraine for years, examined how these patients responded to nasal surgery.[31] The 299 migraine patients underwent various procedures, including sphenoidectomy, middle turbinate resection, ethmoidectomy, and septoplasty. The authors reported an astounding success rate of 90%, with 79% of their patients becoming "permanently asymptomatic." Yet timelines for follow-up were lacking, making the conclusions uninterpretable as to true long-term measurable outcomes.[31]

In a careful and systematic review of the literature published in 2013, Patel and colleagues concluded that the patients most likely to benefit from directed nasal surgery to remove contact points included those who have clearly identified contact points, have failed adequate therapy aimed at migraine, have otherwise normal endoscopy and CT scan, and have had a previous positive response to local anesthetic applied to the contact point.[6] It should be cautioned that surgery may not render the patient headache- or facial pain free. Furthermore, although the risk of surgical complications for these procedures is low, "it is not zero".[6]

For all patients presenting with "sinus headache," a thorough history should be obtained, and include headache pattern, onset, hormonal milestones, and hormonal medications. Examination of the head and neck and the cranial nerves, and otolaryngologic evaluation, should be performed. If criteria establishing migraine are revealed, acute and preventive therapies (see *Chapters 8* and *9*) may be initiated.

MOTION SICKNESS

Association between childhood-onset motion sickness and the subsequent development of migraine has been established in numerous studies.[32] Of 650 adult migraineurs, 60% were reported by Pearce to have suffered from motion sickness during childhood.

Motion sickness includes autonomic and cognitive signs and symptoms in moving environments, including motor vehicles, aircraft simulators, and watercraft. Symptoms may include dizziness, nausea, vomiting, increased salivation, headache, and diaphoresis.[33]

Childhood Equivalents in Migraine

Fenichel and colleagues first described attacks in two young siblings with vertigo, who later developed migraine.[34] These brief attacks of disequilibrium or vertigo, nystagmus, and often vomiting are episodic, and may be recurrent, often for months to years. "Benign paroxysmal vertigo" is the term used for these attacks, and is viewed as a childhood equivalent of migraine. Some authors suggest it is an early manifestation of vestibular migraine.[35,36]

Role of Hormonal Factors

Factors such as environment, sea conditions, car trips, amusement park rides, and others may trigger motion sickness in susceptible individuals. But what of the underlying susceptibility to such triggering events? Genetic, hormonal, and endocrinological status appear to be the underpinnings for such susceptibilities. Gender differences in simulator sickness were noted among military personnel.[37] Motion sickness, migraine, and episodes of vertigo have been linked to female menstrual cycles.[38] Vertigo presenting as a symptom of migraine during perimenopause has been reported.[39]

Migraine and Vertigo

In 1873, the English physician, Edward Living, reported that 6 of 60 migraine patients had spontaneous attacks of vertigo.[40] During the intervening 141 years this association has continued to be clinically observed, but mechanisms underlying the association have remained speculative.

In perhaps the most intriguing study to date, functional MRI has revealed evidence for abnormal thalamic functional response to vestibular stimulation in patients with vestibular migraine.[41] They found that caloric vestibular stimulation activated cortical and subcortical areas, known to be involved in vestibular processing, in both vestibular migraineurs and healthy control patients. However, the vestibular migraine patients showed significantly increased left mediodorsal thalamic activation in response to an ipsilateral vestibular stimulation, relative to controls and patients with migraine without aura. They further showed that the magnitude of the left thalamic activation was uniquely correlated with the frequency of migraine attacks in patients with vestibular migraine. Russo and other investigators postulate that the role of mediodorsal thalamus in vestibular migraine could "reflect the involvement of a dysfunctional vestibulo-thalamocortical network, which overlaps with the migraine circuit", a concept originally proposed by Balaban and colleagues.[42] They feel this altered thalamic processing likely reflects an interictal underlying hyper-responsiveness to external stimuli based on abnormal resting intrathalamic inhibition that may be playing a role during vestibular migraine attacks as well.

The search for a genetic link between migraine and vertigo has been ongoing for several decades and has been dominated by the less common syndromes of episodic ataxia, type 6, including both vertigo and migraine. No confirmed shared genes have, to date, found a genetic locus to explain the increased prevalence between migraine and vertigo, including specific vertigo syndromes such as motion sickness, paroxysmal vertigo of childhood, benign paroxysmal positional vertigo, and Ménière's disease.

During the past two decades, genetic defects of ion channels have been identified as the cause of various paroxysmal neurological disorders. While no specific gene has yet been found, spreading depression may play the critical role in vestibular migraine when cortical and thalamocortical areas are involved which are known to process vestibular information. Many well-established neurotransmitters involved in the pathogenesis of migraine (calcitonin gene-related peptide, serotonin, dopamine, norepinephrine) are also known to modulate vestibular neuronal activity.

Statistically, both migraine and vertigo are common disorders in populations, but random chance would suggest a much lower increased comorbidity than has been confirmed both in selected patient groups and in populations. Interestingly, no study reported in the literature has failed to confirm an association.

At times, confusing the association may be the term "dizziness" itself. Individuals use various terms when asked about dizziness, i.e., spinning, disequilibrium, lightheadedness, swimmy-headedness. Colorful descriptions are often used by patients, and may include the sense of "standing at the end of a bouncing diving board," "whirls going around the room," and "walking on a skateboard," to name but a few. True vertigo is traditionally considered as being illusory movement, but the separators linguistically are complex indeed, and cultural barriers may not aid diagnosis. Mild vestibular dysfunction may be experienced as dizziness rather than vertigo, and may prove misleading to clinicians if the history is taken early in a patient's course. Perhaps an effective strategy may be simply to ask the patient to describe the sensation, timing from onset to end — if it ends — and provocative factors, along with any associated symptoms. For women, careful detail about hormonal status, relation to menstrual cycles and ovulation, as well as use and initiation/stoppage of any hormonal medications will also be crucial to establishing pattern and subsequent timing for preemptive and acute treatments.

Dizziness includes both the vertiginous and non-vestibular symptoms of movement. Eliminating the more obvious complaints of pre-syncope and syncope, alterations in loss of consciousness tend to leave the remainder of symptoms for evaluation in relation to migraine. The former may well be related to migraine, but presumably through different mechanisms, as these may occur with or without migraine.

With vertigo being more common in patients with migraine than in patients with tension-type headache, and then headache-free controls, the statistical epidemiological evidence argues for an increased comorbidity that would be predicted by chance alone. The lifetime prevalence of migraine being 16% and the lifetime prevalence of vertigo of 7% would yield an expected comorbidity of 1.1% in the general population. However, more recent population-based studies have shown actual comorbidity at 3.2%, with odds ratios of 3.8 for the association in migraineurs versus non-migraineurs after adjustment for age and sex.[43,44]

At the examination, start with a thorough patient history, especially including any childhood history of headache or motion sickness. Allow the patient sufficient time to reflect on childhood, adolescent, and early adulthood patterns of headache and vestibular symptoms. For women, determine menarche, exposure to and duration of oral contraceptive pills and/or any other hormonal manipulations, and their effect or impact on any headache or vestibular symptoms, miscarriages, pregnancies, and lactation. Determine if any of these factors correlated with vertigo and headache, since "migraine" may only be used by the patient after eliminating her "sinus headache," "menstrual headache," or "stress headache," and be reserved just for the most severe attacks of headache. A lifetime history is important, since migraine appearing in late perimenopause in a migraineur with only the previous pattern of moderate, undiagnosed, menstrual headache may have been forgotten, or she may not have realized her previous headache pattern.

Other considerations to discuss with the patient include non-hormonal triggers such as the amount of "screen time" (computer, tablet, phone), grocery shopping, boating, jet skiing, cruises, train rides, car rides, flights, rollercoaster rides, biking, horseback riding, weather changes (i.e., changes in the barometric pressure, weather fronts, winds, changes in altitude or climate) for both vertigo and headache. Another important factor to discuss is patterns of medication use, including acute, subtle, or not-so-subtle changes in other medications. Many patients do not consider hormonal therapies or supplements, and sometimes even over-the-counter products, as medications, and are unaware of their perils and impact.

Create a timeline for attacks, vestibular or headache alone, and suggest timing of all patterns and ways to think about the temporal sequence to identify any association between vertigo and headache, if not obvious. Always include associated symptoms, such as parasympathetic activation, which may precede the vertigo, headache, or both. Ask specifically about nasal drainage, ocular tearing, or nasal congestion. Convince the patient that a diary of signs and symptoms is crucial to effective treatment and management.

VESTIBULAR MIGRAINE

Descriptions of vertigo associated with migraine are multiple, and are compounded by variable use between neurologists, otolaryngologists, allergists, neurotologists, and headache specialists. The terms migraine-associated vertigo, benign recurrent vertigo, migrainous vertigo, migraine-associated vestibulopathy, migraine-associated dizziness, basilar-type migraine, and definite vestibular migraine have all been used.

Currently, proposed criteria are being clinically evaluated and final criteria for the entity of vestibular migraine should appear in the next ICHD. The need for a working classification is based on the absence from the current ICHD, except for the appearance of vertigo as a migrainous feature in migraine with brainstem aura (formerly basilar migraine).[1]

Under current guidelines, the diagnosis of migraine with brainstem aura requires at least two symptoms referable to posterior circulation lasting between 5 and 60 minutes, followed by the headache of migraine. Based on these restrictions, Neuhauser and Lempert,[35] among others, note that this leaves 90% of patients with vestibular migraine without a classifiable diagnosis. Patients with isolated vertigo, even if fulfilling the duration for aura, and if followed by headache, remain without a category of diagnosis. Recently released ICHD-III beta criteria from the IHS and the Bárány Society for vestibular migraine likely will aid in more systematic approaches to research, diagnosis and treatment, if adopted (Box 15.3).[1]

Long-Term Follow-up of Clinical Symptoms

In a thoughtful analysis of their patients, 10 years following the original diagnosis, Radtke and colleagues reassessed the evolution of vestibular migraine and vestibulocochlear function in 61 of 127 eligible patients with definite vestibular migraine.[45] The majority of the patients continued to have recurrent vertigo (87%) at a median follow-up of 9 years. They reported that of these 61 patients, the frequency of vertigo was actually increased in 29%, unchanged in 16%, and reduced in 56%. The impact of vertigo remained severe in 21%, moderate in 43%, and mild in 36%.

Positional nystagmus was the most common finding on examination, in 28%, and interictal ocular motor

---

BOX 15.3

ICHD-III BETA CRITERIA FROM THE IHS AND THE BARANY SOCIETY FOR VESTIBULAR MIGRAINE

A1.6.5 Vestibular Migraine

A. At least five episodes fulfilling criteria C and D.

B. A current or past history of 1.1 *Migraine without aura* or 1.2 *Migraine with aura*.

C. Vestibular symptoms of moderate or severe intensity, lasting between 5 minutes and 72 hours.

D. At least 50% of episodes are associated with at least one of the following three migrainous features:

1. Headache with at least two of the following four characteristics:
   a. unilateral location
   b. pulsating quality
   c. moderate or severe intensity
   d. aggravation by routine physical activity
2. Photophobia and phonophobia
3. Visual aura.

E. Not better accounted for by another ICHD-III diagnosis
   a. or by another vestibular disorder.

abnormalities actually increased from 16% at time of diagnosis to 41% at follow-up. However, only one of nine patients with ocular motor abnormalities at initial evaluation had symptoms on follow-up. Mild persistent unsteadiness was reported in 18% of patients. Cochlear symptoms over the intervening years progressed from 15% initially to 49% at the time of follow-up. Mild bilateral sensorineural hearing loss had also developed in 11 patients during the intervening years.

These findings, in a closely observed population of vestibular migraineurs, are concerning, as they suggest persistence of symptoms, often with significant impact on patient functioning. While interictal ocular motor abnormalities may show some variation over time, vestibulo-cochlear dysfunction is slower in these patients with vestibular migraine.[45] As such, interictal central-type peripheral nystagmus may be useful in distinguishing vestibular migraine from peripheral vestibular disorders, such as Ménière's disease.[45] These findings may facilitate a role in the usefulness of clinical examination which is currently lacking in the evaluation of vestibular migraine patients.

Examination

Cranial nerve testing (observing for nystagmus), balance testing with tandem walk, and Romberg testing should be undertaken on all patients with a history suggestive of vertigo and migraine.

Caloric testing has proven to yield inconsistent results in vestibular migraine. Audiometric testing should be pursued if Ménière's disease is being considered as a dual diagnosis. If any hearing loss is reported, magnetic resonance imaging should be considered. A full neurological examination should be performed on all patients.

Although no studies currently exist to confirm or deny, clinical experience would at least seem to suggest that the earlier these patients are diagnosed and treated, the more years of maltreatment and mistreatment may be avoided. Prevention of transformation into chronicity for vestibular migraine, for "sinus headache," and perhaps for chronic migraine or chronic vertigo, if underlying hyperexcitability can be modulated, would be the appealing hope for long-suffering patients and their vulnerable families.

Treatment

Taghdiri and colleagues recently reported cinnarizine, an L-type calcium channel blocker, to be safe and effective in reducing both the headache and vertigo in patients with vestibular migraine or migraine with brainstem aura (with vertigo as aura).[46] Their study findings are limited by small size and design — retrospective and open label — but cinnarizine's known inhibition of vestibular hair cells led to its choice for prophylaxis.[47,48]

The results of reduction in mean frequency of migraine per month, and reduction in mean duration and median intensity of migraine headaches per month, were particularly interesting. Further, Taghdiri found that the vestibular migraine group showed greater decreases in mean frequency and intensity of migraine attacks compared with the basilar type migraine group.[46]

Preventive treatments for vestibular migraine are lacking evidence for effective prophylactic therapy in large randomized placebo-controlled trials. A retrospective study of 100 patients reported that patients on medical prophylaxis receiving medications (beta-blockers, topiramate, valproate, lamotrigine, amitriptyline, and flunarizine) had reduced episodic vertigo attacks in comparison with another group not on prophylaxis.[49] Another study, by Bisdorff, showed lamotrigine to significantly reduce the mean vertigo frequency per month, but it did not produce a significant change in mean headache frequency per month.[50]

Study results have been mixed, with some investigators reporting little improvement and others reporting dramatic improvement in some patients. Reploeg and Goebel reported 72% of 81 patients with amelioration of non-specific migraine-associated vertigo attacks with tricyclic antidepressants or beta-blockers.[51]

For acute attacks of vestibular migraine, ergotamine, sumatriptan, and vestibular suppressants (promethazine, meclizine, prochlorperazine, or benzodiazepines) are recommended.[51,52] Although strong evidence-based recommendations are lacking for preventive treatment, common medications used by our group and others include propranolol, betaxolol, acetazolamide, topiramate, zonisamide, and nortriptyline.[51] Stepwise treatment using low-dose tricyclic antidepressants and a beta-blocker may also be useful. Identification and avoidance of food and beverage triggers are warranted. Adequate hydration and sleep are also critical to management. Lastly, for women whose condition has evolved to vestibular migraine, hormonal stabilization should be attempted, if feasible.

References

1. Headache Classification Committee of the International Headache Society (IHS). The International Classification of Headache Disorders, 3rd edition (beta version). *Cephalalgia*. 2013;33:629–808.
2. Mehle ME. What do we know about rhinogenic headache? The otolaryngologist's challenge. *Otolaryngol Clin North Am*. 2014;47(2):255–264.
3. Mehle ME, Schreiber CP. Sinus headache, migraine, and the otolaryngologist. *Otolaryngology*. 2005;133(4):489–496.
4. Rozen TD. Intranasal contact point headache: missing the "point" on brain MRI. *Neurology*. 2009;72(12):1107.
5. Mohebbi A, Memari F, Mohebbi S. Endonasal endoscopic management of contact point headache and diagnostic criteria. *Headache*. 2010;50(2):242–248.

6. Patel ZM, Kennedy DW, Setzen M, Poetker DM, DelGaudio JM. "Sinus headache": rhinogenic headache or migraine? An evidence-based guide to diagnosis and treatment. *Int Forum Allergy Rhinol.* 2013;3(3):221–230.

7. Al-Hashel JY, Ahmed SF, Alroughani R, Goadsby PJ. Migraine misdiagnosis as a sinusitis, a delay that can last for many years. *J Headache Pain.* 2013;14:97.

8. Eross E, Dodick D, Eross M. The Sinus, Allergy and Migraine Study (SAMS). *Headache.* 2007;47(2):213–224.

9. Schreiber CP, Hutchinson S, Webster CJ, Ames M, Richardson MS, Powers C. Prevalence of migraine in patients with a history of self-reported or physician-diagnosed "sinus" headache. *Arch Intern Med.* 2004;164(16):1769–1772.

10. Hammond SR, Danta G. Occipital neuralgia. *Clin Exp Neurol.* 1978;15:258–270.

11. Bogduk N. Local anesthetic blocks of the second cervical ganglion: a technique with application in occipital headache. *Cephalalgia.* 1981;1(1):41–50.

12. Mason III JO, Katz B, Greene HH. Severe ocular pain secondary to occipital neuralgia following vitrectomy surgery. *Retina.* 2004;24(3):458–459.

13. Kuhn WF, Kuhn SC, Gilberstadt H. Occipital neuralgias: clinical recognition of a complicated headache. A case series and literature review. *J Orofac Pain.* 1997;11(2):158–165.

14. Taylor M, Silva S, Cottrell C. Botulinum toxin type-A (BOTOX) in the treatment of occipital neuralgia: a pilot study. *Headache.* 2008;48(10):1476–1481.

15. Kapural L, Stillman M, Kapural M, McIntyre P, Guirguis M, Mekhail N. Botulinum toxin occipital nerve block for the treatment of severe occipital neuralgia: a case series. *Pain Practice.* 2007;7(4):337–340.

16. Vanelderen P, Lataster A, Levy R, Mekhail N, van Kleef M, Van Zundert J. 8. Occipital neuralgia. *Pain Practice.* 2010;10(2):137–144.

17. Hupp WS, Firriolo FJ. Cranial neuralgias. *Dent Clin North Am.* 2013;57(3):481–495.

18. Obermann M, Katsarava Z. Update on trigeminal neuralgia. *Expert Rev Neurother.* 2009;9(3):323–329.

19. Shahien R, Beiruti K. Preventive agents for migraine: focus on the antiepileptic drugs. *J Cent Nerv Syst Dis.* 2012;4:37–49.

20. Young RF, Vermeulen SS, Grimm P, Blasko J, Posewitz A. Gamma Knife radiosurgery for treatment of trigeminal neuralgia: idiopathic and tumor related. *Neurology.* 1997;48(3):608–614.

21. Rozen TD, Capobianco DJ, Dalessio DJ. Cranial neuralgias and atypical facial pain. In: Silberstein SD, Lipton RB, Dalessio DJ, eds. *Wolff's Headache and Other Head Pain.* 7th ed. New York, NY: Oxford University Press; 2001:509–524.

22. Raskin NH. Facial Pain. In: Raskin NH, ed. *Headache.* New York, NY: Churchill Livingstone; 1988:333–374.

23. Barbanti P, Fabbrini G, Pesare M, Vanacore N, Cerbo R. Unilateral cranial autonomic symptoms in migraine. *Cephalalgia.* 2002;22(4):256–259.

24. Perry BF, Login IS, Kountakis SE. Nonrhinologic headache in a tertiary rhinology practice. *Otolaryngology.* 2004;130(4):449–452.

25. Mehle ME, Kremer PS. Sinus CT scan findings in "sinus headache" migraineurs. *Headache.* 2008;48(1):67–71.

26. Schreiber CP, Cady RK, Billings C. Oral sumatriptan for self-described "sinus" headache. *Cephalalgia.* 2001;21:298.

27. Kari E, DelGaudio JM. Treatment of sinus headache as migraine: the diagnostic utility of triptans. *Laryngoscope.* 2008;118(12):2235–2239.

28. Dadgarnia MH, Atighechi S, Baradaranfar MH. The response to sodium valproate of patients with sinus headaches with normal endoscopic and CT findings. *Eur Arch Otorhinolaryngol.* 2010;267 (3):375–379.

29. Ishkanian G, Blumenthal H, Webster CJ, Richardson MS, Ames M. Efficacy of sumatriptan tablets in migraineurs self-described or physician-diagnosed as having sinus headache: a randomized, double-blind, placebo-controlled study. *Clin Ther.* 2007;29(1):99–109.

30. Wolff HG. *Headache and Other Head Pain.* New York, NY: Oxford University Press; 1948.

31. Novak VJ, Makek M. Pathogenesis and surgical treatment of migraine and neurovascular headaches with rhinogenic trigger. *Head Neck.* 1992;14(6):467–472.

32. Pearce J. General review. Some aetiological factors in migraine. In: Cumings JN, ed. *Background in Migraine. Fourth Migraine Symposium.* London, UK: Heinemann Medical; 1971:1–7.

33. Furman JM, Marcus DA. Migraine and motion sensitivity. *Continuum.* 2012;18(5 Neuro-otology):1102–1117.

34. Fenichel GM. Migraine as a cause of benign paroxysmal vertigo of childhood. *J Pediatr.* 1967;71(1):114–115.

35. Neuhauser H, Lempert T. Vestibular migraine. *Neurol Clin.* 2009;27(2):379–391.

36. Abu-Arafeh I, Russell G. Paroxysmal vertigo as a migraine equivalent in children: a population-based study. *Cephalalgia.* 1995;15(1):22–25 [discussion 24].

37. Kennedy RS, Lanham DS, Massey CJ, Drexler JM, Lilienthal MG. Gender differences in simulator sickness incidence: implications for military virtual reality systems. *SAFE J.* 1995;25:69–76.

38. Grunfeld EA, Price C, Goadsby PJ, Gresty MA. Motion sickness, migraine, and menstruation in mariners. *Lancet.* 1998;351:1106.

39. Park JH, Viirre E. Vestibular migraine may be an important cause of dizziness/vertigo in perimenopausal period. *Med Hypotheses.* 2010;75:409–414.

40. Liveing E. *On Megrim: Sick Headache and Some Allied Health Disorders: a Contribution to the Pathology of Nerve Storms.* London, UK: J. & A. Churchill; 1873.

41. Russo A, Marcelli V, Esposito F, Corvino V, Marcuccio L, Giannone A, et al. Abnormal thalamic function in patients with vestibular migraine. *Neurology.* 2014;82(23):2120–2126.

42. Balaban CD. Migraine, vertigo and migrainous vertigo: Links between vestibular and pain mechanisms. *J Vestib Res.* 2011;21 (6):315–321.

43. Neuhauser HK, Radtke A, von Brevern M, Feldmann M, Lezius F, Ziese T, et al. Migrainous vertigo: prevalence and impact on quality of life. *Neurology.* 2006;67(6):1028–1033.

44. Neuhauser HK, Radtke A, von Brevern M, Lezius F, Feldmann M, Lempert T. Burden of dizziness and vertigo in the community. *Arch Intern Med.* 2008;168(19):2118–2124.

45. Radtke A, von Brevern M, Neuhauser H, Hottenrott T, Lempert T. Vestibular migraine: long-term follow-up of clinical symptoms and vestibulo-cochlear findings. *Neurology.* 2012;79(15): 1607–1614.

46. Taghdiri F, Togha M, Razeghi Jahromi S, Refaeian F. Cinnarizine for the prophylaxis of migraine associated vertigo: a retrospective study. *SpringerPlus.* 2014;3:231.

47. Arab SF, Duwel P, Jungling E, Westhofen M, Luckhoff A. Inhibition of voltage-gated calcium currents in type II vestibular hair cells by cinnarizine. *Naunyn Schmiedebergs Arch Pharmacol.* 2004;369(6):570–575.

48. Pianese CP, Hidalgo LO, Gonzalez RH, Madrid CE, Ponce JE, Ramirez AM, et al. New approaches to the management of peripheral vertigo: efficacy and safety of two calcium antagonists in a 12-week, multinational, double-blind study. *Otol Neurotol.* 2002;23(3):357–363.

49. Baier B, Winkenwerder E, Dieterich M. "Vestibular migraine": effects of prophylactic therapy with various drugs. A retrospective study. *J Neurol.* 2009;256(3):436–442.

50. Bisdorff AR. Treatment of migraine related vertigo with lamotrigine an observational study. *Bull Soc Sci Med Grand Duche Luxemb.* 2004;2:103–108.

51. Reploeg MD, Goebel JA. Migraine-associated dizziness: patient characteristics and management options. *Otol Neurotol.* 2002;23 (3):364–371.

52. Bisdorff AR. Management of vestibular migraine. *Ther Adv Neurol Disord.* 2011;4(3):183–191.

16

Cervicogenic Headache

Wade M. Cooper[1] and Amit K. Masih[2]

[1]Department of Neurology, University of Michigan, Ann Arbor, Michigan, USA
[2]Department of Neurology, Michigan State University, East Lansing, Michigan, USA

CASE STUDY

A 47-year-old police officer presented with a 9-month history of daily right-sided periorbital headache of moderately severe intensity with photophobia and phonophobia. He complained of a daily right-sided cervicalgia and a less intense occipital headache. The patient reported a motor vehicle accident with minor trauma 1 year prior to headache onset. He denied any history of migraine. On physical examination, the patient had occipital notch and cervical facet joint tenderness to palpation, with restricted cervical rotation to the right which escalated his headache intensity. MRI of the cervical spine identified arthropathy of the C2/3 facet. The headache was completely relieved after a series of occipital nerve anesthetic blockades.

Diagnosis: Cervicogenic headache.

THE RELATIONSHIP OF HEADACHE AND NECK PAIN AS A MANIFESTATION OF NECK DISORDERS

Neck pain and cervical myofascial pain are common features of the primary headache disorders. In migraine, 64% of patients report neck pain as a component of their pain experience. Neck pain is most common during the migraine attack itself; however, approximately one-third of patients experience neck pain during the migraine prodrome or postdrome.[1] Neck pain during migraine is commonly unilateral (59%) and almost always ipsilateral to the side of the headache.[2] In this study, patients described their neck pain during migraine as "tightness" (69%), stiffness (17%), or throbbing (5%). Similar to tension-type headache, migraine patients have been found to have a higher degree of cervical myofascial trigger-point tenderness than non-migraine comparison groups.[3]

"Cervicogenic headache" refers to head pain that is generated by, or referred by, cervical structures. It frequently occurs in association with a traumatic injury or identifiable secondary process such as osteoarthritis. However, it may also occur without an identifiable inciting event. Cervicogenic headache may be challenging to identify, as neck disorders may function as a "trigger" to primary headache syndromes, such as migraine, tension-type headache, or hemicrania continua. Therefore, cervicogenic headache often shares clinical features of these primary headache syndromes.

In the general population, cervicogenic headache has a prevalence of between 0.4% and 2.5%. In patients with chronic migraine evaluated at tertiary specialty clinics, the prevalence has been demonstrated to be as high at 20%.[4] Similar to migraine, cervicogenic headache is approximately four times more common in women with a mean age of 43 years. Additionally, cervicogenic headache has similar declines in quality of life measures as migraine, with larger reductions specific to physical functioning.[4]

CERVICOGENIC HEADACHE DIAGNOSIS

The International Classification of Headache Disorders 3rd edition (beta) provides the description of cervicogenic headache as "Headache caused by a disorder of the cervical spine and its component bony, disc and/or soft tissue elements, usually but not invariably accompanied by neck pain".[5] The diagnostic criteria for cerviocogenic headache as accepted by the International Headache Society are listed in Box 16.1.

S. Diamond, R. K. Cady, M. L. Diamond & V. T. Martin (Eds):
Headache and Migraine Biology and Management.

DOI: http://dx.doi.org/10.1016/B978-0-12-800901-7.00016-1

DIAGNOSTIC CRITERIA FOR CERVICOGENIC HEADACHE

Description

Headache caused by a disorder of the cervical spine and its component bony, disc and/or soft tissue elements, usually but not invariably accompanied by neck pain.

Diagnostic Criteria

Any headache fulfilling criterion C
Clinical, laboratory and/or imaging evidence of a disorder or lesion within the cervical spine or soft tissues of the neck, known to be able to cause headache
Evidence of causation demonstrated by at least two of the following:

Headache has developed in temporal relation to the onset of the cervical disorder or appearance of the lesion
Headache has significantly improved or resolved in parallel with improvement in or resolution of the cervical disorder or lesion
Cervical range of motion is reduced and headache is made significantly worse by provocative maneuvers
Headache is abolished following diagnostic blockade of a cervical structure or its nerve supply
Not better accounted for by another ICHD-III beta diagnosis.

The International Classification of Headache Disorders, 3rd Edition (beta version)

This classification requires the presence of headache (criterion A), presence of a cervical disorder or lesion known to contribute to headache (criterion B), and evidence of causation (criterion C). Of interest, the clinical characteristics of the headache are not specified, further acknowledging that cervicogenic headache may present with the same phenotype as migraine, tension-type headache, or other primary headache.

Evidence of causation has been a debated concept in cervicogenic headache. Controlled diagnostic nerve blocks contribute to the accurate diagnosis of cervicogenic headache.[6] However, the diagnostic criteria adopted by the International Headache Society do not require relief of headache from diagnostic nerve blockade.[5] For example, in a patient with cervical facet arthropathy at several levels, a medial branch block of the C3 and C4 dorsal rami may not completely eliminate headache. Conversely, peripheral nerve blocks, such as occipital nerve blocks, have been reported to be effective for several primary headache disorders, such as cluster headache and migraine.[7] Additionally, occipital nerve block duration is typically longer than the duration of the anesthetic blockade, at times lasting more than 4 weeks.[8]

Anatomical Concepts

Cervicogenic headache is a consequence of the convergence from nociceptive afferents from the cervical spinal nerves to the trigeminal nucleus caudalis and central pain pathways that are described for migraine. Anatomical understanding of the cervical spine and its relationship to surrounding structures provides an excellent model for cervicogenic headache. The cervical spine consists of the seven bony rings between the skull base and the thoracic vertebrae. Each cervical vertebrate consists of the vertebral foramen formed by the body and vertebral arch, and the smaller transverse foramen. The spinous process and transverse processes provide attachments for muscle insertion related to neck extension and flexion as well as, to a lesser extent, rotation. Beginning at C2, the upper and lower cervical vertebrae are articulated laterally by the zygapophyseal joint (facet joint), allowing primarily for lateral rotation but also contributing to neck extension and flexion. These are small synovial joints that provide nociception through the medial branches of the cervical dorsal root.

Cervicogenic headache is considered to be related to the C1–C3 segments of the spine (Figure 16.1).

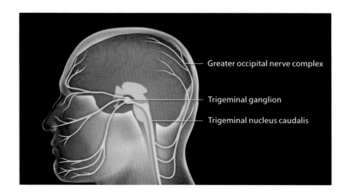

FIGURE 16.1 The trigeminal nucleus caudalis descends to the cervical cord and includes innervation to the occipital scalp and cervical structures.

In human cadaver studies the trigeminal nucleus caudalis (TCN) terminates at the C3—4 level of the spinal cord, but there is substantial anatomical variation for the termination of the TCN that ranges as high as C1 and as low as C7. Based on the anatomical variation, a cervical lesion of C5 may provoke cervicogenic headache in some patients and not in others. A study by Diener and colleagues supported this concept, and concluded that cervical disc prolapse below the level of C4 may contribute to cervicogenic headache in a subset of patients.[9]

Head pain has been demonstrated in human subjects following electrical stimulation of C1 or by noxious stimulation of the greater occipital nerve.[10] Additionally, stimuli to the C2—3 intervertebral disc or C2—3 cervical facet joints may produce headache.[11,12] The C1 nerve root innervates the posterior fossa and portions of the dens. There is evidence that direct stimulation of this nerve root refers pain to the orbit.[13] The second cervical nerve root is the main contributor to the greater occipital nerve. Due to the unique anatomy of C1 and C2 relative to the other cervical vertebrae, the greater occipital nerve has no cervical facet to protect it from injury. Cadaver studies have demonstrated that direct compression of the greater occipital nerve may be achieved between the atlas and axis in extension, and by the muscles and connective tissue the nerve passes through on its way to the occipital scalp in flexion. The C2—3 cervical facet joint is the best studied and generally accepted as a key contributor to cervicogenic headache. Of interest, the most common area of cervical facet arthrosis is at the level of C4—5; however, cervical arthropathy is most severe at C2—3.[14] This joint receives innervation partly from the third occipital nerve, and C2—3 spinal nerve medial branches.

Cervical myofascial pain has been associated with migraine, tension-type headache, and cervicogenic headache. Most anatomical texts refer to the suboccipital musculature as relatively isolated muscles that originate from the tubercle of the atlas and insert in the medial aspect of the inferior nuchal line of the occiput. However, recent anatomical studies have confirmed the presence of a connective tissue bridge between deep cervical extensor muscles and the cervical dura. This connective tissue bridge is found in most if not all specimens, and includes muscles such as the rectus captius posterior minor, rectus capitus posterior major, and oblicus capitus major, and histologically represents tendon, muscle, or fascia. It has been proposed that traumatic injury or primary dystonia may cause spasm of these muscles with subsequent traction of the cervical dura, resulting in cervicogenic headache.[15—17] Due to the minimal contributions these muscles make to large movements of the neck, patients

with this anatomical issue may present with normal range of motion or minimal restrictions to fine head positioning. This can result in delayed diagnosis or misdiagnosis.

CLINICAL CHARACTERISTICS

Patients with cervicogenic headache may present with a wide range of symptoms. Neck pain is common, and usually described as constant, dull, non-throbbing pain that is more prominent on the same side as the headache. However, neck pain is not required for diagnosis, as some patients do not experience pain below the occiput. Confounding matters, patients are frequently focused on headache, and may not verbalize neck pain until prompted by the examiner.

Most commonly, the headache is one-sided and occipital in location, but cervicogenic headache can be located periorbitally or in the temporal, parietal, and frontal areas of the head. It may also be bilateral in location, but if so it typically begins unilaterally and expands over time or with increasing intensity to become holocranial. Patients presenting with holocranial headache may, with a careful history, describe an initial one-sided and occipital location that has gradually progressed over months or years to become a non-focal headache. As discussed previously, with cervicogenic headache there is activation of the TCN from cervical inputs and, as such, cervicogenic headache may have migraine features, including photophobia, phonophobia, nausea, or emesis. Cervicogenic headache shares other migraine features, such as worsening with activity, or a throbbing sensation. Clinical features of trigeminal autonomic dysfunction have also been associated with cervicogenic headache, including lid ptosis, ipsilateral lacrimation, rhinorrhea, and facial fullness.[18] Additionally, nasal edema evidenced on MRI has been shown to resolve following ipsilateral occipital nerve block.[19]

Antonaci and colleagues summarized the most common clinical features of cervicogenic headache to include pain radiation to the arm or shoulder, various duration or fluctuation of continuous pain, moderate and non-throbbing pain, and history of neck trauma (Box 16.2).[20] Patients with cervicogenic headache may develop protective behaviors that limit neck activity and avoid certain neck postures. At times they may be aware of these adjustments, but these may be reflected in behaviors such as elaborate ways of arranging pillows or sleeping environments to protect neck posture during sleep. They may reposition themselves in the examination room to avoid neck rotation. Some may describe a change in daily activities to reduce neck

---

BOX 16.2

LIMITED DIFFERENTIAL DIAGNOSES FOR CERVICOGENIC HEADACHE

Cervical facet arthropathy   Infection – disc, dura, abscess
Cervical disc disease   Cervical cord or spine tumor
Cervical dystonia   Posterior fossa tumor (innervated by cervical nerves)
Myofascial dysfunction/whiplash   Occipital neuritis
Carotid/vertebral artery aneurysm

---

strain, such as no longer checking the blind spot while driving an automobile. Avoidance of neck postures or activities due to concern of worsening headache should alert the examiner to the potential of a cervicogenic headache. A key feature of cervicogenic headache is reduced cervical range of motion, which is reflected in the diagnostic criteria for cervicogenic headache.[5] Neck motion restriction is correlated to headache frequency and associated disability. Patients with cervicogenic headache typically show at least 10 degrees of restriction on the affected side.[21,22]

Patients with cervicogenic headache may have restricted range of motion or abnormal posture, possibly as a protection mechanism to prevent further pain. A reflexive response to spine pain may include spasm of the cervical paraspinal muscles, which may contribute to abnormal posture or motion restriction seen in cervicogenic headache. Myofascial trigger-point tenderness may be present in the suboccipital, cervical, and shoulder regions, which at times refer pain to the head upon stimulation. Rarely, patients may report radicular symptoms favoring the side of headache. Similarly, sensory dysfunction of the occipital scalp or head may be described. For example, a patient may describe occipital numbness or a feeling of water dripping down the occiput when the greater occipital nerve is compromised.

EVALUATION OF CERVICOGENIC HEADACHE

The most important aspect for an accurate diagnosis of cervicogenic headache is a thorough history and physical examination. A comprehensive history, review of systems, and physical examination will assess for an underlying structural disorder or systemic disease.[23] The history should include questions regarding previous head or neck trauma. Direct head trauma may contribute to cervicogenic headache, as head trauma almost invariably impacts the cervical

structures. History of migraine is also important, as those who have a predisposition to migraine are more likely to develop cervicogenic headache than are those not prone to migraine.[24] Patients with cervicogenic headache may report neck pain with passive motion, or report joint crepitus (described as "neck cracking" or "grinding") with neck rotation or side bending. Diagnostic evaluations such as serologies and imaging are helpful to exclude secondary pathologies, and may confirm a cervical lesion. In a study of inflammation markers in cervicogenic headache, Martelletti found elevations in interleukin-1 beta (IL-1β) and tumor necrosis factor alpha (TNF-α) that were elevated compared with healthy controls. They were also higher than in patients with migraine without aura.[25] However, these serologies are not generally used in clinical practice, as they lack specificity for cervicogenic headache and add to medical expense. Vitamin B12 and erythrocyte sedimentation rate may be helpful, as well as other diagnostic evaluations directed by the history and examination to exclude systemic illnesses that may affect the cervical spine, muscles, or vasculature. Systemic illnesses associated with secondary cervicogenic headache include rheumatoid arthritis, Sjögren's syndrome, thyroid and parathyroid dysfunction, and systemic lupus erythematosis. Rheumatoid arthritis is specifically associated with advanced high cervical facet arthropathy and elevated risk for atlantoaxial subluxation. Increased peripheral joint inflammation may be prognostic for risk of subluxation.[26,27] As almost any structural lesion in the cervical region can generate a nociceptive signal, and therefore activate the trigeminal nucleus caudalis, the differential for cervicogenic headache causes is quite large. A limited differential is listed in Box 16.2.

Diagnostic imaging, including magnetic resonance imaging (MRI) and computed tomography (CT), may assist in the diagnosis of cervicogenic headache.[28] Radiologic studies of the spine may be helpful in assessing the degree of cervical facet joint degenerative changes, odontoid structure and stability, spinal fracture, or cervical stenosis. One small study reported no demonstrable differences in the appearance of cervical spine structures on MRI scans when 24 patients with clinical features of cervicogenic headache were compared with 20 control subjects.[29] Cervical disc bulging, which was the primary focus of the study, was reported equally in both groups (45.5% versus 45.0%, respectively). Unfortunately, this study did not include cervical facet evaluation. Cervical spine MRI or CT adequately demonstrate cervical facet hypertrophy when examined in the sagittal or axial plane. Facet hypertrophy has the appearance of blurred facet margins and an irregular, distended facet capsule (Figure 16.2). Diagnostic imaging studies may be useful in excluding soft tissue structures as a source of cervicogenic headache, such as vertebral artery dissection, aneurysm, or myelopathy. An increase in the diameter of the neck extensors has been evaluated with MRI, with some authors finding cervical acceleration/deceleration (whiplash) injury association with rectus capitus posterior minor and major enlargement. This pseudohypertrophy is thought to represent an injury to the muscle, with increased MRI signal of fat within the affected muscles. Although some authors suggest that cervicogenic headache from whiplash resolves within 12 months, the clinical course may vary substantially.[30]

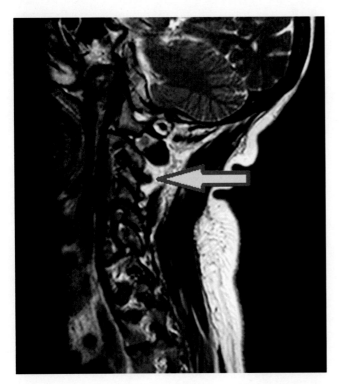

FIGURE 16.2 Sagittal T2 weighted MRI showing cervical facet hypertrophy at C2/3 (arrow).

TREATMENT OF CERVICOGENIC HEADACHE

Cervicogenic headache is a complex multidimensional disorder with numerous contributions from various anatomical and physiological components. Successful treatment of cervicogenic headache includes an awareness of the different therapeutic tools, and clinical judgment to match patients with optimal treatment. A comprehensive approach to cervicogenic headache frequently involves physical modalities, pharmacologic interventions, and interventional treatments. Surgical options are typically employed only if other treatment approaches are not successful. Research on the treatment of cervicogenic headache is limited due to the complexity of the disorder, challenges with diagnostic criteria, and lack of standardization in interventional approaches. There are no FDA approved pharmacologic therapies or consensus treatment algorithms.

Physical Modalities

Physiotherapy is the most commonly prescribed treatment for cervicogenic headache, being requested or performed in 75% of patients with a diagnosis of cervicogenic headache.[1] It is a reasonable approach for cervicogenic headache, considering the substantial involvement of the musculoskeletal system and its relationship to the cervical spine. There are more published reports and studies of physical treatments for cervicogenic headache than for medication or interventional procedures.[31] The effectiveness of manual therapies and exercise on cervicogenic headache does not appear to be related to age, chronicity, or gender.[32] Systematic reviews of physical modalities have demonstrated a gap in the quantity and quality of research citing lack of placebo control or limited sample size. One study showed a physical exercise program and myofascial release to be equally effective in cervicogenic headache.[33] Combining the two did not seem to provide any additional efficacy; however, both demonstrated reduced frequency and severity of headache compared with standardized undirected physician care. In this study, 76% of patients had a 50% reduction in headache frequency at 7 weeks, and 72% of patients had reduced frequency at 12 months, suggesting long-term benefits from manual therapies.[34] In the authors' experience, the skill-set of the provider of manual therapy is directly related to patient outcome.

Pharmacological Therapies

To date, there have been no large randomized placebo-controlled studies examining preventive medication for cervicogenic headache. In a study of cervicogenic headache, morphine was found to be of minimal benefit, with substantial concern for progression of headache pattern.[35] Additionally, ergotamine was found to be ineffective. One study of six patients suggested benefit from infliximab based on elevated TNF-α seen in cervicogenic headache, and suspected cervical joint inflammation.[36] These results were not followed up with a larger study, and it is not common practice. Several medications have been proposed as possibly helpful considering the benefits in other headache syndromes, such as migraine or hemicrania continua. Cervicogenic headache is recognized as a secondary headache syndrome, suggesting a peripheral source of nociception generating headache escalation through the trigeminal nucleus caudalis. Therefore, medication that targets peripheral nociception, similar to other neuropathic pain syndromes such as diabetic neuropathy, may be clinically effective in cervicogenic headache. As cervicogenic headache includes activation of the trigeminal nucleaus caudalis, medications that are effective in migraine prevention or neuropathic pain may reasonably be considered.

Anti-Epileptic Drugs

Neuronal membrane stabilizing medications, known as anti-epileptic drugs (AEDs), are presumed to suppress both peripheral and central nerve pain pathways.[37] Gabapentin has an FDA indication for post-herpetic neuralgia, and is commonly prescribed in neuropathic pain syndromes such as occipital neuralgia and cervical radiculopathy. However, gabapentin failed to show benefit in multiple prevention trials for episodic migraine.[38] Pregabalin has an FDA indication for fibromyalgia and is also commonly prescribed for other centralized pain syndromes and, at times, off-label for chronic daily headache. Although there are no published clinical trials for pregabalin in episodic migraine, pregabalin may be beneficial in cervicogenic headache. Both gabapentin and pregabalin have shown increased restorative sleep patterns,[39,40] which also may play a role in chronic pain syndromes such as cervicogenic headache.

Topiramate is FDA approved for prevention of episodic migraine, with some modest benefit shown for chronic migraine. It has not been effectively studied in cervicogenic headache. Topiramate is the best studied of the neuronal membrane stabilizers for prevention of primary headache syndromes. It is generally considered to have its effect through electrolyte channel stabilizing, and recently has been shown to have a mechanism of action at the glutamate pathway. Specifically, topiramate's mechanism of kainite receptor antagonism has been identified in the glutamate system as part of the trigeminothalamic pathway.[41] Glutamate is an important excitatory amino acid and believed to be involved in the pathophysiology of migraine and other pain disorders. Glutamate related receptors such as the N-methyl-D-aspartate (NMDA) receptor are located in key pain processing structures, including the dorsal root ganglion, spinal cord, and thalamus.[42]

Antidepressant Medications

Many chronic pain syndromes, such as fibromyalgia, neuropathic pain, and migraine, have been associated with dysfunction of the norepinephrine and, to a lesser extent, serotonin systems. Tricyclic antidepressant (TCA) medications such as nortriptyline, amitriptyline, and doxepin are commonly used to prevent migraine and treat other pain syndromes. Although the exact mechanism of benefit is uncertain, antidepressant medications do not appear to function by treating underlying depression. TCAs and other antidepressant medications demonstrate improvement in pain earlier than expected for depression, and their benefit appears independent of the presence of depression.[43] The serotonin and norepinephrine reuptake inhibitors (SNRIs), such as duloxetine and venlafaxine, have been observed to be effective in chronic pain states such as fibromyalgia and neuropathic pain. A recent study demonstrated duloxetine at a target dose of 120 mg per day to be effective in episodic migraine prevention.[44] Anecdotal evidence suggests benefit of SNRI class medications in cervicogenic headache. Selective serotonin reuptake inhibitor (SSRI) class medications have a limited role in migraine prevention and have not been evaluated for cervicogenic headache.

Greater Occipital Nerve Blockade

The greater occipital nerves and lesser occipital nerves originate from mostly C2, with variable contributions from the C1, C3, and C4 nerve roots. They share innervation of the occipital scalp, and are extensions of the cervical dorsal roots that also innervate the zygapophyseal joints of the high cervical spine. Anesthetic blockade is thought to have its clinical effect by reducing pain afferents to the cervical trigeminal nucleus caudalis. One key concept is that the response to occipital nerve block does not prove the occipital nerve itself is the source of headache. Occipital nerve blockade may provide clinical benefit through secondary cervical cord and brainstem modulation of headache. Furthermore, occipital nerve block procedures have been shown to reduce trigeminal autonomic features seen in the nasal structures.[19] Occipital nerve blockade may be helpful in several different headache types, including migraine, cluster headache, and occipital neuralgia.[45] Associated features of cervicogenic headache, including cervical range of motion, neck pain, and dizziness, have been shown to improve following occipital nerve block procedures.[46] Afridi and colleagues found that a 3-mL injection of 2% lidocaine and 80 mg methylprednisolone provided a median of 30 days' benefit in patients with chronic migraine, and that the presence of occipital nerve tenderness to palpation was predictive of favorable response.[8]

Occipital nerve block procedures are commonly performed in cervicogenic headache as a diagnostic procedure. Clinical response to interventional procedures is part of the diagnostic criteria for cervicogenic headache.[5] The type of anesthestic, inclusion of steroid, and quantity of injectate varies substantially by provider. An expert consensus recommendation statement has recently been published by the American Headache Society.[7] Unilateral or bilateral occipital nerve block procedures are relatively simple and safe to perform. The patient is typically seated, with the neck fully flexed. The region of the greater occipital notch is palpated, with the point of maximal tenderness assisting with localization. Using sterile technique, a 25-gauge or 27-gauge needle is used to deliver anesthetic solution at a depth of between 3 and 5 mm, with care not to inadvertently inject into the occipital artery (Figures 16.3, 16.4).

Research supporting the use of occipital nerve blockade in cervicogenic headache is limited due to sample size and lack of controls. One study examined 50 patients that received either greater or lesser occipital nerve injection, or placebo. Anesthetic injection was significant in reducing headache-related pain compared with placebo. The duration of benefit in this study trended longer with subsequent blocks,

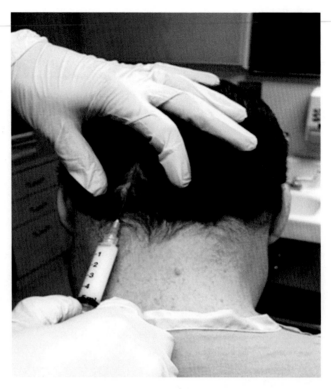

FIGURE 16.4 Patient positioned for occipital nerve block procedure.

suggesting a summation response over time with repeated treatments.[47] Occipital nerve blocks may be similar in efficacy for cervicogenic headache as compared with cervical medial branch blocks outcomes, as noted by Inan and colleagues in their study of 28 patients.[48]

Cervical Medial Branch Neurotomy

The most commonly cited anatomic sources of cervicogenic headache are the C2/3 and C3/4 cervical zygapophyseal joints. Long-term relief for cervicogenic headache stemming from these regions may be obtained by performing a neurotomy of the sensory nerve branches to these regions, including the third occipital nerve (innervates the C2/3 joint) and the C3 and C4 dorsal rami (innervates the C3/4 joint). Radiofrequency ablation is a neurotomy technique whereby an electrode is inserted perpendicular to the target and an electrical current disables the nerve. The nerve sheath is intended to remain intact such that healing over time results in return of nerve activity without aberrant regeneration or resultant neuropathic pain. Duration of radiofrequency ablation varies substantially, but is generally thought to last more than 6 months. Repeating a radiofrequency ablation

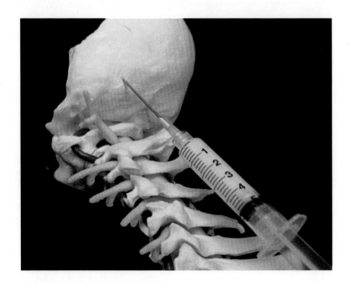

FIGURE 16.3 Placement of needle for occipital nerve block procedure.

procedure is safe and effective, and can maintain headache improvement for more than 2 years.[49]

For optimal results, it is essential to conduct controlled diagnostic blocks prior to long-term treatment with radiofrequency ablation. The ideal outcome is complete cessation of pain following diagnostic block. At times, the duration of pain relief follows the expected duration of anesthetic action. However, as discussed earlier in the case of occipital nerve block procedures, anesthetic nerve blocks without steroid may last weeks or months based on the secondary modulation of centralized pain processes. Therefore, diagnostic blockade is most important to assure the correct location for radiofrequency ablation. A randomized placebo-controlled study of radiofrequency neurotomy in cervicogenic headache showed statistically significant responses that separated it from placebo.[50]

Cervical radiofrequency neurotomy can be effective for cervicogenic headache from zygapophyseal joints once an accurate diagnosis has been established. For patients who have complete relief of cervicogenic headache following third occipital nerve anesthestic blockade, radiofrequency ablation provided 88% of the study patients with complete relief. The duration of response had a median of 297 days, with substantial variance in length of response between patients.[51] Similarly, in patients with clinical response to diagnostic C3/4 medial branch block procedures, radiofrequency ablation of the same area provided 70% of the patients with at least 75% improvement in their headache at 12 months.[49] There have been some studies that have not shown as robust a response for cervical neurotomy; however, these studies were not as stringent with diagnostic block criteria, and some focused on targeting lower cervical zygapophyseal joints.[52,53] Cervical neurotomy appears to be well tolerated. A recent study by Hamer and Purath reported 92.5% of patients who received this procedure for cervicogenic headache would be willing to have it repeated if their clinical symptoms returned.[54]

Surgical Intervention

Surgical intervention for cervicogenic headache has not been adequately studied. For patients with cervical facet arthropathy, conservative therapy including a combination of manual treatments, medication, and nerve blockade is typically pursued first. Jansen reported that well-selected patients with long-lasting and severe cervicogenic headache not responsive to other treatments may benefit from surgical intervention. His case series included anterior discectomy and fusion at typically two levels identified by examination, diagnostic imaging, and response to diagnostic blocks. In this study of 60 patients, a favorable result, including reduction of pain by 50% over 1 year, was seen in approximately 60% of patients, with approximately one-third of patients showing secondary deterioration within 1 year.[55] Similarly, in patients with C2/3 disc disease not responsive to conservative approaches, a cervical fusion of C2 to C3 via bone graft or similar approach may be considered.[56]

Other Modalities

There are several other treatment options for cervicogenic headache that are commonly performed but have limited research to support their use. Transcutaneous electrical nerve stimulation (TENS) is a non-invasive approach that utilizes high frequency, low intensity electrical impulses for pain therapy. In one study, 80% of patients with cervicogenic headache with TENS reported greater than 60% reduction in headache at 1 month.[57] Longitudinal response beyond 1 month was not reported in this study. A task force for neck pain evaluated TENS and concluded that there was no clinical benefit for TENS in neck pain compared with placebo.[58]

Extensive studies for onabutulinumtoxinA in chronic migraine have shown a small but statistically significant benefit compared with placebo.[59] This prompted onabotulinumtoxinA to receive an FDA approved indication for treatment in chronic migraine. Its mechanism is unclear for the prevention of headache.[60] Unfortunately, when onabotulinumtoxinA was studied for benefit in cervicogenic headache, it was found to be no more effective than placebo.[61] Additionally, a Cochrane review found botulinum toxin to have no statisitically significant or clinically relevant effect in either chronic cervical pain or cervicogenic headache.[62]

CONCLUSION

Cervicogenic headache is a relatively common chronic headache disorder that is often unrecognized. It is commonly encountered in headache specialty clinics, and at times is misdiagnosed as migraine or tension-type headache. Cervicogenic headache typically presents with occipital headache that radiates to one side with underlying cervical pain. Physical examination may show restricted range of motion to the affected side, and focal tenderness to cervical facets or other cervical structures. Diagnostic imaging is commonly normal or non-specific, but may identify areas of clinical concern such as C2/3 or C3/4 facet hypertrophy. Physiotherapy may be an effective treatment.

Although there are no FDA approved medication treatments for prevention of cervicogenic headache, medications such as gabapentin, topiramate, or duloxetine may have a clinical role. Interventional treatments such as occipital nerve anesthetic blockade or cervical neurotomy may be considered. Early diagnosis and management can significantly decrease the duration, cost, and disability of this often challenging pain disorder.

References

1. Blau JN, MacGregor EA. Migraine and the neck. *Headache.* 1994;34:88–90.
2. Tfeld-Hansen P, Lous I, Olesen J. Prevalence and significance of muscle tenderness during common migraine attacks. *Headache.* 1981;21:49–54.
3. Marcus D, Scharff L, Mercer MA, Turk DC. Musculoskeletal abnormalities in chronic headache: a controlled comparison of headache diagnostic groups. *Headache.* 1999;39:21–27.
4. Van Suijlekom HA, Lame I, Stomp-Van den Berg SG, Kessels AG, Weber WE. Quality of life of patients with cervicogenic headache: a comparison with control subjects and patients with migraine or tension-type headache. *Headache.* 2003;43:1034–1041.
5. Headache Classification Committee of the International Headache Society. The international classification of headache disorders: 3rd Edition Beta. *Cephalalgia.* 2013;33:629–808.
6. Bogduk N. The neck and headaches. *Neurol Clin.* 2014;32:471–487.
7. Expert consensus recommendations for the performance of peripheral nerve blocks for headaches – a narrative review. *Headache.* 2013;53:437–446.
8. Afridi SK, Shields KG, Bhola R, Goadsby PJ. Greater occipital nerve injections in primary headache syndromes – prolonged effects from a single injection. *Pain.* 2006;122:126–129.
9. Diener HC, Kaminski M, Stappert G, Stolke D, Schoch B. Lower cervical disc prolapse may cause cervicogenic headache: prospective study in patients undergoing surgery. *Cephalalgia.* 2007;27:1050–1054.
10. Piovesan EJ, Kowacs PA, Tatsui CE, Lange MC, Ribas LC, Weneck LC. Referred pain after painful stimulation of the greater occipital nerve in humans: evidence of convergence of cervical afferences on trigeminal nuclei. *Cephalalgia.* 2001;21:107–109.
11. Schellhas KP, Smith MD, Gundry CR, Polleil SR. Cervical discogenic pain: prospective correlation of magnetic resonance imaging and discography in asymptomatic subjects and pain sufferers. *Spine.* 1996;21:300–312.
12. Dwyer A, Aprill C, Bogduk N. Cervical zygapophysial joint pain patterns I: a study in normal volunteers. *Spine.* 1990;15:453–457.
13. Kerr FW. A mechanism to account for frontal headache in cases of posterior fossa tumors. *J Neurosurg.* 1962;18:605–609.
14. Lee MJ, Riew DK. The presence of cervical facet arthrosis: an osseus study in a cadaveric population. *Spine J.* 2009;9:711–714.
15. Kahkeshani K, Ward PJ. Connection between the spinal dura mater and suboccipital musculature: evidence for the myodural bridge and a route for its dissection – a review. *Clin Anat.* 2012;25:415–422.
16. Zumpano MP, Hartwell S, Jagos CS. Soft tissue connection between rectus capitis posterior minor and the posterior atlantooccipital membrane: a cadaveric study. *Clin Anat.* 2006;19:522–527.
17. Pontell ME, Scali F, Marshall E, Enix D. The obliquus capitis inferior myo-dural bridge. *Clin Anat.* 2013;26(4):450–454.
18. Biondi DM. Cervicogenic headache: a review of diagnostic and treatment strategies. *J Am Osteopath Assoc.* 2005;105:S16–S22.
19. Cooper WM. Resolution of trigeminal mediated nasal edema following greater occipital nerve blockade. *Headache.* 2008;48(2):278–279.
20. Antonaci F, Ghirmai S, Bono S, Sandrini G, Nappi G. Cervicogenic headache: evaluation of the original diagnostic criteria. *Cephalalgia.* 2001;21:573–583.
21. Vavrek D, Haas M, Peterson D. Physical examination and self reported pain outcomes from a randomized trial on chronic cervicogenic headache. *J Manipulative Physiol Ther.* 2010;33:338–348.
22. Hall TM, Briffa K, Hopper D, Robinson K. Comparative analysis and diagnostic accuracy of the cervical flexion–rotation test. *J Headache Pain.* 2010;11:391–397.
23. Pfaffenrath V, Dandekar R, Pollma W. Cervicogenic headache – the clinical picture, radiological findings and hypothesis on its pathophysiology. *Headache.* 1987;27:495–499.
24. Stovner LJ, Schrader H, Mickeviciene D, Surkiene D, Sand T. Headache after concussion. *Eur J Neurol.* 2009;16(1):112–120.
25. Martelletti P. Proinflammatory pathways in cervicogenic headache. *Clin Exp Rheumatol.* 2000;18(2):S33–S38.
26. Schwartz N, Mitnick HJ, Nowatsky J. Headaches related to rheumatologic disease. *Rheumatology.* 2013;17:381–392.
27. Neva MH, Isomaki P, Hannonen P, Kauppi M, Krishnan E, Sokka T. Early and extensive erosiveness in peripheral joints predicts atlantoaxial subluxations in pateints with rheumatoid arthritis. *Arthritis Rheumatol.* 2003;48:1808–1813.
28. Fredriksen TA, Fougner R, Tangerud A, Sjaastad O. Cervicogenic headache: radiological investigations concerning headache. *Cephalalgia.* 1989;9:139–146.
29. Coskun O, Ucler S, Karakurum B, Atasoy HT, Yildirim T, Ozkan S, et al. Magnetic resonance imaging of patients with cervicogenic headache. *Cephalalgia.* 2003;23:842–845.
30. Drottning M, Staff PH, Sjaastad O. Cervicogenic headache after whiplash injury. *Cephalalgia.* 1997;17:288–289.
31. Halderman S, Dagenais S. Choosing a treatment for cervicogenic headache: when? what? how much? *Spine J.* 2010;10:169–171.
32. Jull GA, Stanton WR. Predictors of responsiveness to physiotherapy management of cervicogenic headache. *Cephalalgia.* 2005;25:101–108.
33. Posadzki P, Ernst E. Systematic reviews of spinal manipulations for headaches: an attempt to clear up the confusion. *Headache.* 2011;51:1419–1425.
34. Jull G, Trott P, Potter H, Zito G, Niere K, Shirley D, et al. A randomized controlled trial of exercise and manipulative therapy for cervicogenic headache. *Spine.* 2002;27:1835–1843.
35. Bovim G, Sjaastad O. Cervicogenic headache: responses to nitroglycerin, oxygen, ergotamine and morphine. *Headache.* 1993;33:249–252.
36. Martelletti P. Inflammatory mechanisms in cervicogenic headache: an integrative review. *Curr Pain Headache Rep.* 2002;6:315–319.
37. Mulleners WM, McCrory DC, Linide M. Antiepileptics in migraine prophylaxis: an updated cochrane review. *Cephalalgia.* 2014;12 August [Epub ahead of print].
38. Linde M, Mulleners WM, Chronicle EP, McCrory DC. Gabapentin or pregabalin for the prophylaxis of episodic migraine in adults. *Cochrane Database Syst Rev.* 2013;24(6).
39. Lo HS, Yang CM, Lo HG, Lee CY, Ting H, Tzang BS. Treatment effects of gabapentin for primary insomnia. *Clin Neuropharmacol.* 2010;33(2):84–90.
40. Roth T, Arnold LM, Garcia-Borreguero D, REsnick M, Clair AG. *Sleep Med Rev.* 2014;18(3):261–271.
41. Andreou AP, Goadsby PJ. Topiramate in the treatment of migraine: a kainate (glutamate) receptor antagonist within the trigeminothalamic pathway. *Cephalalgia.* 2011;31(13):1343–1349.

42. Huang L, Bocek M, Jordan JK, Sheehan AH. Memantine for the prevention of primary headache disorders. *Ann Pharmacolther.* 2014;48(11):1507–1511.

43. Paneraj AE, Monza G, Movilia P, Bianchi M, Francucci BM, Tiengo M. A randomized within patient, cross-over placebo-controlled trial on the efficacy and tolerability of the tricyclic antidepressants chlorimipramine and nortriptyline in central pain. *Acta Neurol Scand.* 1990;82:34–38.

44. Young WB, Bradley KC, Anjum MW, Gabelline Myers C. Duloextine prophylaxis for episodic migraine in persons without depression: a prospective study. *Headache.* 2013;53(9):1430–1437.

45. Wcibelt S, Andress-Rothrock D, King W, Rothrock J. Suboccipital nerve blocks for suppression of chronic migraine: safety, efficacy, and predictors of outcome. *Headache.* 2010;50:1041–1044.

46. Baron EP, Cherian N, Tepper SJ. Role of greater occipital nerve blocks and trigger point injections for patients with dizziness and headache. *Neurologist.* 2011;17:312–317.

47. Naja ZM, El-Rajab M, Al-Tannir MA, Ziade FM, Tawfik OM. Repetitive occipital nerve blockade for cervicogenic headache: expanded case report of 47 adults. *Pain Practicioner.* 2006;6:278–284.

48. Inan N, Ceyhan A, Inan L, Kavaklioglu O, Alptekin A, Unal N. C2/C3 nerve blocks and greater occipital nerve blocks in cervicogenic headache treatment. *Funct Neurol.* 2011;16:239–243.

49. Barnsley L. Percutaneous radiofrequency neurotomy for chronic neck pain: outcomes in a series of consecutive patients. *Pain Med.* 2005;6:282–286.

50. Haspeslagh SR, Van Suijlekom HA, Lame IE, Kessels AG, van Kleef M, Weber WE. Randomised controlled trial of cervical radiofrequency lesions as a treatment for cervicogenic headache. *BMC Anesthesiol.* 2006;6:1.

51. Lee JB, Park JY, Park J, Lim DJ, Kim SD, Chung HS. Clinical efficacy of radiofrequency cervical zygapophyseal neurotomy in patients with chronic cervicogenic headache. *J Korean Med Sci.* 2007;22:326–329.

52. Govind J, King W, Bailey B, Bogduk N. Radiofrequency neurotomy for the treatment of third occipital headache. *J Neurol Neurosurg Psychiatry.* 2003;74:88–93.

53. Stovner LJ, Kolstad F, Helde G. Radiofrequency denervation of facet joints C2–C6 in cervicogenic headache: a randomized, double blind, sham-controlled study. *Cephalalgia.* 2004;24:821–830.

54. Hamer JF, Purath TA. Response of cervicogenic headache and occipital neuralgia to radio frequency ablation of the C2 dorsal root ganglion and/or third occipital nerve. *Headache.* 2014;54:500–510.

55. Jansen J. Surgical treatment of cervicogenic headache. *Cephalalgia.* 2008;28(1):41–44.

56. Schofferman J, Garges K, Goldthwaite N, Koestler M, Libby E. Upper cervical anterior diskectomy and fusion improves discogenic cervical headaches. *Spine.* 2002;27:2240–2244.

57. Farina S, Granella F, Malferrari G, Manzoni GC. Headache and cervical spine disorders: classification and treatment with transcutaneous electrical nerve stimulation. *Headache.* 1986;26:431–433.

58. Haldeman S, Carroll L, Cassidy JD. Bone joint decade 2000–2010 task force on neck pain and its associated disorders. *Spine.* 2008;33(S4):S5–S7.

59. Aurora SK, Winner P, Freeman MC, Spierings EL, Heiring JO, DeGryse RE, et al. OnabotulinumtoxinA for treatment of chronic migraine: pooled analyses of the 56-week PREEMPT clinical program. *Headache.* 2011;51(9):1358–1373.

60. Durham PL, Cady R. Insights into the mechanism of onabotulinumtoxinA in chronic migraine. *Headache.* 2011;51 (10):1573–1577.

61. Linde M, Hagen K, Salveson O, Gravdahl GD, Helde G, Stovner LJ. OnabotulinumtoxinA treatment of cervicogenic headache: a randomised double-blind placebo controlled crossover study. *Cephalalgia.* 2011;31:797–807.

62. Langevin P, Peloso PM, Lowcock J, Nolan M, Weber J, Gross A, et al. Botulinum toxin for subacute chronic neck pain. *Cochrane Database Syst Rev.* 2011;7.

17

Headache in Children and Adolescents

Jack Gladstein[1], Howard S. Jacobs[2], and A. David Rothner[3]

[1]Department of Pediatric Neurology, and Headache Program, University of Maryland School of Medicine, Baltimore, Maryland, USA [2]Department of Neurology, Nationwide Children's Hospital, and Ohio State University, Columbus, Ohio, USA [3]Center for Pediatric Neurology, Cleveland Clinic Main Campus, Cleveland, Ohio, USA

INTRODUCTION

It is important to study pediatric headache for a number of reasons. Just as in adults, significant disability and pain is associated with headache in young people. Children and adolescents with headache suffer when they miss days of school, or go to school but cannot concentrate.[1] Parents need to take time off work to care for their child who has a bad headache. In addition to disability, we know that most adults with headache trace the beginnings of their headache history to their younger years.[2] We feel that adequate treatment as a youngster may prevent the ravages of chronic disease. The principles of early intervention, identifying triggers, and treating comorbidities may be vital in creating normalcy in a youngster's turbulent adolescent years.[3]

We have learned that differences exist between childhood and adult headaches in duration of pain, time to chronification, symptom description, and ease of identifying psychiatric comorbidities.[4,5] Further, migraine equivalents that present in childhood will cause significant concern for parents. These children grow up to be migraine sufferers. Earlier identification and treatment for cyclic vomiting, benign paroxysmal torticollis, and benign paroxysmal vertigo will help those families obtain the help that their children need to cope with these scary, yet treatable migraine precursors.[6]

HISTORICAL PERSPECTIVE

In 1962, Bo Bille reported on an epidemiologic review of schoolchildren in Uppsala, Sweden. His classic paper taught us that as children age, the proportion of individuals with non-migrainous headaches increases.[7] He also found the prevalence of headache to be equal in boys and girls before puberty, with a more female predominance after puberty.[7]

Early attempts at classification of pediatric headache reflected criteria in the adult literature. Prensky[8] relied upon descriptors such as throbbing, intermittent headache, etc., but did not quantify frequency. No account for severity or localization was presented. As in the adult literature, it was difficult to compare groups of subjects using these criteria.[9] Prensky observed, however, that pediatric pain is more often bilateral and short-lived. This information facilitated IHS modifications when those criteria were established. IHS I[10] provided a uniform definition of migraine that allowed for better quantification of migraine. Criteria, however, were limited to adults. Using adult criteria excluded many youngsters from drug trials because headache duration was required to be 4 hours, and unilateral pain was a major criterion. Maytal[11] and others modified adult criteria for children and enabled us to better diagnose migraine in children. The current ICHD-II criteria[12] provide a detailed classification of headache; however, this classification captures only 62% of pediatric subjects.[13,14]

Classification of chronic daily headache has been controversial in both the adult and the pediatric literature.[4,15] Attempts to modify Silberstein's criteria by Holden,[4] and then via a multicenter approach led by Koenig, demonstrated that in children there was less medication overuse and a shorter period of transformation from episodic to chronic headache, as compared with adults.[5]

We have learned that drug trials are still fraught with problems. It has been difficult to show spread

S. Diamond, R. K. Cady, M. L. Diamond & V. T. Martin (Eds):
Headache and Migraine Biology and Management.

DOI: http://dx.doi.org/10.1016/B978-0-12-800901-7.00017-3

between placebo and study drug, which has made it arduous to acquire pediatric indications for excellent medications.[16] Nevertheless, properly designed triptan trials have been undertaken in adolescents.[17–29] Trials of preventative medications have been mostly inconclusive,[30–46] although topiramate trials using modern classification criteria were successful.[47–49] Evaluation and treatment of youngsters with headache presenting to the ER has been studied through the work of Li,[50] Kabbouche,[51] and Lewis.[52]

Classic articles by Holroyd[53] have encouraged examination for psychiatric comorbidities. The works of Zeltzer[54] and McGrath[55] provide frameworks for self-efficacy and pain management for both the youngster and his or her parents.

EPIDEMIOLOGY AND PATHOPHYSIOLOGY

Headache is a frequent complaint in the pediatric population. Using ICHD-II criteria,[12] which may in fact be overly limiting for pediatric patients, Abu-Arafeh and colleagues state that the prevalence of headache in children up to 20 years of age is approximately 58%, with a 1:1.5 male to female ratio.[56] The incidence of migraine headaches in this population is 7.7%; 6.0% in males and 9.7% in females. When divided by age groups, in those under 14 years of age, 4.7% of males and 7.0% of females suffer from migraines; in those over 14, the incidence rises to 6.0% of males and 9.7% of females.

Bigal and Lipton further stratified headache and migraine by age, finding that 3–8% of 3-year-olds complain of headache, increasing to 19.5% at 5 years of age, 37–51.5% at 7, and 57–82% in patients aged 8–15 years.[2] In their study, the prevalence of migraine in prepubertal children is higher in boys than in girls, but this reverses post-puberty. The peak incidence of migraine with aura in boys is 5 years of age (6.6:1000), and migraine without aura is 10–11 years of age (10:1000). In girls, migraine with aura peaks at 12–13 years (14.1:1000), and migraine without aura peaks at 14–17 years (18.9/1000).

In patients presenting to a pediatric headache center, approximately 35% complained of >15 days/month of headache. In the diagnostic criteria for chronic daily headache (CDH), 22% (63% of CDH patients) experienced 30 days of headache and 6.7% (19.5%) had continuous headache. The incidence of adolescent patients suffering from CDH in the general pediatric population has been reported to be as high as 3.5% using the PedMIDAS questionnaire.[57,58]

When socioeconomics is considered, studies have shown that migraine incidence is inversely related to socioeconomic status.[59,60] In a study by Stevens and colleagues, headache was more frequently diagnosed in pediatric patients receiving medical assistance than in those on private insurance, divided equally between black and white patients. Significantly, despite the diagnosis, less than 40% of those patients received evidence-based therapy.[61]

After an exhaustive literature search, we were not able to find articles that specifically discuss pediatric migraine physiology, so we defer to earlier chapters in this volume to discuss the topic in more detail. Investigations on the role of genetics in childhood migraine are in the nascent stages. Russell showed that migraine both with and without aura has genetic inheritance patterns. Specifically, first-degree relatives of patients who experience migraine without aura have a two-fold chance of suffering the condition, whereas relatives of patients with migraine with aura have a four-fold chance of experiencing the same condition.[62] Although not generalizable to typical migraine, genes have been identified in rare migraine conditions, such as familial hemiplegic migraine.[63,64] The MTHFR polymorphism is increased in patients with migraine with aura, but not in migraine without aura.[65] Polymorphisms in the serotonin transporter gene (5HTTLPR) are also associated with migraine with aura.[66] At present we do not have adequate knowledge of gene therapy in youngsters, but a time can be anticipated when treatment medications and doses can be directed based upon a child's specific genetic information.

CLINICAL APPROACH

Evaluation of a child with headache includes a thorough history, and careful pediatric and neurologic examinations, which support a thoughtful differential diagnosis.[67] The age of the patient, the acuity of the problem, the presence or absence of neurologic symptoms and signs, and the location of the evaluation will modify the approach. Laboratory testing is based on the differential diagnosis, and in primary headaches is often unnecessary. The American Academy of Pediatrics has established guidelines for evaluation of children and adolescents with headache.[68] Routine laboratory testing and lumbar puncture are not recommended. Routine EEG is not recommended for the evaluation of headache, and is not helpful in distinguishing primary versus secondary headache or between migraine and other primary headache disorders. The guideline suggests that incidental abnormalities unrelated to headache symptoms are reported in 16% of patients undergoing routine neuroimaging. Some of these abnormalities include arachnoid cysts,

Chiari malformations (Figures 17.1–17.3),[69] sinus disease, and vascular malformations. Routine neuroimaging is not recommended in the evaluation of headache in children.[70]

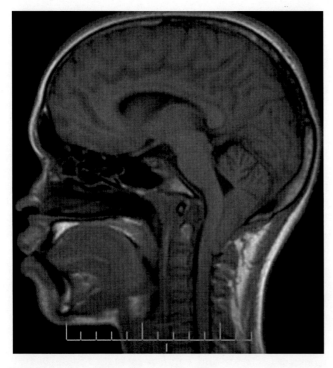

FIGURE 17.1   Chiari malformation (asymptomatic in the majority of patients). The headache history determines the correct diagnosis. The child often provides the most useful information. One must assess the course of the headache over time. Four patterns will be noted (Figure 17.2).[69]

Chronic non-progressive headaches have been present for weeks to months and are essentially remaining the same. After a careful evaluation, an underlying etiology is seldom found. The physician should carefully ascertain the presence of symptoms of increased intracranial pressure, progressive neurologic symptoms, and the presence of neurologic abnormalities. The family history is critical in determining the diagnosis, since, in the majority of children with migraine, a positive family history of migraine is reported.[71] The review of systems, the ascertainment of comorbid medical conditions, the use of medication and/or recreational drugs, educational issues, and psychosocial issues often provide additional useful information.

Childhood periodic syndromes occur often, and provide a history that allows them to be identified. The ICHD-III category includes cyclic vomiting syndrome, abdominal migraine, benign paroxysmal vertigo of childhood, and benign paroxysmal torticollis.[72,73] These diagnoses are of exclusion. Inborn errors of metabolism may mimic these disorders. A thorough evaluation for these disorders is necessary to avoid missing treatable causes.

Vital signs including blood pressure, height, weight, and temperature should be noted. The examiner should pay close attention to the skin, searching for *café au lait* macules, rashes, petechiae, and bruises. The neurologic examination must be thorough, and include examinations of both the fundi and the neck.[67]

At the completion of the history and physical examination, the examiner should have a tentative diagnosis based upon the pattern of the headaches, as well as the patient's symptoms and signs. The patient with a

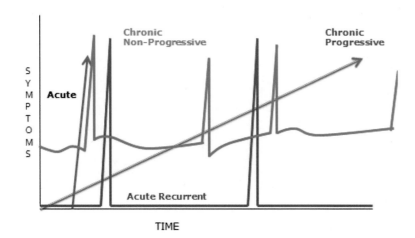

FIGURE   17.2 **Time-courses   of   headache.** The acute headache describes a headache without a previous history of headache. The evaluation is determined by associated symptoms and signs. Acute recurrent headaches indicate headaches that occur acutely but recur over time. The most common type of acute recurrent headache is migraine. Chronic progressive headaches have been present for several weeks to several months, and are worsening over time. If neurologic symptoms and/or signs are present, an organic process should be considered (Figure 17.3). *Adapted from Rothner.*[69]

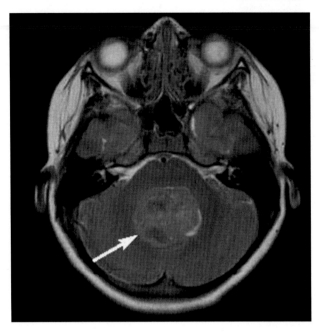

FIGURE 17.3 A posterior fossa tumor, which would likely have presented with a chronic progressive headache.

primary headache disorder should have a completely normal neurological examination. If neurologic symptoms or an abnormal neurologic exam is present, an underlying organic disorder should be ruled out.

The choice of laboratory tests depends on the differential diagnosis. In primary headaches, routine laboratory work and/or scanning is wasteful and not helpful. Indeed, the presence of unrelated abnormalities can often delay the correct diagnosis. A review of the AAN guidelines for the evaluation of a child with headache should be helpful.[68]

TREATMENT

The principles underlying the treatment of a pediatric/adolescent patient with a headache include establishing the diagnosis, educating the patient and the patient's family regarding the nature of the disorder, and providing confident reassurance that the patient has been adequately examined and no underlying neurological or structural abnormality found[13,74] (Box 17.1).

The role of stress should be discussed in detail, especially if the patient has chronic non-progressive headaches.[53–55] Details of school absences, medication overuse, and other comorbidities should be discussed. Important factors in the overall treatment program for headache patients include a regular schedule, 8 hours of restorative sleep, and adequate hydration. The patient should have an adequate diet and avoid missing meals, should not overuse over-the-counter medication, and should participate in an adequate exercise program.[75] Every effort should be made to keep the patient in school.[76] Some clinicians advocate a migraine diet that excludes caffeine, chocolate, luncheon meats, aged cheeses, and monosodium glutamate (MSG).[13,77] Comorbidities include obesity,[78] excessive medication use,[79] anxiety, depression,[80] obsessive-compulsive disorders,[81] sleep disorders, and eating disorders.[82] Parental over-concern and over-participation should be noted and discussed.[83]

Once the basic issues have been addressed, consideration is given to using pharmacologic therapy. The use of medication is divided between rescue medication to be used in an acute situation, and preventive medication when indicated.

Acute treatment of migraine is based upon the principles of early intervention to avoid cutaneous allodynia,[84] and using high doses of oral medications to combat gastroparesis.[85] Adjunctive medications to treat nausea[86] and anxiety[87] may be added if needed. Due to the paucity of controlled trials in this area, variation from the experts exists on when to institute over-the-counter drugs (such as ibuprofen[88] and acetaminophen[89]), and when to use the migraine-specific triptans.[90] The triptans can be administered intranasally, orally, or by subcutaneous injection. The dose and choice of the triptans need to be adjusted based on the age and size of the patient, and the treating physician's preference and comfort level. Side effects from the triptans are well studied. The triptans are not FDA approved in the younger age group because it has been difficult to demonstrate a difference in efficacy between placebo and active drug.[16] Narcotics and narcotic analgesics,[91] as well as barbiturates,[92] should not be used in most circumstances. Sedation often relieves the headache if acute management fails.[93]

Pharmaco-prophylaxis is usually indicated if migraines occur at least three to four times a month and cause significant impairment in the patient's daily function or quality of life. The goal is to reduce frequency of migraines to less than one to two attacks a month. The general approach to pharmaco-prophylaxis is to start at the lowest dose and titrate upwards as needed. Before beginning any preventive therapy, it is important to set realistic expectations with the patient and the family, because any preventive therapy may take at least 8–12 weeks to show recognizable effect.[28,94]

Various agents have been tried in the prophylaxis of pediatric migraines, including tricyclic antidepressants (amitriptyline and nortriptyline), trazodone, beta-blockers (propranolol), calcium channel blockers (verapamil and flunarizine), antiserotonergics (cyproheptadine), gabapentin, and anti-epileptics.[26]

BOX 17.1

ALGORITHM FOR HEADACHE IN CHILDREN

I. Acute Headaches
   A. Check for strep and viruses
   B. If sudden and dramatic, rule out subarachnoid hemorrhage with CT and LP

II. Acute Recurrent

Autonomic symptoms − Migraine
   A. Fluids, consistent sleep, regular meals, stress reduction
   B. Treat at first twinge with high doses of analgesics/triptans
   C. Assess disability
      1. Absenteeism − no. of days of school missed
      2. Presenteeism − no. of days in school but head on desk and not learning
      3. If >2 days per week, consider either, or/and:
         a. Prevention medication − selection based on comorbidities
         b. Complementary treatment − yoga, hypnosis, biofeedback, cognitive behavioral therapy, acupuncture

Autonomic symptoms − Tension-type headaches
   A. Absenteeism − no. of days of school missed
   B. Presenteeism − no. of days in school but head on desk and not learning
   C. If >2 days per week, consider either, or/and:
      1. Prevention medication − selection based on comorbidities
      2. Complementary treatment −
         a. Yoga, hypnosis, biofeedback, cognitive behavioral therapy, acupuncture
         b. Note absence of treatment at first twinge, as there are no autonomic symptoms

III. Chronic Progressive
   A. Worrisome pattern for intracranial mass or benign intracranial hypertension
   B. Needs work-up to include MRI and possible LP if MRI is negative

IV. Chronic Non-Progressive
   A. Usually transforms from episodic migraine or tension-type
   B. Often accompanied by comorbidities
      1. Depression
      2. Anxiety
      3. Insomnia
      4. Dizziness
      5. Medication overuse
      6. School underperformance
   C. Requires multilayered approach
   D. Often requires referral to tertiary headache clinic

Based on Cochrane systematic review of 20 randomized trials, flunarizine has demonstrated efficacy in the prophylaxis of chronic pediatric migraine. This compound is not available in the US. It is associated with daytime sedation and weight gain. Administering it during the early evening can avert daytime sleepiness.[1,27,34−37] Anti-epileptics such as topiramate, 100 mg, have been shown to be effective for reduction of headache frequency, severity, and duration.[38,47,48] The most common side effects include cognitive impairment and weight loss.[1,34] Divalproate 15−45 mg/kg per day has demonstrated efficacy in reducing frequency as well as severity of pediatric migraine, with a 50% headache reduction seen. Side effects were dizziness, drowsiness, and increase in appetite.[32,33]

Amitryptiline,[30,31] cyproheptadine,[30] levetiracetam,[28] and zonisamide[29] have been studied in pediatric migraine, but further evidence is needed to establish efficacy. Conflicting data exist for the efficacy of propranolol.[39−42] Trials have not established efficacy of nimodipine,[95] timolol, papaverine, pizotifen,[44] trazodone,[96] clonidine,[45,46] metoclopramide, and domperidone.[30] In the absence of evidence, clinicians have used their own preferences and experiences to guide treatment options.

Before starting therapy, care must be taken to identify potential side effects of these medications that may affect certain groups of children more than others. For instance, valproic acid may be teratogenic and have other toxic effects, and therefore may not be suitable for adolescent girls. Beta-blockers may not be ideal agents for children with asthma or atopy. Further, athletes using beta-blockers may drop their exercise tolerance. Conversely, antiserotonergics such as cyproheptadine may be particularly useful in children with comorbid environmental allergies.[41]

Treatment of the patient who visits the emergency department depends on the history and the physical examination. An organic process must be excluded. Once serious pathology has been excluded, the use of intravenous medications to hydrate and sedate, and to decrease nausea and vomiting, are standard initial steps. Intravenous ketorolac can be used for analgesia.[97] Intravenous fluids, corticosteroids, nonsteroidal anti-inflammatory drugs (NSAIDs), magnesium, valproate, and dihydroergotamine (DHE) have been used successfully.[51,98–101] Narcotic analgesics should be avoided.[102–104]

WHAT HAPPENS TO OUR PATIENTS AS THEY GROW UP?

Healthcare practitioners who treat children know that one of the realities of their specialty is that patients inevitably grow up and move from their practices to be cared for by physicians who take care of "big people." But what happens to these headache patients? What is the expected natural course of the problem — do they get better? Do they get worse?

Antonacci and colleagues published an excellent review of the literature on headache evolution in 2014, covering approximately the past 20 years.[105] In 1995, Dooley and Bagnell looked at the headache progression of 77 pediatric patients during the decade from 1983 to 1993.[106] In the initial year,

18 patients experienced tension-type headaches (TTH), 49 migraine without aura (MwoA), 5 migraine with aura (MwA), and 5 classified as "other." Ten years later, 22 suffered from TTH, 26 from MwoA, and 7 from MwA, and 22 were headache free. The breakdown is reflected in Table 17.1.[106]

Patients with TTH were 50% likely to be headache free after 10 years, compared with 20% of the patients with migraine and none of the patients who initially presented with migraine with aura. Almost half of the patients who initially presented with migraine continued to experience migraine. Interestingly, in this study only one of the five patients with migraine with aura continued to have MwA after 10 years. In total, 80% of the patients felt that they had improved over the 10-year span (84% male, 71% female). Later studies have confirmed the likelihood of improvement being greater in TTH than migraine, especially in male patients.[107,108] The age of onset was shown to be the only significant predictor — the older the age of onset, the more likely it was that the headache would follow a favorable course.[109]

There is concern that children and adolescents who suffer from headache are at increased risk of developing depression and/or anxiety over time.[80] However, evidence shows that we can be cautiously optimistic that such patients are likely to show improvement over time.

WHAT THE FUTURE HOLDS FOR PEDIATRIC HEADACHE

One of the biggest hurdles to be overcome in pediatric headache care is the paucity of drugs approved by the Food and Drug Administration (FDA). Currently, only rizatriptan is approved for ages 6 years and above, and almotriptan is approved for ages 12 years and above. Despite its widespread use, oral sumatriptan is not FDA approved for pediatric use.

In 2013, sumatriptan in a patch formulation was released for adult use. Since many migraineurs suffer

TABLE 17.1   The Headache Progression of 77 Pediatric Patients, 1983–1993

	1993			
1983	No headache	TTH	MwoA	MwA
TTH (18)	9 (50%)	7 (39%)	2 (11%)	0
MwoA (49)	10 (20.4%)	11 (22.4%)	23 (46.9%)	5 (10.2%)
MwA (5)	0	3 (60%)	1 (20%)	1 (20%)
other (5)	3 (60%)	1 (20%)	0	1 (20%)

TTH, tension-type headache; MwoA, migraine without aura; MwA, migraine with aura.
Adapted from Dooley and Bagnell.[106]

nausea with migraine attacks, this unique delivery system may offer a more tolerable injection-free method of treatment for adults and pediatric patients. However, FDA approval is not likely to be granted in the near future. DHE in an inhaled form (MAP0004) is in the final stages of study and will likely soon be available. In many emergency departments, IV DHE is part of the therapy for status migrainosus. Tepper and colleagues have shown that unlike triptans, the efficacy of orally inhaled DHE is not reduced once central sensitization has occurred.[110] This difference is quite significant. One could potentially consider using inhaled DHE in the home setting to try and prevent an ER visit for status migrainosus.

Calcium gene-related peptide (CGRP) related therapy for acute migraine is currently the target of extensive research. CGRP is released by both peripheral and central trigeminal ganglion neurons, and is the main mediator of pain signals from the trigeminal nerve. If CGRP is blocked, cortical spreading depression is significantly inhibited.[111] Five CGRP receptor antagonists have shown efficacy: one of these agents could not be produced in an oral form, two others were discontinued due to liver toxicity, and the two remaining agents (BI44370A and BMS-927711) have shown promise as possible acute migraine therapies in the future.[112]

No drugs have received FDA approval for treatment of chronic migraine in the pediatric patient. Amitriptyline and topiramate are frequently used in this setting, but as "off-label" therapy. Hershey and Powers are currently conducting the multicenter Children and Adolescent Migraine Prevention (CHAMP) Study, comparing amitriptyline and topiramate versus placebo and versus each other, in the hope of establishing the usefulness of these drugs for chronic migraine. The study should also determine which drug should be first-line therapy.

In the adult population, Botox has received a lot of attention as the only approved treatment for chronic migraine. In a small study in adolescents, Chan and colleagues showed that Botox has the potential for benefit in pediatric patients, but further study was recommended.[113] A double-blinded study is being attempted, but few parents are willing to allow their child to possibly receive 31 injections of a placebo.

Beyond therapies one can hope that in the near future research will provide a better understanding of the relationship between migraine and GI disorders such as inflammatory bowel disease and celiac disease (if a relationship exists at all). Other autonomic disorders, such as postural orthostatic tachycardia syndrome (POTS), may be involved.

More needs to be discovered about the genetics of headache: who gets them and why; why some of our patients quickly progress to chronic migraine but most do not; and why medications have such variable results when treating headaches with the same phenotype.

Finally, more evidence-based therapy is needed, whether for acute migraine, migraine prophylaxis, or ER/hospital treatment. Many centers rely on their own tried and true therapies and pathways. It is hoped that these interventions will be compared in a prospective manner to decide the best courses of care for our patients.

References

1. Termine C, Ozge A, Antonaci F, Natriashvili S, Guidetti V, Wober-Bingol C. Overview of diagnosis and management of paediatric headache. Part II: therapeutic management. *J Headache Pain*. 2011;12(1):25–34.
2. Bigal ME, Lipton RB. The epidemiology, burden and comorbidities of migraine. *Neurol Clin*. 2009;27:321–334.
3. Jacobs H, Gladstein J. Pediatric headache: a clinical review. *Headache*. 2012;52(2):333–339.
4. Holden EW, Gladstein J, Trulsen MA, Wall MA. Chronic daily headache in children and adolescents. *Headache*. 1994;34:508–514.
5. Koenig MA, Gladstein J, McCarter RJ, Hershey AD, Wasiewski W. Chronic daily headache in children and adolescents presenting to tertiary headache clinics. Pediatric Committee of the American Headache Society. *Headache*. 2002;42(6):491–500.
6. Al-Twaijri NA, Shevell MI. Pediatric migraine equivalents: occurrence and clinical features in practice. *Pediatric Neurology*. 2002;26(5):365–368.
7. Bille B. Migraine in pediatric school children. *Acta Paediatric Scand*. 1962;51(Suppl. 136):1–151.
8. Prensky AL. Migraine and migraine variants in pediatric patients. *Ped Clin North America*. 1976;23(3):461–471.
9. Ad Hoc Committee on Classification of Headache. National institute of neurologic diseases and blindness. Classification of headache. *Neurology*. 1962;12:378–380.
10. Headache Classification Committee of the international Headache Society. Classification and diagnostic criteria for headache disorders, cranial neuralgias and facial pain. *Cephalalgia*. 1988;8(suppl 7):1–96.
11. Maytal J, Young M, Shechter A, Lipton RB. Pediatric migraine and the IHS criteria. *Neurology*. 1997;48(3):602–607.
12. The International Classification of Headache Disorders: 2nd edition. *Cephalalgia*. 2004;24(suppl 1): 9–160.
13. Hershey AD. Pediatric headache: update on recent research. *Headache*. 2012;52(2):327–332.
14. Hershey A, Winner P, Kabbouche MA. Use of the ICHD-II criteria in the diagnosis of pediatric migraine. *Headache*. 2005;45:11.
15. Silberstein SD, Lipton RL, Solomon S, Mathew NT. Classification of daily and near daily headaches. Proposed revisions to the IHS criteria. *Headache*. 1994;34:1–7.
16. Lewis DW, Winner P, Wasiewski W. The placebo responder rate in children and adolescents. *Headache*. 2005;45(3):232–239.
17. Winner P, Rothner AD, Saper J, Nett R, Asgharnejad M, Laurenza A, et al. A randomized, double-blind, placebo-controlled study of sumatriptan nasal spray in the treatment of acute migraine in adolescents. *Pediatrics*. 2000;106(5):989–997.
18. Major PW, Grubisa HS, Thie NM. Triptans for treatment of acute pediatric migraine: a systematic literature review. *Pediatr Neurol*. 2003;29(5):425–429.
19. MacDonald JT. Treatment of juvenile migraine with subcutaneous sumatriptan. *Headache*. 1994;34(10):581–582.

20. Linder SL. Subcutaneous sumatriptan in the clinical setting: the first 50 consecutive patients with acute migraine in a pediatric neurology office practice. *Headache*. 1996;36(7):419–422.

21. Ahonen K, Hamaleinen L, Rantala H, Hoppa K. Nasal sumatriptan is effective in treatment of migraine attacks in children: A randomized trial. *Neurology*. 2004;62(6):883–887.

22. Linder SL, Dowson AJ. Zolmitriptan provides effective migraine relief in adolescents. *Int J Clin Pract*. 2000;54(7):466–469.

23. Evers S, Rahmann A, Kraemer C, Kurlemann G, Debus O, Husstedt IW, et al. Treatment of childhood migraine attacks with oral zolmitriptan and ibuprofen. *Neurology*. 2006;67 (3):497–499.

24. Anttila P. Tension-type headache in childhood and adolescence. *Lancet Neurol*. 2006;5(3):268–274.

25. Steiner TJ, Lange R, Voelker M. Aspirin in episodic tension-type headache: placebo-controlled dose-ranging comparison with paracetamol. *Cephalalgia*. 2003;23(1):59–66.

26. Lewis DW. Headaches in children and adolescents. *Curr Probl Pediatr Adolesc Health Care*. 2007;37(6):207–246.

27. Victor S, Ryan SW. Drugs for preventing migraine headaches in children. *Cochrane Database Syst Rev*. 2003;(4):CD002761.

28. Miller GS. Efficacy and safety of levetiracetam in pediatric migraine. *Headache*. 2004;44(3):238–243.

29. Pakalnis A, Kring D. Zonisamide prophylaxis in refractory pediatric headache. *Headache*. 2006;46(5):804–807.

30. Lewis DW, Diamond S, Scott D, Jones V. Prophylactic treatment of pediatric migraine. *Headache*. 2004;44(3):230–237.

31. Hershey AD, Powers SW, Bentti AL, Degrauw TJ. Effectiveness of amitriptyline in the prophylactic management of childhood headaches. *Headache*. 2000;40(7):539–549.

32. Caruso JM, Brown WD, Exil G, Gascon GC. The efficacy of divalproex sodium in the prophylactic treatment of children with migraine. *Headache*. 2000;40(8):672–676.

33. Serdaroglu G, Erhan E, Tekgul H, Oskel F, Erermis S, Uyar M, et al. Sodium valproate prophylaxis in childhood migraine. *Headache*. 2002;42(8):819–822.

34. Lewis D, Ashwal S, Hershey A, Hirtz D, Yonker M, Silberstein S, et al. Practice parameter: pharmacological treatment of migraine headache in children and adolescents: report of the American Academy of Neurology Quality Standards Subcommittee and the Practice Committee of the Child Neurology Society. *Neurology*. 2004;63(12):2215–2224.

35. Sorge F, DeSimone R, Marano E, Nolano M, Orefice G, Carrieri P. Flunarizine in prophylaxis of childhood migraine. A double-blind, placebo-controlled, crossover study. *Cephalalgia*. 1988;8 (1):1–6.

36. Lutschg J, Vassella F. The treatment of juvenile migraine using flunarizine or propranolol. *Schweiz Med Wochenschr*. 1990;120 (46):1731–1736.

37. Sorge F, Marano E. Flunarizine v. placebo in childhood migraine. A double-blind study. *Cephalalgia*. 1985;5(suppl 2):145–148.

38. Lewis D, Winner P, Saper J, Ness S, Poberejan E, Wang S, et al. Randomized, double-blind, placebo-controlled study to evaluate the efficacy and safety of topiramate for migraine prevention in pediatric subjects 12 to 17 years of age. *Pediatrics*. 2009;123 (3):924–934.

39. Forsythe WI, Gillies D, Sills MA. Propanolol ("Inderal") in the treatment of childhood migraine. *Dev Med Child Neurol*. 1984;26 (6):737–741.

40. Ludvigsson J. Propranolol used in prophylaxis of migraine in children. *Acta Neurol Scand*. 1974;50(1):109–115.

41. Damen L, Bruijn J, Verhagen AP, Berger MY, Passchier J, Koes BW. Prophylactic treatment of migraine in children. Part 2. A systematic review of pharmacological trials. *Cephalalgia*. 2006;26(5):497–505.

42. Olness K, MacDonald JT, Uden DL. Comparison of self-hypnosis and propranolol in the treatment of juvenile classic migraine. *Pediatrics*. 1987;79(4):593–597.

43. Battistella PA, Ruffilli R, Viero F, Bendagli B, Condini A. A placebo-controlled crossover trial of nimodipine in pediatric migraine. *Headache*. 1990;30(5):264–268.

44. Gillies D, Sills M, Forsythe I. Pizotifen (Sanomigran) in childhood migraine. A double-blind controlled trial. *Eur Neurol*. 1986;25(1):32–35.

45. Sillanpaa M. Clonidine prophylaxis of childhood migraine and other vascular headache. A double blind study of 57 children. *Headache*. 1977;17(1):28–31.

46. Sills M, Congdon P, Forsythe I. Clonidine and childhood migraine: a pilot and double-blind study. *Dev Med Child Neurol*. 1982;24(6):837–841.

47. Lakshmi CV, Singhi P, Malhi P, Ray M. Topiramate in the prophylaxis of pediatric migraine: a double-blind placebo-controlled trial. *J Child Neurol*. 2007;22(7):829–835.

48. Hershey AD, Powers SW, Vockell AL, LeCates S, Kabbouche M. Effectiveness of topiramate in the prevention of childhood headaches. *Headache*. 2002;42(8):810–818.

49. Brandes JL, Saper JR, Diamond M, Couch JR, Lewis DW, Schmitt J, et al.; MIGR-002 Study Group. Topiramate for migraine prevention. A randomized controlled trial. *JAMA*. 2004;291(8):965–973.

50. Li K, Nagelberg J, Maytal J. Headaches in a pediatric emergency department: etiology, imaging, and treatment. *Headache*. 2000;40 (1):25–30.

51. Kabbouche MA, Linder SL. Acute treatment of pediatric headache in the emergency department and inpatient setting. *Pediatric Annals*. 2005;34(6):466–471.

52. Lewis DW, Qureshi F. Acute headache in children and adolescents presenting to the emergency department. *Headache*. 2000;40 (3):200–203.

53. Holroyd KA, Drew JB. Behavioral approaches to the treatment of migraine. *Seminars in Neurology*. 2006;26(2):199–207.

54. Zeltzer JK, Schlank CB. *Conquering Your Child's Chronic Pain*. New York, NY: Harper Collins; 2004.

55. McGrath PA. *Pain in Children. Nature, Assessment and Treatment*. New York, NY: Guilford Press; 1990.

56. Abu-Arafeh I, Razak S, Sivaraman B, Graham C. Prevalence of headache and migraine in children and adolescents: a systematic review of population-based studies. *Dev Med Child Neurol*. 2010;52(12):1088–1097.

57. Lipton R, Manack A, Ricci J, Manack A, Ricci JA, Chee E, et al. Prevalence and burden of chronic migraine in adolescents: results of the chronic daily headache in adolescent study (C-dAS). *Headache*. 2011;51(5):707–712.

58. Hershey AD, Powers SW, Vockell AL, LeCates S, Kabbouche MA, Maynard MK. PedMIDAS development of a questionnaire to assess disability of migraines in children. *Neurology*. 2001;57:2034–2039.

59. Stang P, Sternfeld B, Sidney S. Migraine headache in a prepaid health plan: ascertainment, demographics, physiological and behavioral factors. *Headache*. 1996;36(2):69–76.

60. Lipton RB, Stewart WF, Diamond S, Diamond ML, Reed M. Prevalence and burden of migraine in the United States: data from the American Migraine Study II. *Headache*. 2001;41 (7):646–657.

61. Stevens J, Harman J, Pakalnis A, Lo W, Prescod J. Sociodemographic differences in diagnosis and treatment of pediatric headache. *J Child Neurol*. 2010;25(4):435–440.

62. Russell MB, Iselius L, Olesen J. Migraine without aura and migraine with aura are inherited disorders. *Cephalalgia*. 1996;16:305–309.

63. Ophoff RA, Terwindt GM, Vergouwe MN, van Eijk R, Oefner PJ, Hoffman SM, et al. Familial hemiplegic migraine and episodic ataxia type-2 are caused by mutations in the Ca^{2+} channel gene CACNL1A4. *Cell.* 1996;87:544−552.

64. De Fusco M, Marconi R, Silvestril L, Atorino L, Rampoldi L, Morgante L, et al. Haploinsufficiency of ATP1A2 enocding the Na^+/K^+ pump alpha2 subunit associated with familial hemiplegic migraine type 2. *Nat Genet.* 2003;33:192−196.

65. Scher A, Terwindt GM, Verschuren WM, Kruit MC, Blom HJ, Kowa H, et al. Migraine and MTHFR C677T genotype in population-based sample. *Ann Neurol.* 2006;59:372−375.

66. Szikagyi A, Boor K, Orosz I, Szantai E, Szekely A, Kalasz H, et al. Contribution of serotonin transporter gene polymorphisms to pediatric migraine. *Headache.* 2006;46:478−485.

67. Rothner AD. Primary care management of headache in children and adolescents. *Fam Practice Certification.* 2002;24 (2):29−45.

68. Lewis DW, Ashwal S, Dahl G, Dorbad D, Hirtz D, Prensky A, et al. Practice parameter for the evaluation of children and adolescents with recurrent headaches. *Neurol.* 2002;59: 490−498.

69. Rothner AD. The evaluation of headache in children and adolescents. *Seminars in Pediatric Neurology.* 1995;2:109−118.

70. Strain JD. ACR appropriateness criteria on headache − child. *J Am Coll Radiol.* 2007;4(10):18−23.

71. Gladstein J. Headache in pediatric patients: diagnosis and treatment. *Topics in Pain Management.* 2007;22(11):1−10.

72. Headache Classification Committee of the International Headache Society (IHS). The International classification of headache disorders, 3rd edition (beta version). *Cephalalgia.* 2013;33(9): 629−808.

73. Rothner AD. Complicated migraine and migraine variants. *Curr Pain Headache Rep.* 2002;6(3):233−239.

74. Rothner AD. *Treatment of Pediatric and Adolescent Headache.* Cleveland Clinic Manual of Headache Therapy. New York, NY: Springer; 2014:209−223.

75. Lewis DW, Yonker M, Winner P, Sowell M. The treatment of pediatric migraine. *Pediatr Ann.* 2005;34(6):448−460.

76. Rach A, Andrasik F. *Psychologic aspects of childhood headache.* Childhood Headache. 2nd edition London, UK: Mac Keith Press; 2013:246−257.

77. Millichap JG, Yee MM. The diet factor in pediatric and adolescent migraine. *Pediatr Neurol.* 2003;28(1):9−15.

78. Hershey AD, Powers SW, Nelson TD, Kabbouche MA, Winner P, Yonker M, et al. Obesity in the pediatric headache population: a multicenter study. *Headache.* 2008;49:170−177.

79. Pakalnis A, Kring D. Chronic daily headache, medication overuse, and obesity in children and adolescents. *J Child Neurol.* 2012;27:577−580.

80. Guidetti V, Galli F, Fabrizi P, Giannantoni AS, Napoli L, Bruni O, et al. Headache and psychiatric comorbidity. Clinical aspects and outcome in an 8 year follow-up study. *Cephalalgia.* 1998;18:455−462.

81. Pakalnis A, Butz C, Splaingard D, Kring D, Fong J. Emotional problems and prevalence of medication overuse in pediatric chronic daily headache. *J Child Neurol.* 2007;22:1356−1359.

82. Seshia SS, Phillips DF, von Baeyer CL. Childhood chronic daily headache. A biopsychosocial perspective. *Dev Med Child Neurol.* 2008;50:541−545.

83. Osterhaus SOL. *Recurrent Headache in Youngsters: Measurement, Behavioral Treatment, Stress and Family Factors.* Amsterdam, The Netherlands: G & FK; 1998.

84. Eidlitz-Markus T, Shuper A, Gorali O, Zeharia A. Migraine and cephalic cutaneous allodynia in pediatric patients. *Headache.* 2007;47(8):1219−1223.

85. Newman LC. Why triptan treatment can fail: Focus on gastrointestinal manisfestations of migraine. *Headache.* 2013;53 (suppl):1−16.

86. Láinez MJA, García-Casado A, Gascón F. Optimal management of severe nausea and vomiting in migraine: improving patient outcomes. *Patient Related Outcome Measures.* 2013;4:61−73.

87. Green Mark W. Headaches: Psychiatric aspects. *Psychiatry for the Neurologist, Neurologic Clinics.* 2011;29(1):65−80.

88. Lewis DW, Kellstein D, Dahl G, Burke B, Frank LM, Toor S, et al. Children's ibuprofen suspension for the acute treatment of pediatric migraine in children. *Headache.* 2002;42: 780−786.

89. Hamalainen ML, Hoppu K, Valkeila E, Santavuori P. Ibuprofen or acetominaphen for the cute treatment of migraine in children − a double blinded, randomized , placebo-controlled, crossover study. *Neurology.* 1997;48:103−107.

90. Evers S. The efficacy of triptans in childhood and adolescent migraine. *Curr Pain Headache Rep.* 2013;17:342−348.

91. Tepper SJ. Opioids should not be used in migraine. *Headache.* 2012;52(suppl):30−34.

92. Tfelt-Hansen PC, Diener HC. Why should American headache and migraine patients still be treated with butalbital-containing medicine? *Headache.* 2012;52(4):672−674.

93. Leung S, Bulloch B, Young C, Yonker M, Hostetler M. Effectiveness of standardized combination therapy for migraine treatment in the Pediatric Emergency Department. *Headache.* 2013;53(3):491−497.

94. Singhi S, Jacobs H, Gladstein J. Pediatric headache: where have we been and where do we need to be? *Headache.* 2014;54 (5):817−829.

95. Battistella PA, Ruffilli R, Moro R, Fabiani M, Bertoli S, Antolini A, et al. A placebo-controlled crossover trial of nimodipine in pediatric migraine. *Headache.* 1990;30(5):264−268.

96. Battistella PA, Ruffilli R, Cernetti R, Pettenazzo A, Baldin L, Bertoli S, et al. A placebo-controlled crossover trial using trazodone in pediatric migraine. *Headache.* 1993;33(1):36−39.

97. Kelley NE, Tepper DE. Rescue therapy for acute migraine Part 3: opioids, nsaids, steroids and post-discharge medications. *Headache.* 2012;52(3):467−482.

98. Kelley NE, Tepper DE. Rescue treatment for acute migraine Part 1: Triptans, DHE, and Magnesium. *Headache.* 2012;52 (1):114−128.

99. Kelley NE, Tepper DE. Rescue treatment for acute migraine Part 2: neuroleptics, antihistamines and others. *Headache.* 2012;52(2):292−306.

100. Callahan N, Raskin N. A controlled study of dihydoergotamine in the treatment of acute headache. *Headache.* 1986;26:168−171.

101. Mathew NT, Kailasam J, Meadors L, Chernyschev O, Gentry P. Intravenous valproate sodium (depacon) aborts migraine rapidly. *Headache.* 2000;40(9):720−723.

102. Richer LP, Laycock K, Millar K, Fitzpatrick E, Khangura S, Bhatt M, et al.; the Pediatric Emergency Research Canada Emergency Department Migraine Group. Treatment of children with migraine in the emergency department: a National Practice Variation Study. *Pediatrics.* 2010;126(1):e150−e155.

103. Szyszkowicz M, Kaplan GG, Grafstein E, Rowe BH. Emergency department visits for migraine and headache: a multi-city study. *Int J Occup Med Environ Health.* 2009;22(3):235−242.

104. Kabbouche MA, Cleves C. Evaluation and management of children and adolescents presenting within an acute setting. *Semin Pediatr Neurol.* 2010;17(2):105−108.

105. Antonaci F, Voiticovschi-Iosob C, Di Stefano AL, Galli F, Ozge A, Balottin. The evolution of headache from childhood to adulthood: a review of the literature. *J Headache and Pain.* 2014;15:15.

106. Dooley J, Bagnell A. The progression and treatment of headaches in children: a ten-year follow-up. *Can J Neurol*. 1995;22: 47–49.

107. Guidetti V, Galli F. Evolution of headache in childhood and adolescence: an 8-year follow-up. *Cephalalgia*. 1998;18:449–454.

108. Kienbacher C, Wober C, Zesch HE, Hafferl-Gattermayer A, Posch M, Karwautz A, et al. Clinical features, classification and prognosis of migraine and tension-type headache in children and adolescents: a long-term follow-up study. *Cephalalgia*. 2006;26:820–830.

109. Hernandez-Latorre M, Roig M. Natural history of migraine in childhood. *Cephalalgia*. 2000;20:573–579.

110. Kori SH, Borland SW, Wang MH, Hu B, Mather NT, Silberstein SD. Efficacy and safety of MAP0004, orally inhaled DHE in treating migraines with and without allodynia. *Headache*. 2012;52(1):37–47.

111. Costa C, Tozzi A, Rainero I, Cupini LM, Calabresi P, Ayata C, et al. Cortical spreading depression as a target for anti-migraine agents. *J Headache Pain*. 2013;14:62.

112. Bigal M. BMS-927711 for the acute treatment of migraine. *Cephalalgia*. 2014;34(2):90–92.

113. Chan VW, McCabe EJ, MacGregor DL. Botox treatment for migraine and chronic daily headache in adolescents. *J Neurosci Nurs*. 2009;41(5):235–243.

18

The Psychiatric Approach to Headache

Robert B. Shulman

Department of Psychiatry, Rush University Medical Center, and Rush Medical College, Chicago, Illinois, USA

INTRODUCTION

Nowhere are the limitations of current psychiatric diagnostic nosology more apparent than at the interface with pain, especially chronic pain. The DSM-IV-Tr did not list pain as a symptom of any mood disorder, and this omission has persisted into the new DSM-V. Anxiety and depression are strikingly marginalized in the list of symptoms required to meet criteria for chronic pain disorder. This segregation of mood and pain translates poorly into clinical and neurobiological reality, where we see a great degree of co-occurrence or comorbidity in our clinical settings. Indeed, there is not only co-occurrence, but a growing body of literature has also implicated comorbid psychiatric illness as contributing to the chronification of pain disorders, especially migraine headache.[1]

Historically, while some of the earliest scholarly works on the topic of headache acknowledged the influence of psychological and behavioral factors, it was not until the late 19th century that Sigmund Freud, a migraine sufferer himself, associated concepts of psychopathology with migraine headache.[2] Harold Wolff, in the late 1930s, first applied systematic observation to the study of these associations in those with migraine who were not also seeking treatment for psychiatric illness.[3] His efforts to integrate knowledge from social and medical sciences were intended to lead to better understanding of headaches and other "psychophysiological" disorders — a foreshadowing of contemporary behavioral medicine.[4] Wolff developed the idea of a "migraine personality" characterized by "feelings of insecurity with tension manifested as inflexibility, conscientiousness, meticulousness, perfectionism, and resentment".[5] However, Wolff's notion of the migraine personality has not stood the test of time, or more scientific inquiry. In fact, an extensive review of the literature, over 300 studies of psychological factors and headache, found little empirical support for personality stereotypes.[6] What has been found, however, through further epidemiological and scientific study is that certain psychiatric disorders occur in the migraine headache population at rates greater than the general population. There appears to be more than a causal relationship, as there are certain shared underlying etiologies.

This chapter will examine issues of comorbidity between migraine headache and psychiatric illness, and the possible neurobiological foundation that underlies this connection, and address treatment concerns for when comorbidity exists. Additionally, although the notion of the migraine personality has settled by the wayside, we will address how those with certain personality disorders may cope with having headache and pain by identifying problematic cognitive patterns in the different personality styles.

MIGRAINE AND PSYCHIATRIC COMORBIDITIES, BEYOND COEXISTENCE

Migraine affects nearly 13% of the adult population in the United States.[7] It appears that psychiatric disorders co-occur in migraine patients at rates significantly higher than those in the general population, with depression and anxiety disorders being particularly common.[8] The relationship between migraine headache and psychiatric illness is complex and, for the most part, unclear. Most data on the topic are largely epidemiological in nature, while a smaller body of recent literature has examined underlying mechanisms and effects on migraine progression.

In data from a prospective longitudinal cohort study of young adults in Zurich, Merikangas and colleagues studied the course and order of onset of comorbid conditions, particularly depression and anxiety disorders with

S. Diamond, R. K. Cady, M. L. Diamond & V. T. Martin (Eds):
Headache and Migraine Biology and Management.

DOI: http://dx.doi.org/10.1016/B978-0-12-800901-7.00018-5

respect to migraine.[9] The findings revealed that the onset of anxiety disorders generally preceded that of migraine, and that the onset of depressive illness generally occurred subsequent to the development of migraine. The two possible sources of this non-random association between migraine and psychiatric illness are that they either share common underlying pathological mechanisms or could be causally related. However, Breslau and colleagues found, from data from the Detroit Area Study of Headache, a three-fold greater lifetime prevalence of major depression in migraineurs than in the general population, and significantly higher in migraine with aura than those without aura. The authors also observed a bidirectional relationship between migraine and depression, and that severe headache signaled an increased risk for major depression but not so the converse. The authors concluded that multiple epidemiological studies suggest that "the migraine—major depression association is unlikely to be a psychological reaction to the demoralizing experience of recurrent headaches and suggest shared causes, an explanation consistent with evidence that similar neurochemical abnormalities might be implicated in both conditions".[10]

So, what is this shared neurochemical link between psychiatric illness or emotion, and pain? First, the concept that emotional pain can be expressed physically is not new. Hippocrates' explanation for the causation of depression was that the condition was due to black bile (Greek: *melan chole*). Sir Henry Maudsley (1835—1918), the pioneering British psychiatrist, is quoted as saying: "The sorrow which has no vent in tears may make other organs weep".[11] Edgar Degas' (1834—1917) portrait entitled *Melancholy* (late 1860s) shows a woman, seemingly doubled over, arms folded across her abdomen with a pained expression on her face (Figure 18.1).

FIGURE 18.1   Edgar Degas (1834—1917). *Melancholy. Reproduced courtesy of The Phillips Collection, Washington, DC.*

The neurochemical link between emotion and pain has begun to be elucidated over the last 50 years as we have gained greater knowledge of brain networks and circuitry function. We know that dysregulation of serotonin (5-HT) and norepinephrine (NE) in the brain is strongly associated with depression and anxiety. Moreover, both 5-HT and NE are key modulatory neurotransmitters involved in regulating somatosensory perception via the somatosensory cortex and limbic system, as well as the descending periaqueductal gray matter tracts in the spinal cord. Those, in turn, stimulate endogenous opioid receptors in the lamina of the dorsal horn of the spinal cord, thus modulating the ascending pain fiber signals to the brain.[12] However, it appears the story goes beyond the monoamines. Functional magnetic resonance imaging (fMRI) studies demonstrate that brain areas (such as the dorsal anterior cingulate) central to the experience of negative affect in response to physical pain also mediate the distress in response to the "pain" of social exclusion.[13] Similar structural and functional changes in the amygdala and hippocampus have been described in major depression and pain disorders.[14,15] Dysfunction of these limbic formations is believed to contribute to abnormalities in neuroendocrine, autonomic, and immune functioning that may further contribute to the generation and/or worsening of mood and pain disorders.[16]

Beyond coexistence, the complex underpinnings of mood disorders and pain can cause neuroplastic changes and alterations in gene expression resulting in central or neurosensitization, a common etiology for chronic pain, depression and anxiety disorders, and post-traumatic stress disorder.[17,18] The model is such that stress or injury act upon genetic vulnerability, resulting in dysregulation of: (1) neural circuitry or (2) neuroendocrine, autonomic, or immune systems; or (3) impact on cellular or subcellular functions such as intracellular signaling, gene transcription, and neurotrophic support (Figure 18.2). The ultimate results are systemic manifestations of pain, and neuropsychiatric emotional, cognitive, behavioral, and physical symptoms.[16] This is, of course, at the heart of the concept of the "limbically augmented pain syndrome" (LAPS) introduced by Rome and Rome in 2000.[19] The distinguishing features of LAPS include alterations in pain perception that are chronic, and treatment refractory in association with disturbances of mood, sleep, energy, libido, memory and/or concentration, behavior, and stress tolerance. Here, the normal arousal produced by acute pain becomes a pathological stress if it persists over time. There is then a failure to adapt, to turn off the mind—body's arousal hormones, resulting in the sensory and affective pain pathways becoming

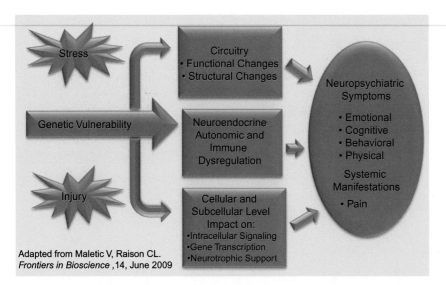

FIGURE 18.2 Stress or injury acting upon genetic vulnerability resulting in a potential myriad of complex neurobiological sequelae. *Adapted from Maletic and Raison.[16]*

sensitized. Ultimately, this results in neuroplastic changes in the corticolimbic system or LAPS.[20] Interestingly, this failure to adapt is also an explanation for how post-traumatic stress disorder (PTSD) develops when the individual, exposed to a life-threatening event, is unable to adapt neurophysiologically to turn off or tune down limbic activation. Rauch and colleagues, looking at male combat veterans using trauma script-driven imagery and PET, showed that the anterior cingulate fails to activate in response to anxiety-provoking stimuli in persons with PTSD.[21,22] This failure to inhibit may contribute to re-experiencing phenomena.

BEYOND COEXISTENCE: COMPLEX NEUROBIOLOGICAL UNDERPINNINGS OF MOOD DISORDERS AND PAIN

Some of the latest models of the effects of chronic stress in pain disorders and depression include discussion of the effects of proinflammatory cytokines. In this model, stress begets a cascade of intracellular events beginning with altered tryptophan metabolism, leading to the activation of microglia support cells, which in turn stimulate NMDA receptors, which result in a decline of neurotrophic support, disrupted neural plasticity, and eventual apoptosis (Figure 18.3).[21]

FIGURE 18.3 The effects of stress altering tryptophan metabolism.

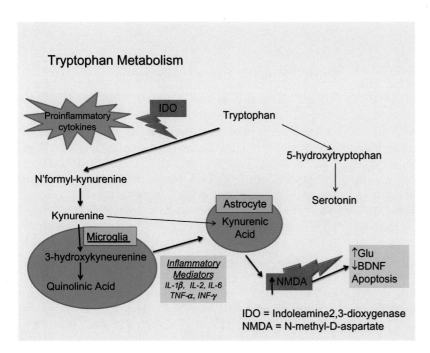

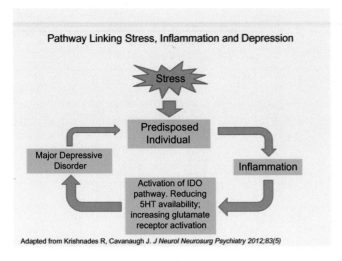

FIGURE 18.4 Stress and inflammation can lead to depression. *Adapted from Krishnades and Cavanaugh.*[23]

Affecting an array of physiological processes in the central nervous system, this may play a significant role in neurosensitization and the chronification of pain and mental illness disease states (Figure 18.4).[23]

PREVALENCE OF PSYCHIATRIC COMORBIDITIES IN MIGRAINE PATIENTS

This text should be read in consultation with Table 18.1.

Mood Disorders

Psychiatric disorders co-occur in migraine patients at rates significantly higher than those observed in the general population, with depression and anxiety disorders being particularly common.[24] In population studies, migraine sufferers are between 2.2 and 4.0 times more likely to suffer major depressive disorder than non-migraineurs, and are also at higher risk for suicide attempts, regardless of depression status.[25] Depression appears to be more common in migraine patients than in any other chronic pain conditions, occurring in 28% of migraineurs,[26] with rates of depression and suicide attempts higher in migraine with aura than in migraine without aura.[27] Although research is limited regarding migraine and bipolar disorder, evidence suggests a similar finding — that migraine patients are three times more likely to suffer from a bipolar spectrum disorder than are non-migraineurs, with this association strongest for migraineurs with aura.[28]

TABLE 18.1 Prevalence of Mental Health Disorders in the United States

Disorder	No. of Adults in US (millions)	% Adult population
Any mood disorder	20.9	9.5
Major depressive disorder	14.8	6.7
Dysthymia	3.3	1.5
Bipolar disorder	5.7	2.6
Schizophrenia	2.4	1.1
Any anxiety disorder	40	18.1
Panic disorder	6	2.7
OCD	2.2	1.0
PTSD	7.7	3.5
Generalized anxiety disorder	6.8	3.1
Agoraphobia	1.8	0.8
Specific phobia	19.2	8.7

Source: National Institute of Mental Health website: www.nimh.nih.gov (April 2014).

Anxiety Disorders

The prevalence of anxiety disorders in migraineurs is almost twice that of major depression, ranging from 51% to 58%.[29] Compared with those without migraine, individuals with migraine are at four to five times greater risk for generalized anxiety disorder (GAD), and five times greater risk for obsessive-compulsive disorder (OCD), and are three to ten times more likely to suffer from panic disorder.[28] As stated earlier, the onset of anxiety often precedes migraine, with the risk of depression increasing subsequently.

Post-Traumatic Stress Disorder

Post-traumatic stress disorder (PTSD) has been studied less frequently, although recent work by Peterlin and colleagues showed that in tertiary, clinic-based studies, 22–30% of headache sufferers fulfilled PTSD criteria.[30] However, in a veteran cohort survey the prevalence of PTSD was even greater than that found in tertiary care clinics, with almost 50% of those with migraine fulfilling criteria for PTSD.[31] Indeed, PTSD was found to be a risk factor for migraine chronification in those migraineurs with depression, and the presence of PTSD in migraine sufferers is associated with significantly greater level of disability as compared with those without PTSD.

One can look further at the relationship between the presentation of pain in primary care and the prevalence of comorbid psychiatric illness. Kurt Kroenke

TABLE 18.2   Relationship Between Common Physical Symptoms and Psychiatric Disorders

Symptom	% all PCP patients with this symptom	% with symptom and anxiety disorder	% with symptom and mood disorder
Joint or limb pain	59%	24%	34%
Back pain	41%	28%	38%
Headache	36%	28%	40%
Chest pain	21%	33%	46%
Abdominal pain	19%	31%	43%

Reproduced from Kroenke et al.,[32] Archives of Family Medicine 1994; 3:774-9, with permission.

examined this relationship in a prospective study occurring in multiple primary care clinics, and found a strong relationship between patients presenting with a chief complaint of pain and a co-occurring psychiatric condition – primarily depression or anxiety.[32] Of those presenting with a complaint of headache (more than one-third of all patients with a complaint of pain), 28% had a co-occurring anxiety disorder while 40% suffered a mood disorder (Table 18.2).

In fact, the chronic headache/pain patient may share symptoms with those with major depression. These symptoms include changes in appetite and weight, loss of energy, sleep disturbance, poor concentration, and psychomotor slowing. The predictors of depression in the pain patient are the frequency that severe pain is experienced, the number of pain areas of the body, pain intensity, and functional disability. Also predictive are psychological factors of low self-esteem, poor coping styles, and poor problem-solving.

PSYCHIATRIC ILLNESS: MAKING THE CORRECT DIAGNOSIS

Given the marked prevalence of psychiatric disorders in the community, especially elevated in the headache population as well as in those who present to primary care with the simple complaint of pain, it makes sense to acquaint oneself with the various psychiatric disorders and their treatments. Indeed, primary care physicians are asked to address psychiatric disorders increasingly for various reasons, including insurance network considerations, managed care, and lack of access to psychiatric specialists – and they have responded in their willingness to treat these disorders.[33] With the development of better tolerated medical treatments for psychiatric illness, more non-psychiatrists have been willing to treat even bipolar disorder in its less severe forms. This has allowed many patients to gain access to effective treatments that may never have happened before the

TABLE 18.3   Pain and Psychiatric Comorbidity

Mood disorders	Anxiety disorders
Depressive disorders	Generalized anxiety
• Major depression	Panic disorder
• Dysthymia	
• Adjustment reaction	Social anxiety and phobias
Bipolar disorders	**Somatoform disorders**
• BP I	Pain, bowel and fatigue syndromes
• BP II	Psychological factors
• Cyclothymia	Neurasthenia
Substance use disorders	**Post-traumatic stress disorder**
Alcohol – RXs	History of abuse as a child
Opiates – OTCs	Traumatic life events
	Often with comorbid mood and sleep disorders and ineffective coping styles

psychopharmacology drug development boom of the past three decades. However, given this willingness of primary care providers, there also comes the responsibility of making the correct diagnosis, especially as regards the distinction between unipolar and bipolar depression (Table 18.3).

The reason to differentiate between unipolar and bipolar depression is related to the important difference in therapeutic medication approach. There is a truism in psychiatry that I try to impart to resident physicians and students, as well as in other educational opportunities – that while depressed patients may well suffer, manic patients will too often get into trouble. In fact, this was once well illustrated in a cartoon in *The New Yorker* magazine of the man standing before the judge in the courtroom: "Your Honor," he says, "now that I've swung back to depression, I'm truly sorry for what I did when I was manic." We, as clinicians, sometimes unwittingly contribute to this situation by giving patients who are depressed, anxious, or in pain antidepressant medication, resulting in triggering manic episodes or accelerating mood cycles.

It is thus important to differentiate between unipolar and bipolar depression. The unipolar depressive may present with more anxiety, appetite disturbance, irritability, anger, and physical complaints. The bipolar depressive may have fewer physical complaints, and present as more flat and empty with psychomotor retardation, hypersomnia and fatigue. Red flags for a bipolar diathesis are: younger age of onset of the condition; occurring in the post-partum period; greater number of suicide attempts; and psychosis. In all depressed patients, a careful history of illness and family history, including substance use disorders, should be obtained. Of note, 29% of bipolar patients experience three episodes of depression before the correct bipolar diagnosis is made.[34] One should always "think bipolar" in depression when there is a cycle induction or acceleration either with the addition of an antidepressant medication or when such occurs with the abrupt discontinuation of antidepressant medication.

CHOOSING THE RIGHT MEDICATION

For many years, psychiatric or centrally acting agents have been a mainstay in the treatment of headache and different chronic pain conditions. I learned early in my training that a little amitriptyline went a long way in the treatment of headache, fibromyalgia, and other neuropathic pain disorders. In fact, the analgesic response seemed to occur more quickly and at lower doses than one would expect if one were treating an episode of major depression. Subsequent data have borne out that tricyclic antidepressants (TCA), venlafaxine, duloxetine, and mirtazapine — all dual-acting serotonergic/noradrenergic agents — can modulate pain.[35] Milnacipran, also a dual-acting norepinephrine/serotonin reuptake inhibitor, is approved in the United States only for certain chronic pain conditions, although it has been used as an antidepressant for years in Europe. Although there are good data for dual-acting antidepressants, the data for the mono-mechanistic serotonin reuptake inhibitors (SRI) are conflicting, tending to do better in open-label trials and less so in double-blinded placebo-controlled studies.[36] In addition to the antidepressants, other centrally acting psychiatric drugs are used in certain pain disorders. The mood stabilizing anti-epileptic drugs (AEDs) and antipsychotics, both first and second generations (FGAs and SGAs), have been used for a variety of conditions. We now have options as we encounter a patient with headache and psychiatric comorbidity, which allow us to choose the class of drug that offers help for the pain while addressing the psychiatric condition as well, either by helping with its treatment, or at least not making it worse.

TABLE 18.4   The Antidepressants

Classical mechanisms	New mechanisms
Tricyclic and heterocyclic antidepressants (TCAs/HCAs)	SRIs: fluoxetine, sertraline, paroxetine, fluvoxamine, citalopram, escitalopram
Monoamine oxidase inhibitors (MAOIs): • MAOI-A: phenylzine, isocarboxazid, tranylcypromine • MAOI-B: selegiline transdermal	SRI(+): vilazodone (5HT1a partial Ag), vortioxetine (multiple MOAs)
	DNRI: buproprion
	SNRIs: venlafaxine, duloxetine, o-desmethylvenlafaxine, milnacipran, levomilnacipran
	SARI: nefazodone, trazodone
	NaSA: mirtazapine
	NRI: atomoxetine (ADHD)

SRI(+), serotonin reuptake inhibitor plus secondary mechanisms of action; DNRI, dopamine/norepinephrine reuptake inhibitor; SNRI, serotonin/norepinephrine reuptake inhibitor; SARI, serotonin receptor (5-HT2) antagonist with reuptake inhibition; NaSA, noradrenergic and serotinergic antidepressant; NRI, norepinephrine reuptake inhibitor.

The Antidepressants

The group of medications known as the antidepressants have multiple indications for use in the United States, predominantly for mood and anxiety disorders, but they are also used for pain, sleep disorders, impulse dysregulation disorders, autism, functional bowel disorders, and others. Depending on the mechanism of action, an effect can either be therapeutic or a side effect. This is, of course, another of those clinical pearls passed on through generations; that one person's side effect is another's mechanism of action. Sometimes, medications will be chosen for secondary mechanism effects such as sedation, antimuscarinic anticholinergic effects, or activation of other circuitry.

At present, there are multiple classes of antidepressant medications, including the early TCAs and monoamine oxidase inhibitors (MAOI). The newer classes of antidepressants had their start with the introduction in the United States of the first selective serotonin reuptake inhibitor (SRI), fluoxetine, in January 1988. Since then, in addition to the SRIs, at least a half dozen other classes of novel agents with various mechanisms of action are available to select (Tables 18.4 and 18.5).

The Anti-Epileptic Mood Stabilizers

In addition to reducing and preventing seizures, the AEDs have broad clinical action in the central nervous system (Table 18.6). Primarily used in psychiatric disorders to treat bipolar disorder and its variants, the AEDs (primarily carbamazepine and divalproex sodium) provided the earlier alternatives to lithium in treating

TABLE 18.5 Other Uses for the Antidepressants

Approved:	Non-approved, but commonly used:
OCD — multiple SRs & clomipramine	Eating disorders
Panic disorder — multiple SRIs & venlafaxine GAD — multiple SRIs, venlafaxine & duloxetine	Impulse control
	Agitation in dementia
Social anxiety — multiple SRIs & venlafaxine PTSD — multiple SRIs	Autistic disorders
	Irritable bowel
PMDD — multiple SRIs	Pain disorders
Nicotine dependence — bupropion	Premature ejaculation
Diabetic neuropathy & fibromyalgia — duloxetine, mulnacipran (FM)	ADHD
	Migraine headache

TABLE 18.6 Main Indications for AEDs — Approved or General Acceptance of Utility[36]

AED	Epilepsy	Bipolar disorder	Neuropathic pain	Migraine
Carbamazepine	✓	✓	✓	
Felbamate	✓			
Gabapentin	✓		✓	✓
Lamotrigine	✓	✓	✓	
Leviteratetam	✓		✓	
Oxcarbazepine	✓		✓	
Pregabalin	✓		✓	
Tiagabine	✓			
Topiramate	✓			✓
Valproate	✓	✓	✓	✓
Vigabatrin	✓			
Zonisamide	✓			

Adapted from Landmark.[37]

mania. Once divalproex sodium was approved for use to treat mania, the manufacturers of other AEDs looked at their agents for similar indications. In addition to epilepsy and bipolar disorder, the AEDs are also used to treat migraine, neuropathic pain, schizophrenia, anxiety, tremor, and numerous other conditions.[37] Interestingly, these disorders are possibly all related to disturbed excitability in the central nervous system and may share pathophysiological processes. The main targets for AEDs in the synapses include enhancement of GABAergic inhibitory neurotransmission, decrease in excitatory neurotransmission directly or via inhibition of voltage-gated ion (sodium and calcium) channels, or interference with intracellular signaling pathways.[38]

The common result of these interventions is a decrease in neuronal excitability. Although each AED has a primary mechanism of action, many have secondary, tertiary, or even quaternary mechanisms of action. Several of these agents have been very important in the treatment of bipolar disorder, and can be considered strongly when migraine and bipolar disorder are comorbid in a patient.

The Antipsychotics

Antipsychotics, formally the first generation (FGA) and more recently the second generation (SGA), have been used in pain disorders; however, systematic studies have been few. Fishbain and colleagues did a review of SGAs in pain disorders (10 scholarly articles in all) and found that the data were generally consistent and that some of the SGAs may have an analgesic effect.[39] However, there were few double-blind placebo-controlled trials, and many of the reports/studies involved less than 50 patients. Still, given the overlay of psychiatric comorbidity with headache and pain disorders, the use of antipsychotics, particularly the SGAs, makes sense, as we will see.

Acute psychotic episodes are associated with neuroinflammation and elevations of cytokines such as IL-1, IL-6, TNF-α, and INF-γ. These inflammatory biomarkers are released by microglia, which are rapidly activated by psychosis and mediate brain tissue damage during psychosis.[40] SGAs can suppress induction of these proinflammatory cytokines, and thus may be neuroprotective in that regard. Interestingly, haloperidol, an FGA, does not appear to have the same effect.

The use of SGAs extends beyond treating only conditions that exhibit psychosis (Table 18.7). They are, for the most part, primarily D2 and 5-HT2 antagonists (with the exception of aripiprazole, which is a partial agonist at D2) (Table 18.8), but that is where the comparison stops and most SGAs have tertiary, quaternary, or beyond, mechanisms of action. All are approved for use in schizophrenia, and many are also approved for use in bipolar disorder, mostly as an anti-manic agent. Some have an approved use in depressive disorders. Additionally, many of these agents are used for other conditions or symptoms, including anxiety and anxiety disorders, PTSD, insomnia, agitation, aggression, impulse dyscontrol syndromes, OCD, and pain disorders.[41]

MIGRAINE, PAIN, AND SUICIDE

Suicide was the fourth leading cause of death in adults, ages 18 to 65, in the United States in 2007, and

TABLE 18.7   Second-Generation Antipsychotic Indications for Use

	Schizophrenia	Bipolar manic/mixed	Bipolar depressed	Major depression as adjunct	Other
Aripiprazole	✓	✓		✓	Agitation (IM)
					Depot
Asenapine	✓	✓			
Clozapine	✓				Suicide prevention in schizophrenia
Iloperidone	✓				
Lurasidone	✓		✓		
Olanzapine	✓	✓	✓ (with fluoxetine)	Treatment-resistant depression (w/fluoxetine)	Agitation (IM)
					Depot
Palliperidone	✓				Schizoaffective d/o
					Depot
Quetiapine	✓	✓	✓	✓ (XR only)	
Risperidone	✓	✓			Depot
Ziprasidone	✓	✓			Agitation (IM)

Depot: long-acting injectable formulation available in the US.

TABLE 18.8   Second-Generation Antipsychotic Mechanisms of Action and Receptor Binding Capacities

Generic	D2	5-HT1A	5-HT1B	5-HT2A	5-HT2C	5-HT7	M1	H1
Aripiprazole	+++ PA	+++ PA	+	++	++	+++	−	++
Asenapine	+++	++ PA	+	++++	++++	++++	+	+++
Clozapine	+	+ PA	+	++	++	++	+++	+++
Iloperidone	+++	++ PA	+	+++	+	++	−	++
Lurasidone	+++	+++ PA		++	+	++++	−	−
Olanzapine	++/+++	−	+	+++	++	+	++	+++
Palliperidone	+++	+ PA	+	++++	++	+++	−	++
Quetiapine	+	+ PA	+	++	+	++	++	+++
Risperidone	+++	+ PA	+	++++	++	+++	−	++
Ziprasidone	+++	++ PA	+++	++++	++	+++	−	++

++++, very strong affinity; +++, strong affinity; ++, moderate affinity; +, weak affinity; −, no affinity; PA, partial agonist.
Adapted from Stahl's Essential Psychopharmacology.[41]

the tenth leading cause over all ages.[42] In addition, it was the third leading cause of death for young people, ages 15 to 24, accounting for 14.4% of all deaths in 2009. The annual adjusted rate of suicide mortality is 11.8 per 100,000 population, while the rate of non-serious suicide attempts is approximately 1800 per 100,000. While most research has focused on psychiatric disorders, there is a small but growing literature suggesting that chronic pain is associated with suicidal behaviors. Ilgen and colleagues looked at non-cancer pain conditions and risk of suicide in a study that extracted data from the National Death Index and treatment records from the Department of Veterans Affairs Healthcare System.[43] The analyses examined the association between baseline clinical diagnoses of pain-related conditions, namely arthritis, migraine, back pain, neuropathy, headache or tension headache, fibromyalgia, and psychogenic pain, and subsequent suicide deaths in a predominantly male population during fiscal years 2006 through 2008 ($n = 4,863,086$, 91.7% male, 8.3% female). Controlling for age, sex, medical conditions, and psychiatric diagnoses, three pain conditions were significantly associated with an increased risk of suicide. The greatest risk was psychogenic pain (1.58 hazard ratio $P < 0.001$) followed by migraine (1.34 hazard ratio $P < 0.005$), and back pain (1.13 hazard ratio $P < 0.001$). Of note, the predominant methods of suicide in this retrospective study were death by firearms (67.9%), followed by poisoning (16.6%). These figures most likely represent the primarily male study population in the Veterans Affairs Healthcare System. However, women represented 30% of the migraine population compared with 8.3% overall. In the migraine population, the method of suicide was more evenly distributed, with 44.1% deaths by firearms and 38.6% deaths by poisoning. The authors concluded that in this older male population, ambiguity related to causes and treatments of psychogenic pain may be the core part of its relationship to suicide, and some biological mechanisms for pain conditions, such as migraine (i.e., serotonergic dysfunction), may be associated with increased risk of suicide, even in subthreshold psychiatric disorders.[43]

Specifically considering migraine and suicide, Breslau and colleagues noted an increased risk of suicide attempts in persons with migraine over a 2-year period as compared with controls. The odds ratio, adjusting for gender, psychiatric disorder, and previous suicide attempt at baseline was 4.43.[44] The authors also found that patients with migraine with aura and major depressive disorder had significantly higher rates of suicide attempts and suicidal ideation than the combined rates of those with just one of these two disease states. However, they found that migraine without aura alone was not associated with increased risk of suicide attempts or suicidal ideation. Fuller-Thomson and colleagues found certain risk factors for suicidal ideation associated with gender among Canadian migraineurs.[45] For females, younger age, being unmarried, being poor, and having greater limitations in activities are associated with greater suicidal ideation. Male migraineurs share three of these risk factors, excluding only being poor. Lastly, severity and chronicity of pain is associated with emotional turmoil and hopelessness.[46] Given the increased risk for suicide attempts and suicidal ideation in chronic pain and migraine headache specifically, it is clear that it is important to screen not only for psychiatric comorbid conditions, but also for the presence of suicidal thoughts alone.

PERSONALITY AND COPING STYLES IN MIGRAINE HEADACHE

The relationship between migraine and personality was historically discussed much more than it was studied. The notion of "migraine personality" first arose from clinical observations. Early studies tended to be of highly selected patients in specialty clinics, and thus had methodology and selection bias. The researchers tended to have basic assumptions that migraineurs shared common personality traits, and that these traits were enduring, and distinguished migraineurs from controls. These traits included such behavioral styles as deliberate, hesitant, insecure, detailed, perfectionistic, sensitive to criticism, deeply frustrated emotionally, rigid, compulsive, competitive, and chronically resentful. Additionally, migraineurs were said to lack warmth, have difficulty making social contacts, be unable to delegate responsibility, protect themselves from intimacy, have troubled relationships with loved ones, be meticulous in appearance, be hostile or irritable, and try to dominate their environment.[3,47,48]

Really, a cacophony of behavioral and personality styles, it seems. These studies, however, were fraught with diagnostic inconsistencies and selection bias, and did not control for any psychiatric comorbidity. Later studies found that data often conflicted and were rarely robust. Migraineurs, if anything, may tend to be higher on scales of neuroticism compared with non-migraineurs with elevations on the MMPI (Minnesota Multiphasic Personality Inventory) areas of depression, hysteria, and hypochondriasis.[49]

Our discussion will focus on personality styles and how they may correlate with problematic cognitive patterns in chronic pain and headache.

Author's note: The following discussion of personality and behavior, while based diagnostically on the DSM, will use terms and explanation based on years of training ... with my father, Bernard H. Shulman, MD, the original psychiatric consultant to the Diamond Headache Clinic's inpatient treatment unit in Chicago. He is the author of many scholarly papers, articles of original thought, and books mostly circumscribed around the Individual Psychology theories of Alfred Adler, and co-founder of the Adler School of Professional Psychology in Chicago. Dr Bernard, as he was fondly known around the office, finally announced on his 91st birthday that he was done seeing patients and retired. A benevolent figure and mentor to generations of clinicians, educators, and students around the world through his writings, lectures, and leadership of the *International Association of Individual Psychology* (1970–1982), he remains adored by his children and grandchildren. And while, like Albert Camus, I owe much of what I know most surely about morality and the duty of man to my years of goalkeeping for various football clubs, I most certainly owe the rest, and more, to my father, Dr Bernard. This opus is dedicated to him.

The term "psychodynamic" refers to patterns of perceiving, relating to, and thinking about the environment and oneself that are exhibited in a wide range of social and personal contexts. In this way, personality can be understood as a way of dealing with life, a disposition, and a stance toward the world. Neurosis, on the other hand, is a way of dealing with a life situation, a coping style, and a way of arranging life how one wants it (thus giving purpose to the neurotic behavior). The personality disorders are broken into clusters reflecting various commonalities amongst them. Cluster A comprises those personalities that appear odd or eccentric and seem to live according to some private logic. Cluster B includes those personality disorders that are dramatic, highly emotional and erratic, and often break rules. Lastly, Cluster C personalities tend to be most fixed in their neurotic behavior. Herein follows a brief primer of the personality disorders describing their behavioral styles and methods. We will then examine how each affects the relationship between the person and pain.

Cluster A Personality Disorders

The Cluster A personality disorders are three: Paranoid, Schizoid, and Schizotypal personalities. The paranoid personality is a pervasive and unwarranted tendency to interpret the actions of people as deliberately demeaning or threatening. The schizoid personality is a pervasive pattern of indifference to social relationships and a restricted range of emotional experience and expression. Finally, the schizotypal personality is a pervasive pattern of peculiarities of ideation, appearances and behavior, and deficits in interpersonal relationships.

Paranoid Personality Disorder

The behavioral traits of the paranoid personality are that they may be distrustful, argumentative, easily slighted, and may bear grudges. They are critical of others, rigid, defensive, and unemotional. They may also be moralistic and make mountains out of molehills. Ultimately, they are keenly aware of power and rank. Their *self* view is that of an innocent victim, a good person surrounded by bad people. Their *fictitious goal* in life is to arrange to not be at fault. Their *method* is to project blame onto others, and their *impairments* include anxiety, occupational difficulties, and problems with authority figures. There is a prevalence of males in this personality disorder.

Schizoid Personality Disorder

The schizoid personality is a loner, cold, and aloof. They may seem self-absorbed or absent-minded, may lack social skills, and neither desire nor enjoy close relationships. Their *self* view is that they have no place in the world and may be easily overwhelmed by stimuli. Their *fictitious goal* is to reach an ivory tower where they are safe. Their *method* is one of safeguarding, avoidance and distancing, and being indifferent and aloof. Their *impairments* are that they lack social skills and rarely experience strong emotions.

Schizotypal Personality Disorder

Individuals with this personality disorder will be suspicious and display odd beliefs and referential thoughts. The schizotypal personality may report clairvoyance or claim telepathic powers, a "sixth sense." They see their *self* as not competent in the consensual world, and thus their *fictitious goal* is to find a safe place. Their *method* is to engage in magical thinking and eccentric behavior, making up their own rules. Their *impairments* include increased social anxiety, inappropriate affect, and decreased interpersonal relatedness. They may experience transient psychosis when under stress. The prevalence of schizotypal personality disorder is increased in first-degree relatives of schizophrenics.

Cluster B Personality Disorders

There are four Cluster B personality disorders: Antisocial, Borderline, Histrionic, and Narcissistic. The antisocial personality demonstrates a pervasive pattern of irresponsible and antisocial behavior. The borderline personality is characterized by a pervasive pattern of instability of self-image, interpersonal relationships, and mood. The histrionic personality shows a pervasive pattern of excessive emotionality and attention-seeking. Lastly, the narcissistic personality is characterized by a pervasive pattern of grandiosity, hypersensitivity to evaluation by others, and lack of empathy.

Antisocial Personality Disorder

One of the two Cluster B personality disorders that appear to have a genetic component, the antisocial personality often begins with signs in childhood, including lying, stealing, truancy, vandalism, fighting, running away from home, and cruelty. Adult signs are the failure to honor obligations, failure to conform to norms, and repeated antisocial acts. The antisocial personality appears unable to tolerate boredom and may be depressed. Individuals may also engage in domestic violence or criminal acts. Their *self*-image is that of a rule-breaker where life is hostile, and they thrive on defiance. Their *fictitious goal* is to successfully defy the world as their highest obligation is to the self, and rules prevent getting their needs met. Their *method* is that "might makes right," and they have a total lack of empathy for others.

Three to four times more common in males, the *impairments* of the antisocial personality are dysphoria, substance abuse, and inability to sustain lasting close, warm, responsible relationships.

Borderline Personality Disorder

The borderline personality is the other personality disorder that has genetic inheritability. With a prevalence of 2% of the population, it accounts for 10% of patients seen in outpatient mental health centers. It is also five times more common in first-degree relatives of those with borderline personality disorder. The characteristics of the borderline personality include a marked and persistent identity disturbance, chronic feelings of emptiness and boredom, and intense unstable personal relationships. The borderline tends to have difficulty tolerating being alone, and will fear abandonment. They are highly impulsive, and may engage in self-mutilating behavior, have recurring suicidal threats, and manipulate others to meet their immediate needs. The borderline also tends toward having an intense unstable affect mood, display inappropriate anger, have perceptual distortions, and under great stress may depersonalize. They see the *self* as justified; since they feel intolerably bad, they are entitled to go by impulse instead of common sense, and entitled to soothe themselves. Their *fictitious goal* is to do whatever they want as they cannot be happy by how others do it. Their *methods* are protean, and include splitting, primitive idealization, projective identification, denial, and devaluation. The borderline personality is not a cooperator. Their *impairments* include affective instability, anxiety and panic, and engaging in self-harmful behaviors. These impairments may cause significant interference in social or occupational functioning. Because the borderline uses others to meet their needs, they can often be the "problematic patient" that medical offices dread.

Histrionic Personality Disorder

Those with histrionic personality want to be the center of attention, and are uncomfortable in situations where they are not. They can display rapidly shifting and shallow expressions of emotions, and their behavior is over-reactive and intensely expressed. They may crave novelty, stimulation, and excitement, and may constantly demand reassurance due to feelings of helplessness or dependence. The histrionic personality is over-concerned with physical attractiveness, and has little or no tolerance for frustration or delayed gratification. Their manner is seductive and flamboyant, though their expressionistic speech will most likely be devoid of depth and lack detail. The *self* for the histrionic personality is that "I am sensitive and everyone should admire and approve of me." Their *fictitious goal* is to be the center of admiring attention without really working for it, like a child. Their *methods* include hypersensitivity, covert manipulation, emotional displays, and claims to be the center of attention. More common in females and in first-degree relatives, the histrionic personality's *impairments* include being controlling or dependent in relationships, impressionable and easily influenced, and overly trusting. Of note in our discussion here is that the histrionic personality may somaticize a great deal as a means of gaining attention.

Narcissistic Personality Disorder

The last of the Cluster B personality disorders, the narcissistic personality is preoccupied with fantasies of unlimited power, success, brilliance, and beauty. With fragile self-esteem, they have chronic feelings of envy for those perceived as being more successful. Like the histrionic, there is the exhibitionist need for constant attention and admiration, and they may constantly fish for compliments. When criticized, the narcissistic personality may react with rage, shame, or humiliation. In romantic relationships, the partner is often treated as an object to bolster their own self-esteem. The narcissist is always measuring: "Am I better, or is this good for me?" The *self* for the narcissist is special and unique, entitled to extraordinary privilege whether earned or not. They have the *fictitious goal* of superiority, and are owed admiration and privilege. The *method* is one of self-enhancement and deprecation of others. Their *impairments* include anger and rage, often depression, disturbed relationships, and a lack of empathy.

Cluster C Personality Disorders

Cluster C personality disorders comprise a divergent group with very fixed behavioral styles. The avoidant personality demonstrates a pervasive pattern of social discomfort, timidity, and fear of negative evaluation. The dependent personality displays a pervasive pattern of submissive behavior. Lastly, the obsessive-compulsive personality is where there is a pervasive pattern of perfectionism and inflexibility.

Avoidant Personality Disorder

The avoidant personality is easily hurt by criticism and devastated by the slightest hint of disapproval. Generally unwilling to enter into relationships, they will avoid activities that involve significant interpersonal contact. The avoidant personality may yearn for acceptance, and is usually distressed by the lack of ability to relate comfortably with others. They view the *self* as feeling inferior and unacceptable, as having deficits, and are frightened by rejection. Their *fictitious goal* is to avoid humiliation — as life is unfair, they must be vigilant. The *method* is avoidance, and their *impairments* are depression, anxiety, and social phobia.

Dependent Personality Disorder

The dependent personality is unable to make everyday decisions without an excessive amount of advice and reassurance from others. They will even allow others to make very important decisions for them. They may feel uncomfortable and helpless when alone, and will go to great lengths to avoid such situations. They can be devastated when relationships end and, like the borderline personality, may be preoccupied with fears of being abandoned. Easily hurt by criticism, the dependent personality will occupy a subordinate position and try to get others to like them. The *self* for the dependent personality is small and weak, not competent, and overwhelmed by life. Their *fictitious goal* is that others are here to take care of them, as they cannot do it themselves. Their *methods* are to use weakness and childlike appeal, self-abasement, and to seek service and protection. More common in women, their *impairments* are a lack of confidence, and they will suffer anxiety and depression if a dependent relationship is threatened.

Obsessive-Compulsive Personality Disorder

The final personality disorder, the obsessive-compulsive, constantly strives for perfection. Their unattainable standards frequently interfere with completion of tasks and projects. The obsessive-compulsive personality will tend to be preoccupied with logic, overly conscientious, moralistic, scrupulous, and judgmental of self and others. They may be intolerant of emotional behavior in others, and decisions are avoided out of fear of making a mistake. There is a strong need to be in control, and their relationships are formal and serious. The obsessive-compulsive personality sees the *self* as responsible if something goes wrong. They will not take risks because they are not sure of the results, as they want to control every life situation. The *fictitious goal* for the obsessive-compulsive personality is perfect control for perfect security. Their *methods* are concern with detail and demand for order as they suppress emotion and use logic. More common in men, their *impairments* are depression and anger, or rumination if not in control.

Influence of Axis II Personality Disorders in Headache

There is no one personality disorder more commonly associated with headache, although we feel that the Cluster B personalities may be more likely to present for treatment. Identification of a personality disorder, or at least a predominance of one personality's traits, should heighten the clinician's awareness of the patient's needs and coping repertoire as well as perhaps the meaning of illness to the patient. However, personality type has been employed mostly in a pejorative sense in most clinical settings to imply that a patient is difficult to treat and does little to facilitate or complement treatment. Looking at the clusters as a whole in headache, we rarely see Cluster A personality disorders, as they tend not to trust the healthcare provider. Cluster C personality disorders will tend to want to actively control treatment or even sabotage it more passively. Cluster B personality disorders are highly reactive, emotional and dramatic, intolerant of discomfort, and unable to delay gratification. This group is more likely to be drug-seeking, and their defensive styles may make treatment very difficult.

Raphael Leo identified eight problematic cognitive patterns in patients with chronic pain[50]:

1. *Catastrophizing*: tendency to view and expect the worst. "I am doomed to have pain and misery forever."
2. *Helplessness*: belief that nothing one does matters. "My doctor says I should exercise, but I know it won't help."
3. *Help-rejecting*: rejection of the efforts of well-meaning others as a means of expressing anger, securing attention, or ongoing support. "I had problems with every medicine you gave me."
4. *Labeling*: ascribing the behavior of a person to a characteristic or nature of a person. "The medicine the doctor gave didn't help me; what a quack!"
5. *Magnification*: exaggerating significance of a negative event. "My pain got worse and I had to leave work early; I am totally disabled."
6. *Overgeneralization*: expanding one adverse event to many or all aspects of one's life. "If this medicine doesn't help me, nothing will."
7. *Personalization*: interpretation that an event or setback is indicative of something about oneself. "Because of the pain, I am a worthless failure."
8. *Selective abstraction*: propensity to attend selectively to negative aspects of one's life while ignoring satisfying or rewarding aspects. "Everything that happens in my life is bad."

Given our understanding of personality styles, we can now begin to attribute Leo's schema of problematic cognitive patterns to each of the personality disorders (Tables 18.9—18.11). In a broad sense, we see only the paranoid personality from Cluster A as demonstrating some of these patterns. The schizoid and

TABLE 18.9 Problematic Cognitive Patterns in Cluster A Personality Disorders

	Paranoid	Schizoid	Schizotypal
Catastrophizing	+++	−	−
Helplessness	+++	−	−
Help-rejecting	++	−	−
Labeling	++++	−	+
Magnification	+	−	+
Overgeneralization	+	+	+
Personalization	−	−	−
Selective abstraction	++++	+	++

++++, very strong; +++, strong; ++, moderate; +, weak; −, none.

TABLE 18.10 Problematic Cognitive Patterns in Cluster B Personality Disorders

	Antisocial	Borderline	Histrionic	Narcissistic
Catastrophizing	+	+++	++++	+++
Helplessness	+	++	++	+
Help-rejecting	++++	++++	++++	+++
Labeling	++++	++++	++	+++
Magnification	++	++++	++++	+++
Overgeneralization	++	++++	+++	++
Personalization	+	++++	+++	+
Selective abstraction	+++	++++	+++	+++

++++, very strong; +++, strong; ++, moderate; +, weak.

TABLE 18.11 Problematic Cognitive Patterns in Cluster C Personality Disorders

	Avoidant	Dependent	Obsessive-compulsive
Catastrophizing	−	++	−
Helplessness	+	++++	−
Help-rejecting	−	+	++
Labeling	−	−	−
Magnification	−	−	+
Overgeneralization	+	−	+
Personalization	+	+	+
Selective abstraction	+	+	+

++++, very strong; ++, moderate; +, weak +, −, none.

schizotypal individuals are probably too untrusting of, or disconnected from, the healthcare provider to even present for treatment. They employ other methods in the service of their fictitious goals. On the other hand, the paranoid personality may engage in several of the problematic cognitive patterns, especially selective abstraction and labeling, as they project blame to avoid being at fault. Cluster C personality disorders also have limited engagement with Leo's problematic cognitive patterns. Like the schizoid and schizotypal personalities, the avoidant personality makes little use of these problematic cognitive patterns, as they suffer anxiety and want no attention drawn to them. The obsessive compulsive personality may reject help but is otherwise too caught up in arranging and controlling the world to achieve security. And the dependent personality utilizes the problematic cognitive pattern of helplessness to a great degree, and maybe catastrophizes to a lesser extent. This is done, of course, in the service of having others care for them.

It is clear with the Cluster B personality disorders that we can see problematic cognitive patterns played out in clinical settings. The antisocial personality may not catastrophize, personalize, or project helplessness much, but they will reject help, label and selectively abstract to meet their need to defy the world. This personality is very likely to use pain as a means to obtain and misuse prescription drugs. The borderline personality is the one that probably uses most, if not all, of the eight problematic cognitive patterns, and each one to a great extent as they desperately want to feel good. This personality feels entitled to self-soothe and will attempt to do so by any means available, including misusing prescription medications. The selective abstraction, among other problematic patterns, can be especially daunting to the clinician, and firm, clear explanation and instruction need to be given to the borderline patient so that there are no misunderstandings. Like the borderline personality, the histrionic patient utilizes most or all of the problematic cognitive patterns to achieve their goal of being the center of attention, but especially the patterns of catastrophizing and help-rejecting. The histrionic will also magnify and overgeneralize to a great extent as they strive for attention. Lastly, the narcissistic personality may reject help, label, magnify, and selectively abstract in their quest for superiority.

Understanding and Managing Personality Disorders in Headache Patients

How can we understand and manage personality styles in headache patients? First, we need to know if the observed behaviors are part of a long-standing pattern of functioning, or a response to having chronic pain. Second, we need to examine whether there are biological, environmental, and/or developmental factors that contribute toward the expression of a particular personality trait characteristic of a disorder — for example, the unstable affect seen in the borderline personality disorder. Third, under the stress of chronic pain, such traits may be heightened and expressed in a manner that might suggest a personality disorder or at least give the clinician clues as to the individual's behavioral and coping style. This is also a phenomenon we see in psychiatry, where the usually well-compensated patient, under the stress of a major depressive episode or other such acute illness, may display traits and coping styles of a personality disorder that they would not otherwise demonstrate.

Once we have identified the problematic cognitive patterns and coping styles of the patient with features of a personality disorder, our management of the headache can follow certain guidelines that should diminish the "headache" that these difficult patients may induce in the clinic. Make a therapeutic alliance, be kind but consistent, and clearly set expectations early — both of what to expect from treatment, and how treatment will be conducted. It is important to establish and maintain boundaries that will invariably be tested. It is also important to validate the patient's concerns by accepting that the pain is real. However, we help further by identifying the affect as unpleasant emotions that may complicate pain management. If we feel anger toward a patient, recognize that is probably the result of our empathy; that is, we are now feeling as the patient feels. At this time, it is especially important to step back and reflect upon the affect and not act upon it. I will often ask the patient, when I am getting angry during an encounter, what it is they are angry about, or reflect that they seem angry, or wonder if they should not be angry about something; this opens lines of communication, and the anger, including mine, tends to defervesce. Lastly, recognize that any discussion you have with the patient is, in fact, an opportunity for psychotherapeutic intervention.

Psychotherapy in Headache and Pain Disorders

If one understands psychotherapy as the reorientation of thought, feeling, and/or perception to the real world, then any interaction the clinician has with a patient is the opportunity for psychotherapy, intended or otherwise. Psychotherapy does not have to occur in the traditional mode of the 50-minute hour, sitting in comfortable chairs chatting about one's life. Indeed, a simple smile and warm greeting can be "psychotherapeutic" if they can momentarily reorient the patient to the consensual world, however briefly. So even the clinician, with the limited time allotted to each patient, can do psychotherapy by recognizing affect as a signal, and help patients identify and label feelings. By identifying the precipitant for the feelings, we can teach the patient that it is better to express with words rather than actions, such as substance abuse. We then determine, with the patient, what can be done with unpleasant feelings, and encourage them to take better care of themselves. Most experienced clinicians, whether in primary care or specialty clinics, learn these techniques and practice them intuitively as the art form of direct patient care.

If there is evidence of greater personality disorder or just a breakdown in the ability to cope with headache and pain, the patient can be referred for more specific psychotherapy as practiced by trained specialists (Box 18.1). In the realm of pain treatment, as in depression, cognitive behavioral therapy has the best data supporting its use. There are also good data supporting the use of adjunctive therapies such as biofeedback and relaxation/imagery training. However, depending on the precipitants to headache and the

---

BOX 18.1

TYPES OF PSYCHOTHERAPY

- Behavior therapy — mitigates excessive problematic pain-associated behaviors
- Cognitive behavioral therapy — cognitive restructuring and coping skills training
- Supportive therapy — warm and empathic, helps reduce patient's distress

- Marital/family/group therapy — addresses system and reduces isolation
- Insight-oriented psychotherapy
- Adjunctive therapies — biofeedback; relaxation and imagery training; hypnosis; vocational rehabilitation.

---

effects headaches may have on life, use of other psychotherapies may be relevant and helpful as well.

SUMMARY: WHAT IS THE PSYCHIATRIC APPROACH TO HEADACHE?

There are few psychiatrists who have taken on the role as primary physician to the headache or chronic pain patient. However, due to the marked evidence of high comorbidities and probable shared underlying pathophysiology, psychiatrists really should take note and become more proficient in understanding pain and its treatments. The goal for psychiatrists and all clinicians in the treatment of headache is, along with symptom reduction, to improve the patient's functional status and ability to perform activities of daily living without significant pain-related interruption or dysfunction. The role psychiatry has to play is to assess for psychiatric comorbidity, both major mental illness and personality disorders, as well as substance misuse, abuse, and dependency. The psychiatrist can then recommend or initiate treatment with the most appropriate agent, based on the comorbid psychiatric illness, and engage the patient supportively in therapy regarding issues of stress, coping styles, and learned behavior. If needed, the psychiatrist can also refer to a specific psychotherapy specialist if indicated, such as for addictions counseling or family therapy. In this way we participate in the care of the headache patient, getting the central nervous system to forget its molecular memories, controlling psychiatric comorbid illness, and contributing to recovery and productive life.

References

1. Smitherman TA, Rains JC, Penzien DB. Psychiatric comorbidities and migraine chronification. *Current Pain Headache Rep.* 2009;13:326−331.
2. Adler C, Adler S, Friedman A. A historical perspective on psychiatric thinking about headache. In: Adler C, Adler S, Packard R, eds. *Psychiatric Aspects of Headache.* Baltimore, MD: Williams & Wilkins; 1987:3−21.
3. Wolff HG. Personality features and reactions of subjects with migraine. *Arch Neurol Psychiatry.* 1937;37(4):895−921.
4. Lake AE, Rains JC, Penzien DB, Lipchik GL. Headache and psychiatric comorbidity: historical context, clinical implications, and research relevance. *Headache.* 2005;45:493−506.
5. Wolff H. *Headache and Other Head Pain.* New York, NY: Oxford University Press; 1948:348
6. Penzien DB, Rains J, Holroyd KA. Psychological assessment of the recurrent headache sufferer. In: Tollison C, Kunkel R, eds. *Headache: Diagnosis and Treatment.* Baltimore, MD: Williams & Wilkins; 1993:39−49.
7. Lipton RB, Stewart WF, Diamond S, Diamond ML, Reed M. Prevalence and burden of migraine in the United States: data from the American Migraine Study II. *Headache.* 2001;41:646−657.
8. Radat F, Swendson J. Psychiatric comorbidity in migraine: a review. *Cephalgia.* 2005;25:165−178.
9. Merikangas KR, Angst J, Isler H. Migraine and psychopathology: results of the Zurich Cohort Study of young adults. *Arch Gen Psychiatry.* 1990;47:849−853.
10. Breslau N, Schultz LR, Stewart WF, Lipton RB, Lucia VC, Welch KM. Headache and major depression: is the association specific to migraine? *Neurology.* 2000;54:308−313.
11. Webster R. *Why Freud Was Wrong: Sin, Science and Psychoanalysis.* London, UK: HarperCollins; 1996.
12. Gallagher RM, Verma S. Managing pain and comorbid depression: a public health challenge. *Semin Clin Neuropsychiatry.* 1999;4(3):203−220.
13. Singer T, Seymour B, O'Doherty J, Kaube H, Dolan RJ, Frith CD. Empathy for pain involves the affective but not sensory components of pain. *Science.* 2004;303(5661):1157−1162.
14. Frodl T, Jäger M, Smajstrlova I, Born C, Bottlender B, Palladino T, et al. Effect of hippocampal and amygdala volumes on clinical outcomes in major depression: a 3-year prospective magnetic resonance imaging study. *J Psychiatry Neurosci.* 2008;33(5):423−430.
15. Emad Y, Ragab Y, Zeinhom F, El-Khouly G, Abou-Zeid A, Rasker JJ. Hippocampus dysfunction may explain symptoms of fibromyalgia. A study with single-voxel magnetic resonance spectroscopy. *J Rheumatol.* 2008;35(7):1371−1377.
16. Maletic V, Raison CL. Neurobiology of depression, fibromyalgia and neuropathic pain. *Front Biosci.* 2009;14(6):5291−5338.
17. Post RM. Kindling and sensitization as models for affective episode recurrence, cyclicity, and tolerance phenomena. *Neurosci Biobehav Rev.* 2007;31(6):858−873.
18. Miller L. Neurosensitization: a model for persistent disability in chronic pain, depression, and posttraumatic stress disorder following injury. *Neurorehabilitation.* 2000;14(1):25−32.
19. Rome HP, Rome JD. Limbically augmented pain syndrome (LAPS): kindling, corticolimbic sensitization, and the convergence of affective and sensory symptoms in chronic pain disorders. *Pain Med.* 2000;1:7−22.
20. Cady R, Farmer K, Dexter JK, Schreiber C. Cosensitization of pain and psychiatric comorbidity in chronic daily headache. *Current Pain Headache Rep.* 2005;4:7−12.
21. Rauch SL, van der Kolk BA, Fisler RE, Alpert NM, Orr SP, Savage CR, et al. A symptom provocation study of posttraumatic stress disorder using positron emission tomography and script-driven imagery. *Arch Gen Psychiatry.* 1996;53(5):380−387.
22. Miller AH, Maletic V, Raison CL. Inflammation and its discontents: the role of cytokines in the pathophysiology of major depression. *Biol Psychiatry.* 2009;65:732−741.
23. Krishnades R, Cavanaugh J. Depression: an inflammatory illness? *J Neurol Neurosurg Psychiatry.* 2012;83(5):495−502.
24. Hamelsky SW, Lipton RB. Psychiatric comorbidity of migraine. *Headache.* 2006;46:1327−1333.
25. Breslau N, Davis GC, Andreski P. Migraine, psychiatric disorders, and suicide attempts: an epidemiologic study of young adults. *Psychiatry Res.* 1991;137:11−23.
26. McWilliam LA, Goodwin RD, Cox BJ. Depression and anxiety associated with three pain conditions: results from a nationally represented sample. *Pain.* 2004;111:77−83.
27. Oedegaard KJ, Neckelmann D, Mykletun A, Dahl AA, Zwart JA, Hagen K, et al. Migraine with and without aura: association with depression and anxiety disorder in a population-based study. The HUNT study. *Cephalgia.* 2006;26(1):1−6.

28. Baskin SM, Smitherman TA. Migraine and psychiatric disorders: comorbidities, mechanisms and clinical applications. *Neurol Sci.* 2009;30(suppl 1):561–565.

29. Breslau N. Psychiatric comorbidity in migraine. *Cephalgia.* 1998;18(suppl 2):56–61.

30. Peterlin BL, Nijjar SN, Tietjen GE. Posttraumatic stress disorder and migraine: epidemiology, sex differences, and potential mechanisms. *Headache.* 2011;51:860–868.

31. Peterlin BL, Tietjen GE, Brandes JL, Rubin SM, Drexler E, Lidicker JR, et al. Posttraumatic stress disorder in migraine. *Headache.* 2009;49:541–551.

32. Kroenke K, Spitzer RL, Williams JBW, Linzer M, Hahn SR, deGruy III FV, et al. Physical symptoms in primary care. Predictors of psychiatric disorders and functional impairment. *Arch Fam Med.* 1994;3(9):774–779.

33. Olfson M, Kroenke K, Wang S, Blanco C. Trends on office-based mental health care provided by psychiatrists and primary care clinicians. *J Clin Psychiatry.* 2014;75(3):247–253.

34. Goodwin FK, Jamison KR. *Manic-Depressive Illness.* New York, NY: Oxford University Press; 1990.

35. Barkin RL, Fawcett J. The management challenges of chronic pain: the role of the antidepressants. *Am J Ther.* 2000;7(11):131–147.

36. Lynch ME. Antidepressants as analgesics: a review of randomized controlled trials. *J Psychiatry Neurosci.* 2001;26(1):30–36.

37. Landmark CL. Targets for antiepileptic drugs in the synapse. *Med Sci Monit.* 2007;13(1):RA1–7.

38. Rogawski MA, Loscher W. The neurobiology of antiepileptic drugs for the treatment of nonepileptic conditions. *Nat Med.* 2004;10:685–692.

39. Fishbain DA, Cutler RB, Lewis J, Cole B, Rosomoff RS, Rosomoff HL. Do the second generation "atypical neuroleptics" have analgesic properties? A structured evidence-based review. *Pain Med.* 2004;5(4):359–365.

40. Drzyzga L, Obuchowicz E, Marcinowska A, Herman ZS. Cytokines in schizophrenia and the effects of antipsychotic drugs. *Brain Behav Immun.* 2006;20(6):532–545.

41. Stahl SM. *Stahl's Essential Psychopharmacology.* 3rd ed. New York, NY: Cambridge University Press; 2008.

42. National Institute of Mental Health. *Statistics.* Bethesda, MD: NIMH. <www.nimh.nih.gov>.

43. Ilgen MA, Kleinberg F, Ignacio RV, Bohnert ASB, Valenstein M, McCarthy JF, et al. Noncancer pain conditions and risk of suicide. *JAMA Psychiatry.* 2013;70(7):692–697.

44. Breslau N, Schultz L, Lipton R, Peterson E, Welch KMA. Migraine headaches and suicide attempts. *Headache.* 2012;52(5):723–731.

45. Fuller-Thomson E, Shrumm M, Brennenstuhl S. Migraine and despair: factors associated with depression and suicidal ideation among Canadian migraineurs in a population-based study. *Depress Res Treat.* 2013;2013:401487.

46. De Filippis S, Erbuto D, Gentili F, Innamorati M, Lester D, Tatarelli R, et al. Mental turmoil, suicide risk, illness perception, and temperament, and their impact on quality of life in chronic daily headache. *J Headache Pain.* 2008;9:349–357.

47. Touraine GA, Draper G. The migraine patient: a constitutional study. *J Nerv Ment Dis.* 1934;80:1–204.

48. Schmidt FN, Carney P, Fitzsimmons G. An empirical assessment of the migraine personality type. *J Pyschosom Res.* 1986;30:189–197.

49. Silberstein SD, Lipton RB, Breslau N. Migraine: association with personality characteristics and psychopathology. *Cephalgia.* 1995;15(5):358–369.

50. Leo RJ. *Pain Management in Psychiatry.* Washington, DC: American Psychiatric Publishing; 2007.

Psychological Approaches to Headache

Elizabeth K. Seng[1] and Steven M. Baskin[2]

[1]Ferkauf Graduate School of Psychology of Yeshiva University, and Albert Einstein College of Medicine of Yeshiva University, New York, New York, USA [2]New England Institute for Neurology and Headache, Stamford, Connecticut, USA

INTRODUCTION

As a group, headache disorders are painful and disabling, impacting the lives of people with headache across multiple domains and reducing their ability to be productive and satisfied members of society. The majority of work examining psychological approaches to headache has occurred within the two most prevalent primary headache disorders: migraine and tension-type headache.

Recent epidemiological studies have demonstrated prevalence rates for episodic migraine of 12% across the United States population, including 18% of women and 6% of men.[1] Chronic migraine affects a smaller proportion of the United States population (1–2%), with a higher proportion of women to men similar to episodic migraine, but can be far more disabling because of its chronic nature, and because it is less responsive to available treatments.[2]

Migraine is associated with high levels of disability, and can interfere with the individual's functioning across domains. Migraine has been ranked as the seventh specific cause of disability worldwide,[3] and accounts for more than 1% of years lived in disability.[4] Migraine occurs during the most productive years of life, and has a higher prevalence in women.[5] Migraine can reduce the capacity of individuals to function in social and work environments both during the period of the headache attack (during which a person with migraine may need to completely withdraw to a dark, quiet room), as well as between attacks, during which time a person with migraine may experience anxiety about potential migraine onset and restrict their activities to avoid migraine onset.[6] A burgeoning literature on stigma suggests that migraine patients report greater disease-related stigma than do those with epilepsy, and

that much of this stigma is related to work-related productivity,[7] and can even change decision-making about taking acute migraine treatment.[8]

Tension-type headache is a painful condition characterized by episodes of mild to moderate bilateral pressing head pain. Most people experience episodes of tension-type headache at some point in their lives.[9] However, a minority of individuals report experiencing tension-type headache on 15 or more days per month, or chronic tension-type headache. As with chronic migraine, chronic tension-type headache is far more disabling than episodic tension-type headache, and can be less responsive to treatment.[9]

Psychological factors have been considered integral to our understanding of headache since the first recorded information about headaches.[10,11] The term "psychological factors" encompasses a broad range of topics that may contribute to headaches including: patient behaviors during and in-between headache attacks; patient thoughts and beliefs about illness and headaches; and, characteristics of the environment in which patients live and work.

This chapter will describe our current understanding of the role of psychological factors in headache disorders. First, we will review the role of trigger factors and their avoidance in the management of headache disorders, as well as the role of behaviors, thoughts, and psychiatric comorbidities in the onset and maintenance of headache disorders. Then, we will examine the contribution of psychological factors to medical treatment of headache disorders, including medication adherence and patient–physician communication. It should be noted that a large proportion of people with migraine manage their headaches without the assistance of prescription medications[1]; thus, those who seek medical consultation likely have a slightly

S. Diamond, R. K. Cady, M. L. Diamond & V. T. Martin (Eds):
Headache and Migraine Biology and Management.

DOI: http://dx.doi.org/10.1016/B978-0-12-800901-7.00019-7

different psychological profile (e.g., feel less control over their headaches), which emphasizes the importance of assessing psychological factors in a treatment-seeking population. Finally, we will describe treatments based on psychological principles that have demonstrated efficacy in headache disorders, including educational strategies, biofeedback and relaxation strategies, and cognitive behavioral treatments, as well as new approaches that are promising avenues for behavioral headache treatment.

TRIGGER FACTORS

Patients often note that encountering certain phenomena or circumstances seems frequently to precipitate, or "trigger," headache episodes. The belief that these factors actually trigger headaches can lead patients to invest time and effort in identifying, avoiding, and mitigating headache triggers. When asked about patient (and healthcare provider) perceptions, most identify a consistent conglomeration of factors that they believe are headache triggers in at least some people, including dietary factors (aged cheese, red wine); situational factors (weather patterns, noxious odors, flickering lights); cognitive factors (stress); and behavioral factors (poor/inconsistent sleep and dietary schedule, physical activity).[12] Healthcare providers often encourage patients to keep track of factors that might trigger their headaches in order to establish their idiosyncratic trigger profile. However, recent evidence suggests that factors commonly thought of as well-established triggers for migraine with aura (strenuous activity and flickering lights) have a low likelihood of actually triggering a migraine with aura (11%) when administered to individuals who report these as salient triggering factors.[13] Further, a study examining fluctuations in three commonly reported triggers (weather, ovarian hormones, and perceived stress) in a daily diary found that the day-to-day variation in these triggers would make it difficult to identify true triggers in an individual person with migraine solely using methods of natural experimentation.[14]

Although identifying unique migraine triggers may be challenging through daily-diary natural experimentation methods, research empirically examining which factors are likely to precipitate migraine episodes may provide useful information that can be broadly applied in clinical situations. Stress, sleep, and some aspects of diet are psychological/behavioral precipitating factors that are modifiable, and therefore have a high clinical utility. It may be helpful for providers to introduce these psychological factors as contributing to reaching the "migraine threshold," a useful concept in migraine education. The migraine threshold can be thought of as a tipping point at which the patient's nervous system generates a migraine. Management of individual psychological "triggering" factors may help people with migraine to avoid reaching the "migraine threshold."

Stress

Stress is one of the perceived triggers most commonly identified across different headache types and clinical populations.[12,15,16] However, the term "stress" is now commonly used in lay-language in a variety of fluid contexts, which might color survey findings. Psychological stress can be defined as the perception that one does not have adequate resources to manage a potentially threatening situation, or stressor.[17] However, lay persons might use the term "stress" to describe a large number of daily hassles (in which "stress" is equated with stressor), or a state of feeling overwhelmed by phenomena in one's life. Thus, in order to meet the definition of psychological stress, the daily hassle or life event (i.e., stressor) must be something that threatens one's typical daily routine or state of being in some way, *and* the individual must have a concern about lack of resources (e.g., psychological resources, time resources, social and environmental support systems) to manage this stressor.[18]

Recent evidence examining day-to-day fluctuations in stress and headache symptoms suggests that stress on a single day may not be related to onset of headache symptoms, but rather stress on one day may impact the presence of headache on the following day.[19,20] In a sample of subjects with chronic migraine or chronic tension-type headache, two consecutive days with elevated stress was associated with the onset of a headache.[20] In a different sample of patients with episodic migraine, a decline in perceived stress from the previous day was associated with onset of migraine, providing evidence for the "let-down" stress phenomenon precipitating migraine episodes.[19]

Stress could impact headache through a variety of mechanisms, but no single mechanism has been convincingly demonstrated in empirical research. The biological reaction of stress, activation of the hypothalamic–pituitary axis and sympathetic nervous system, could be associated with migraine onset.[18] From a behavioral perspective, stress on one day might increase the presence of more proximal behavioral precipitating headache factors, such as poor sleep or diet.[19] It has also been posited that the prodromal symptoms of headache itself might change the perception of one's ability to successfully cope with one's constant levels of daily hassles. In other words, headache might change one's perceived ability to manage the daily hassles that commonly arise in one's life.[21]

These are each plausible mechanisms that deserve continued study.

The impact of stress reaches beyond attacks, and may be associated with chronification of migraine (in which migraine attack frequency increases from episodic migraine to meet criteria for chronic migraine).[22] A population-based study found that greater numbers of major life changes were associated with migraine chronification.[23] Again, more research must be conducted to ascertain mechanisms through which stress influences headache chronification. Stress may play a role in headache chronification through influencing other factors potentially more proximal to chronification (medication overuse, consumption of caffeine), or more directly through physiological processes such as central sensitization.[18] Further, additional work is needed to determine whether intervening with stress can prevent headache chronification in individuals with headache and high levels of stress.

Interventions that focus on stress management have demonstrated efficacy to treat both migraine and tension-type headaches.[24,25] Stress management training typically incorporates both behavioral and cognitive interventions. Behavioral interventions specifically to manage stress might include relaxation techniques (progressive relaxation, deep breathing, autogenic training, imagery), biofeedback, and increasing activities with stress-reducing properties, such as exercise. Cognitive aspects of stress management interventions might include awareness of when one's stress level is increasing, identification of thoughts that are initiating and maintaining high levels of stress, and modifying maladaptive patterns of thinking to increase a sense of self-efficacy to manage life stressors.

Sleep

Sleep plays an important role in headache. Primary headache disorders such as migraine, tension-type headache, and cluster headache are influenced by sleep. Disturbances in the normal pattern of sleep are commonly perceived to be headache triggers: going to bed later than usual, getting up earlier than usual, oversleeping, or disruptions in sleep pattern due to work or jet lag.[26] Empirical evidence suggests that two consecutive nights of sleep disturbance have been associated with onset of migraine and tension-type headache, and sleep disturbance can build upon daily stress to precipitate headache onset.[20]

Several patterns of primary headache disorders, including morning headaches, headaches that awaken the patient from sleep, and chronic daily headaches, are associated with sleep disorders.[27] Insomnia is the most prevalent sleep disorder in chronic migraine and tension-type headache. Sleep apnea commonly occurs in individuals with cluster headache and chronic daily headache patterns. Further, sleep apnea is associated with a secondary headache disorder — sleep apnea headache — that occurs within the context of a diagnosis of sleep apnea, characterized by bilateral pressing headaches that present on awakening, resolve within 30 minutes, occur on more than 15 days per month, and resolve upon successful treatment of sleep apnea.[28] Circadian rhythm disorders are common in people with migraine, and recent evidence suggests that a genetic mutation associated with severe familial advanced sleep phase may be associated with migraine with aura.[29]

Few studies have examined the effect of sleep treatment on headache, although existing evidence suggests that treating sleep disturbances may improve some headaches. Treating sleep apnea with continuous positive airway pressure (CPAP) can improve sleep apnea headache.[30] One study demonstrated that in veterans from the war in Iraq and Afghanistan presenting with comorbid headache and insomnia, headache days and severity improved when veterans were treated for insomnia with education and pharmacotherapy.[31] Cognitive behavioral treatments have demonstrated efficacy to treat sleep disturbances, with the strongest evidence for the treatment of insomnia.[32] Cognitive behavioral treatments for insomnia reduce conditioned arousal related to the sleep environment, ineffective or harmful habits developed in an effort to improve sleep, and sleep-related worry. Common elements of cognitive behavioral therapy for insomnia include: stimulus control (e.g., only go to bed when truly sleepy, maintain a regular rise time, reduce napping); reduce sleep-interfering activation (e.g., undertake relaxation techniques, avoid stimulants and exercise later in the day); and sleep restriction. A recent trial demonstrated that a 12-week cognitive behavioral sleep intervention improved headache symptoms in individuals with chronic migraine.[33] It provides promising preliminary evidence supporting the use of these interventions in headache sufferers.

Diet and Obesity

Given the inconsistent evidence regarding foods perceived to trigger headache, and the sheer number of foods thought to trigger headache, simply avoiding these foods is unlikely to improve headaches in an individual patient. Additionally, this strategy runs the risk of exacerbating headaches through reinforcing maladaptive avoidance coping.[26] In one double-blind study, chocolate was no more likely than carob (a chocolate-like substance) to trigger a migraine across all participants, regardless of whether participants believed chocolate to be a migraine trigger.[34] Dietary recommendations appear to be most effective when individualized. For example, a randomized clinical cross-over trial showed that a diet eliminating foods to which people with migraine demonstrated an immune reaction (elevated IgG antibodies) reduced headache days and migraine attacks.[35]

Obesity is prevalent among people with migraine[36] and is associated with higher headache frequency and severity among people with migraine.[37] Obesity is further associated with the change of migraine (but not tension-type headache) from an episodic to a chronic condition.[38] Behavioral interventions that incorporate monitoring and modification of diet and physical activity can be efficacious to modify weight. However, changes in eating patterns (and in particular fasting) are known to exacerbate migraine.[39] A promising recent randomized clinical trial demonstrated that a 12-month weight-loss program in adolescents with migraine was associated with decreases in migraine frequency, intensity, and disability; further reductions in body mass index were associated with migraine outcomes.[40] More research should examine the impact of weight-loss interventions on migraine.

PERSONALITY TRAITS AND MIGRAINE

There have been numerous hypotheses concerning the role of personality traits that may predispose to headache or be characteristic of people with migraine. Harold Wolff developed the notion of "the migraine personality," depicting people with migraine as rigid, conscientious, meticulous, perfectionistic, and obsessive.[11] This early literature had a significant selection bias, as it was based on non-standardized assessment and cross-sectional designs that did not distinguish between causes and effects of migraine. These early studies also did not control for headache frequency, substance use, psychiatric comorbidity, or disability. Later epidemiological studies found a moderate relationship between migraine and the personality trait of neuroticism, or susceptibility to experience negative affect, in both clinic- and population-based studies.[41-43] This higher level of neuroticism remained after controlling for comorbid anxiety and depression.[43] However, many psychological questionnaires with temporal qualifiers that reflect general health concerns, fatigue, social withdrawal, and generalized distress could result in inflated neuroticism scale scores because the headache disorder itself could produce this symptomatology. Therefore, it is possible that differences in neuroticism scale scores on personality questionnaires may reflect frequency of head pain as

well as related somatic issues rather than a stable personality trait. Later studies showed that neuroticism seems to be unrelated to headache frequency, intensity, or duration,[44,45] except for one study with women with migraine where there was a strong positive correlation between headache duration and neuroticism.[46]

Neuroticism is the personality trait that has been most consistently associated with migraine. Harm avoidance shares some properties with neuroticism, and is characterized by behavioral inhibition, excessive fear and worry, introversion, and pessimism. Subjects with migraine endorsed higher levels of harm avoidance than non-migraine controls on the Temperament and Character Inventory.[47-50] These studies showed inconsistent results on the traits of self-directness and persistence. Catastrophizing is a construct where individuals show an exaggerated appraisal of the negative consequences of pain, often exhibiting magnification, rumination, and helplessness. Holroyd and colleagues have showed that this psychological response to migraine is associated with impaired functioning and poor quality of life — independent of migraine characteristics and comorbid psychopathology.[51]

The Minnesota Multiphasic Personality Inventory (MMPI) is the most widely used personality questionnaire and instrument to measure adult psychopathology. Several MMPI studies have shown that chronic migraine and chronic tension-type headache, with and without medication overuse, have significant elevations on the hypochondriasis, depression, and hysteria scales (neurotic triad), as well as social introversion, as compared with episodic migraine. The frequency of headache seems to be more indicative of "personality traits" than is the actual headache diagnosis.[52-55] A recent Italian study showed that patients experiencing chronic daily headache with psychiatric comorbidity had significantly higher MMPI-2 scores (than chronic daily headache patients without psychiatric comorbidity) on all scales of the neurotic triad. Medication overuse was not a significant factor.[56] Galli and associates have shown that medication overuse headache (MOH) patients and pure substance abusers do not share dependency characteristics on the MMPI-2.[57] There may be two separate profiles in medication overuse headache patients with a more complex group exhibiting significant dependency features and who are more likely to abuse opioid analgesics.[58,59]

PSYCHIATRIC COMORBIDITY

Comorbidity is the presence of any additional coexisting disorder in an individual with a particular index disease; a more than coincidental association between two conditions. A variety of reviews have shown a significant relationship between migraine and psychiatric disorders.[60–62] Psychiatric comorbidities may complicate differential diagnosis; increase medical costs; affect adherence to treatment regimens; impact quality of life; contribute to increased headache-related disability; and impact the course of migraine, in some cases leading to headache chronification.[63–68]

In population studies, subjects with migraine are between 2.2 and 4.0 times more likely to suffer from major depressive disorder (MDD) than people without migraine.[61,69,70] A bidirectional relationship has been consistently demonstrated between migraine and major depression, in longitudinal studies, with either disorder increasing the risk for the other, suggesting a possible shared neurobiology.[70] Potential mechanisms underlying this comorbidity remain largely unexplored, and include serotonergic and other neurotransmitter dysfunction; dysregulation of the HPA axis; proinflammatory cytokines; central sensitization processes; and fluctuations in ovarian hormones.[71] Two population-based studies found an overall prevalence of MDD of 28% in the migraine sample, and only 12% in two other chronic pain states.[72,73] There is about a 2.5- to 3-fold higher relationship between migraine and bipolar spectrum disorders.[74–78] Bipolar spectrum disorder, MDD, recurrent depression, and suicide attempts exhibit a stronger relationship for migraine with aura than for migraine without aura.[75–80] Chronic migraine has higher associated depression than episodic migraine,[81–84] and there is also an increased prevalence in MOH, with the psychiatric disorder often preceding the medication overuse.[85,86]

The lifetime prevalence of anxiety disorders in people with migraine (ranging from 51% to 58%) is almost twice that of MDD.[82,87] Anxiety disorders may complicate migraine more than depression, with greater long-term persistence, greater headache-related disability, and reduced satisfaction with acute therapies.[65,68,88] The onset of anxiety generally precedes the onset of episodic migraine, whereas the onset of major depression follows the onset of migraine.[76,89] Anxiety may appear in childhood, followed by episodic migraine then depression, and then transformation to chronic migraine. Compared with individuals without migraine, individuals with migraine have about a four- to five-fold greater risk for generalized anxiety disorder, five-fold greater risk for obsessive-compulsive disorder, and approximately 3.75-fold greater risk for panic disorder.[61,66,72,75,76,90–92] A stronger relationship is noted between panic disorder and transformed and chronic migraine, with lifetime prevalence between 25% and 30%.[83,91] Panic disorder is also more common in migraine with aura[92] and, similar to the depression data, has a bidirectional relationship with migraine.[93]

Childhood maltreatment, including physical, emotional, and sexual abuse, is associated with migraine. Headache clinic studies have shown 25–30% of migraine patients reporting a history of physical or sexual abuse. Headaches in individuals who suffered from childhood maltreatment tend to be more disabling, and more likely to "transform" from episodic to chronic variants.[94–96] A higher prevalence of post-traumatic stress disorder in migraine sufferers has been noted in general population surveys, tertiary-care headache clinic studies, and military surveys. In tertiary care centers, approximately 22–30% of headache sufferers meet post-traumatic stress disorder diagnostic criteria.[97,98] Approximately 25% of the patients in an inpatient refractory daily headache population suffered from a personality disorder.[99] In one clinical study, a diagnosis of borderline personality disorder was associated with more pervasive headache, high headache-related disability, lower probability of responding to standard preventive pharmacological therapy, and greater risk for medication overuse.[100]

Although comorbid depression and anxiety are associated with poorer functioning and quality of life in people with migraine,[71,92] individuals with migraine and comorbid depression and/or anxiety can experience significant improvement in migraine over the course of comprehensive treatment.[101] Behavioral/psychological treatments for migraine as well as anxiety and depression share similar goals. Psychological treatments for depression focus on changing behaviors, cognitive patterns, and social interactions that may be conducive to the development of depression.[102,103] Psychological treatments for migraine as well as depression-specific psychotherapies attempt to increase self-efficacy. Patients learn ways to better regulate their pain and mood state while increasing their functionality and problem-solving skills. Many headache patients who have comorbid depression tend to be helpless, hopeless, discouraged, and non-adherent to headache treatment recommendations. In both migraine and mood disorders, it is important to maintain consistent biological rhythms, activate behavior, modify self-defeating thinking, and develop better coping skills.

Headache patients with comorbid anxiety also pose significant treatment challenges. Anxiety disorders are typically accompanied by fear-based thoughts of danger and vulnerability. Avoidance of feared stimuli reduces discomfort but maintains the danger belief that the fear stimuli and anxiety symptoms themselves are harmful.[103] Patients with anxiety disorders tend to overestimate the likelihood of the occurrence of a negative event, such as migraine, and perceive situations as more catastrophic, threatening, and unmanageable than the objective reality.

Panic disorder is a chronic condition similar to migraine, with episodic attacks of high impairment and interictal worry of future attacks. Headache patients with anxiety disorders often develop conditioned anticipatory fear of somatic sensations that are perceived as "warning signals" of unpredictable severe pain that is similar to interoceptive panic conditioning.[92,104] Patients may treat fear or what they think is a migraine prodrome with medication, believing that they will pre-emptively avoid migraine. Medication reduces their emotional distress and "prevents" the migraine – a powerful avoidance learning condition process. Many migraine patients, similar to panic disorder sufferers, are exquisitely sensitive to bodily signals and have high anxiety sensitivity in which they fear benign anxiety-related physical sensations because they believe that they will have catastrophic consequences.[92,105] Patients may become exquisitely sensitive to perceived headache triggers. It is helpful when treating those headache patients with significant anxiety to provide opportunities for exposure to feared somatic symptoms (interoceptive exposure) so that they can challenge their danger cognitions and negative predictions about these internal sensations. Patients learn to better tolerate migraine prodromes, limit any avoidance responses, and make rational therapy decisions.

PSYCHOLOGICAL FACTORS IN MEDICAL TREATMENT OF HEADACHE DISORDERS

Medication Adherence

At the most basic level, medications are only effective if patients take them. However, particularly for headache medication, simply "taking" a medication is insufficient knowledge as to whether a patient is taking that medication *optimally*. Medication adherence refers to the extent to which a patient engages in the medication-taking behaviors required to optimally manage headaches.[106] For headaches, medication adherence can take place in a preventive manner (on a fixed schedule) and in an acute manner (in the context of a headache episode).

Keeping medication on hand is essential to adherence with both preventive and acute headache medication. However, barriers to having a medication on hand when needed do occur. For preventive medication, individuals with headache may experience difficulty in remembering to take their medication, or may have low motivation to take their medication, if they are not currently experiencing a headache. With acute medication, patients with headache describe difficulty keeping medication with them at all times, particularly

when their schedule is disrupted (which is also when a headache is most likely to occur).[8] Further, some headache sufferers do not have medication on hand because they did not refill their prescription. For example, one pharmacy utilization study in Israel ($n = 1498$) found that over half of the participants (56.1%) purchased triptans only once within a 6- to 18-month period.[107]

Adherence to preventive therapy, beyond filling prescriptions and having medication on hand, requires incorporating taking the preventive medication into the patient's daily routine. Across chronic diseases, only 50–75% of patients consistently use medication prescribed on a fixed schedule.[108] Adherence to taking medications on a fixed schedule is associated with less complex medication regimens, fewer daily doses, oral rather than other routes of administration, lower risk of side effects, and higher perceived need for the medication.

Medication adherence with acute headache medications, or medications taken on an as-needed basis to treat acute headache attacks, requires a complex, iterative decision-making process based on dynamic headache symptoms and circumstantial demands.[109] Patients and providers have described a series of interdependent behaviors required for optimal use of acute medication, including distinguishing between headache types, choosing what type of medication to take, taking medication early, repeating doses as needed, and limiting overuse of acute headache medication.[8] Further, multiple steps of individually tailoring the acute headache medication regimen is common in clinical practice.[11] Thus, the "rules" regarding optimal use of acute headache medication are likely to change over the course of treatment, which can be cognitively demanding for the person with headache to remember during a painful headache episode.

During a migraine episode, using a migraine-specific medication initially is often associated with better outcomes.[110,111] However, many patients do not use a migraine-specific medication initially but instead rely on non-specific pain medications.[112] Further, between 40% and 85% of individuals with headache often wait to take acute medication (especially migraine-specific medication) until the headache is moderate or severe, rather than taking the medication earlier when the pain is mild (which is associated with better treatment outcomes).[111,113,114] Patients describe concerns about side effects and perceived dangerousness of prescription (as opposed to over-the-counter) medication as contributing to this delay. However, patients also describe difficulty identifying an acute headache as a migraine (as opposed to a tension-type headache) early during the course of the headache.[8] Patients also may wait to take migraine-specific medication as a strategy to prevent overuse of acute headache medication, or because they have a limited supply.

Headache sufferers taking acute medications must be particularly careful to avoid medication overuse headaches. Medication overuse headache refers to the phenomenon of headaches actually increasing in frequency and severity when acute headache medications are taken frequently. Therefore, it is generally recommended that headache sufferers use acute medications no more than 2 or 3 days per week to prevent medication overuse headache.[10] Estimates suggest that at least 4% of patients, with headache overall, fail to limit their use of acute headache medications, placing them at high risk for medication overuse headache.[115,116] This phenomenon is significantly more common in tertiary care.[117]

Comprehensive multidisciplinary treatments for primary headaches, as well as medication overuse headache, commonly incorporate education and motivation enhancement strategies to bolster acute medication adherence.[118] Brief targeted educational and motivation enhancement strategies alone, such as nurse education or between-visit motivational phone calls, have also demonstrated some efficacy to modify specific acute medication adherence behaviors (e.g., taking medication early, or reducing medication overuse).[113,119]

Several steps can be taken to increase medication adherence in patients with headaches.[118] Specifically, assessing how individuals with headache take their medications can provide valuable information for the clinical assessment, and allow for targeted interventions. Involving the patient in treatment planning can ensure that the medication regimen recommended is realistic and aligns with patient goals for treatment. Reducing the complexity of the treatment regimen is an important first step in improving adherence to both preventive and acute medications. Consistent education across providers tailored to the individual patient's problems with medication adherence is essential, and retention can be enhanced by providing written materials at an appropriate literacy level. Additionally, incorporating family members or other significant individuals in the patient's life can be useful to support medication adherence.

Motivation to take headache medications may vary with episodic exacerbations of headache symptoms; therefore, motivation enhancement strategies are warranted. These strategies may include linking medication adherence to desired goals in the patient's life, and developing specific and realistic plans for implementing changes in medication. Providers should consider strategies that are appropriate to a patient's level of readiness to change. For example, if a patient is unaware that medication overuse is causing problems in his or her life, non-judgmental education strategies may be beneficial. However, if a patient is well aware of the negative consequences of medication overuse and has a strong desire to reduce medication use, strategies to build patient self-efficacy and provide specific realistic goals would be warranted. Reinforcing successful medication adherence can enhance motivation.

Patient—Physician Communication

Perceived empathy from one's physician has been associated with adherence with behavioral and pharmacological treatment recommendations and a decrease in migraine disability and symptoms over 3 months.[120] Both patients and physicians identified high quality communication in the patient—physician relationship to be a key factor in adherence with acute headache medications.[8] However, both patients and physicians have idiosyncratic assumptions when entering a new encounter. These assumptions can color interpretation of communication, and can leave both parties with an incomplete or inaccurate representation of what the other was attempting to convey. Patients and physicians often have differing impressions of number of migraine days and disability due to headache[121] as well as aberrant medication-taking.[122]

Patient—physician communication can be improved through several measures. The first is through standardizing methods of gaining information regarding headache. Utilization of surveys about headache symptoms and quality of life (such as the Migraine Disability Assessment or MIDAS),[123] and standard headache diaries,[124] can provide a starting point for conversations about migraine symptoms and improve communication about migraine.[124,125] The second method is through improving communication techniques to enhance understanding between the physician and patient.[126] One useful technique is the "ask—tell—ask" method of communication, which involves asking an open-ended question, relaying in simple terms information pertinent to the patient's treatment, and then, to ensure understanding, requesting that the patient rephrase what was just communicated. The American Migraine Communication Study II demonstrated that training physicians in using open-ended questions and the "ask—tell—ask" method of relaying treatment information can be effective in improving communication.[127]

PSYCHOLOGICAL APPROACHES TO TREATING HEADACHE DISORDERS

Non-pharmacological approaches to treating headache disorders are appropriate for a wide range of patients, including patients who are pregnant or planning to become pregnant, patients for whom medications are contraindicated, patients who have not responded adequately to pharmacological interventions, and patients who prefer not to take medications. Relaxation training, biofeedback training, and cognitive behavioral therapy have a robust evidence base, and are recommended in treatment guidelines for both migraine[128] and tension-type headache.[129] These treatments can be combined with each other for effective management of migraine[24] and chronic tension-type headache.[25]

Biofeedback and Relaxation-Based Therapies

Biofeedback refers to any treatment in which biological processes (most often those associated with sympathetic arousal) are brought to the awareness of the patient.[130] This awareness gives patients enhanced understanding about the factors that contribute to changes in their bodies, and, with training, can provide an opportunity to increase the patient's control over these processes.[131] The physiological feedback is often paired with training in methods of relaxation. Considerable evidence from randomized clinical trials exists to support the efficacy of biofeedback to treat migraine and tension-type headache, with estimates of improvement in headache of between 35% and 60%.[132,133]

The purpose of biofeedback is to assist patients in learning to control their autonomic nervous system, which, in people with headache, is typically over-excitable. In a biofeedback session patients are presented with computerized feedback, typically an audible tone or visual indicator, about physiological responses (e.g., peripheral skin temperature, muscle tension, heart rate variability) not typically under the patient's voluntary control. Recent research suggests that the main benefit of biofeedback is to help patients learn how psychological factors impact their bodies in demonstrable ways. Thus, biofeedback appears to impact headaches through increasing a sense of self-efficacy in managing headaches, as well as through any physiological changes.[132,134,135] For example, a patient may relate a story about a work-related stressor that occurred during the past week, along with a relaxing evening spent with a friend, and then notice the changes in physiological responses between these two stories. Patients are encouraged to experiment with different thoughts and techniques to modify those responses. Although we recommend doing biofeedback with a trained behavioral healthcare professional when possible, a resource for biofeedback can be found at the HeadacheCareCenter™ website (www.headachecare.com/biofeedback.html).

Relaxation training refers to the utilization of cognitive and behavioral techniques to voluntarily reduce sympathetic arousal in-the-moment. In headache patients, common relaxation techniques include diaphragmatic breathing, progressive muscle relaxation, autogenic training, and imagery/visualization.[136,137] Diaphragmatic breathing involves engaging diaphragm muscles in taking slow, measured breaths, and is occasionally combined with counting during in-breaths and out-breaths. Progressive muscle relaxation is a technique, initially developed by Edmund Jacobson in the 1930s, designed to teach individuals to discriminate between tense and relaxed muscles, with the intent of inducing physiological relaxation.[138] As patients improve their skills they may be able to combine muscle groups, or forego the muscle tension component altogether (which is termed "body scanning"). Autogenic training refers to imagining one's limbs as warm and/or heavy, and is commonly paired with hand-warming biofeedback. Imagery/visualization involves creating a rich, immersive environment in one's mind, and allowing all five senses to absorb and interact with this environment. Relaxation techniques can be utilized in short and long versions, and can be combined for maximum effectiveness in an individual patient. Relaxation techniques are often paired with biofeedback, which provides a rationale for their use, and individualized feedback to observe the efficacy of various relaxation techniques.[136,139]

Biofeedback and relaxation strategies are typically taught in sessions with clinicians, but the primary emphasis of the treatment is consistent home practice of the skill, to be utilized both in a daily manner to prevent headaches and in an acute manner during a headache to increase coping.[140] Biofeedback and relaxation training teach skills that individuals with headache can continue to utilize on a regular basis, with benefits that extend beyond the treatment period.[130]

Cognitive Behavioral Approaches

For headache disorders, cognitive behavioral therapy (CBT) attempts to modify both thinking and behavior patterns to improve quality of life.[137,141,142] Behavioral components typically include keeping a diary of headache symptoms and potentially relevant behaviors (such as medication adherence, sleep, stress, and diet); setting goals to modify these behaviors; and utilization of the relaxation and biofeedback techniques discussed above. Cognitive components include improving a sense of self-efficacy to manage headaches, and identifying and challenging maladaptive thought patterns. Self-monitoring of cognitive factors by keeping a record of situations, thoughts, and emotions that co-occur with headaches is necessary to identify maladaptive thought patterns.[143] A common maladaptive thought pattern in migraine is catastrophizing, or focusing on and expecting the worst possible outcome of a situation. Catastrophizing has been associated with higher headache-related disability,[144] and can be reduced effectively with CBT for headache.[142]

Cognitive behavioral therapy for migraine is considered to have Grade A evidence, based on findings from randomized clinical trials.[128] In one trial, CBT reduced headache activity by an average of 68%, compared with 56% for biofeedback and 20% in the control group.[145] A recent trial demonstrated that a 10-session CBT protocol reduced migraine symptoms and migraine-related disability in children with chronic migraine compared with 10 sessions of headache education.[146] Further evidence suggests that the combination of CBT and preventive drug therapy for headaches may be more effective than either treatment alone. In a trial of a minimal contact CBT for severe migraine (comprising four monthly sessions with a workbook and between-session phone calls), the combination of preventive drug treatment was more effective in reducing migraine frequency and migraine-related disability compared with CBT alone, preventive medication alone, or placebo.[24] In another trial of minimum contact CBT for chronic tension-type headache (three monthly sessions with a workbook and between-session phone calls), a combination of CBT and a preventive medication was more likely to produce clinically significant (larger than 50%) reductions in headache index than CBT alone, preventive medication alone, or placebo.[25]

CBT is time- and cost-effective compared with pharmacological preventive treatments.[147] It can be successfully delivered at home or with minimal therapist contact through technology-augmented interventions such as the telephone or Internet, and can be tailored to meet the needs of individual persons with headache.[24,148,149] Although CBT has demonstrated efficacy, a large proportion of participants in trials (roughly 40%) may not experience clinically significant reductions in migraine symptoms.[150] However, CBT may have greater effects on other factors related to quality of life, including functional and social disability, psychiatric symptoms (depression and anxiety), and adaptive thoughts and beliefs (e.g., self-efficacy).[151] For example, a small trial examining 10-week CBT for headache and focused on reducing catastrophizing ($n = 34$) did not find differences between changes in headache frequency and intensity compared with a wait-list control ($n = 11$); however, the CBT group reported significantly greater reductions in headache-related catastrophizing and anxiety, and increases in self-efficacy to manage headaches.[142] Thus, CBT may be useful in improving

quality of life in individuals with headaches, even for the proportion of those who do not experience clinically significant reductions in headache symptoms.

CONCLUSION

Understanding the influence of psychological factors on a patient's headache presentation can improve quality of care and, in turn, the patient's quality of life. Modification of psychological factors to improve quality of life can occur within the clinic visits with the physician, or through treatments provided by psychologists and other behavioral healthcare providers. Healthcare providers can educate patients to normalize known headache triggers, particularly stress, sleep, and diet, which can help patients gain control over headache onset, exacerbation, and chronification. For patients experiencing difficulty managing these triggers, a course of behavioral treatment to address such triggers (e.g., cognitive behavioral therapy to normalize sleep or to decrease stress) may be warranted. Treatment of psychiatric comorbidities of headache disorders, most commonly depression and anxiety disorders, can also improve the quality of life of patients with headaches. Psychological interventions can also improve the care of headache patients through improving medication adherence and communication with providers. Psychological treatments, including relaxation, biofeedback, and cognitive behavioral therapies, have demonstrated efficacy for the treatment of headache disorders. Referral to psychologists and other behavioral healthcare providers trained in these interventions can improve the care of many patients with headaches.

References

1. Lipton R, Bigal M, Diamond M, Freitag F, Reed M, Stewart W. Migraine prevalence, disease burden, and the need for preventive therapy. *Neurology*. 2007;68:343–349.
2. Buse D, Manack A, Serrano D, Reed M, Varon S, Turkel C, et al. Headache impact of chronic and episodic migraine: results from the American Migraine Prevalence and Prevention study. *Headache*. 2012;52(1):3–17.
3. Martelletti P, Birbeck GL, Katsarava Z, Jensen RH, Stovner LJ, Steiner TJ. The global burden of disease survey 2010, lifting the burden and thinking outside-the-box on headache disorders. *J Headache Pain*. 2013;14(1):13.
4. Leonardi M, Steiner TJ, Scher AT, Lipton RB. The global burden of migraine: measuring disability in headache disorders with WHO's classification of functioning, disability and health (ICF). *J Headache Pain*. 2005;6(6):429–440.
5. Smitherman TA, Burch R, Sheikh H, Loder E. The prevalence, impact, and treatment of migraine and severe headaches in the United States: a review of statistics from national surveillance studies. *Headache*. 2013;53(3):427–436.
6. Brandes JL. The migraine cycle: patient burden of migraine during and between migraine attacks. *Headache*. 2008;48(3):430–441.

7. Young WB, Park JE, Tian IX, Kempner J. The stigma of migraine. *PLoS ONE*. 2013;8(1):e54074.

8. Seng EK, Holroyd KA. Optimal use of acute headache medication: a qualitative examination of behaviors and barriers to their performance. *Headache*. 2013;53(9):1438−1450.

9. Rosen NL. Psychological issues in the evaluation and treatment of tension-type headache. *Curr Pain Headache Rep*. 2012;16 (6):545−553.

10. Silberstein S, Lipton R, Goadsby P. *Headache in Clinical Practice*. London, UK: Martin Dunitz; 2002.

11. Wolff HG. *Headache and other Head Pain*. New York, NY: Oxford University Press; 1948.

12. Andress-Rothrock D, King W, Rothrock J. An analysis of migraine triggers in a clinic-based population. *Headache*. 2010;50 (8):1366−1370.

13. Hougaard A, Amin FM, Hauge AW, Ashina M, Olesen J. Provocation of migraine with aura using natural trigger factors. *Neurology*. 2013;80(5):428−431.

14. Houle TT, Turner DP. Natural experimentation is a challenging method for identifying headache triggers. *Headache*. 2013;53 (4):636−643.

15. Hauge AW, Kirchmann M, Olesen J. Trigger factors in migraine with aura. *Cephalalgia*. 2010;30(3):346−353.

16. Martin PR. Behavioral management of migraine headache triggers: learning to cope with triggers. *Curr Pain Headache Rep*. 2010;14(3):221−227.

17. Coyne J, Holroyd K. Stress, coping and illness: a transactional perspective. In: Millon T, Green C, Meagher R, eds. *Handbook of Health Care Clinical Psychology*. New York, NY: Plenum; 1982.

18. Nash J, Thebarge R. Understanding psychological stress, its biological processes, and impact on primary headache. *Headache*. 2006;46(9):1377−1386.

19. Lipton RB, Buse DC, Hall CB, Tennen H, Defreitas TA, Borkowski TM, et al. Reduction in perceived stress as a migraine trigger: testing the "let-down headache" hypothesis. *Neurology*. 2014;82(16):1395−1401.

20. Houle TT, Butschek RA, Turner DP, Smitherman TA, Rains JC, Penzien DB. Stress and sleep duration predict headache severity in chronic headache sufferers. *Pain*. 2012;153(12):2432−2440.

21. Goadsby PJ. Stress and migraine: something expected, something unexpected. *Neurology*. 2014;82(16):1388−1389.

22. Bigal ME, Lipton RB. What predicts the change from episodic to chronic migraine? *Curr Opin Neurol*. 2009;22(3):269−276.

23. Scher A, Stewart W, Buse D, Krantz DL, Lipton RB. Major life changes before and after the onset of chronic daily headache: a population based study. *Cephalalgia*. 2008;28:868−876.

24. Holroyd KA, Cottrell CK, O'Donnell FJ, Cordingley GE, Drew JB, Carlson BW, et al. Effect of preventive (beta blocker) treatment, behavioural migraine management, or their combination on outcomes of optimised acute treatment in frequent migraine: randomised controlled trial. *BMJ*. 2010;341:c4871.

25. Holroyd KA, O'Donnell FJ, Stensland M, Lipchik GL, Cordingley GE, Carlson B. Management of chronic tension-type headache with tricyclic antidepressant medication, stress-management therapy, and their combination: a randomized controlled trial. *JAMA*. 2001;285(17):2208−2215.

26. Martin PR, MacLeod C. Behavioral management of headache triggers: avoidance of triggers is an inadequate strategy. *Clin Psychol Rev*. 2009;29(6):483−495.

27. Rains JC, Poceta JS. Headache and sleep disorders: review and clinical implications for headache management. *Headache*. 2006; 46(9):1344−1363.

28. Headache Classification Committee of the International Headache Society. The International Classification of Headache Disorders, 3rd edition (beta version). *Cephalalgia*. 2013;33(9): 629−808.

29. Xu Y, Padiath QS, Shapiro RE, Jones CR, Wu SC, Saigoh N, et al. Functional consequences of a CKIdelta mutation causing familial advanced sleep phase syndrome. *Nature*. 2005;434(7033):640−644.

30. Rains J, Poceta JS. Sleep-related headache syndromes. In: Avidan A, ed. *Seminars in Neurology*. Vol 25. New York, NY: Thieme Medical Publishers; 2006:69−80.

31. Ruff RL, Ruff SS, Wang XF. Improving sleep: initial headache treatment in OIF/OEF veterans with blast-induced mild traumatic brain injury. *J Rehabil Res Dev*. 2009;46(9):1071−1084.

32. Babson KA, Feldner MT, Badour CL. Cognitive behavioral therapy for sleep disorders. *Psychiatr Clin North Am*. 2010;33 (3):629−640.

33. Calhoun A, Ford S. Behavioral sleep modifications may revert transformed migraine to episodic migraine. *Headache*. 2007;47 (8):1178−1183.

34. Marcus DA, Scharff L, Turk D, Gourley LM. A double-blind provocative study of chocolate as a trigger of headache. *Cephalalgia*. 1997;17(8):855−862.

35. Alpay K, Ertas M, Orhan EK, Ustay DK, Lieners C, Baykan B. Diet restriction in migraine, based on IgG against foods: a clinical double-blind, randomised, cross-over trial. *Cephalalgia*. 2010;30(7):829−837.

36. Peterlin BL, Rapoport AM, Kurth T. Migraine and obesity: epidemiology, mechanisms, and implications. *Headache*. 2010;50 (4):631−648.

37. Bond DS, Roth J, Nash JM, Wing RR. Migraine and obesity: epidemiology, possible mechanisms and the potential role of weight loss treatment. *Obes Rev*. 2011;12(5):e362−e371.

38. Bigal ME, Lipton RB. Obesity is a risk factor for transformed migraine but not chronic tension-type headache. *Neurology*. 2006;67(2):252−257.

39. Torelli P, Manzoni GC. Fasting headache. *Curr Pain Headache Rep*. 2010;14(4):284−291.

40. Verrotti A, Agostinelli S, D'Egidio C, Di Fonzo A, Carotenuto M, Parisi P, et al. Impact of a weight loss program on migraine in obese adolescents. *Eur J Neurol*. 2013;20(2):394−397.

41. Silberstein SD, Lipton RB, Breslau N. Migraine: association with personality characteristics and psychopathology. *Cephalalgia*. 1995;15:358−369.

42. Brandt J, Celentano D, Stewart W, Linet M, Folstein M. Personality and emotional disorder in a community sample of migraine headache sufferers. *Am J Psychiatry*. 1990;147:303−308.

43. Breslau N, Andreski P. Migraine, personality, and psychiatric comorbidity. *Headache*. 1995;35:382−386.

44. Persson B. Growth environment and personality in adult migraineurs and their migraine-free siblings. *Headache*. 1997;37 (3):159−168.

45. Cao M, Zhang S, Wang K, Wang Y, Wang W. Personality traits in migraine and tension-type headaches: a five-factor model study. *Psychopathology*. 2002;35(4):254−258.

46. Huber D, Henrich G. Personality traits and stress sensitivity in migraine patients. *Behav Med*. 2003;29(1):4−13.

47. Abbate-Daga G, Fassino S, Lo Giudice R, Rainero I, Gramaglia C, Marech L, et al. Anger, depression and personality dimensions in patients with migraine without aura. *Psychother Psychosom*. 2007;76(2):122−128.

48. Mongini F, Fassino S, Rota E, Deregibus A, Levi M, Monticone D, et al. The temperament and character inventory in women with migraine. *J Headache Pain*. 2005;6(4):247−249.

49. Sanchez-Roman S, Tellez-Zenteno JF, Zermeno-Phols F, Garcia-Ramos G, Velazquez A, Derry P, et al. Personality in patients with migraine evaluated with the "Temperament and Character Inventory". *J Headache Pain*. 2007;8(2):94−104.

50. Davis RE, Smitherman TA, Baskin SM. Personality traits, personality disorders, and migraine: a review. *Neurol Sci*. 2013; 34(suppl 1):S7−S10.

51. Holroyd K, Drew J, Cottrell C, Romanek K, Heh V. Impaired functioning and quality of life in severe migraine: the role of catastrophizing and associated symptoms. *Cephalalgia*. 2007;27:1156–1165.

52. Bigal ME, Sheftell FD, Rapoport AM, Tepper SJ, Weeks R, Baskin SM. MMPI personality profiles in patients with primary chronic daily headache: a casecontrol study. *Neurol Sci*. 2003;24(3):103–110.

53. Karakurum B, Soylu O, Karatas M, Giray S, Tan M, Arlier Z, et al. Personality, depression, and anxiety as risk factors for chronic migraine. *Int J Neurosci*. 2004;114(11):1391–1399.

54. Mongini F, Rota E, Deregibus A, Mura F, Francia Germani A, Mongini T. A comparative analysis of personality profile and muscle tenderness between chronic migraine and chronic tension-type headache. *Neurol Sci*. 2005;26(4):203–207.

55. Weeks R, Baskin S, Rapoport A, Sheftell F, Arrowsmith F. A comparison of MMPI personality data and frontalis electromyographic readings in migraine and combination headache patients. *Headache*. 1983;23(2):75–82.

56. Rausa M, Cevoli S, Sancisi E, Grimaldi D, Pollutri G, Casoria M, et al. Personality traits in chronic daily headache patients with and without psychiatric comorbidity: an observational study in a tertiary care headache center. *J Headache Pain*. 2013;14(1):22.

57. Galli F, Pozzi G, Frustaci A, Allena M, Anastasi S, Chirumbolo A, et al. Differences in the personality profile of medication-overuse headache sufferers and drug addict patients: a comparative study using MMPI-2. *Headache*. 2011;51(8):1212–1227.

58. Radat F, Lanteri-Minet M. What is the role of dependence-related behavior in medication-overuse headache?. *Headache*. 2010;50(10):1597–1611.

59. Da Silva AN, Lake III AE. Clinical aspects of medication overuse headaches. *Headache*. 2014;54(1):211–217.

60. Radat F, Swendsen J. Psychiatric comorbidity in migraine: a review. *Cephalalgia*. 2005;25:165–178.

61. Hamelsky SW, Lipton RB. Psychiatric comorbidity of migraine. *Headache*. 2006;46(9):1327–1333.

62. Baskin S, Lipchik G, Smitherman T. Mood and anxiety disorders in chronic headache. *Headache*. 2006;46(s3):S76–S87.

63. Guidetti V, Galli F, Fabrizi P, Giannantoni AS, Napoli L, Bruni O, et al. Headache and psychiatric comorbidity: clinical aspects and outcome in an 8-year follow-up study. *Cephalalgia*. 1998;18:455–462.

64. Pesa J, Lage MJ. The medical costs of migraine and comorbid anxiety and depression. *Headache*. 2004;44(6):562–570.

65. Lanteri-Minet M, Radat F, Chautard MH, Lucas C. Anxiety and depression associated with migraine: influence on migraine subjects' disability and quality of life, and acute migraine management. *Pain*. 2005;118:319–326.

66. Bigal ME, Lipton RB. Modifiable risk factors for migraine progression. *Headache*. 2006;46(9):1334–1343.

67. Scher A, Midgette L, Lipton R. Risk factors for headache chronification. *Headache*. 2008;48:16–25.

68. Smitherman TA, Penzien DB, Maizels M. Anxiety disorders and migraine intractability and progression. *Curr Pain Headache Rep*. 2008;12(3):224–229.

69. Breslau N, Lipton RB, Stewart WF, Schultz LR, Welch KMA. Comorbidity of migraine and depression. *Neurology*. 2003;60:1308–1312.

70. Breslau N, Schultz LR, Stewart WF, Lipton RB, Lucia V, Welch KMA. Headache and major depression: is the association specific to migraine? *Neurology*. 2000;54:308–313.

71. Baskin SM, Smitherman TA. Migraine and psychiatric disorders: comorbidities, mechanisms, and clinical applications. *Neurol Sci*. 2009;30(suppl 1):S61–S65.

72. McWilliams LA, Goodwin RD, Cox BJ. Depression and anxiety associated with three pain conditions: results from a nationally representative sample. *Pain*. 2004;111(1–2):77–83.

73. Patel NV, Bigal ME, Kolodner KB, Leotta C, Lafata JE, Lipton RB. Prevalence and impact of migraine and probable migraine in a health plan. *Neurology*. 2004;63(8):1432–1438.

74. Jette N, Patten S, Williams J, Becker W, Wiebe S. Comorbidity of migraine and psychiatric disorders—a national population-based study. *Headache*. 2008;48(4):501–516.

75. Breslau N, Davis GC, Andreski P. Migraine, psychiatric disorders, and suicide attempts: an epidemiologic study of young adults. *Psychiatry Res*. 1991;37(1):11–23.

76. Merikangas KR, Angst J, Isler H. Migraine and psychopathology: results of the Zurich cohort study of young adults. *Arch Gen Psychiatry*. 1990;47:849–852.

77. Merikangas KR, Merikangas JR, Angst J. Headache syndromes and psychiatric disorders: association and family transmission. *J Psychiatr Res*. 1993;27:197–210.

78. Ratcliffe GE, Enns MW, Jacobi F, Belik SL, Sareen J. The relationship between migraine and mental disorders in a population-based sample. *Gen Hosp Psychiatry*. 2009;31(1):14–19.

79. Fasmer OB, Oedegaard KJ. Clinical characteristics of patients with major affective disorders and comorbid migraine. *World J Biol Psychiatry*. 2001;2(3):149–155.

80. Oedegaard KJ, Neckelmann D, Mykletun A, Dahl AA, Zwart JA, Hagen K, et al. Migraine with and without aura: association with depression and anxiety disorder in a population-based study. The HUNT Study. *Cephalalgia*. 2006;26(1):1–6.

81. Verri AP, Cecchini AP, Galli C, Granella F, Sandrini G, Nappi G. Psychiatric comorbidity in chronic daily headache. *Cephalalgia*. 1998;18(suppl 21):45–49.

82. Zwart JA, Dyb G, Hagen K, Odegard KJ, Dahl AA, Bovim G, et al. Depression and anxiety disorders associated with headache frequency. The Nord-Trondelag Health Study. *Eur J Neurol*. 2003;10(2):147–152.

83. Juang K, Wang S, Fuh J. Comorbidity of depressive and anxiety disorders in chronic daily headache and its subtypes. *Headache*. 2000;40(10):818–823.

84. Buse DC, Manack A, Serrano D, Turkel C, Lipton RB. Sociodemographic and comorbidity profiles of chronic migraine and episodic migraine sufferers. *J Neurol Neurosurg Psychiatry*. 2010;81(4):428–432.

85. Atasoy HT, Atasoy N, Unal AE, Emre U, Sumer M. Psychiatric comorbidity in medication overuse headache patients with pre-existing headache type of episodic tension-type headache. *Eur J Pain*. 2005;9(3):285–291.

86. Radat F, Creac'h C, Swendsen JD, Lafittau M, Irachabal S, Dousset V, et al. Psychiatric comorbidity in the evolution from migraine to medication overuse headache. *Cephalalgia*. 2005;25(7):519–522.

87. Breslau N. Psychiatric comorbidity in migraine. *Cephalalgia*. 1998;18(suppl 22):56–58 [discussion 8–61].

88. Guidetti V, Galli F. Evolution of headache in childhood and adolescence: an 8-year follow-up. *Cephalalgia*. 1998;18:449–454.

89. Mercante JP, Peres MF, Bernik MA. Primary headaches in patients with generalized anxiety disorder. *J Headache Pain*. 2011;12(3):331–338.

90. Saunders K, Merikangas K, Low NC, Von Korff M, Kessler RC. Impact of comorbidity on headache-related disability. *Neurology*. 2008;70(7):538–547.

91. Wang SJ, Juang KD, Fuh JL, Lu SR. Psychiatric comorbidity and suicide risk in adolescents with chronic daily headache. *Neurology*. 2007;68(18):1468–1473.

92. Smitherman TA, Kolivas ED, Bailey JR. Panic disorder and migraine: comorbidity, mechanisms, and clinical implications. *Headache*. 2013;53(1):23–45.

93. Breslau N, Schultz LR, Stewart WF, Lipton R, Welch KM. Headache types and panic disorder: directionality and specificity. *Neurology*. 2001;56(3):350–354.

94. Tietjen GE, Brandes JL, Peterlin BL, Eloff A, Dafer RM, Stein MR, et al. Childhood maltreatment and migraine (part I). Prevalence and adult revictimization: a multicenter headache clinic survey. *Headache*. 2010;50(1):20−31.

95. Tietjen GE, Brandes JL, Peterlin BL, Eloff A, Dafer RM, Stein MR, et al. Childhood maltreatment and migraine (part II). Emotional abuse as a risk factor for headache chronification. *Headache*. 2010;50(1):32−41.

96. Tietjen GE, Peterlin BL. Childhood abuse and migraine: epidemiology, sex differences, and potential mechanisms. *Headache*. 2011;51(6):869−879.

97. Peterlin BL, Tietjen GE, Brandes JL, Rubin SM, Drexler E, Lidicker JR, et al. Post-traumatic stress disorder in migraine. *Headache*. 2009;49(4):541−551.

98. Peterlin BL, Nijjar SS, Tietjen GE. Post-traumatic stress disorder and migraine: epidemiology, sex differences, and potential mechanisms. *Headache*. 2011;51(6):860−868.

99. Lake AE, Saper JR, Madden SF, Kreeger C. Comprehensive inpatient treatment for intractable migraine: a prospective long-term outcome study. *Headache*. 1993;33:55−62.

100. Rothrock J, Lopez I, Zweilfer R, Andress-Rothrock D, Drinkard R, Walters N. Borderline personality disorder and migraine. *Headache*. 2007;47(1):22−26.

101. Seng EK, Holroyd KA. Psychiatric comorbidity and response to preventative therapy in the treatment of severe migraine trial. *Cephalalgia*. 2012;32(5):390−400.

102. Lipchik GL, Smitherman TA, Penzien DB, Holroyd KA. Basic principles and techniques of cognitive-behavioral therapies for comorbid psychiatric symptoms among headache patients. *Headache*. 2006;46(s3):S119−S132.

103. Smitherman TA, Maizels M, Penzien DB. Headache chronification: screening and behavioral management of comorbid depressive and anxiety disorders. *Headache*. 2008;48(1):45−50.

104. De Cort K, Griez E, Buchler M, Schruers K. The role of "interoceptive" fear conditioning in the development of panic disorder. *Behav Ther*. 2012;43(1):203−215.

105. Norton PJ, Asmundson GJ. Anxiety sensitivity, fear, and avoidance behavior in headache pain. *Pain*. 2004;111(1−2):218−223.

106. Katic BJ, Krause SJ, Tepper SJ, Hu HX, Bigal ME. Adherence to acute migraine medication: what does it mean, why does it matter? *Headache*. 2010;50(1):117−129.

107. Ifergane G, Wirguin I, Shvartzman P. Triptans − why once? *Headache*. 2006;46(8):1261−1263.

108. Dunbar-Jacob J, Erlen JA, Schlenk EA, Ryan CM, Sereika SM, Doswell WM. Adherence in chronic disease. *Annu Rev Nurs Res*. 2000;18:48−90.

109. Peters M, Abu-Saad HH, Vydelingum V, Dowson A, Murphy M. Patients' decision-making for migraine and chronic daily headache management. A qualitative study. *Cephalalgia*. 2003;23(8):833−841.

110. D'Amico D, Moschiano F, Bussone G. Early treatment of migraine attacks with triptans: a strategy to enhance outcomes and patient satisfaction? *Expert Rev Neurother*. 2006;6(7):1087−1097.

111. Foley K, Cady R, Martin V, Adelman J, Diamond M, Bell C, et al. Treating early versus treating mild: timing of migraine prescription medications among patients with diagnosed migraine. *Headache*. 2005;45(5):538−545.

112. Malik SN, Hopkins M, Young WB, Silberstein SD. Acute migraine treatment: patterns of use and satisfaction in a clinical population. *Headache*. 2006;46(5):773−780.

113. Cady RK, Farmer K, Beach ME, Tarrasch J. Nurse-based education: an office-based comparative model for education of migraine patients. *Headache*. 2008;48(4):564−569.

114. Gallagher R, Kunkel R. Migraine medication attributes important for patient compliance: concerns about side effects may delay treatment. *Headache*. 2003;43(43):36−43.

115. Ottervanger J, Valkenburg H, Grobbee D, Pingel J, Theofanous J, Jackson D, et al. Pattern of sumatriptan use and overuse in general practice. *Eur J Clin Pharmacol*. 1996;50(5):353−355.

116. Packard RC, O'Connell P. Medication compliance among headache patients. *Headache*. 1986;26:416−419.

117. Meskunas CA, Tepper SJ, Rapoport AM, Sheftell FD, Bigal ME. Medications associated with probable medication overuse headache reported in a tertiary headache center over a 15-year period. *Headache*. 2006;46(5):766−772.

118. Rains J, Penzien D, Lipchik G. Behavioral facilitation of medical treatment for headache part II: theoretical models and behavioral strategies for improving adherence. *Headache*. 2006;46(9):1395−1403.

119. Stevens J, Hayes J, Pakalnis A. A randomized trial of telephone-based motivational interviewing for adolescent chronic headache with medication overuse. *Cephalalgia*. 2014;34(6):446−454.

120. Attar HS, Chandramani S. Impact of physician empathy on migraine disability and migraineur compliance. *Ann Indian Acad Neurol*. 2012;15(suppl 1):S89−S94.

121. Buse D, Rupnow M, Lipton R. Assessing and managing all aspects of migraine: migraine attacks, migraine-related functional impairment, common comorbidities, and quality of life. *Mayo Clin Proc*. 2009;84(5):422−435.

122. Weaver MF, Bond DS, Arnold BL, Waterhouse E, Towne A. Aberrant drug-taking behaviors and headache: patient versus physician report. *Am J Health Behav*. 2006;30(5):475−482.

123. Lipton RB, Stewart WF, Sawyer J, Edmeads J. Clinical utility of an instrument assessing migraine disability: the Migraine Disability Assessment (MIDAS) Questionnaire. *Headache*. 2001;41:854−861.

124. Baos V, Ester F, Castellanos A, Nocea G, Caloto MT, Gerth WC. Use of a structured migraine diary improves patient and physician communication about migraine disability and treatment outcomes. *Int J Clin Pract*. 2005;59(3):281−286.

125. Edmeads J, Lainez JM, Brandes JL, Schoenen J, Freitag F. Potential of the Migraine Disability Assessment (MIDAS) Questionnaire as a public health initiative and in clinical practice. *Neurology*. 2001;56(6 suppl 1):S29−S34.

126. Manzoni GC, Torelli P. The patient−physician relationship in the approach to therapeutic management. *Neurol Sci*. 2007;28(suppl 2):S130−S133.

127. Buse DC, Lipton RB. Facilitating communication with patients for improved migraine outcomes. *Curr Pain Headache Rep*. 2008;12(3):230−236.

128. Silberstein S. Practice parameter: evidence-based guidelines for migraine headache (an evidence-based review). *Neurology*. 2000;55(6):754−762.

129. Bendtsen L, Bigal ME, Cerbo R, Diener HC, Holroyd K, Lampl C, et al. Guidelines for controlled trials of drugs in tension-type headache: second edition. *Cephalalgia*. 2010;30(1):1−16.

130. Andrasik F. Biofeedback in headache: an overview of approaches and evidence. *Cleve Clin J Med*. 2010;77(suppl 3):S72−S76.

131. Nicholson RA, Buse DC, Andrasik F, Lipton RB. Nonpharmacologic treatments for migraine and tension-type headache: how to choose and when to use. *Curr Treat Options Neurol*. 2011;13:28−40.

132. Nestoriuc Y, Rief W, Martin A. Meta-analysis of biofeedback for tension-type headache: efficacy, specificity, and treatment moderators. *J Consult Clin Psychol*. 2008;76(3):379−396.

133. Nestoriuc Y, Martin A. Efficacy of biofeedback for migraine: a meta-analysis. *Pain*. 2007;128(1–2):111–127.

134. Holroyd KA, Penzien DB, Hursey K, Tobin D, Rogers L, Holm J, et al. Change mechanisms in EMG biofeedback training: cognitive changes underlying improvements in tension headache. *J Consult Clin Psychol*. 1984;52(6):1039–1053.

135. Mizener D, Thomas M, Billings R. Cognitive changes of migraineurs receiving biofeedback training. *Headache*. 1988;28(5): 339–343.

136. Schwartz MS, Andrasik F. *Biofeedback: A Practitioner's Guide*. 3rd ed. New York, NY: Guilford Press; 2003.

137. Lipchik G, Holroyd K, Nash J. Cognitive-behavioral management of recurrent headache disorders: a minimal-therapist contact approach. In: Turk D, Gatchel R, eds. *Psychological Approaches to Pain Management*. 2nd ed. New York, NY: Guilford Press; 2002:356–389.

138. Jacobson E. *Progressive Relaxation*. Chicago, IL: University of Chicago Press; 1938.

139. Nestoriuc Y, Martin A, Rief W, Andrasik F. Biofeedback treatment for headache disorders: a comprehensive efficacy review. *Appl Psychophysiol Biofeedback*. 2008;33(3):125–140.

140. Rains JC, Penzien DB, McCrory DC, Gray RN. Behavioral headache treatment: history, review of the empirical literature, and methodological critique. *Headache*. 2005;45(S2):92–109.

141. Beck A. Cognitive approaches to stress. In: Lehrer P, Woolfolk R, eds. *Principles and Practice of Stress Management*. New York, NY: The Guilford Press; 1993:333–371.

142. Thorn BE, Pence LB, Ward LC, Kilgo G, Clements KL, Cross TH, et al. A randomized clinical trial of targeted cognitive behavioral treatment to reduce catastrophizing in chronic headache sufferers. *J Pain*. 2007;8(12):938–949.

143. Holroyd K, Chen Y. A hand-held computer headache diary program: monitoring headaches, medication use, and disability in real time. *Cephalalgia*. 2000;36:123.

144. Holroyd K, Cottrell C, Drew J. Catastrophizing response to migraines predicts impaired functioning and quality-of-life in frequent migraine. *J Pain*. 2006;7:S71.

145. Martin PR, Forsyth MR, Reece J. Cognitive behavioral therapy versus temporal pulse amplitude biofeedback training for recurrent headache. *Behav Ther*. 2007;38(4):350–363.

146. Powers SW, Kashikar-Zuck SM, Allen JR, LeCates SL, Slater SK, Zafar M, et al. Cognitive behavioral therapy plus amitriptyline for chronic migraine in children and adolescents: a randomized clinical trial. *JAMA*. 2013;310(24):2622–2630.

147. Schafer AM, Rains JC, Penzien DB, Groban L, Smitherman TA, Houle TT. Direct costs of preventive headache treatments: comparison of behavioral and pharmacologic approaches. *Headache*. 2011;51(6):985–991.

148. Nicholson R, Nash J, Andrasik F. A self-administered behavioral intervention using tailored messages for migraine. *Headache*. 2005;45(5):1124–1139.

149. Trautmann E, Kroner-Herwig B. A randomized controlled trial of Internet-based self-help training for recurrent headache in childhood and adolescence. *Behav Res Ther*. 2010;48(1):28–37.

150. Andrasik F. What does the evidence show? Efficacy of behavioural treatments for recurrent headaches in adults. *Neurol Sci*. 2007;28(suppl 2):S70–S77.

151. Weeks RE. Behavioral management of headache. *Tech Reg Anesth Pain Manag*. 2009;13(1):50–57.

20

Too Much of a Good Thing: Medication Overuse Headache

Duren Michael Ready

Headache Clinic, Baylor Scott & White, Temple, Texas, USA

INTRODUCTION

Many factors are involved in the progression of episodic migraine to chronic migraine.[1] One of the most significant perpetuating factors for migraine progression is the overuse of analgesic medication.[2] Medication overuse headaches, colloquially known as rebound headaches, result from the frequent use of abortive medications. The use of "rebound" may lead to confusion as people typically associate rebound with the return of an acute headache once the medication has worn off, instead of the intended increased frequency of attacks. This chapter will review the history, classification, pathogenesis, behavior, and, treatment of medication overuse headache. The use of opiate medication in headache will also be discussed, and how to manage it in appropriately selected patients.

All individuals with chronic migraine should be assessed for medication overuse.[3,4] In the primary care setting, the incidence of medication overuse headache (MOH) is estimated to be 21%.[5] MOH is the most common secondary headache seen in clinical practice, representing upwards of 70% of the patients seen in a tertiary headache clinic.[6]

Typically, MOH occurs because the patients were just following the doctors' orders.[7] Remember that medication overuse is only one of the factors associated with migraine progression. A successful resolution of medication overuse may not lead to a return to the baseline headache frequency without first searching out and addressing additional perpetuating factors.

Medication overuse is known to blunt the effectiveness of preventive interventions.[6] Few preventive therapies have demonstrated benefit in the face of MOH. It may also be influenced by cultural aspects.[7] Similar to the disease of addiction, MOH may best be understood as a chronic, relapsing, remitting disorder with many distinct influences. It is also seen in children, who appear to be at higher risk because of their age.[8] Patients may also be at risk due to adverse events of the medications they are overusing.

Identifying a single responsible agent may be difficult, as patients will frequently engage in polypharmacy to treat their headaches.[9] Some physicians have attempted to avoid provoking MOH by treating their patients on alternating days with opiates and butalbital-containing medications. These strategies are not effective, as the number of days of acute medication usage are additive.

Almost every acute headache medication has been associated at some point in time with migraine progression, although all have differing propensities. Opiates and barbiturates have been shown to induce progression with a lesser frequency of the days used.[10–12] Triptans appear to provoke migraine progression at a much more rapid pace than other medications, but, after withdrawal, triptan-induced MOH seems to revert to a baseline level of headache more rapidly.[13] Non-steroidal anti-inflammatory drugs (NSAIDs) seem to present a mixed picture of protection and progression. NSAIDS appear to be protective with headache frequencies of 10 days a month or less.[13] Low-dose aspirin seems to be protective in men but not in women. While the earliest reports of MOH were associated with ergotamine compounds, it is generally believed that an ergotamine derivative, dihydroergotamine (DHE), does not induce progression; however, it has been reported to do so in a few instances.[14]

While almost all acute medication may induce progression in isolation or while under favorable circumstances, it does not appear that all medications induce MOH by the same pathways.[14]

S. Diamond, R. K. Cady, M. L. Diamond & V. T. Martin (Eds):
Headache and Migraine Biology and Management.

DOI: http://dx.doi.org/10.1016/B978-0-12-800901-7.00020-3

HISTORY/BACKGROUND

The paradoxical effect of analgesic medications increasing headache frequency and intensity has long been reported. Initially, MOH was called "ergotamine overuse headache" or "drug-induced headache." In 1982, progressive headaches were linked with overusing simple analgesics.[15] Over time the most common term for progressive headaches became "transformed migraine," with overusing medication labeled as the leading reason for progression.[16]

The increasing awareness of this condition in headache clinics led to the development of formal criteria for chronic migraine by Silberstein and colleagues.[17]

They reported that medication overuse was one factor associated with migraine progression. Today, transformed migraine is known as chronic migraine with multiple identifiable mutable and non-mutable risk factors for progression.[7] Of these risk factors, medication overuse is thought to be the most important factor in migraine progression.[18,19]

The International Classification of Headache Disorders (ICHD-III beta), now in its third edition, defines the criteria for MOH as a headache that occurs in an individual with pre-existing primary headache who (in the presence of medication overuse) develops a new type of headache or a marked worsening of their pre-existing headache (Table 20.1).[20] It further

TABLE 20.1   International Headache Society International Classification of Headache Disorders (ICHD-III) Criteria for Medication Overuse Headache

	Medication responsible for overuse headache	Diagnostic criteria	Average monthly days used/months	Comments
8.2.1	Ergotamine	A, B, C	≥10 days/month × 3 months	No minimum dose defined
8.2.2	Triptan	A, B, C	≥10 days/month × 3 months	Most rapid inducer of MOH
8.2.3	Simple analgesic 8.2.3.1 APAP 8.2.3.2 ASA 8.2.3.3 NSAID	A, B, C	≥15 days/month × 3 months	
8.2.4	Opioids	A, B, C	≥10 days/month × 3 months	Lipton/Bigal identified 8 days/month as the level with a greater effect in men Individuals using opiates for other pain condition can induce MOH These patients tend to have higher relapse rates following withdrawal
8.2.5	Combination analgesic	A, B, C	≥10 days/month × 3 months	Butalbital is believed to induce progression with as few as 5 days/month
8.2.6	MOH attributed to multiple drug classes not individually overused	A, B, C	≥10 days/month × 3 months of any of the above medications	
8.2.7	MOH attributed to unverified overuse of multiple drug classes	A, B, C	Regular overuse of other medications for greater than 3 months	Needs to have regular use greater than 10 days/month × 3 months AND the classification of the overused medication cannot be reliably established
8.2.8	MOH attributed to other medication	A, B, C	A. Regular overuse, on 10 days per month for >3 months, of one or more medications other than those described above, taken for acute or symptomatic treatment of headache.	

Diagnostic criteria:
A. Headache occurring on 15 days per month in a patient with a pre-existing headache disorder.
B. Regular overuse for >3 months of one or more drugs that can be taken for acute and/or symptomatic treatment of headache.
C. Not better accounted for by another ICHD-III diagnosis.

states that individuals who meet the criteria for chronic migraine and MOH should be given both diagnoses. Past ICHD criteria required that headaches improve after withdrawal of the offending medication, but this requirement is no longer present in the ICHD-III beta criteria. The latest edition also established constitutes "overuse" based on what medication is used.

PATHOPHYSIOLOGY

It is important to keep an open mind when trying to determine the pathway for migraine progression. As there are many pathways for the progression from episodic to chronic migraine, of which medication overuse is but one, it is wise to remember Lord Osler's admonition that it is more important to know "what kind of person has the disease than what disease the person has".[21] Individuals may arrive at MOH by one or multiple pathways. They may develop chronic migraine or MOH slowly, singularly over time, or acting in concert to bring about a more rapid progression. The final diagnosis of MOH might not be as important as identifying which pathway the patient took to develop MOH. Patients' attitude regarding their medication usage potentially provides clues as to which pathway has led to progression. Additionally, there are some patients who overuse medications but do not develop MOH, and if we draw the premature conclusion that MOH is the "usual suspect" we may delay the ultimate regression back to episodic migraine.

In migraine patients, daily triptan exposure downregulates receptors in multiple cortical regions by disrupting the 5-HT system. Chronic acetaminophen use upregulates cerebral cortex pronociceptive 5-HT2A receptors.[22] Continual use of opiates is frequently associated with increased pain. One explanation for this is the phenomenon known as opioid-induced hyperalgesia (OIH) — the paradoxical worsening of pain with continuing or increasing opiate usage.[23] Many of the features seen in OIH are also observed in MOH. In both cases, the increasing pain involves upregulation of CGRP and the hyperexcitability of dorsal horn neurons.[24,25] Opiates are thought to induce hyperalgesia through glial cell activation via TLR-4 stimulation. Emerging evidence associates chronic pain with increased TLR-4 sensitivity.[25] Animal models demonstrate diminished 5-HT levels, increased cortical spreading depression susceptibility, and increased CGRP expression in the trigeminal ganglia with cortical spreading depression-induced increased calcitonin gene-related peptide (CGRP) release. The 5-HT system dysfunction facilitates trigeminal nociception. Frequent exposure to the same substance changes the expression and sensitization of the receptors.[26]

Trigeminal system function is substantially disrupted by the loss of central modulating control. This disruption may occur as decreasing nociceptive inhibition, facilitating nociception or augmenting central sensitization itself. This derangement in the endogenous pain modulating system increases pain perception and likely reinforces medication usage.

Medication overuse produces an increase excitability of cerebral cortex and trigeminal neurons. This hyperexcitability renders the migraine brain more susceptible to cortical spreading depression, and facilitates central and peripheral sensitization through the trigeminal system.[14]

The hyperexcitability (sensitization) is believed to be the same process that is taking place in migraine. The repetitive activation of hypersensitive neurons of the trigeminovascular pathway produces biologic and functional changes in wide dynamic nociceptive neurons in the nucleus caudalis, resulting in a reduced firing threshold, expansion of receptive fields, and the clinical correlates of cutaneous allodynia.[14]

Central Sensitization Pathway

Central sensitization may be a pathway for the development of MOH, or it may represent a final common pathway where influences combine to drive migraine progression. It may be induced by frequent and inadequately treated pain.[27,28] Over time, the continued pain state lowers the threshold centrally for pain activation and increases the sensory receptive fields, resulting in progression to chronic migraine. Central sensitization has been shown to change the structure and function of the trigeminal nucleus caudalis. These changes decrease platelet serotonin and increase 5-H2A excitatory receptors.[29] One observed alteration produced by chronic migraine is the increased iron deposition around the periaqueductal gray, representing an alteration of the central pain pathways impairing descending pain modulation.[30–32]

Central sensitization (allodynia) is a risk factor for migraine progression. It may be induced by opiates,[33] triptans,[34] or headache frequency.[35] It impairs descending pain modulation from the brainstem rostral ventral medial medulla, limiting individuals' ability to tolerate nociceptive input.[36] It is believed that continuous trigeminovascular neuronal activation in turn activates descending modulating pain pathways, culminating in functional impairment by free radical-induced neuronal cell damage in the periaqueductal gray (PAG), and ultimately damaging the migraine generator. MRI imaging demonstrating iron deposition in the PAG is consistent with this hypothesis.

Nervous system changes as measured by CSF orexin-A and corticotrophin-releasing hormone have been found to correlate with monthly medication usage. These findings were validated by self-reported dependency measures. This identifies changes to the nervous system that correlate to medication intake.[37]

NEUROIMAGING

Imaging has provided a clue into the neuroplastic changes and a possible reversal. PET imaging, before and after discontinuation of the overuse of medication, looks for changes in the thalamus, anterior cingulate gyrus, and insula/ventral striatum. In the right inferior parietal lobule, it demonstrated diminished glucose uptake/hypometabolism while overusing medication. The cerebellar vermis was found to be hypermetabolic. After withdrawal, cerebral regions recovered to near baseline metabolic status. One region that did not change was the orbital frontal cortex, which demonstrated further metabolic decrease.

GENETIC BASIS FOR MEDICATION OVERUSE HEADACHE

Epidemiological studies support the hypothesis of genetic contributions to MOH.[38] A family history of MOH appears to impart a three-fold greater likelihood that the patient will develop it.[39] Additionally, a hypothesized link has been identified between substance abuse/dependence and MOH. Patients with MOH are also more likely to have relatives with a substance abuse disorder even if the patients do not (except for their overused medication). This potentially represents a familial vulnerability to the disease of addiction that increases the risk of loss of control of analgesic and acute migraine medications.

The genetic basis for migraine progression was furthered by a recent study where a specific biomarker was associated with headache improvement following successful weaning.[25] Over 1300 patients, indistinguishable at intake with 39.1% overusing medication, were withdrawn from their offending substance. After weaning, 44.5% of the patients had a greater than 50% reduction in headache frequency. No change was observed in 41.6%. Blood genomic expression patterns were obtained in 33 patients, with 19 of those patients overusing analgesics. A unique genomic expression pattern was identified in patients who were likely to have headache improvement upon withdrawal, suggesting that blood genomic patterns can accurately identify patients who respond to withdrawal. It would appear that these two initially indistinguishable groups progressed to MOH via differing pathways.[40]

PSYCHOLOGICAL/BEHAVIORAL ASPECTS OF MEDICATION OVERUSE HEADACHE

Psychiatric comorbidities are more prevalent in MOH, with the onset of the behavioral disorders preceding MOH progression. These psychological states appear to be important in the development and perpetuation of medication overuse. Such factors include bipolar disorder, personality disorders, anxiety, depression, and obsessional drug-taking and/or dependence-related behaviors,[41] and these alone may compel medication overuse.[42] Psychiatric and substance abuse disorders appear to be less common in those who overuse triptans.[43,44] Patients who overuse butalbital and opiates are more likely to have comorbid psychopathology (irrespective of the presence or absence of a substance abuse disorder).[43,45,46]

Medication overuse has also been likened to addiction (compulsion to use with a persistent desire, or failure to reduce in the face of harm).[42,47] MOH patients are at greater risk for substance abuse disorders than episodic migraineurs, thereby supporting the idea that MOH is part of the spectrum of addictive disorders.[42] A large percentage of patients with MOH will meet the DSM-IV criteria for substance abuse. Leeds Dependency Questionnaire responses in individuals who were unable to withdraw from medication had compulsive dependence subscores similar to those seen in addicts.[48] Prior to withdrawal all subjects believed their medication use was needed to maintain function, and without it a meaningful life was unlikely. In subjects able to withdraw from the overused medication, an attitude change regarding medication use was accomplished with proper headache management and subsequent improvement in their condition. Those who were unable to withdraw persisted in beliefs that are seen in substance abuse disorders.

Another behavioral pathway may involve a patient's attitudes towards their medication usage.

Patient's attitudes may reflect the different behavioral pathways for MOH. One study found that patients who viewed their medications as indispensable, and believed that their lives would be unbearable without their medication, focused more on their pain and less on the amount of medication they used.[37] They also had difficulty accepting a primary headache diagnosis, holding skeptical beliefs about daily prophylactic medications, and had become resigned with daily acute medications. They failed to see the irony of their willingness to engage daily acute medication and reluctance to commit to preventive therapy. They saw their medication usage as a direct result of the increasingly frequent headaches, and not the influence the acute medication was having on progression.[49]

One small study showed a deficit in decision-making in those who overuse psychoactive medications such as opiates.[37] Individuals with MOH showed significant difference between episodic migraineurs and controls, in behavioral dependence, depression, anxiety, and catastrophizing.[50]

CLINICAL PRESENTATION OF MEDICATION OVERUSE HEADACHE

Successful medical practice requires the ability to recognize patterns. While all individuals with medication overuse carry a comorbid diagnosis of chronic migraine, not all individuals with chronic migraine overutilize medication. It is important to remember that medication overuse does not universally lead to migraine progression. When MOH does occur, it tends to do so slowly, over many months to years.[7,51] The number of average monthly medication days seems to be a more important factor for progression than the number of average monthly doses taken.[38] When a headache transforms rapidly from episodic to chronic (especially progressing to daily), the patient must be carefully evaluated for secondary headaches and addiction.

MOH should be considered in any patient with chronic migraine. Additionally, other daily headaches must be ruled out. A differential diagnosis for MOH should include chronic migraine, new daily persistent headache, hemicranias continua, high- or low-pressure headaches, and secondary headaches.

Three factors are considered necessary and sufficient for MOH development: a history of migraine, an elevated baseline frequency, and overuse of acute medication(s).[6] Clinical risk factors for the development of medication overuse include the number of headache days, the average days of monthly medication use, the number of years of medication use, the baseline headache frequency at the start of the year, the number of physicians consulted, and the number of different medications used.

As medication overuse induces transformation, the clinical presentation of the headache and related comorbidities frequently changes. Sleep worsens and becomes non-restorative, and psychiatric comorbidities worsen. Often, an increase in neck pain and autonomic signs such as vasomotor instability occurs. The headache may develop a circadian rhythm, often awakening individuals with an early-morning headache (as the medication is wearing off).[52]

During the examination, risk factors for migraine progression and medication overuse should be identified (Box 20.1). The interview should include the patient's thoughts and practices regarding their headaches, the use of acute medication, and their prognosis. The fear of an impending headache attack has been associated with medication overuse.[54] Patterns of beliefs have been identified in individuals who are able to recover from medication overuse as being those who overused in order to overcome their headaches, and that the medication was essential to return to normal functioning. In this group, the drug-dependence behavior appeared to reflect a consequence of their headache. In this study, the Leeds Dependency Questionnaire was utilized and the non-responders group demonstrated multiple similarities to individuals suffering from the disease of addiction.[46,47]

If the patient overuses one medication, he or she may be at risk for overusing other medications. The physician should inquire about alcohol and tobacco usage, as these may be the first signs of someone who is at risk for a substance abuse disorder. Tobacco usage has been shown to be a risk factor for aberrant drug-taking behavior.[55] If possible, a report on the patient's

---

BOX 20.1

PATIENT QUESTIONS FOR MEDICATION OVERUSE HEADACHE

Can you tell when you're going to have a headache?
　If so, what do you do to try to prevent it?
　If you use a medication, when and how do you take it?
Are you taking prescription or over-the-counter medications for other painful conditions?
How often do you take medications, even though they don't seem to work for you?
When do you use caffeine during the day, and how much?

How often do you use substances (alcohol, cannabis, drugs, and prescription medications) for recreational purposes?
Do you or any family members have a substance abuse disorder or addiction?
If I were to obtain a report of scheduled prescription medications that you have used in the last year, what medications would be on that list?
　How often do you take those medications?

Modified from Young et al.[53]

prescription drug use should be obtained. The risk of addiction is greatly underestimated by both clinicians and patients. When headaches progress, individuals are likely to see their increasing medication usage as a necessary response to the worsening headaches without realizing that their increasing pain is often a sign of analgesic withdrawal.

DETOXIFICATION FROM MEDICATION OVERUSE HEADACHE

Restoration to the episodic migraine pattern requires withdrawing the offending medication. Although withdrawal may be the most important step, it is imperative to address all migraine progression risk factors. Failure to do so will likely influence or unnecessarily complicate treatment.

Improvement may take weeks to months, and the goal is to return the attacks to an episodic pattern. Headache diaries provide important clues, as change may be slow and not noticed. A reduction in headache intensity and duration may be noticed before a reduction in headache days.

Effective treatment plans require addressing patient expectations, and encouraging patience and vigilance. Patients must be prepared for prolonged treatment, as therapy requires the commitment of physician and patient resources. The patient should prepare for a short-term increase in their headache pain, which will undoubtedly affect their productivity and quality of life for a while.[6] Some physicians will consider short-term disability (if available) for the patient to allow a greater focus towards recovery.

Behavioral management techniques should be actively engaged during recovery. This requires preparation prior to the withdrawal date. These techniques are more effective when they are used regularly. Behavioral treatments consistently improve preventive and abortive medication effectiveness.[6,52]

We must remember that we are dealing with pain. It is instinctive to perceive pain primarily as a product of nociception. In MOH, suffering is a product of the neuroplastic changes stimulated by continued medication usage, nociception, and the accompanying emotional response. At a fundamental level, what is perceived is pain, and it should be considered as such. Pain is a perception of a nociceptive input and an accompanying emotional response. The brain is hardwired to interpret pain as a threat to survival, and it will reflexively respond to such. The accompanying emotional response to the nociception requires behavioral and not pharmacological therapies. Often patients refuse to accept the value of these treatments because they have "real pain." In these situations, it is often best to avoid using the word "pain," but to divide the concept into signal and suffering. Humans are capable of tolerating a great deal of pain (signal). What we do not do well is suffering. Behavioral treatments target suffering.

Behavioral management tools help blunt the emotional response/stress response to pain, thereby diminishing the suffering. It is also beneficial to state the overwhelming evidence that it is appropriate for all migraine patients with significant disability to use behavioral interventions, often under the direction of an appropriately interested and trained mental health professional. Frequently, even in the presence of severe disability, patients will deny any significant stress. It is helpful to point out that their headache pain is a stress, and that their headache comorbidities and factors for headache progression are all stressors. It is these stressors that drive progression, as stress in the simplest terms is anything that acts to provoke a response. A useful analogy for pain is found in electricity. In electricity, the volts represent nociception and amps the emotional response – hence, emotion amplifies the signal.

Patients should understand the essential role of behavioral interventions. Behavioral interventions allow patients to learn how to do for themselves what the medication is doing for them. This shifts the locus of control from the physician to the patient, fostering self-efficacy. Patients should understand that the only one who can do anything for their headaches is themselves.

PATIENT EDUCATION FOR MEDICATION OVERUSE HEADACHE

It is through migraine and MOH education that the patient can visualize a picture of regression back to an episodic headache pattern – what others have done to get better. This picture of restoration, like a jigsaw puzzle, may have many pieces, and often many pieces must be present and in place at the same time before we can clearly see the image of recovery. Patients must be challenged with this idea by asking, "What are you going to do to prepare for a successful transformation from chronic to episodic migraine?" For optimal outcomes, patients must know how they got here (chronic migraine) and what is keeping them here (medication overuse and migraine comorbidities). Many of the acute medications are so insidious that patients are seduced by their charms. Knowledge that the overused medication is part of the problem is a prerequisite for embracing the difficult path of withdrawal, regression, restoration, and recovery.

Patients must understand the risk factors for migraine progression and the development of MOH, and how they are applicable to their own experience. They should know that these risk factors for progression lower their migraine threshold, triggering greater attacks over time. The necessity of engagement with education can be enhanced by using the "as if" principle: if you want a characteristic, act as if you already have it! Ask them: "What have people like you done to gain control of their migraine?" then suggest they go out and do the same. Highlight that individuals who have gained control of their headaches rarely continuously overuse their medications.

Reserving judgment, a clinician should be empathic if not sympathetic to the sufferer. We are unlikely to successfully change behaviors by blaming and shaming when, under most circumstances, the patients were "just following doctors' orders." This shared vision and understanding of their circumstances creates an environment in which the patients are able to "buy in" to an agreed-upon treatment plan.[6,56]

Frequently, patients are unwilling to accept the absence of a secondary cause for their headache. It is difficult to comprehend that a high degree of pain is not caused by a serious threat to the patient's health. This belief cannot be changed if patients are not educated about their headache problem. They should understand that migraine will not prevent the development of a secondary headache disorder. Secondary headaches are uncommon, and have a distinct pattern[57] that almost always differs from migraine. Patients must realize that migraine alone is sufficient to explain their continued pain and disability. Reassuring them that there is not an ominous threat to their health can be very beneficial. If the headaches are chronic and not yet daily, simply asking "Where does the tumor go during the times

when you don't have a headache?" can often provide insight to the point you're making.

It is very easy for the physician to use medical jargon, or to speak in a nuanced fashion. For patients, the physician's words may have very different meanings. We must ensure that the patients understand what we mean when we say overuse, abuse, dependence, and addiction.[58]

Patients do not intuitively understand medication overuse. They may respond that "I'm only taking them when I have a headache." Patients do not believe they are overusing their medications, but that these medications are essential to reduce suffering and improve their ability to function.[47] The reality must be presented in terms the patient is able to understand. We can explain the development of MOH as an unfortunate consequence in individuals who are desperately seeking escape from a disabling pain, and believing that they are doing the right thing. It is a classic example of "too much of a good thing."

Patients should understand that their headaches did not transform overnight, and are unlikely to resolve rapidly. Regression to episodic attacks takes, minimally, 3 months or longer if opiates or butalbital are the overused medication (Figure 20.1).[59] This point should be discussed with the patients to better identify their expectations of treatment. If they are anticipating a rapid regression, they will unlikely tolerate any treatment plan, especially if it requires not treating acute headaches.

Patients should be aware that headaches worsen in the short term after the overused medication is withdrawn.[60] This knowledge may also have a therapeutic benefit. Humans are able to tolerate a great deal of

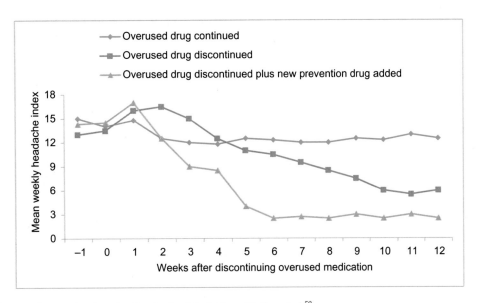

FIGURE 20.1   Pattern of regression to episodic attacks. *Adapted from Mathew et al.*[59]

pain; what we do not do well is suffering. When we perceive that pain is without purpose, or unending, the result is suffering. If we understand that the increase in headaches is expected and is a part of the usual course of healing, patients are better able to tolerate the discomfort. Bridge therapy (see below) may help to suppress this exacerbation.

Family members and significant others also need to be educated about the process, and avoid focusing not on their loved one's pain but on recognizing the non-pain aspects of the patient's life. What we pay attention to grows, and we no longer want to grow the pain.

Initially, the physician and the patient are likely to have competing interests. The physician often has a view of the condition over time with the goal of guiding the return to episodic. The patient has a priority of resolving the suffering of each individual attack. It might be helpful to frame the treatment plan as if we were preparing for war. Strategies are made over time with the knowledge that resources are limited. If all resources are dedicated to a single focus, it will likely fail. In this "migraine war," there will be many complications, there will be pain, and there will be doubt. We must work wisely to maximize achievements, minimize losses, and expend our resources appropriately.[52]

Disagreements regarding acute medication limits are common. Both parties must agree on these limits, and when exceptions to these limits are allowed. Physicians understand that these medications, in the absence of apparent benefit, are perpetuating the migraine progression potentially to the point of no return, and, in the case of habituating substances, physicians may actually believe that the medications are contributing to drug abuse. Patients are facing the prospect that if they are unable to control their usage of pain medications, they will suffer more.

TREATMENT OF MEDICATION OVERUSE HEADACHE

After the decision to wean the offending medication is made, the next step is to determine which path to follow to withdrawal — outpatient slow, fast infusion, or inpatient treatment. By not stopping the overused medication(s), the likelihood of the headaches improving spontaneously is exceedingly low. Conversely, successful weaning is directly associated with improved headache outcomes.[52] Meta-analyses have demonstrated that withdrawal alone may induce regression in 45–60% of cases. When headache prophylaxis was added to successful weaning, 72–85% of the patients demonstrated improvement at 1 year.[52]

Outpatient treatment of MOH requires active engagement[6,61] of both patient and physician. Frequent office visits are often needed. The advantage of outpatient treatment is its lower costs; however, it is also not as effective. One small randomized European trial did, however, document similar outcomes at 2 years, but no opiate users were included in the outpatient group and only one in the inpatient group.[53] These outcomes should be considered with caution.[62] Outpatient care requires a very motivated patient who is acutely aware of the goals of treatment. Outpatient treatment may be further subdivided into two categories: slow wean or fast wean. In a slow wean, the patient reduces the number of pills from his or her daily dose by one pill every week. For example, if an individual is taking hydrocodone 60 mg a day in six tablets, during the first week this should be reduced to 50 mg, in the subsequent week to 40 mg, in the next week to 30 mg, and so on until the medication is discontinued. Patients on short-acting medications that provoked the MOH may be transitioned to similar longer-acting agents, and then the dose titrated down. An example would be transitioning the patient from sumatriptan 200 mg a day to naratriptan 5 mg a day. An individual on short-acting opiates may be transitioned to a milligram equivalent of a longer-acting agent. Individuals overusing butalbital may be weaned using phenobarbital.

Preventive medications may be started before initiating withdrawal. Frequently, with migraine, a patient's preventive medication should be started at a low dose and slowly titrated up in an attempt to minimize adverse reactions. It is important that the patient should not conclude that the treatment has failed if there is no improvement in the baseline headache. Depending upon the medication that provoked the MOH, restoration of the hypersensitive nervous system to an episodic pattern may take several months.[57,63] If a rapid wean is planned, a quit date should be scheduled. On that date, the patient is to stop the offending medication, and must be prepared for this cessation. I advise patients to envision the time to the quit date as one of the most important journeys of their life.

A rapid wean is obviously not recommended for individuals who are likely to have significant withdrawal symptoms, such as those taking butalbital medications more than twice a day. In order to limit the anticipated worsening of the baseline headache disorder, a "bridge therapy" may be initiated, which may run between 5 and 14 days (Box 20.2).

For either outpatient plan, limits on acute medication therapy need to be established at the onset, and the patient should not use the medication that provoked the MOH. When possible, it is reasonable to consider using simple analgesics or polypharmacy for the acute treatment of headache, based on pre-established limits.

BOX 20.2

BRIDGE THERAPIES
(Do not use the offending medication as a part of the bridge)

Non-steroidal bridge therapy

Naproxen 500 mg twice daily for 7–10 days.
Ketorolac 60 mg IM twice daily for 5 days
(should use PPI or H2 blocker if using ketorolac).

Steroidal bridge therapy

Dexamethasone 4 mg BID × 7 days or 4 mg
BID × 4 days, then 4 mg daily × 4 days, then stop.
Dose packs typically are not effective.
Prednisone taper 60 mg daily for 2 days, then 40 mg
daily for 2 days, then 20 mg daily for 3 days,
then stop.
Not advisable to use NSAID in combination with
steroids (increased risk of GI side effects). May use
steroids first then follow with a bridge NSAID
therapy.

Triptan bridge therapy

Short acting: sumatriptan and 25 mg three times a
day for 10 days or until the patient is pain free for
24 hours.

Long acting: naratriptan 2.5 mg twice daily for
1 week.

Ergotamines bridge therapy

Dihydroergotamine 1 mg sub Q twice daily for 7–10
days (likely to be most effective).
Dihydroergotamine nasal spray twice to three times
daily for 7–10 days (more available than injections
but less effective).
Methylergonovine 0.2 mg 1–2 pills three times a day
for up to 14 days.

Bridge polypharmacy/adjunctive medications

Prochlorperazine 10 mg ± diphenhydramine
25–50 mg TID.
Metoclopramide 10 mg ± diphenhydramine
25–50 mg TID (do not take with prochlorperazine).
Olanzapine 10–20-mg PO QHS times 5–7 days. Start
dosing with 10 mg, repeating the dose in 1 hour if
not sleepy. A 20-mg dose is often needed if there is
significant anxiety or insomnia.

Infusion Center Withdrawal

An outpatient infusion center may be considered a hybrid of oral outpatient treatment and inpatient intravenous treatment. It is often reserved for individuals who do not have the flexibility for an inpatient admission, or who have failed the oral outpatient treatment plan. Typically, these infusions use twice-daily dosing of DHE with a self-administered subcutaneous dose of DHE in the evening, for 5 days. Administration of subcutaneous DHE via a portable pump has been reported.[53]

Inpatient Treatment of Medication Overuse Headache

The principal benefit of inpatient treatment is a greater chance of successfully withdrawing the offending medication. The controlled environment allows for management of withdrawal and pain, and provides multidisciplinary treatment.[62]

Inpatient therapy should be considered if the MOH has been ongoing for greater than 5 years, or when opiates or butalbital are responsible for progression to MOH.[13] Inpatient therapy may also be considered based upon the patient's past experience. The patient may have failed outpatient withdrawal or had previously successful inpatient withdrawal.

Inpatient therapy is primarily achieved through intravenous medications. Whichever intravenous therapy is used, continuing the offending medication or others that provoke MOH is likely to hinder the resolution of the baseline headaches. The principal IV medication is dihydroergotamine (DHE). It should be administered at the maximum subnauseating dose (after premedication with an anti-emetic) up to 1 mg. This dose is repeated three times per day until the headache is resolved or no further improvement is demonstrated. DHE should be administered for at least 36 hours (four doses) if no visible improvement is demonstrated. Tolerability can be enhanced by administering slowly (1 mg/50 mL NS over 30 minutes) or continuously over 24 hours.[64]

It is not uncommon for a headache to "spike" during inpatient treatment, although this does not represent a failure of the principal IV therapy but merely an overlay of an acute headache or an eruption of a migraine triggered by withdrawal of the offending medication. While the baseline headache disorder is being treated with intravenous medications, other medications may be used to treat the spike in the headache. Inpatient intravenous treatment allows for faster and more nimble pain control during withdrawal. The inpatient time is also used to teach the patient pain management skills, healthy habits, and lifestyle changes that enhance success.

Appropriate Opiate Use in Chronic Migraine

Many physicians may not feel comfortable prescribing opiates for migraine patients, as this is rarely appropriate. However, "rarely appropriate" does not mean never appropriate. Criteria, and several clinical experiences regarding opiates in headache, have been published.[65,66]

Migraine opiate therapy is practiced as episodic use, continuous use, or a hybrid where opiates are used daily on a scheduled basis and additional medications are used on an "as needed" basis. When considering using the latter, it is imperative to define a patient's monthly usage, as it is frequently underestimated. Asking the specific question: "How many pills do you need for 30 days?" can often provide valuable information.

If episodic opiate use is selected, its purpose is for migraine rescue and to decrease utilization of the emergency department. For this purpose, limits of 7 or less days per month may be reasonable, as long as these limits are not exceeded over time and opiates are not used for other pain conditions. It is possible that episodic opiate use may have advantages for some patients who are unable to tolerate migraine-specific therapies because of a comorbid health condition.

Continuous opiate therapy for migraine has shown limited benefit. One study demonstrated that few patients show functional improvement with continuous opiate usage.[67] They further reported that large numbers of patients continued to use daily opiates in the absence of demonstrable benefits.

Continuous usage of opiates for any condition should never be undertaken without a discussion about how these medications will be started, used, and, if necessary, ultimately discontinued. Remembering the patients' perspective, by the time they reach this stage, they likely believe these medications are essential for their survival. Any obstacles a clinician places between them and the medication might be viewed as a threat. This is where the overlap between MOH and addiction is obvious. However, even in these patients there may be a place for continuous opiates in order to improve a patient's functioning. These patients did not ask for the physiology and circumstances that brought them to MOH; they are deserving of our care, and it is incumbent upon us to use our best judgment to determine what the best care will be.

For continuous opiate therapy, it is important to treat the patient not only as a headache patient but also as a chronic pain patient. In chronic pain, as in headache, there are serious discussions regarding the appropriateness of continuous opiates. In all cases, there is consensus that the success of continuous opiate therapy hinges on patient selection. For a patient to be considered an appropriate candidate for opiates, one requirement is the ability to follow the physicians' orders. Patients unable to do so are not appropriate candidates. Additionally, if the patient increases opiate usage without authorization, and after an appropriate review a new secondary headache has been ruled out, the medication should be stopped. The drug is likely responsible for the exacerbation, or the medication has unmasked the disease of addiction. Either of the two previous reasons demonstrates that this patient is an inappropriate candidate for opiates. In general, patients with an axis II diagnosis should not be started on continuous opiates. Other groups do not improve on continuous opiates, particularly those with serious psychological comorbidities. However, it is important to understand that some disorders, such as depression, may be a product of the pain. The depression may spontaneously improve once the pain has been properly treated. It may be useful to ask a very pointed question: "Considering that the use of continuous opiates is not benign and is associated with many severe and substantial risks, and that so few people do well on chronic opiates, what makes you believe that your experience will be different?"

Box 20.3 lists patient factors that indicate a better outcome with the use of opiates, while Table 20.2[68] provides indications of those patients who are less likely to do well on opiates.

Recently, evidence-based guidelines for chronic opioid therapy in chronic non-cancer pain have been published by the American Pain Society and the American Academy of Pain Medicine (Box 20.4). It should be pointed out that these evidence-based guidelines all carry strong recommendations, but their evidence base is simultaneously rated as "weak".

Tools are available to screen for the disease of addiction and potential aberrant drug-taking behavior, including the Opiate Risk Tool and the Screener and Opioid Assessment for Patients with Pain (SOAPP). Of the two, the SOAPP is believed to have a greater predictive value, although any sufficiently motivated patients are able to defeat the screen. State registries, if available, should be used to determine from whom the patient is receiving controlled substances.

Overall, long-acting therapies have not proven superior to short-acting therapies. The principal benefit of the short-acting opiates is cost. However, the disadvantage is that repetitive dosing throughout the day reinforces

---

BOX 20.3

PATIENT FACTORS ASSOCIATED WITH BETTER OUTCOMES WITH OPIATES

1. Individuals are generally goal oriented and adherent to medical regimens.
2. They are functional.
3. They take full responsibility for health outcomes and their role in multimodality treatment.
4. They understand concepts in opioid use such as tolerance, dependency, and addiction.

5. There is an absence of severe chronic psychopathology.
6. There is an absence of serious personality disorder.
7. They rarely overuse medication.
8. There is no history of illegal drug or alcohol use.

Reproduced from Gatchel.[69]

TABLE 20.2   Indications for Patients that are Less Likely to do Well on Opiates

Risk factors for opiates abuse	Patient behaviors associated with development of problems managing opioid intake
Personal history of a substance abuse disorder	Poly substance abuse
Family history of a substance abuse disorder	Focus on opioids
Young age	Non-functional status due to pain
History of preadolescent sexual abuse	Exaggeration of pain
Mental illness	Cigarette dependency
Psychological stress	Social patterns of drug use
Lack of a 12-step program	Legal problems
Poor social support	
History of repeated drug EtOH rehab	
Chronic pain syndrome	
Unclear etiology for pain	

Adapted from Webster.[68]

BOX 20.4

APS/AAPM GUIDELINES AND UNIVERSAL PRECAUTIONS FOR OPIATE PRESCRIBING

1. Make a diagnosis with appropriate differential following a comprehensive evaluation.
2. Perform a psychological assessment, including addiction risk and stratification.
3. Obtain informed consent for the risks and benefits of the treatment.
4. Have a treatment agreement in place outlining expectations for starting, continuing, and stopping medications*.
5. Make a pre- and post-intervention assessment of pain level and function.
6. Start an appropriate trial of opioid therapy with or without adjunctive medication.

7. Reassess pain score and functioning at every appointment.
8. Regularly assess the "A"s of Pain Medicine (Analgesia, Activities of daily living, Adverse side effects, and Aberrant drug-taking behaviours; "Adherence" and "Affect" [observed mood] might also be added).
9. Perform urine toxicology*.
10. Periodically review pain diagnosis and comorbidities, including addictive disorders.
11. Documentation.

Considered optional in APS/AAPM 2009 Guidelines.

---

operant behavioral conditioning. Long-acting opiates may allow for a greater baseline level of pain control, but be insufficient for spikes in pain. This might be used to motivate the patient to work on other pain management strategies. If long-acting opiates are chosen, then it is necessary to predetermine what (if any) medications will be allowed for the use of "breakthrough" pain.

Whatever limits are established for opioid therapy, patients must understand that if the limit is exceeded, the drug will no longer be prescribed. When a previously effective treatment plan is no longer effective, the physician should determine why. Over time, any patient will probably develop tolerance to a particular opiate, and report that it is not as effective as previously. However, tolerance to the analgesic properties develops slowly over time. Tolerance to the euphoric effects develops rapidly. If a patient requires more than the pre-agreed upon amount of medication, evaluation for worsening of the baseline (medication overuse) headache, the development of a secondary headache, or the disease of addiction is required. If there are any questions, consultation with an addictionologist may be considered. Referral of the patient, for co-management or for total management, to an addiction specialist who also treats pain is another option for consideration.

MEDICATION OVERUSE PEARLS

MOH is commonly seen in primary care and very common in tertiary care headache clinics.

- Most patients with MOH will improve with withdrawal of the overused medication, but it may take weeks to months to achieve this improvement.
- When dealing with treatment failures, the physician must pursue prevention more aggressively rather than return to prescribing the overused medication.
- A short course of steroids might benefit patients in whom headaches are worsening during a withdrawal of the offending medication.
- For patients on 150 mg/day or more of butalbital, transition to phenobarbital at 30 mg per 100 mg of butalbital and titrate down 30 mg/day until discontinued.
- A DHE infusion, as either an inpatient or outpatient, may benefit those who have failed outpatient tapering of the overused medication.

References

1. Ashina S, Lipton RB, Bigal ME. Treatment of comorbidities of chronic daily headache. *Curr Treat Options Neurol.* 2008;10: 36–43.
2. Bigal ME, Sheftell FD, Rapoport AM, Tepper SJ, Lipton RB. Chronic daily headache: identification of factors associated with induction and transformation. *Headache.* 2002;42:575–581.
3. Rapoport A, Stang P, Gutterman DL, Cady R, Markley H, Weeks R, et al. Analgesic rebound headache in clinical practice: data from a physician survey. *Headache.* 1996;36:14–19.
4. Mathew NT. Transformed migraine, analgesic rebound, and other chronic daily headaches. *Neurol Clin.* 1997;15:167–186.
5. Von Korff M, Galer BS, Stang P. Chronic use of symptomatic headache medications. *Pain.* 1995;62:179–186.
6. Tepper SJ, Tepper DE. Treatment of medication overuse headache. In: Tepper SJ, Tepper DE, eds. *The Cleveland Clinic Manual of Headache Therapy.* New York, NY: Springer; 2011: 153–166.
7. Diener HC, Silberstein SD. Medication overuse headaches. In: Olesen J, Goadsby PJ, Ramadan NM, Tfelt-Hansen P, Welch KMA, eds. *The Headaches.* 3rd ed. Philadelphia, PA: Lippincott Williams & Wilkins; 2005:971–980.
8. DeVries A, Koch T, Wall E, Getchius T, Chi W, Rosenberg A. Opioid use among adolescent patients treated for headache. *J Adolesc Health.* 2014;55(1):128–133.
9. Rapoport A, Weeks R, Sheftell F. Analgesic rebound headache: theoretical and practical implications. In: Olesen J, Tfelf-Hansen P, Jensen K, eds. *Proceedings of the Second International Headache Congress.* Copenhagen: International Headache Society; 1985: 448–449.
10. Wilkinson SM, Becker WJ, Heine JA. Opiate use to control bowel motility may induce chronic daily headache in patients with migraine. *Headache.* 2001;41:303–309.
11. Bigal ME, Serrano D, Buse D, Scher AI, Stewart WF, Lipton RB. Migraine medications and evolution from episodic to chronic migraine: a longitudinal population-based study. *Headache.* 2008;48:1157–1168.
12. Bahra A, Walsh M, Menon S, Goadsby PJ. Does chronic daily headache arise de novo in association with regular use of analgesics? *Headache.* 2003;43:179–190.
13. Diener HC, Limmroth V, Katsarava Z. Medication-overuse headache. In: Goadsby PJ, Silberstein SD, Dodick DW, eds. *Chronic Daily Headache for Clinicians.* Hamilton, ON: BC Decker; 2005: 117–127.
14. Dodick DW, Freitag F. Evidence-based understanding of medication-overuse headache: clinical implications. *Headache.* 2006;46(suppl 4):S202–S211.
15. Kudrow L. Pradoxical effects of frequent analgesic use. In: Critchley M, Friedman A, Goroni S, Sicuteri F, eds. *Advances in Neurology.* 33. New York, NY: Raven Press; 1982:335–341.
16. Mathew NT, Stubits E, Nigam MP. Transformation of episodic migraine into daily headache: analysis of factors. *Headache.* 1982;22:66–68.
17. Silberstein SD, Lipton RB, Solomon S, Mathew NT. Classification of daily and near-daily headaches: proposed revisions to the IHS criteria. *Headache.* 1994;34:1–7.
18. Zwart JA, Dyb G, Hagen K, Svebak S, Holmen J. Analgesic use: a predictor of chronic pain and medication overuse headache: the Head-HUNT Study. *Neurology.* 2003;61:160–164.
19. Boes CJ, Capobianco DJ. Chronic migraine and medicaton-overuse headache through the ages. *Cephalalgia.* 2005;25:378–390.
20. Headache Classification Committee of the International Headache Society (IHS). The International Classification of Headache Disorders, 3rd edition (beta version). *Cephalalgia.* 2013;33:629–808.
21. Gyles C. On the cusp of a paradigm shift in medicine? *Can Vet J.* 2009;50:1221–1222.
22. Supornsilpchai W, le Grand SM, Srikiatkhachorn A. Involvement of pro-nociceptive 5-HT2A receptor in the pathogenesis of medication-overuse headache. *Headache.* 2010;50:185–197.

23. Johnson JL, Hutchinson MR, Williams DB, Rolan P. Medication-overuse headache and opioid-induced hyperalgesia. A review of mechanisms, a neuroimmune hypothesis, and a novel approach to treatment. *Cephalalgia*. 2013;33:52–64.

24. Boes CJ, Black DF, Dodick DW. Pathophysiology and management of transformed migraine and medication overuse headache. *Semin Neurol*. 2006;26:232–241.

25. LeGrand S, Saengiaroentham C, Supronsilpchai W, Srkiatkhachorn A. The increase in the calcitonin gene-related peptide immunoreactivity following the CSD activation in the serotonin depleted state. *J Headache Pain*. 2010;11:S136–S137.

26. Srikiatkachorn A, Le Grand SM, Supornsilpchai W, Storer RJ. Pathophysiology of medication overuse headache – An update. *Headache Curr*. 2014;54:204–210.

27. Baranauskas G, Nistri A. Sensitization of pain pathways in the spinal cord: cellular mechanisms. *Prog Neurobiol*. 1998;54:349–365.

28. Staud R, Smitherman ML. Peripheral and central sensitization in fibromyalgia: pathogenetic role. *Curr Pain Headache Rep*. 2002; 6:259–266.

29. Srikiathachorn A. Chronic daily headache: a scientist's perspective. *Headache*. 2002;42:532–553.

30. Burstein R, Cutrer MF, Yarnitsky D. The development of cutaneous allodynia during a migraine attack: clinical evidence for the sequential recruitment of spinal and supraspinal nociceptive neurons in migraine. *Brain*. 2000;123(pt 8):1703–1709.

31. Srikiathachorn A, Maneesri S, Govitrapong P, Kasantikul V. Derangement of serotonin system in migrainous patients with analgesic abuse headache: clues from platelets. *Headache*. 1998; 38:43–49.

32. Srikiatkhachorn A, Tarasbu N, Govitrapong P. Acetaminophen-induced anti-nociception via central 5-HT2A receptors. *Neurochem Int*. 1999;34:491–498.

33. Jakubowski M, Levy D, Goor-Aryeh I, Collins B, Bajwa Z, Burstein R. Terminating migraine with allodynia and ongoing central sensitization using parenteral administration of COX1/COX2 inhibitors. *Headache*. 2005;45:850–861.

34. Burstein R, Collins B, Jakubowski M. Defeating migraine pain with triptans: a race against the development of cutaneous allodynia. *Ann Neurol*. 2004;55:19–26.

35. Bigal ME, Ashina S, Burstein R, Reed ML, Buse D, Serrano D, et al. Prevalence and characteristics of allodynia in headache sufferers: a population study. *Neurology*. 2008;70:1525–1533.

36. Welch KM, Nagesh V, Aurora SK, Gelman N. Periaqueductal gray matter dysfunction in migraine: cause of the burden of illness? *Headache*. 2001;41:629–637.

37. Fumal A, Laurey S, DiClemente I, Boly M, Bohotin V, Vandenheede M, et al. Orbitofrontal cortex involvement in chronic analgesic overuse headache evolving from episodic migraine. *Brain*. 2006;129:543–550.

38. Evers S, Marziniak M. Clinical features, pathophysiology, and treatment of medication-overuse headache. *Lancet Neurol*. 2010;9:391–401.

39. Cevoli S, Sancisi E, Grimaldi D, Pierangeli G, Zanigni S, Nicodemo M, et al. Family history for chronic headache and drug overuse as a risk factor for headache chronification. *Headache*. 2009;49:412–418.

40. Hershey AD, Burdine D, Kabbouche MA, Powers SW. Genomic expression patterns in medication overuse headaches. *Cephalalgia*. 2011;31:161–171.

41. Saper J, Hamel R, Lake III AE. Medication overuse headache (MOH) is a biobehavioural disorder. *Cephalalgia*. 2005;25: 545–546.

42. Radat F, Creac'h C, Swendsen J, Swendsen JD, Lafittau M, Irachabal S, et al. Psychiatric comorbidity in the evolution from migraine to medication overuse headache. *Cephalalgia*. 2005;25: 519–522.

43. Lake III AE. Placebo, chronic daily headache, and pain: ten points to ponder. *Curr Pain Headache Rep*. 2006;10:4–6.

44. Pozo-Rosich P, Oshinsky M. Effects of dihydroergotamine (DHE) on central sensitization of neurons in the trigeminal nucleus caudalis. *Proc Am Acad Neurol*. 2005;5 [Abstract].

45. Saper JR, Dodick D, Gladstone JP. Management of chronic daily headache: challenges in clinical practice. *Headache*. 2005;45 (suppl 1):S74–S85.

46. Saper JR. Chronic daily headache: a clinician's perspective. *Headache*. 2002;42:538–542.

47. Ferrari A, Cicero AF, Bertolini A, Leone S, Pasciullo G, Sternieri E. Need for analgesics/drugs of abuse: a comparison between headache patients by the Leeds Dependence Questionnaire (LDQ). *Cephalalgia*. 2006;26:187–193.

48. Corbelli I, Caproni S, Eusebi P, Sarchielli P. Drug dependence behaviour and outcome of medication overuse headache after treatment. *J Headache Pain*. 2012;13:653–660.

49. Jenssen P, Jakobsson A, Hensing F, Linde M, Moore CD, Hodenrud T. Holding on to the indispensable medication. A grounded theory on medication use from the perspective of persons with medication overuse headache. *J Headache Pain*. 2013;14:43.

50. Radat F, Chanraud S, DiScala G, Dousset V, Allard M. Psychological and neuropsychological correlates of dependence-related behaviour in medication overuse headaches: a one year follow-up study. *J Headache Pain*. 2013;14:59.

51. Limmroth V, Katsarava Z, Fritsche G, Przywara S, Diener HC. Features of medication overuse headache following overuse of different acute headache drugs. *Neurology*. 2002;59:1011–1014.

52. Tepper SJ, Tepper DE. Medication overuse headache in refractory migraine and its treatment. In: Schulman EA, Levin M, Lake AE, Loder E, eds. *Refractory Migraine: Mechanisms and Management*. New York, NY: Oxford; 2010:136–159.

53. Young WB, Silberstein SD, Nahas SJ, Marmura MJ. *Medication overuse: diagnosis and treatment. Jefferson Headache Manual*. New York, NY: Demos Medical; 2010:97–104

54. Saadah HA. Headache fear. *J Oklahoma State Med Assoc*. 1997;90:179–184.

55. Dhingra LK, Passik SD. Smoking and aberrant behavior in chronic pain patients. *Pract Pain Manage*. 2007;<www.practicalpainmanagement.com/resources/smoking-aberrant-behavior-chronic-pain:patients> Accessed 01.05.14.

56. Lake AE. Medication overuse headache: biobehavioral issues and solutions. *Headache*. 2006;46:S88–S97.

57. Martin VT. The diagnostic evaluation of secondary headache disorders. *Headache*. 2011;51:346–352.

58. Warner JS. Prolonged recovery from rebound headache. *Headache*. 2001;41:817–822.

59. Mathew NT, Jurman R, Perez F. Drug induced refractory headache – clinical features and management. *Headache*. 1990;30: 634–638.

60. Lipchik GL, Nash JM. Cognitive-behavioral issues in the treatment and management of chronic daily headache. *Curr Pain Headache Rep*. 2002;6:473–479.

61. Warner JS. Time required for improvement of an analgesic rebound headache. *Headache*. 1998;38:229–230.

62. Ford RG, Ford KT. Continuous intravenous dihydroergotamine in the treatment of intractable headache. *Headache*. 1997;37: 129–136.

63. Stillman M. Commentary on perspectives on outpatient intravenous dihydroergotamine for probable medication overuse headache. *Headache Care*. 2006;3:51.

64. Saper JR, Lake AE, Bain PA, Stillman MJ, Rothrock JF, Mathew NT, et al. A practice guide for continuous opioid therapy for refractory daily headache: patient selection, physician requirements, and treatment monitoring. *Headache*. 2010;50:1175–1193.

65. Robbins L. Long acting opioids for refractory chronic migraine. *Pract Pain Manage*. 2009;9:45–54.

66. Saper JR, Lake AE. Continuous opioid therapy (COT) is rarely advisable for refractory chronic daily headache: limited efficacy, risks, and proposed guidelines. *Headache*. 2008;48:838–849.

67. Chou R, Fanciullo GJ, Fine PG, Adler JA, Ballantyne JC, Davies P, et al. American Pain Society — American Academy of Pain Medicine opioids guidelines panel. *J Pain*. 2009;10:113–130.

68. Webster LR. Determining the risk of opioid abuse. Emerging solutions in pain CE-accredited monograph. *Pract Pain Manage*. 2006;6(1).

69. Gatchel, RJ. *Clinical Essentials of Pain Management*. American Psychological Association. 2005, APA, Washington DC, p. 199.

21

Presentation of Headache in the Emergency Department and its Triage

Benjamin W. Friedman

Albert Einstein College of Medicine, Bronx, New York, USA

INTRODUCTION

Nearly 4 million patients annually present to emergency departments (EDs) in the United States for management of headache.[1] Most of these patients have a benign headache etiology. A small number will have a malignant secondary cause of headache that must be diagnosed expeditiously. The twin challenges for ED physicians are to provide both an accurate diagnosis and effective treatment in a timely manner.

THE ROLE OF THE EMERGENCY DEPARTMENT

Emergency physicians play a variety of roles in headache management. At one extreme, the ED is the ideal medical destination for patients with true headache emergencies. EDs have the capability to intake and stabilize patients rapidly. For patients presenting with devastating secondary headaches, emergency clinicians can control the airway, administer therapeutic agents, obtain diagnostic testing rapidly, and coordinate definitive care. At the other extreme is the ED's role as a medical safety net. Patients without financial resources or access to health care can obtain medical care in the ED — federal law mandates treatment of all comers. Thus, emergency clinicians commonly encounter patients with undiagnosed primary headache disorders and associated comorbidities. Although these patients should be referred to an appropriate outpatient setting for ongoing care, the emergency physician can provide acute therapeutics, as well as a diagnosis and headache-specific education.

For patients with established primary headache diagnoses, the ED is a place of last resort to obtain headache relief at all hours of the day and night. On the one hand, the ED is the type of location least conducive to migraine treatment — many EDs are congested, bright, and noisy spaces. Wait times can be substantial, and subjecting oneself to a history and physical exam can be taxing within the throes of an acute migraine. On the other hand, the ED is capable of delivering high-flow oxygen or administering potent parenteral medications while providing cardiopulmonary monitoring. Well-functioning healthcare systems should divert primary headache patients to locations that can optimize their comfort while treating refractory headaches expeditiously.

Writing in the pre-triptan era of migraine management, one observer commented that two types of patients present to the ED for management of headache.[2] The first, sufferers of the "first or worst syndrome," present to the ED out of concern about the significance of their pain. These patients seek assurance that they do not have a pathological cause of headache, such as brain tumor or aneurysm. The treatment of their pain is straightforward once they are convinced of the benign nature of their headaches. The second group, patients with the "last straw syndrome," are more difficult to treat. These patients have been unable to obtain freedom from their headaches, and are thus characterized by frustration, exhaustion, and despair. Psychiatric comorbidities and medication overuse are common in this group. These patients are often managed poorly in the ED. Ideally, they should be referred to comprehensive headache or pain management programs at which their complex needs can be addressed thoroughly and in an ongoing multidisciplinary manner.

S. Diamond, R. K. Cady, M. L. Diamond & V. T. Martin (Eds):
Headache and Migraine Biology and Management.

DOI: http://dx.doi.org/10.1016/B978-0-12-800901-7.00021-5

EPIDEMIOLOGY OF HEADACHE IN THE EMERGENCY DEPARTMENT

In the ED, frequency of secondary headache has ranged from 23% to 38%, and frequency of primary headaches from 55% to 77%.[3-5] The frequency of migraine is generally four- to five-fold greater than that of tension-type headache.[3,4]

Despite rigorous evaluations, several authors have been unable to provide a specific diagnosis to every acute headache patient. Frequency of "unable to diagnose" has ranged from 17% to 33% of ED headache patients.[3,4,6] Barriers to providing a classification-based diagnosis to every patient have included prolonged duration of headache, and an infrequent number of similar prior headaches.

From a population perspective, use of an ED for management of migraine is uncommon. When surveyed, 5% of US migraineurs reported one to three visits to an ED or urgent care facility within the previous 12 months, while 1% reported more than three visits. Nearly 94% of US migraineurs report no use of an ED or an urgent care facility within the previous 12 months.[7] Among patients with tension-type headache, use of the ED is even less common – 3% report one to three visits in the previous 12 months, and 1% report frequent visits. Of American women, only 1% stated that an emergency physician was the type of physician seen most frequently for migraine, while even fewer males reported the same.[8] In a sample of Medicaid migraineurs from Georgia in the US, only 3% had a migraine-related ED visit within a 12-month period.[9] In this sample, nearly 70% of those who used the ED used it only once in a 12-month period while 4% had five or more ED visits.

Risk factors for ED use have been described. Patients with lower socioeconomic status and worse underlying illness have a greater propensity for ED use.[7] Depression is associated with ED use while insurance is protective against a visit. After controlling for sociodemographic characteristics in a Georgia Medicaid population, receipt of an opioid prescription was associated with ED use. However, obtaining butalbital-containing prescriptions was not associated with ED use.[9] In this sample, a higher hospital density was associated with ED use, while a higher density of general practitioners was protective against an ED visit.

DIAGNOSIS

Emergency department clinicians must first differentiate between primary and secondary headaches, and then provide a specific diagnosis within these categories. Typically, emergency physicians worry about malignant secondary headaches and consider it mandatory to exclude these headaches from the differential diagnosis.

Clinical features can be used to determine an individual patient's risk for malignant secondary headache. For patients with a typical recurrence of an episodic headache disorder, diagnosis should be relatively straightforward. In contrast, new onset focal neurological deficits and altered mental status place patients at higher risk of poor outcomes. In these latter patients, a malignant cause of headache should be assumed and pursued until excluded from the differential diagnosis. Among patients with a new headache type but normal cognition and no focal deficits, malignant causes of headache are not common but should be considered (Figure 21.1). In the following paragraphs, we present a brief description of the 10 more common causes of malignant secondary headaches that may present to an ED, and specific features of these headaches that should be apparent in the patient's history and physical exam.

1. *Aneurysmal subarachnoid hemorrhage*

   Autopsy studies demonstrate that nearly 2% of the population harbor intracranial aneurysms.[10] Rupture can cause devastating neurologic impairment or death. Most patients with aneurysmal subarachnoid hemorrhage (aSAH) report abrupt onset headache.[11] Many describe it as the worst headache of their life.[12] The diagnosis is readily apparent when patients arrive in the ED with an acute onset headache and focal deficits. Patients presenting with a headache and a normal sensorium and no focal deficit represent more of a diagnostic challenge. Among neurologically intact patients who present with an abrupt onset headache or a headache associated with syncope, a validated clinical decision rule can help determine the need for further testing. Specifically, those patients without any of the following characteristics are considered safe for discharge without diagnostic work-up: (a) age greater than 40 years; (b) neck pain or stiffness; (c) witnessed loss of consciousness; (d) onset during exertion; (e) thunderclap headaches; (f) limited neck flexion on exam.[13] There is an imperative to diagnose this catastrophic disease rapidly because ruptured aneurysms are at risk of re-bleeding.[14] Cerebrovascular aneurysms are readily amendable to surgical repair. Thus, patients who are diagnosed while neurologically intact are much more likely to be discharged from the hospital neurologically intact.

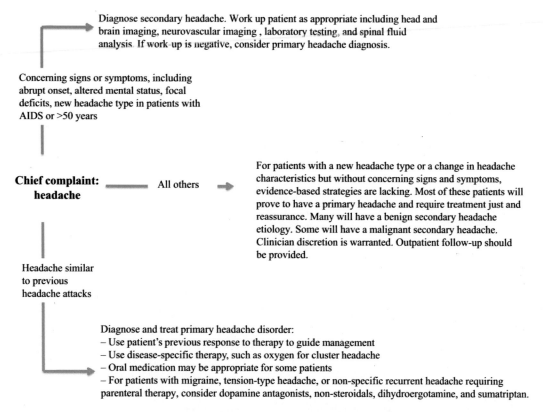

Diagnose secondary headache. Work up patient as appropriate including head and brain imaging, neurovascular imaging , laboratory testing, and spinal fluid analysis. If work-up is negative, consider primary headache diagnosis.

Concerning signs or symptoms, including abrupt onset, altered mental status, focal deficits, new headache type in patients with AIDS or >50 years

Chief complaint: headache ——— All others →

For patients with a new headache type or a change in headache characteristics but without concerning signs and symptoms, evidence-based strategies are lacking. Most of these patients will prove to have a primary headache and require treatment just and reassurance. Many will have a benign secondary headache etiology. Some will have a malignant secondary headache. Clinician discretion is warranted. Outpatient follow-up should be provided.

Headache similar to previous headache attacks

Diagnose and treat primary headache disorder:
– Use patient's previous response to therapy to guide management
– Use disease-specific therapy, such as oxygen for cluster headache
– Oral medication may be appropriate for some patients
– For patients with migraine, tension-type headache, or non-specific recurrent headache requiring parenteral therapy, consider dopamine antagonists, non-steroidals, dihydroergotamine, and sumatriptan.

FIGURE 21.1   An algorithmic approach to headache management in the emergency department.

2. Traumatic head injury

Altered mental status or focal neurological deficits following head trauma warrant consideration of intracranial hemorrhage, such as epidural or subdural hematomas, which may be amenable to neurosurgical intervention. In the setting of trauma, subarachnoid hemorrhage is generally a less concerning finding. Among head trauma patients, mechanism of injury and presenting signs and symptoms are predictive of outcome. Older age and use of anticoagulation or antiplatelet agents place patients at higher risk of clinically important bleeds and warrant consideration of prolonged observation. Persistently abnormal Glasgow Coma Scale (GCS) scores are associated with worse brain injury. Normal GCS scores are more likely to result in a post-concussive syndrome consisting of headaches, dizziness, memory disturbances, or alterations in mood or personality. Among patients with head trauma and normal mental status, various clinical decision rules can be used to predict the need for diagnostic imaging (Table 21.1).[18–20]

3. Brain tumor

The majority of patients with intracranial tumors report headaches at the time of diagnosis, though as many as 40% of patients with intracranial tumors will not complain of headache.[21] Brain metastases of non-CNS malignancies are more common than primary CNS tumors. Rarely, the symptoms of the brain metastasis are the presenting manifestation of a non-CNS cancer.[21] Brain tumors cause focal headaches, usually anatomically related to tumor location, and generalized headaches due to increased intracranial pressure, which can develop if the tumor bulk occludes the flow of CSF.[22] Headaches associated with brain tumors are quite variable in character. Typically, the headaches are intermittent, of moderate intensity, develop and resolve within hours, and are not specific in location.[23–25] Often, the headaches fulfill diagnostic criteria for tension-type headache or migraine.[26] A personal susceptibility to headache or family history of headaches may increase the risk of headaches associated with brain tumors. Emergency clinicians cannot rely on the classic description of a progressive severe headache — worse upon awakening and associated with nausea and vomiting — to exclude this diagnosis.

4. Giant cell arteritis (GCA)

Among elderly patients presenting to medical attention with a new headache, giant cell arteritis

TABLE 21.1   Clinical Decision Rules to Exclude Clinically Significant Injuries in Patients with Minor Head Trauma. The Presence of One of These Features Mandates Head CT

New Orleans Criteria[15]	Canadian CT Head Rule[16]
Headache	Failure to reach GCS 15 within 2 hours
Vomiting	Suspected open skull fracture
Seizure	Sign of basilar skull fracture
Intoxication	Vomiting > two episodes
Short-term memory deficit	Age > 65 years
Age > 60 years	Amnesia before impact > 30 minutes
Injury above the clavicles	Dangerous mechanism of injury

Test characteristics for clinically important findings in an independent sample[17]:

NOC – Sensitivity 97.7% (95% CI: 92.1–99.4), Specificity 5.5% (95% CI: 2.6–8.7)
CCHR – Sensitivity 87.2% (95% CI: 82.5–90.9%), Specificity 39.3 (CI: 36.6–42.0%)

is as common as migraine.[27] It is a systemic large vessel vasculitis with a predilection for branches of the aortic arch, especially the carotid and vertebral arteries. Although it can affect patients as young as the sixth decade of life, GCA is predominantly a disease of the geriatric population — incidence of disease is much less in middle-aged patients. Headache is the most common symptom, and vision loss or ischemic stroke is the most feared. Other common symptoms include fatigue, weight loss, fever, scalp tenderness, and, jaw claudication. Formal diagnostic criteria require three of the following five features: (a) age at onset ≥ 50 years; (b) new onset or new type of headache; (c) temporal artery tenderness or decreased pulsation; (d) erythrocyte sedimentation rate ≥ 50 mm/h; or, (e) abnormal artery biopsy.[28] A low index of suspicion is required for all patients aged 50 years or older who present with a new headache type. Corticosteroids are the treatment of choice for giant cell arteritis, and, particularly in the setting of visual symptoms, treatment should be initiated immediately rather than waiting for confirmatory biopsy.[29]

5. *Venous sinus thrombosis*

This potentially lethal disease is difficult to diagnose because it is uncommon and because symptoms, at least initially, are often subtle. Before widespread use of emergent MRI, patient with venous sinus thrombosis (VST) typically remained undiagnosed for 3 days after presentation to the hospital. VST may cause a thunderclap headache, but more typically causes an indolent, gradual onset headache. This is a thrombotic disease, so typical thromboembolic risk factors should be sought during the patient interview. Late pregnancy and the post-partum period place patients at higher risk, as does use of oral contraceptives. Patients with recent facial or neurosurgical interventions are also at risk. The cavernous sinus is involved uncommonly, but typically presents more overtly with proptosis, facial edema, ocular palsies, and visual deficits.[30]

6. *Cervical artery dissection*

Carotid artery dissection is more common than vertebral dissection, although the latter is more likely to present with neurological symptoms other than headache.[31] Tears occur in the arterial wall, resulting in an intramural hematoma which can lead to blood flow obstruction, stenosis, or aneurysmal dilatation. Cervical artery dissection may be spontaneous or traumatic. Preceding hyperextension or rotation of the neck are proximate causes of traumatic dissection. Valsalva that occurs with coughing, vomiting, and heavy lifting may also contribute to dissection. The classical presentation of cervical artery dissection involves headache, neck pain, and signs of brain ischemia. Among patients with carotid artery dissection, headache is often the initial presenting symptom and can precede other neurologic findings by hours to days. Patients may experience cranial nerve palsies, transient monocular blindness, tinnitus, or distortion of taste. A partial Horner's syndrome, consisting of ptosis and meiosis, in the setting of neck or facial pain, is highly concerning for carotid dissection. In contrast, patients with vertebral dissection often experience occipital headache which may be either unilateral or bilateral, in addition to neck pain, vertigo, nausea, vomiting, unilateral facial numbness, and unsteady gait.[32]

7. *Meningitis*

The headache of meningitis is non-specific. The constellation of signs and symptoms rather than the headache itself remains key to diagnosis.[33] Fever and alterations in mental status are often present in patients with bacterial meningitis. The absence of fever, neck stiffness, and altered mental status practically eliminates bacterial meningitis from the differential diagnosis, although the individual components of this triad in isolation have poor predictive ability.[34] Jolt accentuation — sharp, sudden head movements — worsens the headache of meningitis. In the US, patients with Lyme meningitis frequently recall the classic erythema migrans rash and often present with a concomitant seventh cranial nerve palsy. In Europe, painful radiculopathy is more common.[35] Cryptococcal meningitis is classically a disease of the immune-compromised, though it may occur in the immune-competent as well. It often presents as an indolent, subacute meningitis that one is tempted to call "tension-type headache." The emergency clinician, therefore, must have a low index of suspicion in patients with CD4 counts less than 100 cells/mm^3. As cryptoccal meningitis progresses, elevated intracranial pressure may cause alterations in consciousness, papilledema, cranial nerve palsies, and seizures.[36]

8. *Pituitary apoplexy*

Pituitary apoplexy — a rare headache etiology — causes ophthalmoplegia, visual field deficits, hormonal collapse, and thunderclap headache. Thus, an examination of the extraocular muscles and visual fields is important for all headache patients. Although a history of pituitary adenoma may be useful for establishing this diagnosis, the apoplectic headache is often the first symptom of the adenoma. Precipitating features such as recent surgery, use of anticoagulants, and recent parturition should be sought during the patient interview.

9. *Hypertensive headache*

Elevated blood pressure and headache have long been linked in the medical literature, although the causative pathway is unclear. Ambulatory blood pressure monitoring studies revealed no association between mild fluctuations in blood pressure and headaches.[37] Similarly, migraine patients who present to an ED with elevated blood pressure are just as likely to improve whether or not their blood pressure improves.

On the other hand, headache and elevated blood pressure are defining features of various malignant syndromes such as pre-eclampsia and posterior reversible encephalopathy syndrome (PRES). Blood pressure is often elevated in patients who present with non-traumatic intracranial hemorrhage. The emergency physician should measure blood pressure in headache patients, and inquire about a history of hypertension and any antihypertensive medication used. Pre-eclampsia requires the patient to be in late stages of pregnancy or the post-partum period, and is usually accompanied by brisk reflexes. PRES, although it may involve regions other than the posterior part of the brain, is accompanied by focal neurological deficits. Intracranial hemorrhage almost always presents with a severe, atypical headache. Assuming the headache is in other ways typical for the patient, elevated blood pressure by itself should not trigger a diagnostic work-up.

10. *Ischemic/hemorrhagic stroke*

Severe and abrupt onset headache typically accompanies intraparenchymal or intraventricular hemorrhage. Abrupt onset headache may accompany ischemic strokes as well, although most patients suffering from acute ischemic stroke present with focal neurological deficits as the predominant presenting symptom. It is uncommon to identify ischemic stroke as the headache etiology in patients who are otherwise neurologically intact. One of four patients with acute ischemic cerebrovascular disease will report a headache. Among patients with ischemic stroke, headache is more common among younger patients and those with a history of migraine.[38]

Once secondary headaches have been excluded, the emergency physician must next provide the patient with a specific primary headache diagnosis. Considering the high prevalence of migraine in the ED setting and the relatively uncommon occurrence of malignant secondary headaches, migraine can often be correctly diagnosed based on specific historical features of the headache or the use of a simple questionnaire. Migraine can be identified using the POUNDing mnemonic: Pulsating quality, duration of 4–72 hOurs, Unilateral, Nausea, functionally Disabling. Patients with three or four of these predictors can be diagnosed as having a migraine headache with high sensitivity and specificity.[39] The combination of functional disability, nausea, and sensitivity to light has a high positive predictive value for a diagnosis of migraine among patients with recurrent episodes of headache.[40]

Cluster headache is uncommon in the ED. Often, it has resolved or improved substantially by the time the patient sees the physician, who therefore needs to inquire about recent headache history. Once secondary headaches and other primary headaches have been eliminated from the differential diagnosis, tension-type headache is often assigned as a diagnosis of exclusion.

DIAGNOSTIC TESTING

Routine laboratory testing of patients with primary headache disorders is unlikely to be of benefit. Patients with migraine may present with intravascular volume depletion. Occasionally, the dehydration is severe enough to cause electrolyte abnormalities. However, for the vast majority of patients, dehydration can be diagnosed clinically and treated appropriately without evaluation of serum chemistries.

A serum or urine pregnancy test is useful confirmatory testing that may influence treatment decisions. Medications used to treat migraine, such as dihydroergotamine and valproic acid, have unfavorable pregnancy ratings and should be avoided if testing reveals early pregnancy.

Laboratory testing may be useful when various secondary headaches are on the differential diagnosis. The combination of a normal erythrocyte sedimentation rate and a normal C-reactive protein eliminates giant cell arteritis from the differential diagnosis, although poor specificity means that positive tests do not confirm the diagnosis. For patients at low risk of venous sinus thrombosis, a normal D-dimer can exclude this disease. However, because sensitivity is only 94%, false negatives may occur.[41] This diagnosis should still be pursued in patients deemed to be at high risk of disease. Similarly, for patients in whom the clinician suspects cryptococcal meningitis, a serum cryptococcal antigen is useful to rule out but not confirm the disease.[36]

The use of advanced neuroimaging studies in US EDs has increased explosively in the 21st century, increasing from 12% of all headache visits to more than 30% of headache visits in 15 years.[42] Because of wide availability and brief test durations, computed tomography (CT) is the go-to imaging modality in EDs across the US. While gradually increasing, use of MRI has historically been much less common. Concerns related to the explosion in use of head imaging include a growing understanding of the harm associated with iatrogenic radiation toxicity, an awareness of the costs to the healthcare system, and the uncertainty associated with discovering incidental findings. The rationale for the increase in ordering has not been explored experimentally. Availability of scanners, encouraging economic forces, and medico-legal concerns likely contribute to this phenomenon. Lack of confidence in a neurological exam probably contributes as well. Finally, a paucity of high quality data is available to identify those headache patients who are at low risk of pathological processes. It is clear that the yield on ED-initiated head imaging is low — national data reveal that more than 95% of head and facial CTs ordered for non-traumatic complaints were unremarkable.[42] In which patients, then, can neuroimaging be avoided?

Authoritative guidelines recommend brain imaging in all patients with "first," "worst," or "changed" headaches.[14] While not an evidence-based practice, this practice is designed as a conservative strategy to avoid missing potentially treatable life- or limb-threatening causes of headache. However, casting a wide net invariably ensnares many normal cases.

Imaging studies are generally unwarranted in patients with a typical recurrence of their underlying primary headache disorder.[43] It is unclear if, and when, patients with an atypical recurrence of their primary headache disorder require imaging. Primary headache disorder patients who require ED treatment usually describe their headache as somewhat different than usual — perhaps the headache has lasted longer, not responded to typical treatments, or been more intense. These so called "changed" headaches are considered higher risk than typical recurrences of primary headache disorders. However, whether the risk is substantial enough to require emergent imaging routinely is unknown. The emergency clinician should determine the extent to which the headache varies from the typical headache, and whether other high risk features, such as abrupt onset or focal neurological deficits, are present. Patients are often better able to evaluate the current headache relative to previous headaches after they have been afforded pain relief.

The limitations of head CT must be understood. For patients with aneurysmal subarachnoid hemorrhage, sensitivity of head CT is related directly to elapsed time since the event. If performed within 6 hours, sensitivity of head CT approaches 100%.[44] Sensitivity declines to 93% if performed within 24 hours and continues to decline from that time, resulting in a sensitivity of no more than 75% after several days have elapsed. Similarly, head CT is considered insufficiently sensitive to exclude acute ischemic stroke and venous sinus thrombosis from the differential diagnosis. Vascular imaging is required to evaluate the cervical arteries for pathology when arterial dissection is suspected.

Lumbar puncture, too, may provide useful information as the clinician excludes various secondary headache etiologies from the differential diagnosis. This procedure is substantially more uncomfortable than phlebotomy or CT scan,[45] particularly in obese patients, and so should not be performed without appropriate justification. Lumbar puncture can provide two useful pieces of information. First, the opening pressure can confirm a high- or low-pressure headache disorder, such as idiopathic intracranial hypertension, and may alert the clinician to subtle diagnoses that present with elevated intracranial pressure, including cryptococcal meningitis and venous sinus thrombosis. Second, the cerebrospinal fluid (CSF) analysis can be used to exclude a variety of ailments, including subarachnoid hemorrhage and meningitis. For patients in whom the clinician suspects subarachnoid hemorrhage despite a normal CT scan, a CSF analysis may be performed to identify xanthochromia — a straw-colored discoloration of the fluid attributable to hemoglobin degradation, which appears in the CSF 6–12 hours after the headache began — and the persistence of red blood cells in the final tube sent for analysis. CSF analysis can also confirm infectious, inflammatory, or malignant causes of meningitis.

APPROACH TO TREATMENT

Once emergency clinicians conclude they are dealing with a primary headache disorder, a vast armamentarium of treatment options is available. Traditionally, emergency physicians must first decide whether they are dealing with migraine, tension-type, or cluster headache. However, a specific diagnosis often is not required prior to acute treatment. Regardless of primary headache diagnosis, many patients are likely to respond to non-specific analgesics such as ketorolac, migraine-specific medication such as sumatriptan, and anti-emetics such as metoclopramide.[46]

Often, a specific primary headache diagnosis is difficult to clarify in the ED, particularly if the patient is in the throes of an acute headache. An accurate headache diagnosis requires both sufficient time to obtain a detailed history, and a comfortable patient who can provide an appropriate perspective on the relationship between the current headache and previous headaches. In this context, it is reasonable to treat primary headaches with an all-purpose parenteral headache medication such as a non-steroidal anti-inflammatory agent (NSAID) or an anti-dopaminergic anti-emetic, and elucidate the full headache history more carefully once the patient is comfortable.

DIFFICULT EMERGENCY DEPARTMENT POPULATIONS

One of the strongest predictors of frequent ED use is severe, chronic, underlying illness.[7] Because management of chronic illness is beyond the purview of most emergency practice, emergency physicians tend to feel less comfortable managing these patients. However, focusing exclusively on short-term goals may not appreciably affect the patient's longer term health outcomes. This is certainly true with migraine patients. Frequent use of an ED is associated with worse underlying migraine disability, so any measures a physician can implement to improve the underlying migraine disorder (such as prescriptions for an effective acute or preventive medication, or referring the patient to an outpatient physician capable of managing migraine) will likely do more good for the patient's long-term health outcomes than any acute intervention.

A commonly cited concern regarding ED management of migraine is the frequent use of parenteral opioids. Although adept at managing acute pain, acute care practitioners are less comfortable managing patients who make frequent use of an ED for acute exacerbations of chronic pain conditions. Chronic migraine is a prototypical example of a condition that is often poorly managed in acute care settings. In 1998, towards the beginning of the triptan era, opioids were used in 50% of all ED migraine visits.[1] Twelve years later, now well into the triptan era, the use of opioids for migraine in the ED has not decreased. Parenteral opioids are familiar and highly effective analgesics; however, they are less desirable for migraine care because they are associated with recurrent ED visits and subsequent refractoriness to standard medications.[47] Also, there is a perception that parenteral opioid medications are associated with a vicious cycle, in which patients begin to rely on increasing doses of opioids. For all of these reasons, opioids should never be used as first-line migraine medications and should not be offered to patients presenting to an ED de novo with an acute migraine refractory to oral treatment.

Whatever their intentions, emergency physicians are often placed in the uncomfortable position of caring for patients who insist on treatment with opioids. Often, patients may insist on a particular regimen of opioids and adjunctive medications. These patients, too, should be offered an alternate non-opioid parenteral medication. Unlike other migraine therapeutics, a prior history of successful response to opioids does not indicate that opioids are the appropriate medication for the patient. Successive doses of parenteral opioids is a treatment strategy that should be offered only in conjunction with a thoughtful overarching treatment plan that includes appropriate oral abortive therapy, preventive medications, and screening for medication overuse headache.

For an occasional patient, successive doses of opioids may be an appropriate therapy; however, it is incumbent upon the ED staff to determine this indication. Patients for whom successive doses of opioids are appropriate include those who attend medical clinics regularly, have a treatment plan in place, and have a healthcare provider who can vouch for their compliance with medical treatment. This process may or may not be possible during the course of a busy clinical shift. In those EDs with an abundance of chronic pain patients who receive most of their care in the ED, problematic patient committees may need to be established. In conjunction with other interested departments and stakeholders, these committees can determine individualized treatment plans for each patient that must be enforced by treatment contracts. Once such a treatment plan exists, it can be implemented department wide. Non-compliance with the treatment plan is grounds for using therapies other than parenteral opioids.

Models of comprehensive headache management programs have been demonstrated to be effective for select, motivated patients. These programs, which provide the multidisciplinary care and close follow-up that complicated headache patients require, can decrease the number of ED visits. One group enrolled HMO patients with migraine or tension-type headache in a headache management program consisting of an educational intervention and appropriate headache care.[47] They were able to demonstrate a decrease in the rate of ED visits, although they were not successful in attracting most ED users to participate in their headache management program. Another group enrolled HMO patients with elevated headache disability scores in a headache management program that included diagnosis, education, and evidence-based treatment.[48] Although only a select group of patients participated in these programs, those who did demonstrated substantial improvement in headache pain and functional scores.[49] Although both groups were able to improve outcomes in participants, many patients could not participate because of failure to meet enrollment criteria, or unwillingness or inability to attend the educational sessions. The challenge remains to implement these programs on a large-scale basis, and recruit frequent ED visitors.

Patients who visit an ED frequently for management of headache are complicated patients. The emergency physician may not have the time or training to manage the complexity of these patients. The management of chronic psychiatric illness, medication overuse headache, and opioid dependence is beyond the scope of most emergency physicians. Yet, discussing these issues with patients and directing them towards follow-up care is well within the scope of emergency practice. Although it is tempting to denigrate frequent ED visitors as drug seekers or malingerers, it is useful to remember the frustration and despair that many of these patients experience. Focusing on long-term rather than short-term goals may make the experience more meaningful for both the physician and the patient.

DISCHARGE CARE

Regardless of treatment received in the ED and regardless of primary headache diagnosis, patients are very likely to suffer a recurrence of their headache within 48 hours of ED discharge.[50] These headaches are often moderate to severe in intensity, and frequently functionally disabling. Therefore, it is incumbent upon the emergency physician to provide the patient with a management strategy to address the post-discharge headache. This plan should be either a prescription for an acute medication, or a next day follow-up with an outpatient physician experienced in headache management.

For patients with migraine, dexamethasone is an underutilized evidence-based treatment.[51] Although the ideal dose and duration of corticosteroid medication are not clear, one parenteral dose of medication is effective at decreasing the incidence of post-discharge headache. This benefit, however, is modest, with a number needed to treat of nearly 10. Unfortunately, it is not clear which patients are most likely to benefit.

Dispensing either one naproxen 500-mg tablet or one sumatriptan 100-mg tablet is modestly and comparably effective for decreasing the severity of headache after ED discharge.[15] Regardless of acute headache diagnosis, one-fourth of discharged patients continue to suffer moderate or severe headache in spite of taking these medications. Instructing patients to combine these medications may result in greater efficacy.

Emergency care is often perceived as a stop-gap measure to improve the acute headache, and then transition the patient to ongoing outpatient care. For many ED patients, the transition to outpatient care does not happen effectively. Considering this aspect, one ED-based randomized trial sought to improve outcomes among migraine patients with worse underlying disease through a multistep intervention.[16] However, the strategy of providing a structured educational session and appropriate acute medications, and offering a complementary headache clinic appointment, failed to improve outcomes more than typical care. This result contrasts with outpatient headache care models, in which comprehensive programs did improve outcomes[17] – which is a humbling lesson for emergency practitioners. Although interventions may help the acute headache, it is essential to refer patients to appropriate outpatient care to improve the underlying headache disorder.

CONCLUSION

Although ED visits for headache are relatively uncommon when viewed from the perspective of the general population, the very large number of headache sufferers in the general population contributes to making headache the fifth most common cause of an ED visit. Emergency physicians must attend to the widely varying needs of the individual headache patient, whether emergent management of a life-threatening secondary headache, or providing a diagnosis, education, referral for a recurrent headache sufferer, or timely and appropriate treatment of a severe headache exacerbation.

References

1. Vinson DR. Treatment patterns of isolated benign headache in US emergency departments. *Ann Emerg Med.* 2002;39:215–222.
2. Edmeads J. Emergency management of headache. *Headache.* 1988;28:675–679.
3. Bigal M, Bordini CA, Speciali JG. Headache in an emergency room in Brazil. *São Paulo Med J.* 2000;118:58–62.
4. Friedman BW, Hochberg ML, Esses D, Corbo J, Toosi B, Meyer RH, et al. Applying the International Classification of Headache Disorders to the emergency department: an assessment of reproducibility and the frequency with which a unique diagnosis can be assigned to every acute headache presentation. *Ann Emerg Med.* 2007;49:409–419:19 e1–9.
5. Leicht MJ. Non-traumatic headache in the emergency department. *Ann Emerg Med.* 1980;9:404–409.
6. Gupta MX, Silberstein SD, Young WB, Hopkins M, Lopez BL, Samsa GP. Less is not more: underutilization of headache medications in a university hospital emergency department. *Headache.* 2007;47:1125–1133.
7. Friedman BW, Serrano D, Reed M, Diamond M, Lipton RB. Use of the emergency department for severe headache. A population-based study. *Headache.* 2009;49:21–30.
8. Lipton RB, Stewart WF, Simon D. Medical consultation for migraine: results from the American Migraine Study. *Headache.* 1998;38:87–96.
9. Kwong WJ. Determinants of migraine emergency department utilization in the Georgia medicaid population. *Headache.* 2007;47:1326–1333.
10. Rinkel GJ, Djibuti M, Algra A, van Gijn J. Prevalence and risk of rupture of intracranial aneurysms: a systematic review. *Stroke.* 1998;29:251–256.
11. Linn FH, Rinkel GJ, Algra A, van Gijn J. Headache characteristics in subarachnoid haemorrhage and benign thunderclap headache. *J Neurol Neurosurg Psychiatry.* 1998;65:791–793.
12. Perry JJ, Stiell IG, Sivilotti ML, Bullard MJ, Lee JS, Eisenhauer M, et al. High risk clinical characteristics for subarachnoid haemorrhage in patients with acute headache: prospective cohort study. *BMJ.* 2010;341:c5204.
13. Perry JJ, Stiell IG, Sivilotti ML, Bullard MJ, Hohl CM, Sutherland J, et al. Clinical decision rules to rule out subarachnoid hemorrhage for acute headache. *JAMA.* 2013;310:1248–1255.
14. Edlow JA, Caplan LR. Avoiding pitfalls in the diagnosis of subarachnoid hemorrhage. *N Engl J Med.* 2000;342:29–36.
15. Friedman BW, Solorzano C, Esses D, Xia S, Hochberg M, Dua N, et al. Treating headache recurrence after emergency department discharge: a randomized controlled trial of naproxen versus sumatriptan. *Ann Emerg Med.* 2010;56:7–17.
16. Friedman BW, Solorzano C, Norton J, Adewumni V, Campbell CM, Esses D, et al. A randomized controlled trial of a comprehensive migraine intervention prior to discharge from an emergency department. *Acad Emerg Med.* 2012;19:1151–1157.
17. Matchar DB, Harpole L, Samsa GP, Jurgelski A, Lipton RB, Silberstein SD, et al. The headache management trial: a randomized study of coordinated care. *Headache.* 2008;48: 1294–1310.
18. Haydel MJ, Preston CA, Mills TJ, Luber S, Blaudeau E, DeBlieux PM. Indications for computed tomography in patients with minor head injury. *N Engl J Med.* 2000;343:100–105.
19. Stiell IG, Wells GA, Vandemheen K, Clement C, Lesiuk H, Laupacis A, et al. The Canadian CT Head Rule for patients with minor head injury. *Lancet.* 2001;357:1391–1396.
20. Smits M, Dippel DW, de Haan GG, Dekker HM, Vos PE, Kool DR, et al. External validation of the Canadian CT Head Rule and the New Orleans Criteria for CT scanning in patients with minor head injury. *JAMA.* 2005;294:1519–1525.
21. Kirby S. Headache and brain tumours. *Cephalalgia.* 2010;30:387–388.
22. Boiardi A, Salmaggi A, Eoli M, Lamperti E, Silvani A. Headache in brain tumours: a symptom to reappraise critically. *Neurol Sci.* 2004;25(suppl 3):S143–S147.
23. Vazquez-Barquero A, Ibanez FJ, Herrera S, Izquierdo JM, Berciano J, Pascual J. Isolated headache as the presenting clinical manifestation of intracranial tumors: a prospective study. *Cephalalgia.* 1994;14:270–272.
24. Valentinis L, Tuniz F, Valent F, Mucchiut M, Little D, Skrap M, et al. Headache attributed to intracranial tumours: a prospective cohort study. *Cephalalgia.* 2010;30:389–398.
25. Schankin CJ, Ferrari U, Reinisch VM, Birnbaum T, Goldbrunner R, Straube A. Characteristics of brain tumour-associated headache. *Cephalalgia.* 2007;27:904–911.
26. Forsyth PA, Posner JB. Headaches in patients with brain tumors: a study of 111 patients. *Neurology.* 1993;43:1678–1683.
27. Solomon GD, Kunkel Jr. RS, Frame J. Demographics of headache in elderly patients. *Headache.* 1990;30:273–276.
28. Hunder GG, Bloch DA, Michel BA, Stevens MB, Arend WP, Calabrese LH, et al. The American College of Rheumatology 1990 criteria for the classification of giant cell arteritis. *Arthritis Rheum.* 1990;33:1122–1128.
29. Salvarani C, Cantini F, Boiardi L, Hunder GG. Polymyalgia rheumatica and giant-cell arteritis. *N Engl J Med.* 2002;347:261–271.
30. Saposnik G, Barinagarrementeria F, Brown Jr. RD, Bushnell CD, Cucchiara B, Cushman M, et al. Diagnosis and management of cerebral venous thrombosis: a statement for healthcare professionals from the American Heart Association/American Stroke Association. *Stroke.* 2011;42:1158–1192.
31. Kim YK, Schulman S. Cervical artery dissection: pathology, epidemiology and management. *Thromb Res.* 2009;123:810–821.
32. Schievink WI. Spontaneous dissection of the carotid and vertebral arteries. *N Engl J Med.* 2001;344:898–906.
33. Gladstone J, Bigal ME. Headaches attributable to infectious diseases. *Curr Pain Headache Rep.* 2010;14:299–308.
34. Attia J, Hatala R, Cook DJ, Wong JG. The rational clinical examination. Does this adult patient have acute meningitis? *JAMA.* 1999;282:175–181.
35. Pachner AR, Steiner I. Lyme neuroborreliosis: infection, immunity, and inflammation. *Lancet Neurol.* 2007;6:544–552.
36. Jackson A, van der Horst C. New insights in the prevention, diagnosis, and treatment of cryptococcal meningitis. *Curr HIV/ AIDS Rep.* 2012;9:267–277.
37. Headache Classification Committee of the International Headache Society (IHS). The International Classification of Headache Disorders, 3rd edition (beta version). *Cephalalgia.* 2013;33:629–808.
38. Tentschert S, Wimmer R, Greisenegger S, Lang W, Lalouschek W. Headache at stroke onset in 2196 patients with ischemic stroke or transient ischemic attack. *Stroke.* 2005;36:e1–e3.

39. Detsky ME, McDonald DR, Baerlocher MO, Tomlinson GA, McCrory DC, Booth CM. Does this patient with headache have a migraine or need neuroimaging? *JAMA*. 2006;296:1274–1283.

40. Lipton RB, Dodick D, Sadovsky R, Kolodner K, Endicott J, Hettiarachchi J, et al. A self-administered screener for migraine in primary care: the ID Migraine validation study. *Neurology*. 2003;61:375–382.

41. Dentali F, Squizzato A, Marchesi C, Bonzini M, Ferro JM, Ageno W. D-dimer testing in the diagnosis of cerebral vein thrombosis: a systematic review and a meta-analysis of the literature. *J Thromb Haemost*. 2012;10:582–589.

42. Gilbert JW, Johnson KM, Larkin GL, Moore CL. Atraumatic headache in US emergency departments: recent trends in CT/MRI utilisation and factors associated with severe intracranial pathology. *Emerg Med J*. 2012;29:576–581.

43. Loder E, Weizenbaum E, Frishberg B, Silberstein S. Choosing wisely in headache medicine: the American Headache Society's list of five things physicians and patients should question. *Headache*. 2013;53:1651–1659.

44. Perry JJ, Stiell IG, Sivilotti ML, Bullard MJ, Emond M, Symington C, et al. Sensitivity of computed tomography performed within six hours of onset of headache for diagnosis of subarachnoid haemorrhage: prospective cohort study. *BMJ*. 2011;343:d4277.

45. Singer AJ, Richman PB, Kowalska A, Thode Jr. HC. Comparison of patient and practitioner assessments of pain from commonly performed emergency department procedures. *Ann Emerg Med*. 1999;33:652–658.

46. Weinman D, Nicastro O, Akala O, Friedman BW. Parenteral treatment of episodic tension-type headache: a systematic review. *Headache*. 2014;54:260–268.

47. Evans RW, Friedman BW. Headache in the emergency department. *Headache*. 2011;51:1276–1278.

48. Maizels M, Saenz V, Wirjo J. Impact of a group-based model of disease management for headache. *Headache*. 2003;43:621–627.

49. Harpole LH, Samsa GP, Jurgelski AE, Shipley JL, Bernstein A, Matchar DB. Headache management program improves outcome for chronic headache. *Headache*. 2003;43:715–724.

50. Friedman BW, Hochberg ML, Esses D, Grosberg BM, Rothberg D, Bernstein B, et al. Recurrence of primary headache disorders after emergency department discharge: frequency and predictors of poor pain and functional outcomes. *Ann Emerg Med*. 2008;52:696–704.

51. Colman I, Friedman BW, Brown MD, Innes GD, Grafstein E, Roberts TE, et al. Parenteral dexamethasone for acute severe migraine headache: meta-analysis of randomised controlled trials for preventing recurrence. *BMJ*. 2008;336:1359–1361.

22

Headache Clinics

Merle L. Diamond

Diamond Headache Clinic, Chicago, Illinois, USA

INTRODUCTION

The concept of a clinic dedicated to the diagnosis and treatment of headache is not new. During World War II, in response to the return of military personnel with traumatic brain injuries and the resulting post-traumatic headaches, Arnold Friedman, MD, with H.H. Merritt and Charles Brenner, founded the first headache clinic in the United States.[1] The unit, officially organized in 1945, was located at Montefiore Medical Center in the Bronx, New York, within the Department of Neurology. For many years this type of dedicated clinic would be limited to academic centers, including Mayo Clinic in Rochester, Minnesota. During the 1930s, Dr Harold G. Wolff did extensive research on migraine at Cornell Medical College in New York City, but again he worked as a member of the neurology staff. In 1950, a headache center was also opened at Faulkner Hospital in Boston by Dr John Graham, who had worked with Dr Wolff at Cornell.

In England, the British Migraine Association (now the Migraine Action Association) founded the first dedicated migraine clinic in 1963 in Bournemouth. This clinic and the association gave the impetus for other sites to open, including the City of London Migraine Clinic, the Princess Margaret Migraine Clinic at Charing Cross Hospital, and a clinic at the National Hospital in Queen Square.[2] In 1954, Professor Federigo Sicuteri was the initial Director of the Headache Clinic at the University of Florence – the first such facility in Italy.[3] Building on the work of T. Dalsgaard-Nielsen, started during the 1940s, Dr Jes Olesen and his colleagues opened the Copenhagen Acute Headache Clinic in 1976.[4] Throughout the world, these dedicated headache centers were opened not only to help headache sufferers through appropriate diagnosis and treatment, but also as research centers which contributed to the many discoveries in headache therapies.

The 1970s has been called "The Era of the Headache Clinic".[1] Dr Seymour Diamond, in 1964, published his first article on depression and headache.[5] Soon, the number of his headache patients grew sufficiently large that he decided to limit his practice to only the treatment of headache. In 1974, the name was officially changed to the Diamond Headache Clinic – the oldest, largest, and most comprehensive private headache clinic in the USA. The clinic, now located on the near north side of Chicago, was established without institutional backing. During that time, this action was unheard of within the medical community. Because of the unique model of the clinic and its originality, reimbursement for services from the insurance carriers was a problem.

A deluge of physicians interested in starting their own headache clinics sought advice from Dr Seymour Diamond, and requested the opportunity to visit the Diamond Headache Clinic. These physicians recognized the success of the clinic and its comprehensive approach to headache management. Unlike other headache centers, the clinic utilized nursing professionals in obtaining the headache histories and handling patient education. A Health History Questionnaire that was developed at Cornell was utilized at the clinic. This simple form was completed by the patient, indicated comorbid issues, and certainly expedited patient appointments. The services of other healthcare professionals were provided at the clinic, including psychologists, dieticians, and pharmacists, and the biofeedback department was established in 1972 as part of Dr Diamond's practice. Certain policies established by Dr Diamond were also adapted by other physician practices. A "no refill" policy facilitated patient compliance in maintaining follow-up appointments and enabling the staff to monitor medication usage. Patients were instructed that they would be able to communicate with their physician during established

S. Diamond, R. K. Cady, M. L. Diamond & V. T. Martin (Eds):
Headache and Migraine Biology and Management.

DOI: http://dx.doi.org/10.1016/B978-0-12-800901-7.00022-7

"calling hours," when the physician would be available for phone calls. A clinic physician was also available, on call, for 24-hour coverage daily.

Although the Diamond Headache Clinic is not affiliated with an academic center, the clinic physicians have maintained faculty positions at medical schools and have welcomed medical students and residents to complete clerkships at the clinic. In 1980, recognizing the need for specialized inpatient headache units, Dr Diamond opened the first Diamond Inpatient Headache Unit at Bethany Methodist Hospital in Chicago. This unit is now located at Saint Joseph Hospital, and is discussed in Chapter 23.

The number of private headache clinics has multiplied during the past 40 years. Some centers are located within multidisciplinary practices and many remain within academic centers, such as the Cleveland Clinic, Mt Sinai (New York), the University of Cincinnati, and the Scripps Clinic. The clinic at Montefiore continues to thrive as a major research center through its affiliation with the Albert Einstein Medical School.

In 2007, the United Council for Subspecialties in Neurology established certification in headache medicine. It also recognized the many fellowships in headache medicine that had been created. Those physicians embarking on a headache fellowship will focus their work and research in outpatient clinics, as well as on those patients admitted to an inpatient unit because of their headaches. In addition to facilitating the diagnosis and care of those experiencing headache, the fellowship programs contribute significantly to current headache research.

For some physicians and healthcare providers, the thought of dedicating their entire practice to the management of headache patients may be daunting. As with Dr Seymour Diamond during the 1960s and early 1970s, the change may be gradual. For headache patients, seeking consultation at a specialized headache center may be a final step in their quest to resolve their headache problems. Patients may have presented with the complaint of headache at their primary care provider (family practice, internal medicine, OB-GYN). The next step may have been a referral to a neurologist, otolaryngologist, or ophthalmologist. Finally, the patient may seek evaluation by a headache specialist at a tertiary headache center.

The patient referred to a headache specialist may have undergone an extensive battery of laboratory tests and scans. Frequently, the patient seen at a headache clinic is experiencing chronic daily headaches or medication overuse headaches. Up to that time, the patient has not received an appropriate diagnosis, and in turn, has not been treated effectively. Hopefully, by undergoing a comprehensive evaluation by physicians experienced in headache medicine, the patient will receive appropriate medications and adjunct therapies, and be able to manage the headache condition.

ESTABLISHMENT OF THE HEADACHE CLINIC

The headache clinic may be a solo practice, part of a multidisciplinary group, or a section in an academic department (typically, neurology). Frequently, the patient attending a headache clinic has been referred by another healthcare practitioner.

In the case of the Diamond Headache Clinic, the initial patients were part of Dr Diamond's original practice. When he decided to establish the clinic, he not only saw the need for the specialized center for patients, but also recognized that as a solo practitioner he needed to dedicate his time and effort to the management of all types of headache. The solo practitioner must rely on his or her own knowledge and experience to continue successful treatment. For some physicians, a solo practice may entail educating their staff, which could include nurses, physician assistants, advance practice nurses, or medical assistants. The staff can assist the physician with the initial interview, utilizing a specialized headache history, and overseeing testing such as the Migraine Disability Assessment (MIDAS) questionnaire. The staff should also be cognizant of the standard therapies for headache treatment and the precautions and side effects of these therapies. A solo practitioner can rely on a knowledgeable staff to triage phone calls regarding headache status and any problems with medications.

At some headache clinics, such as in the current organization of the Diamond Headache Clinic, several practitioners will be on staff, including physicians and physician assistants experienced in the management of headache. These healthcare practitioners (HCPs) may represent many disciplines, including neurology, internal medicine, and pain specialists. Some of the physicians may have received certification in headache medicine from the United Council of Neurological Subspecialties (UCNS) or the Certificate of Added Qualification from the National Headache Foundation. With a larger staff, a higher number of patients can be managed and the burden of responsibility does not fall on one person. A larger staff also enables the HCPs to share call, and provide each patient with access to emergency care on a 24/7 basis.

For those centers within an academic setting, the multidisciplinary approach to headache management is certainly facilitated by a large referral base within the medical center. Patients have access to all departments within the facility, and the practitioners are able

---

BOX 22.1

MULTIDISCIPLINARY TEAM AT THE HEADACHE CLINIC

- Physicians
- Physician assistants/Advance practice nurses
- Registered nurses
- Nursing/medical assistants
- Psychologists
- Addiction counselors
- Acupuncturists
- Dieticians

- Pharmacists
- Physical therapists
- Occupational therapists
- Recreational/activity therapists (art therapy, pet therapy)
- Massage therapists
- Biofeedback technicians

---

to arrange contiguous consultations. This type of headache clinic may also have ready access to other allied health professionals, including nutritionists, physical therapists, occupational therapists, acupuncturists, massage therapists, biofeedback technicians, etc. (Box 22.1).

All forms of the headache clinic — solo practice, group practice, and academic setting — can be successful. For headache patients, it is the continuity of care that will facilitate their headache management. Communication between the various disciplines is essential to provide overall care of the patient.

STAFFING OF THE HEADACHE CLINIC

Whatever the type of practice, the mission of the clinic should be apparent to all staff members and the patients. At the start, patient education can be an excellent motivator for successful treatment. The goal of the individual's treatment, as well as the clinic in general, should be accessible through the website, brochures, and during a patient's visit.

The structure of the clinic staff is dependent on the type of practice. In a solo practice, the physician is also the administrator, managing the financials as well as human resources, and establishing medical care of the patient. In a group practice, a practice administrator will oversee the non-medical issues involved in the day-to-day activities of the practice. Finally, in an academic setting, the headache clinic will be subject to the structure and regulations of the medical center in general.

In all types of headache clinic, the receptionists are important to the success of the practice. Often, a receptionist is the first contact with a new patient. They can address any questions the patient may have about an initial visit, and assure patients about their expectations. To ensure facilitation of the new patient's appointment, an intake form may be utilized (Figure 22.1) This screening form will help determine which HCP should evaluate the patient, and what material the new patient should bring to the initial visit. A receptionist or billing clerk can confirm insurance coverage, copays, deductibles, and accounting procedures. Handling of phone calls, mail, or emails may also be transferred to a capable receptionist, who can distribute queries to the appropriate HCP or staff member.

Medical assistants can be vital in ensuring good patient flow. In addition to monitoring vital signs and weight, and itemizing complaints, the medical assistant can prepare rooms for parenteral treatments and procedures. The medical assistant can provide patient education information and instruct the patient on office procedures, such as appointments, phone calls, and refills.

Registered nurses can be especially effective in the triage of patient calls. These licensed professionals can assist with authorizations on prescriptions, pre-authorization on inpatient admissions, and general patient education. The nurse can assist the physicians with parenteral therapies, such as IV histamine desensitization, IV dihydroergotamine (DHE), and other therapeutic modalities. During the infusion therapy, educational materials may be discussed with the patient. These materials may be written or computer-based.

Physician assistants and advance practice nurses can independently evaluate patients and be especially helpful to the solo practitioner. They will consult with the physician about changes in therapy, and offer suggestions for treatment. The scope of these HCPs varies from state to state, and the physician should determine guidelines for their role within the headache clinic. The PAs and NPs may also receive training in specific therapies, such as Botox injections or nerve blocks. Their role may be greatly enhanced if an infusion center is also established within the headache clinic.

NHA Intake Coordination Form   Rev 6/14   Referred By: WOM   PT   MD   WEB   PST   PT   NHF

Patient Name: _____   SS#: _____
         Last name   First name   DOB: ____/____/_____

Insurance Holder Name: _____   DOB: ____/____/_____
         Last name   First name   SS#: _____

Relation to Pt: _____   Contact #: _____   Chart#: _____

1). Insurance Co: _____ Payor ID: _____   Cust. Svc: _____

Address to send claims: _____ City: _____ State: _____ Zip: _____

Policy#_____ Group:_____ Effective date: _____/_____/_____

2). Insurance Co: _____ Payor ID: _____   Cust. Svc: _____

Address to send claims: _____ City: _____ State: _____ Zip: _____

Policy#_____ Group:_____ Effective date: _____/_____/_____

Ins. Holder Name: _____ DOB: _____ SS# _____ Rel to Pt: _____

Predictor for Inpatient   Call By: _____to CANCEL
☐Yes   ☐No   Have you been hospitalized for your headaches in the last year?
☐Yes   ☐No   In the last 6 months, have you been to the ER for your headaches at least once a month?
☐Yes   ☐No   Recently, have you missed school, work or social activities at least once a week due to headaches?
☐Yes   ☐No   Are you on 5 or more prescription medications for your headaches?
☐Yes   ☐No   Have you been taking rescue medications to stop the headaches daily for over six months?
☐Yes   ☐No   Have you seen more that 2 doctors for your headache problems?
☐Yes   ☐No   Do you have a headache more than 15 days out of the month?
☐Yes   ☐No   Have you had an MRI of the head? When? ____/____ Normal or Abnormal
☐Yes   ☐No   Have you been diagnosed as Bipolar or Borderline?   NH form EM or M _____

Appt Date: ___/___/_____   **Arrival time:** _____   **Patient to see:** _____   **MD Request** ___ **Rep:**_____

Advanced Program Fee: ☐Accepted   ☐Declined   ☐Other: _____   CX policy reviewed? ☐Y ☐N
                               Cancellation FEE:

Benefits verified by: _____   Date: _____   Does pre-existing apply: Y   N   If yes, until when:_____

Insurance Type
☐PPO-Name: _____   ☐In ☐Out   ☐HMO- Is referral needed?   ☐Yes   ☐No
☐POS-Is referral needed?   ☐Yes ☐No   Referral obtained?   ☐Yes   ☐No
☐Indemnity   Process if inpatient? _____

Outpatient Benefits (Office visit)
Co-insurance_____% Copay $____ Deductible met? Y N
Deductible $_____   Amount met $_____

Diagnostic Benefits (Labs)
Co-insurance_____%   Deductible met? Y N
Deductible $_____   Amount met $_____

Inpatient Benefits
Co-insurance_____%   Deductible met? Y   N
Deductible $_____   Amount met $_____
OOP $ _____   Met $ _____
Is pre-certification needed for inpatient?   Y   N
Pre-Cert phone #: _____

Pre-Cert (RQI) needed for MRI?   ☐Yes   ☐No
Is Pre-Cert (RQI) needed for MRA?   ☐Yes   ☐No
MRI/MRA pre-cert phone #:_____
_____

Confirmation
Date:___/_____/_____   Initials: _____
Referral policy reviewed?   ☐Yes   ☐No
Benefits explained to patient?   ☐Yes   ☐No

FIGURE 22.1   NHA Intake Coordination Form.

In a clinic affiliated with a large medical center, the availability of allied health professionals may enhance the program. Classes in nutrition may be easily scheduled, as well as specific referrals to a dietician for discussion of tyramine-free diets, menu planning, and other pertinent issues for the headache patient.

Referrals to the staff pharmacist or pharmacy classes may also be offered in the academic setting. The presence of a staff pharmacist will greatly enhance patient education.

In any of the settings, biofeedback may be offered, and a biofeedback technician may be added to the staff. A full-time biofeedback technician will provide the option of utilizing this therapy for all eligible patients and enhance the program's treatment choices. Other alternative therapies, including acupuncture and massage therapy, can also be offered at the outpatient headache center.

At some headache clinics, the services of a psychologist may be considered. These practitioners can provide on-site counseling services as well as psychological testing. The psychologist, in consultation with the physician and other HCPs, can help establish a treatment plan for patients requiring counseling. The psychologist may also assist in referring the patient for outside services, including family counseling and addiction rehabilitation.

PHYSICAL PLANT OF THE HEADACHE CLINIC

Headache clinics may be located in stand-alone facilities, in a medical office building, or as a department within a large medical center. Although there are variations on the structure of the clinics, some aspects will be common to all forms.

If possible, the clinic may have separate interview rooms for new patients. The headache history could be obtained in that area, as well as allowing patients adequate time to complete forms and questionnaires to be included in their patient record. Histories may also be obtained in examination rooms. The important factor in designing a specialized clinic is to facilitate patient flow.

The furniture in the examination rooms should be simple, as complicated procedures will not be undertaken at this facility. A basic exam table will suffice, as will a small desk for the computer monitor and keyboard. Chairs should be provided for the patient and whoever accompanies the patient to the visit. The physician or staff member can utilize a stool or chair.

Soft, non-fluorescent lighting should be utilized in the headache clinic. Many headache patients are sensitive to light during a headache, and bright lights may precipitate a headache. Ideally, the exam rooms should be soundproof. Loud noises can be irritating to the patient in the midst of a headache attack. Also, because of HIPAA regulations, it is essential to prevent a patient overhearing conversations in another exam room. The color of wall paints/wallpapers should be muted — again, to prevent exacerbation of an acute headache.

Staff workrooms should be located so as not to engender noise or lack of privacy. The workroom will be stocked with the drugs used for parenteral administration. It can also be the area where the physicians give instructions to staff members and consult with their colleagues. A workroom should also be provided for the staff members handling phone calls, including refill requests. Each of these areas will need computer monitors and keyboards for charting.

The addition of an infusion center may be considered for the headache clinic. For those patients who visit the clinic during an acute headache, parenteral agents can be administered. Patients receiving this form of therapy can be carefully monitored by the staff. The creation of an infusion center may comprise assigning one or two examination rooms for prolonged therapy, or there may be actual design of dedicated rooms for infusion. The availability of the infusion center will assist patients in decreasing emergency department visits. The use of reclining chairs or beds is recommended for the infusion center.

One or two rooms may be dedicated for procedures such as acupuncture, Botox injections, or nerve blocks. Other rooms may be assigned for biofeedback therapy or massage therapy. A dedicated room for biofeedback training should be in the quietest area of the clinic, to facilitate relaxation of the patient.

REIMBURSEMENT ISSUES

Whatever the setting, receiving payment for services is vital in order to continue the economic health of the practice. Whether a solo practitioner or in a group practice, payment histories must be monitored for the various insurance agencies and Medicare. The practitioner and/or the practice manager need to be proficient with CPT coding and how it applies to headache medicine.[6] The physician must be cognizant of changes in the International Classification of Headaches[7] as well as the coding requirements.

The introduction of Electronic Medical Records (EMRs) has eased the use of coding, particularly if using a billing module. For any practice, the selection of an EMR system is dependent on the needs of the practice. If the clinic is affiliated with a large academic center, the system will need to be compatible with the network's EMR system. The EMR system has created uniform records, and can be of great benefit in a group practice when one of the physicians needs to review a chart of another HCP's patient. EMR systems, if used appropriately, should not negatively impact patient/physician dialog. In this author's practice, the positioning of the computer screen/keyboard is selected to allow the physician to continue face-to-face interviews with the patient. Observing the patient during the headache history and later in follow-up visits is essential in completing a thorough examination. For example, when asking about personal, family, work, or school relationships, it is important to notice the patient's demeanor and response to questioning. Viewing the patient's reactions or lack of affect will certainly help in establishing an appropriate diagnosis.

From the patient's perspective, a face-to-face interview will enhance communication with the HCP. If the physician's attention is directed to a computer instead of the patient, all further lines of communication and trust can be impacted. This initial visit will define the rest of the physician/patient relationship, and the success or failure of the treatment.

For the headache clinic or any other clinical setting, the EMR system should be customized for each practice. The information that practitioners and their colleagues require in establishing diagnosis and treatment selection should be included in the records. The use of EMR systems in headache management is discussed in Chapter 3.

The information gathered in the records will be utilized in the reimbursement process, as well as gaining pre-authorization for therapies such as Botox. The availability of the records for staff members will help them maneuver the process of dealing with insurance companies. The information gleaned from these records will also facilitate the completion of disability applications for patients and assist patients with school/employment issues (for example, return-to-work forms).

When reviewing reimbursement issues, the physicians and/or practice manager should also review the soundness of the coding used and the procedures provided at the headache clinic. For instance, should patients be referred for laboratory testing to a hospital or testing facility? Or would it benefit the clinic financially to include its own laboratory for blood-drawing or urinalysis? This issue will be moot if the clinic is located within an academic center, but can impact the revenue of a free-standing clinic. The majority of practices will not have their own scanning facilities. Referral for CT and MRI is usually handled in a hospital or independent scanning facility. Pre-authorization for scanning may be done by the clinic staff, utilizing the information in the EMRs.

Equitable reimbursement for services may require negotiation with the insurance companies. During negotiations, it is important to identify reimbursement rates dependent on the specialty. A visit with the staff neurologist may be paid at a higher rate than that assigned to internal medicine. Negotiations may need to be accomplished on an annual basis.

It is also important to have insurance information on each patient prior to the initial visit. With managed care, deductibles have increased and these should be collected at the time of the visit. Staff should be aware of copays and collect them at the visit. For procedures such as nerve blocks and Botox injections, a policy for deposits should be established. These deposits may also be considered for patients who will be hospitalized.

If any one factor can be emphasized in reimbursement issues, it is preparation. The staff should be aware of the patient's insurance policy prior to the initial visit. It not only helps with reimbursement; it will also alleviate patients' anxiety about billing if they are fully aware of their fiscal responsibility. The clinic's financial policy should be posted on its website and made available to new patients prior to the visit. For follow-up visits, if a financial issue needs to be discussed, there should be an area established for such a conversation. A patient's financial issues should not be discussed at the reception desk, in order to prevent other patients from overhearing the conversation.

Preparation also involves careful recording of the patient's history, progress, and treatment in the EMRs. Again, the completeness of these records will facilitate any claim processing and pre-authorizations.

At the Diamond Headache Clinic, the practice manager is proactive in containing costs. This process may involve consolidating services and utilizing preferred vendors for drugs and medical supplies. It also involves the review of expenditures for medications and supplies in order to equitably set charges for the clinic's services. Again, this review should be done on an annual basis. Negotiating with vendors to obtain the lowest price for drugs and other supplies is essential.

MARKETING THE HEADACHE CLINIC

For those clinics affiliated with a large academic center, marketing is usually assumed by the medical center. However, the solo practitioner and group practice may need to depend on their website for "getting the word out." The clinic website will probably be the first point of contact with a new patient.

The website should be easy to use, and provide practical information on the clinic and its staff. Brief biographies of essential personnel (physicians and other HCPs) should be included on the site. Other information should include "what to expect at the first visit," and financial policy. Goals of treatment and educational information should be easily obtained. Forms such as headache diaries, diets, and relaxation exercises can be included, and make printing easy. Directions to the clinic, important phone numbers and email addresses, and refill information should be highlighted.

If the physician is a member of the National Headache Foundation, an interested patient can visit its website (www.headaches.org) to access the "Physician Finder" and find the member listing. On a state-to-state basis, the Physician Finder provides names and addresses of physician members of the NHF who are interested in headache medicine. If the physician has certification in headache medicine, this information will be indicated.

Most headache clinics rely on referrals from other physicians. In addition to the clinic's website, the HCPs at the headache clinic should make their colleagues aware of the clinic and its activities. This may be accomplished through hospital or network affiliations. Practitioners may work through the education office at the medical center or local medical school to be available to lecture to colleagues regarding headache medicine. They may also work through their local medical society, placing announcements in the society journal or newsletter regarding their practice. Networking at local society meetings or at continuing medical education courses may be helpful in making colleagues aware of the practice. Communication with the referring physician is vital to maintaining a good relationship with colleagues. At the Diamond Headache Clinic, it has been a long-term practice to send thank-you letters to the referring physician, and detail the findings of the history and plans for treatment.

Advertising in phone directories, which may have been routine during the past century, has negligible results. Social networking may provide better results, and the clinic may consider creating a Facebook page. This option can offer interested patients a resource regarding the facility.

The key to marketing the headache clinic is being proactive. Headache specialists are often contacted for interviews or lectures to patient groups. Practitioners may consider affiliating with a network which could provide referrals from fellow network HCPs, or joining a professional organization (American Headache Society) or attending a postgraduate course to meet other HCPs interested in headache medicine.

Continuing education in headache medicine is essential to be aware of updates and changes in the specialty. It is also beneficial for new staff members to obtain new knowledge about the intricacies of diagnosis and treatment. At the Diamond Headache Clinic we organize three courses throughout the year which not only assist our colleagues in learning about headache but also make them cognizant about the clinic and its experience. Many organizations offer continuing medical education in headache, and information about these courses is easily accessible on the Internet.

To remain updated on headache therapies, theories, and tests, it is important to read the various headache journals that are available. The American Headache Society publishes the journal *Headache*, which reviews current research. A similar journal, *Cephalalgia*, is published by the International Headache Society. Important articles on headache medicine can also be found in scholarly, peer-reviewed journals. For patients, the National Headache Foundation publishes a quarterly magazine, *Head Wise*. Upon request, multiple copies of this magazine will be sent to physician offices for distribution to their patients. Patients will also be able to access various blogs and social networking sites that focus on headache issues.

THE PATIENT ATTENDING THE HEADACHE CLINIC

As stated previously, the patient consulting a headache specialist has probably seen several physicians, been treated with various medications, and undergone various tests and scans. Typically, the patient has been referred by another physician and has been refractory to prior therapy.

Often, the patient is experiencing chronic daily headache — including chronic migraine and medication overuse headache. A thorough headache history with appropriate testing is essential for these patients to receive the correct diagnosis. The headache specialist will be familiar with treatment approaches that can be attempted in patients who have not responded to earlier trials of medications. As discussed in Chapter 23, patients who are overusing analgesics, caffeine-containing drugs, or triptans should be considered for admission to a specialized inpatient headache unit for withdrawal of the drug.

Patients with comorbid issues may be referred to a specialized headache clinic. For those with concomitant illness that could be complicated by headache therapies, inpatient admission may be necessary to initiate therapy. Collaborative efforts with the primary care provider and the headache specialist will facilitate appropriate treatment that should not interfere with concomitant therapy. Also, the headache specialist will be aware of those agents used for comorbid disease which could aggravate a headache condition.

Occasionally, a patient with recent onset headaches will visit the specialized headache unit. These patients and/or their physicians are aware of the headache clinic and its treatment successes. Patients with recent onset headaches must undergo thorough evaluation with a careful history, laboratory testing, and scanning.

Primary care providers and emergency department physicians, if aware of the headache clinic, will refer those patients who are frequent visitors to the ED. It is essential that these patients receive a correct diagnosis and start preventive therapy for headache control.

Patients with atypical headache patterns, or presenting with difficult-to-treat headaches, may seek help at a specialized clinic. For example, cluster headaches may not be recognized by a primary care physician and ineffective treatment prescribed. A headache specialist will be knowledgeable about this type of headache, and will be able to establish the correct diagnosis and prescribe appropriate medications. Patients experiencing posttraumatic headaches present with their own complicated history, and a headache specialist will be aware of the testing and treatment options for these patients.

For whatever reason patients have sought services at a headache clinic, they must be reassured that the HCP is interested in helping them control their headache condition and prescribe the most beneficial therapy. Effective communication is essential for successful therapy. Patients should be encouraged to participate in their treatment and be compliant with the various therapies.

At the initial visit, patients should receive instructions on what is expected from them. For example, they should be advised on hours in which they can contact a staff member or physician. If the clinic is not open on weekends, patients should be advised that refills will not be prescribed on Saturday and Sunday, and they should plan accordingly. For some practices, patients will be charged for after-hour calls. Patients should be instructed on emergency care for their headaches if they are not able to travel to the clinic on short notice.

CONCLUSION

Although the initial establishment of a specialized headache clinic occurred in 1945, the years between 1970 and 1980 saw an upsurge in the number of these dedicated centers. This interest in limiting practices to headache medicine is probably due to the many advances occurring in headache research since the 1950s. Headache clinics are not confined to academic centers but may be solo practices, within group practices dedicated to headache management, or as part of a large multidisciplinary practice. A variety of specialties may undertake a dedicated practice in headache medicine. Allied health providers, including physician assistants and advance practice nurses, may comprise the staff of these headache centers.

For headache patients, referral to one of these dedicated centers may be the final stage of finding relief for their headache condition. The headache clinic offers hope, as the healthcare providers have experience in headache medicine and are interested in helping patients to manage their headache condition. In the future, the number of headache clinics and headache specialists will continue to grow and increase the options for successful headache management.

References

1. Solomon S, Diamond S, Mathew N, Loder E. American headache through the decades: 1950 to 2008. *Headache.* 2008;48:671–677.
2. Surgery Door. Berkhamsted, UK: Intuition Communication Ltd. <www.surgerydoor.co.uk/advice/living-with/migraine/clinics/> Accessed 20.06.14.
3. Fanciullacci M. In memory of Federigo Sicuteri (1920–2003), a headache medicine pioneer. *Cephalalgia.* 2004;24:1090–1091.
4. Tfelt-Hansen P. History of headache research in Denmark. *Cephalalgia.* 2001;21:748–752.
5. Diamond S. Depressive headaches. *Headache.* 1964;4:255–259.
6. Freitag FG, Black SB. *Establishing a Headache Center.* Hamilton, ON: Baxter Publishing; 2011.
7. Headache Classification Committee of the International Headache Society (IHS). The International Classification of Headache Disorders, 3rd edition (beta version). *Cephalalgia.* 2013;33:629–808.

23

Inpatient Treatment of Headaches

Alexander Feoktistov

Diamond Headache Clinic, Chicago, Illinois, USA

INTRODUCTION

Treatment of headache disorders has drastically improved over the past few decades due to the discovery of new medications and treatment modalities. For acute pain relief, triptans and ergotamine-containing medications as well as non-steroidal anti-inflammatory drugs (NSAIDs) and opioids are still in use, and may be found beneficial in certain clinical scenarios. We have gradually expanded the pool of preventative or prophylactic medications that may provide stability and pain control for prolonged periods. Interventional treatment modalities (cervical facet medial branch nerve blocks and radiofrequency ablations, cervical epidural steroid injections, sphenopalatine ganglion blocks and ablations, etc.) are being actively utilized in the treatment of certain headache disorders. During the past several years, the field of neuromodulation has also been rapidly evolving, potentially offering novel solutions in intractable headache management (occipital nerve stimulation, sphenopalatine nerve stimulations, etc.).

Despite these advances in treatment, some patients will fail to respond to traditional treatment and require more sophisticated and perhaps intense approaches, such as inpatient treatment. The purpose of inpatient treatment is to provide rapid pain relief and associated symptom control, change the course of preventative treatment, and find improved rescue solutions. Generally, effective and targeted treatment is difficult to accomplish in emergency department settings or a general medical floor. Frequently, admission to a specialized headache unit is warranted (Figure 23.1). The mission of a dedicated headache unit is to provide effective, flexible, and continuous treatment for patients who have failed outpatient treatment modalities and whose functional ability has been significantly compromised. In general, we found that up to one-third of all new patients referred to the Diamond Headache Clinic appears to be refractory to outpatient treatment modalities. It is critical to recognize the need for inpatient treatment in a timely manner and institute appropriate referral or therapy.

INDICATIONS FOR INPATIENT HEADACHE TREATMENT

Factors that are important in making the decision for inpatient headache treatment include the intractability of the patient's pain, pain-related disability, the patient's previous treatment success or failure, and experience and refractoriness to established treatment modalities. Overall treatment outcome may be affected by factors such as dependence on opioids, barbiturates, or benzodiazepines; certain types of medication overuse headache; and other comorbid medical and psychological/psychiatric conditions. In a survey of 174 physicians involved in the care of headache patients, at least 50% reported attempts of inpatient treatment of patients with refractory headaches.[1]

The most common reason for admission was to undergo detoxification programs from opioids and barbiturates. Other frequent reasons for inpatient treatment included a change in the headache pattern, clinically significant nausea and vomiting, need for frequent parenteral medications for pain control, presence of complicated migraine, failed acute outpatient therapy, prolonged pain state, and medication overuse headache. Although medication overuse is a very common and frequently quite challenging clinical entity, not every patient with medication overuse headache requires inpatient treatment. For example, a patient with recent onset of simple analgesic overuse (the patient has been overusing over-the-counter medications for 3–6 months) may respond well to an

S. Diamond, R. K. Cady, M. L. Diamond & V. T. Martin (Eds):
Headache and Migraine Biology and Management.

DOI: http://dx.doi.org/10.1016/B978-0-12-800901-7.00023-9

FIGURE 23.1   Headache inpatient unit.

outpatient treatment modality. Thorough education about medication overuse as well as clarifications on medication usage limits must be provided. To improve patient compliance, discontinuation of the "offending" medication may be insufficient, and an alternative treatment plan should be instituted.

In those cases in which the patient has been overusing large quantities of over-the-counter medications, caffeine, opioids, barbiturates, and benzodiazepines, inpatient treatment will most likely be required. These patients need to be closely monitored in the milieu of a dedicated inpatient unit for signs of withdrawal. Seizures may be seen in the patient withdrawing from barbiturates and benzodiazepines. Also, patients with signs of dehydration and electrolyte abnormalities as well as those with unstable vital signs would benefit from an inpatient stay. Patients with suspected organic etiology, such as infection, brain tumor, cerebral aneurysm, or subarachnoid hemorrhage, should also be considered for treatment in a monitored setting. Most of the patients with complications of migraine, such as status migraine, usually require an acute and rather intense treatment approach, and would benefit from an inpatient treatment program as well.

The presence of comorbid medical and psychiatric conditions may also complicate treatment and negatively affect outcome. It has been estimated that psychological and psychiatric conditions occur more often in migraine sufferers than in the general population.[2] Recognition and balanced management of these conditions, in addition to intensive headache care, are required for a successful treatment outcome.

About 2000 new headache patients are referred to the Diamond Headache Clinic each year, and 30% of these patients require an inpatient treatment program. The median age of patients being admitted is 39 years, and on average a headache history spanning 12 years is reported. In reviewing these patients' headache intractability and response to treatment, we found that the typical patient admitted to our inpatient headache unit has been averaging 5.7 visits to the emergency department per year (unpublished information). Although the duration of inpatient treatment programs varies, depending on individual clinical situations (presence of medication overuse, type of headache, presence of comorbid medical conditions, and overall response to treatment), the average duration of the admission is 5 days.

ADMISSION CRITERIA

Admission criteria for inpatient management of intractable headaches have been proposed over the years. These criteria are represented in Box 23.1.

ADVANTAGES OF INPATIENT TREATMENT

What advantage does inpatient treatment have as compared to an outpatient treatment program? First, treatment in a specialized inpatient unit should emphasize a comprehensive multidisciplinary approach that consists of detoxification therapy (when appropriate), acute and prophylactic pharmacologic therapy, and associated symptom management. Treatment may also include non-pharmacologic treatment modalities, such as biofeedback, acupuncture, physical therapy, exercise programs, psychological counseling and psychotherapy, stress management, and dietary and lifestyle modifications. An important aspect of inpatient treatment is the daily clinical rounds performed by the attending physicians who are familiar with the patients' histories. During the clinical rounds, important discussion and evaluation occurs that may ultimately lead to necessary treatment adjustments. Throughout the admission, patients remain under continued monitoring by experienced nursing staff. At our inpatient unit, most nurses have been trained to care for patients with intractable headaches, and are familiar with our treatment regimens and protocols, which facilitate and improve patient care. Because the treatment of these headache patients is individualized, we suggest that nursing staff become familiar with the treatment modalities and clinical parameters that require monitoring.

TREATMENT

Detoxification

One of the most challenging steps in the treatment process is detoxification. Detoxification therapy, which is usually the initial stage of inpatient therapy, is indicated for patients with medication overuse headache.

> **BOX 23.1**
>
> ## ADMISSION CRITERIA OF THE NATIONAL HEADACHE FOUNDATION FOR INPATIENT TREATMENT OF HEADACHE[2]
>
> - Severe dehydration, for which inpatient parenteral therapy may be necessary.
> - Diagnostic suspicion (confirmed by appropriate diagnostic testing) of organic etiology, such as an infectious disorder involving the central nervous system (e.g., brain abscess, meningitis), acute vascular compromise (e.g., aneurysm, subarachnoid hemorrhage), structural disorder with accompanying symptoms (e.g., brain tumor).
> - Prolonged unrelenting headache with associated symptoms, such as nausea and vomiting, which, if allowed to continue, would pose a further threat to the patient's welfare.
> - Status migraine or dependence on analgesics, ergots, opiates, barbiturates, or tranquilizers.
> - Pain that is accompanied by serious adverse reactions or complications from therapy — continued use of such therapy aggravates or induces further illness.
> - Pain that occurs in the presence of significant medical disease, but appropriate treatment of headache symptoms aggravates or induces further illness.
> - Failed outpatient detoxification, for which inpatient pain and psychiatric management may be necessary.
> - Intractable and chronic cluster headache, for which inpatient administration of histamine or dihydroergotamine (DHE) may be necessary.
> - Treatment requiring co-pharmacy with drugs that may cause a drug interaction, thus necessitating careful observation (e.g., monoamine oxidase inhibitors and beta-blockers).

It consists of discontinuation of the medications that the patient was overusing, and management of any withdrawal symptoms. Selecting the type of detoxification methods to be used depends on each patient, the medication that the patient was overusing, and the patient's past personal treatment experience and overall clinical presentation. Detoxification represents a critical time that requires close monitoring and observation to recognize withdrawal symptoms with potentially serious consequences. Patients who have been overusing over-the-counter medications, NSAIDs, triptans, ergotamine- or caffeine-containing medications, or using small doses of opioids or barbiturates for a brief interval, may safely undergo abrupt cessation of the "offending" medications in the milieu of a dedicated inpatient unit. Withdrawal symptoms can be closely monitored and adequately managed. Patients who have been using large quantities of opioids, barbiturates, or benzodiazepines, or who present with signs of dependence, should undergo gradual discontinuation of the medications with close monitoring of withdrawal symptoms. Occasionally, short-acting barbiturates may be temporarily replaced with longer-acting medications (such as phenobarbital) and then tapered gradually over several days or weeks, depending on the patient's symptoms.

The most frequent opioid-related withdrawal symptoms include anxiety, rhinorrhea, muscle aches and cramps, tremor, irritability, insomnia, nightmares, mydriasis, nausea, and vomiting. These symptoms usually subside over 5—10 days. Depending on the severity of the withdrawal symptoms, the dose of the opioid could be gradually reduced every few days, or the patient could be switched from a short-acting opioid to a long-acting opioid (such as methadone or buprenorphine), and the dose then gradually tapered. In clinically unstable patients whose clinical condition may change rapidly, it is prudent to use short-acting opioids and reduce the dose gradually while closely monitoring for signs of withdrawal. Clonidine may be useful in those patients taking low-dose opioids, as it reduces catecholamine release and may help reduce withdrawal symptoms.[3]

The most common withdrawal symptoms from barbiturates (butalbital) include anxiety, insomnia, nausea, vomiting, tachycardia, and seizures. To manage these symptoms, we suggest stopping butalbital-containing medications and starting phenobarbital, 30 mg, for every 100 mg of butalbital used, and tapering the dose gradually by 30 mg per day every 3—4 days, depending on the patient's clinical condition and presence of withdrawal symptoms.[4]

Pharmacological Treatment

As most patients (including those with medication overuse headache) are experiencing an underlying "primary" headache disorder, the treatment should not be limited by detoxification therapy. Management must also target the coexisting intractable headache (migraine, cluster headache, new daily persistent headache). In patients with medication overuse headache, the primary management goal remains the discontinuation and avoidance of the medication that

---

BOX 23.2

DIAMOND HEADACHE INPATIENT UNIT DHE-45 PROTOCOL

- Premedicate the patient with:
  Trimethobenzamide (200 mg IM or 300 mg PO) *or*
  Promethazine (50 mg IM or PO) *or*
  Ondansetron (8 mg IV) *or*
  Metoclopromide (10 mg IM or PO).
- At 30 minutes after parenteral or 60 minutes after oral administration of above, administer first dose of dihydroergotamine, 0.5 mg, diluted in 50 ml of normal saline by slow IV infusion over 30 minutes.

- Subsequent doses should be given every 8 hours for a total of 9 doses.
- Doses 2 through 9 should be administered intravenously using 1 mg of dihydroergotamine diluted in 50 ml of normal saline.
- Vital signs should be monitored and adverse effects assessed before each dose and 30 minutes after each dose.
- Adverse events may be reduced by infusing DHE-45 at a slower rate over 2 hours.

---

caused the "rebound" headache. Patients with chronic daily headaches with migraine features (chronic migraine, medication overuse headache, new daily persistent headache) or patients with intractable cluster headaches may benefit from the use of intravenous dihydroergotamine (DHE).[5] The most commonly used protocol for DHE administration was developed by Raskin.[6] Based on our experience, we have modified intravenous DHE protocol according to our patients' needs (Box 23.2). The protocol consists of slow (over 30−60 minutes) intravenous infusion of DHE 0.5−1 mg every 8 hours for a total of nine doses. As nausea appears to be the most common side effect of DHE therapy, all patients receive premedication with an anti-emetic drug (ondansetron, trimethobenzamide, promethazine, etc). The patient's vital signs need to be monitored closely during DHE infusion. The dose should be withheld if the vital signs are not in normal range.

In addition to the DHE protocol, patients may benefit from intravenous infusion of ketorolac, 30 mg, every 8−12 hours. Because the majority of the patients with chronic daily headaches have coexisting chronic myofascial pain syndrome, parenteral administration of the muscle relaxant orphenadrine, 30−60 mg, every 12 hours, may be effective. Patients receiving muscle relaxants should be routinely monitored for sedation. We have also found intravenous administration of diphenhydramine, 25−100 mg, every 6−12 hours may be beneficial in select patients. Patients receiving intravenous (IV) diphenhydramine should be monitored closely for sedation and signs of arrhythmia, impaired coordination, and urinary retention.

Other treatment modalities that could be utilized during inpatient treatment include administration of magnesium sulfate, 500−1000 mg IV, every 8−12 hours, or valproate sodium, 500−750 mg IV, every 8−12 hours as needed. Patients who failed to respond to the abovementioned treatment options may benefit from parenteral corticosteroids (methylprednisolone, 80 mg IM, single injection). Suppression of the hypothalamic–pituitary–adrenal axis should be considered, especially in patients with recent corticosteroid exposure.

Patients may also respond to IV administration of droperidol, 0.625−2.5 mg, usually combined with diphenhydramine, 25−50 mg IV, to help reduce the risk of extrapyramidal symptoms and akathisia. Since droperidol may contribute to QT prolongation, it is imperative to assess the ECG prior to initiation of treatment, as well as 2−3 hours after completion of the treatment. We also suggest discontinuing or avoiding treatment with droperidol in patients with known QT prolongation. Monitoring the ECG for QT prolongation increases its importance, considering that multiple preventative medications are being used in chronic headache treatment (tricyclic antidepressants, etc.) and may also contribute to QT prolongation.

On any given day, the patient may be exposed to several of the aforementioned medications. We suggest alternating different medications every 6−8 hours in an attempt to affect a variety of aspects of chronic headache pathogenesis. In those cases, it is imperative to monitor for potential drug-to-drug interaction and adverse reactions.

Patients with intractable chronic cluster headaches who do not respond to the abovementioned treatment may benefit from a course of intravenous histamine desensitization. Histamine desensitization was originally described by Horton in 1939.[7] Another trial of intravenous histamine desensitization, conducted in 1997 at the Diamond Headache Clinic, was beneficial in patients with intractable cluster headaches.[8] Histamine desensitization protocol consists of intravenous infusion of 21 doses of histamine phosphate (Box 23.3). This type of treatment requires a prolonged hospital stay, since most patients are only able to tolerate two

---

BOX 23.3

HISTAMINE DESENSITIZATION PROTOCOL

- Twenty-one bags are prepared containing histamine phosphate in 250 mL of normal saline or dextrose in water.
- The first dose contains 2.75 mg of histamine phosphate.
- Doses 2 through 21 contain 5.5 mg of histamine phosphate.
- On the first hospital day, administer only a single dose of histamine phosphate solution.
- On the second and succeeding days, administer histamine phosphate at titrated rates of infusion based on occurrence of cluster headache and drug tolerance.

- Initial infusion rate is 10 drops per minute.
- Infusion rate may be titrated upward every 15 to 30 minutes in increments of 10 drops per minute.
- Occurrence of a cluster headache while receiving an infusion should be followed by an immediate decrease in the infusion rate to 10 drops per minute until the cluster headache attack ceases.
- Adverse effects of flashing, dyspepsia, and nasal stuffiness are an indication of excess rate of infusion.
- When the patient has been free of cluster headache for 24 hours, infusion rate may be increased to accommodate a third dose per day of histamine phosphate.

---

doses of histamine phosphate within a 24-hour period. Once the patient becomes free of cluster headaches for 24 hours, the rate of the histamine infusion may be increased to accommodate a third dose per day, which may shorten the duration of treatment.

Interventional Treatment Modalities

Interventional treatment modalities may be considered in patients who present with chronic upper back and neck pain (due to spinal canal stenosis, degenerative spine disease, radiculopathy, facet arthropathy), coexisting facial pain/atypical facial pain, or trigeminal or occipital neuralgia. These conditions are believed to represent a contributing factor to the patient's chronic headache disorder. If the patient fails to respond to standard medical management, interventional treatment modalities may be utilized, including facet medial branch nerve blocks and radio frequency ablations, cervical epidural steroid injections, trigeminal ganglion and sphenoplatine ganglion blocks, trigger point injections, and occipital nerve blocks. Procedures such as cervical facet medial branch blocks, cervical epidural steroid injections, and trigeminal and sphenopalatine ganglion blocks are performed in the procedure or surgical suites, under mild sedation — either under conscious sedation or under monitored anesthesia care — to provide a safe and comfortable environment for the patent. We utilize fluoroscopy imaging to achieve better accuracy and improve the safety of the procedures. Cervical facet medial branch nerve block (the most commonly utilized procedure) is a purely diagnostic modality which helps

not only to identify the source of the neck pain (which in turn may contribute to headache) but also to narrow the site to two to three levels of the cervical spine. Those patients who achieve significant pain relief (in the neck and upper back) from the initial series of diagnostic cervical facet medial branch nerve blocks may be candidates for therapeutic cervical facet medial branch radiofrequency neurotomy, which aims to provide long-term (at least 6 months) pain relief. It is important to recognize, though, that at this point these procedures are not considered a standard of care for patients with migraine or tension-type headaches. These procedures instead represent important tools in managing chronic intractable neck and upper back pain that frequently contributes to the patient's headache disorders. Also, these procedures are frequently used to treat patients with chronic cervical facet arthropathy, cervical spinal stenosis, and cervical radiculopathy. Thus, accurate patient selection is important.

All patients undergoing cervical spinal interventions are required to participate in physical therapy. Addressing the chronic upper spine and neck pain may provide significant improvement in the patient's clinical condition, and improve overall headache management and outcome. Additionally, if there is a need for consultation with other specialists (cardiologist, neurologist, endocrinologist, neurosurgeon, etc.) the inpatient unit affords easy referral in a timely fashion and adjustments in treatment instituted.

Non-Pharmacological Treatment

In addition to the intense pharmacological treatment and detoxification therapy described above, patients

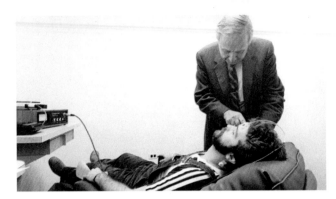

FIGURE 23.2   Biofeedback session in progress.

also are being exposed to a variety of non-pharmacological treatment modalities such as biofeedback and acupuncture. We widely utilize physical and recreational therapy in our inpatient program. Patients participate in art therapy and craft sessions, stress management, and relaxation exercises and biofeedback (Figures 23.2). A typical program is illustrated in Table 23.1.

These non-pharmacological approaches are a vital part of the multidisciplinary inpatient treatment program, and help patients not only to alter their pain behavior but also to develop new pain-coping strategies. Art therapy has been helpful in enabling patients to transform the emotional aspects of their pain and pain perception into more discrete visual and cognitive objects. Additionally, during inpatient treatment, each patient undergoes evaluation by a team of psychologists who provide invaluable information about potential psychological factors contributing to the patient's chronic pain condition as well as help develop new pain- and stress-coping strategies.

At the inpatient unit, we conduct educational pharmacology classes designed specifically for patients in terms and language that are easy to understand, and also to assist patients in gaining a better understanding of our treatment philosophy. They quickly grasp the idea of medication overuse. During this class they learn about the actions of specific medications, potential side effects, and expected results. We believe that instructing patients to understand the reasons and purpose of their treatment enables them to become more comfortable with the treatment process, adhere to the regimen, and be more compliant in the long term.

During the hospital stay, the physical therapy team evaluates patients and assists in adjusting their posture. The individual exercise program helps patients to better manage and prevent myofascial pain, which is frequently experienced by those with chronic headaches.

All patients admitted to the inpatient headache unit participate in a targeted diet class, conducted by a dietician, in which the importance of diet adjustments and modifications is discussed as applied to headache management. Family members are encouraged to participate in some educational activities so that patients will have additional support once they return home. We believe that providing education to patients and their families during the admission dramatically improves adherence to the program and increases overall success of the treatment.

Each week an educational headache class is conducted by a staff physician, during which patients may ask a variety of questions related to headache management, prognosis, etc. Patients are also encouraged to participate in discussion groups chaired by a staff psychologist. These groups provide an opportunity for patients to share their headache stories and experiences, and allow them to appreciate, even more so, that they are not alone in their headache experiences.

Finally, by the end of the admission, each patient undergoes routine discharge planning. Each patient should leave the hospital with a clear understanding of the goals of treatment and the resources that are available for his or her recovery. About 1 or 2 days prior to discharge, a member of the pharmacy staff will visit the patient to discuss available outpatient medication options.

It is important that the patient establishes realistic short- and long-term goals for therapy before discharge. Also, the patient should schedule a follow-up appointment and receive discharge instructions in order to ensure continuity of care.

Overall, an inpatient treatment program for patients with intractable headaches appears to be beneficial. Also, these programs demonstrate reduced disability and long-term healthcare costs.[9] Based on the results of several studies, inpatient headache treatment proved to be effective in more than 50% of patients with intractable headaches.[5,10]

Although individual portions of the described treatment plan could be accomplished on an outpatient basis, inpatient therapy will require less time and effort from the patient. Attempting outpatient therapy may cause limited compliance, and possibly some patients will be lost to follow-up. The advantage of inpatient treatment programs is the availability of the pharmacological and non-pharmacological treatment, as well as educational modalities, to the patient on a daily basis. All of these regimens can be administered in one setting, which facilitates adherence to a treatment program and often dramatically improves the prognosis.[10,11]

TABLE 23.1   Inpatient Unit Activities Schedule

Time	Sunday	Monday	Tuesday	Wednesday	Thursday	Friday	Saturday
8 am	Breakfast	Breakfast	Breakfast	Breakfast	Breakfast	Breakfast	Breakfast
9 am	Music relaxation	Orientation group	Orientation group	Yoga	Orientation group	Music relaxation	
10 am		Pharmacy class	10.30–11 am Diet class	Orientation group	Pharmacy class	Orientation group	Tai chi/Search for serenity
11 am		Stress management		Mindfulness	Morning stroll	Yoga	Soothing spa
12 noon	QUIET TIME	QUIET TIME	QUIET TIME	QUIET TIME	QUIET TIME	QUIET TIME	QUIET TIME
1 pm		Alternate headache treatment	Coping skills	Spirituality group	Card games/ casino fun	Diet class	
			Pediatric unit only: Creative kids				
2 pm	Classic movie matinee with popcorn	Yoga		Assertiveness	Care-giving group	Support group	2.30 pm Wellness à la carte
							Activity cart rounds by volunteers/activity therapy staff
3 pm		High tea: get to know one another	Exercising with headaches	High tea: get to know one another	3.30 pm Art expression, jewelry making	How to have better posture	Saturday social
4 pm			Crafts and hobbys				
5 pm	Dinner	Dinner	Dinner	Dinner	Dinner	Dinner	5 pm Saturday night dinner club
							Informal dinner gathering
6 pm	5.30 pm Email, Google, Wi-fi fun	5.30 pm Email, Google, Wi-fi fun	5.30 pm Email, Google, Wi-fi fun	5.30 pm Email, Google, Wi-fi fun	5.30 pm Email, Google, Wi-fi fun	5.30 pm Email, Google, Wi-fi fun	5.30 pm Email, Google, Wi-fi fun
7 pm	Open rec	Open rec	Wheel of Fortune or Jeopardy		New release movie night	Open rec	Open rec

References

1. Rapoport A, Stang P, Gutterman DL. Analgesic rebound headache in clinical practice: data from a physician survey. *Headache*. 1996;36:14–19.
2. Hamelsky SW, Lipton RB. Psychiatric comorbidity of migraine. *Headache*. 2006;46:1327–1333.
3. Gold MS, Pottash AC, Sweeney DR, Kleber HD. Opiate withdrawal using clonidine. A safe, effective, and rapid nonopiate treatment. *JAMA*. 1980;243(4):343.
4. Hayner G, Galloway G, Wiehl WO. Haight Ashbury free clinics' drug detoxification protocols: part 3: benzodiazepines and other sedative-hypnotics. *J Psychoactive Drugs*. 1993;25(4):331–335.
5. Silberstein SD, Silberstein JR. Chronic daily headache: long-term prognosis following inpatient treatment with repetitive IV DHE. *Headache*. 1992;32:439–445.
6. Raskin NH. Repetitive intravenous dihydroergotamine as therapy for intractable migraine. *Neurology*. 1986;36:995–997.
7. Horton BT, MacLean AR, Craig W. A new syndrome of vascular headache: results of treatment with histamine: preliminary report. *Proc Staff Meet Mayo Clin*. 1939;14:257–260.
8. Freitag FG, Diamond S, Bambhvani IV S. Histamine in chronic cluster headaches. *Clin Pharmacol Ther*. 1997;61:141.
9. Freitag FG, Lyss H, Nissan GR. Migraine disability, healthcare utilization, and expenditures following treatment in a tertiary headache center. *Proc (Bayl Univ Med Cent)*. 2013;26 (4):363–367.
10. Lake AE, Saper JR, Madden SF, Kreeger MA. Comprehensive inpatient treatment for intractable migraine: a prospective long-term outcome study. *Headache*. 1993;33(2):55–62.
11. Diamond S, Freitag FG, Maliszewski M. Inpatient treatment of headache: long-term results. *Headache*. 1986;26:189–197.

24

Newer Research and its Significance

Vincent T. Martin

Department of Internal Medicine, University of Cincinnati, Cincinnati, Ohio, USA

INTRODUCTION

During recent years, there have been many advances in the field of headache. These advances have changed our understanding of the epidemiology, genetics, neuroimaging, pathophysiology, and treatment of headache disorders. The bulk of the past research has been conducted on migraine headache. Therefore, this chapter will focus on new discoveries within the area of migraine headache and their potential significance.

EPIDEMIOLOGY

Most of the recent epidemiologic studies have focused on chronic migraine. Chronic migraine is more common in women than men, and in those of low socioeconomic status.[1,2] It has been associated with four times more headache-related disability than episodic migraine.[1] The prevalence ratios (PRs) are highest in women between the ages of 40 and 49 years (PR = 4.2; 95% confidence interval [CI]: 2.7–6.1) as compared with adolescent girls.[1] High-frequency migraine (i.e., ≥ 10 headache days per month) is more common in women during the perimenopausal stage than the premenopausal time period.[3] These data could suggest that the hormonal changes occurring during the perimenopause might contribute to the development of chronic migraine in some women.

Chronic migraine does not occur in isolation, and is more likely to be associated with comorbid medical and psychiatric disorders than is episodic migraine. Buse and colleagues[4] found that compared with episodic migraineurs, those with chronic migraines were more likely to suffer from chronic pain (odds ratio [OR] = 2.5; 95% CI: 2.1–3.0), depression (OR = 2.0; 95% CI: 1.7–2.3), anxiety (OR = 1.80; 95% CI: 1.5–2.2), asthma (OR = 1.5; 95% CI: 1.3–1.8), hay fever (OR = 1.5;

95% CI: 1.3–1.8), and stroke (OR = 1.7; 95% CI: 1.1–2.5). Blumenfeld and colleagues[2] also reported that non-headache pain conditions, psychiatric disorders, and vascular disease events were more common in chronic than episodic migraineurs. There are several possible explanations for these findings. First, one comorbid illness could cause the development of another, which may represent a "kindling process." For example, chronic pain disorders might sensitize thalamic neurons and further increase the risk of chronic migraine. Second, shared genetic or environment factors could predispose an individual to both disorders (e.g., smoking, air pollution, etc.). Third, comorbid disorders may not directly cause chronic migraine, but could be confounded or mediated by another factor associated with the comorbid disorder. For example, the medications used for asthma (e.g. beta-2 agonists) could trigger migraine attacks and increase the risk for chronic migraine. Regardless of the underlying explanation for these associations, they may provide important clues to the pathogenesis of chronic migraine. In addition, identifying chronic migraine patients with specific comorbid medical conditions may provide more uniform subgroups to perform genetic testing.

A number of predisposing factors have recently been identified that lead to the transformation from episodic to chronic migraine/headache. The most significant risk factor for chronic headache is medication overuse, increasing its likelihood by 7- to 19-fold.[5,6] Another study[7] found that depression at baseline was found to increase the risk of new onset chronic migraine by 1.7-fold in patients with episodic migraine. Caffeine consumption, snoring, obesity, traumatic life events, and asthma have also been shown to increase the risk of chronic migraine or chronic daily headache.[8–12] The potential significance of these findings is that modification and/or treatment of these risk factors may decrease the transformation of episodic to chronic migraine.

S. Diamond, R. K. Cady, M. L. Diamond & V. T. Martin (Eds):
Headache and Migraine Biology and Management.

DOI: http://dx.doi.org/10.1016/B978-0-12-800901-7.00024-0

NEUROIMAGING

Recent studies using functional MRI and H$_2$O-activation PET have elucidated the importance of central mechanisms in the pathophysiology of migraine. Weiller and colleagues[13] were the first to report activation of the rostral pons during a spontaneous migraine attack, and also demonstrated that the activation remained with relief of the headache after administration of sumatriptan. They also found activation of other cortical areas commonly involved with pain in general (e.g., the cingulate, auditory, and visual association cortices). Since the rostral pons is not commonly activated in other pain disorders, it was termed the "migraine generator." A more recent PET study[14] confirmed brainstem activation during a spontaneous acute migraine attack, but also noted hypothalamic activation. Maniyar and colleagues[15] induced migraine with nitroglycerin and performed H$_2$O-activation PET scans during the migraine prodrome. They reported activation of the posterolateral hypothalamus, midbrain tegmental area, periaqueductal gray and dorsal pons, as well as the occipital, temporal, and prefrontal cortices. It was postulated that early activation of the hypothalamus may explain many of the prodromal symptoms experienced by migraine patients (e.g., yawning, polyuria, thirst). Therefore, recent neuroimaging studies implicate activation of the hypothalamus during the prodromal and headache phases of the migraine attack.

A recent study[16] suggests the threshold for activation of the spinal trigeminal nucleus may vary with time and possibly predispose to migraine. Stankewitz and colleagues[16] performed functional MRI in migraineurs and controls after administration of intranasal ammonia during pre-ictal (i.e., 72 hours prior to an attack), inter-ictal (i.e., between attacks, but not within 72 hours of the next attack), and ictal (i.e., during an attack) time periods. Intranasal ammonia has been shown in past studies to activate V1 and V2 afferents of the trigeminal nerve in the nares. The researchers found greater activation of the spinal trigeminal nucleus after ammonia administration during the pre-ictal as compared with inter-ictal and ictal time periods in migraineurs. The intensity of its activation was inversely correlated with the time to onset of the next migraine attack ($R^2 = -0.50$; $P < 0.001$). Interestingly, the dorsal pons was only activated during an attack, possibly suggesting that the vulnerability of the trigeminal nucleus to migraine could vary over time and that trigeminal activation occurs prior to that of the dorsal pons. This might also indicate that activation of the dorsal pons is more of a secondary event and not the "migraine generator."

Enhanced activation of affective brain regions may occur in migraine patients. Eck and colleagues[17] subjected migraine patients and controls to verbal pain descriptors (words that have a negative valence, such as excruciating) and performed functional MRI to determine which areas of the brain were activated. They found increased activation of the left anterior insula in the brains of migraineurs as compared with controls. The left anterior insula is involved in the affective processing of pain, possibly suggesting enhanced recruitment of the affective areas of the brain with unpleasant sensory stimuli in migraineurs.

A number of studies have reported reductions in gray matter in the anterior cingulate cortex and insular cortex of migraine patients as compared with controls.[18–21] This finding has been observed in a number of chronic pain syndromes other than migraine, and the reductions in gray volume occur in areas of the brain involved in the pain matrix. The degree of reduction in gray matter is generally correlated with the duration of illness.[22,23] It is unknown whether these changes are the cause or the result of the chronic pain condition, but one study[24] found that gray matter changes were found to regress after joint replacement in patients with hip osteoarthritis when the chronic hip pain abated postoperatively. These data would argue that the changes are likely the consequence of the chronic pain condition.

Cortical thickness is increased in the somatosensory cortex of migraineurs as compared with controls.[25] The cause may be a result of repeated activation of these brain areas with chronic pain. A recent study[26] demonstrated a positive correlation between thickness of the somatosensory cortex and enhanced pain sensitivity to thermal stimuli applied to the volar surface of the forearm. These results suggest that the brains of chronic pain patients are plastic and can be modulated by chronic pain states.

Magnetic resonance imaging (MRI) in migraineurs has demonstrated increased white matter abnormalities, as compared with controls.[27,28] These abnormalities are more likely to occur in women than men.[27,28] Other risk factors include elevated homocysteine levels, greater duration of migraine, higher migraine frequency, and those individuals with thyroid dysfunction.[29] A recent study[30] found that frequent syncope (more than five lifetime events) was strongly associated with white matter lesions in migraine patients (OR = 2.7; 95% CI: 1.3–5.5). White matter lesions may result from ischemia, endothelial dysfunction, microemboli through right to left shunts, or hyperhomocysteinemia. The Cerebral Abnormalities in Migraine (CAMERA-II) study[28] demonstrated that white matter abnormalities at baseline were not associated with cognitive decline at the 9-year follow-up. Therefore, the clinical significance of white matter abnormalities is unknown, and really does not influence the clinical management of most migraine patients. However, extensive burdens of white matter disease can be encountered in a number of small vessel vasculopathies, such as cerebral autosomal dominant arteriopathy with subcortical infarcts (CASASIL); COL4-A1 mutations (encodes the α-subunit of type IV collagen); hereditary endotheliopathy with retinopathy, nephropathy, and stroke (HERNS); and, cerebral autosomal recessive arteriopathy with subcortical infarcts and leukoencephalopathy (CARASIL).[31]

Functional MRI can also afford us a glimpse into which brain regions are connected with others, which has been termed "resting state functional connectivity (RSFC)." RSFC differs between migraineurs and controls within certain brain regions. Mainero and colleagues[32] reported greater RSFC in migraineurs as compared with controls between the periaqueductal gray (PAG) and several other brain regions involved in pain processing. Migraineurs with severe allodynia may also have stronger RSFC between the PAG and other pain processing areas than those migraineurs without allodynia.[33] In addition, altered RSFC has been identified between the amydala, insula, basal ganglia, and other pain processing areas in migraineurs.[34-37] Maleki and colleagues[38] found differences in the RSFC between the basal ganglia and other brain regions in migraineurs with low and high attack frequencies (Figure 24.1). These data suggest that "brain wiring" varies between migraineurs and controls, as well as between migraineurs with more and less frequent attacks.

GENETICS

Some of the greatest advances in the headache field have evolved in the area of genetics. Earlier studies had identified three genetic mutations (i.e., CACN1A [alpha-1 subunit calcium channel], ATPA-2 [ATPase Na/K transporting, alpha-2a polypeptide], and SCNA1A [sodium channel, voltage-gated, type 1, alpha subunit]) for familial hemiplegic migraine, a rare autosomal dominant genetic disorder that is characterized by hemiplegic auras. The proteins encoded by these genes form ion channels that regulate the flow of ions in neurons and glia. A recent study also found a fourth genetic mutation associated with familial hemiplegic migraine. Riannt and colleagues[39] reported a PRRT2 (proline-rich transmembrane protein 2) mutation in 4 of 101 patients who had been diagnosed with hemiplegic migraine. The gene product of the PRRT2 gene is thought to interact with SNAP-25 proteins, which are important in the function of Ca 2.1 calcium channels.

The genetic mutations linked to familial hemiplegic migraine have not been found in the more common types of migraine.[40] Polymorphisms of gene coding for serotonin transporters, dopamine beta hydroxylase, and dopamine receptors have been over-represented in migraineurs.[41-44] Two past meta-analyses[45,46] reported an increased prevalence of migraine with aura with the TT variant of the methylenetetrahydrofolate reductase gene (MTHFR C677T). This polymorphism impairs the conversion of 5,10-methylenetetrahydrofolate to 5-methyltetrahydrofolate and leads to high serum levels of homocysteine. High levels of homocysteine cause vascular damage and increase the risk for vascular events (e.g., myocardial infarction, stroke). Studies linking migraine with estrogen receptor polymorphisms have been conflicting with both positive[47-50] and negative[51,52] results.

Genome-wide association studies (GWAS) offer the advantage of discovering novel genetic loci linked to migraine that are not preselected as performed in candidate gene studies. Anttila and colleagues[53] found that the prevalence of migraine was increased in those subjects with marker rs1835740. This genetic marker is located near the MTDH (astrocyte elevated gene 1) and the PGCP (plasma glutamate carboxypeptidase) genes, which are involved in glutamate homeostasis. This association was strongest in those with aura. Chasman and colleagues[54] performed a GWAS in 23,230 women from the Women's Genome Health Study. They identified three additional markers at rs11172113, rs2651899, and rs10166942 that are in close proximity to the LRP1 (low density lipoprotein receptor related-protein 1), PRDM16 (PR domain containing 16), and TRPM8 (transient receptor potential melastatin 8) genes, respectively. The LP1 protein interacts with N-methyl-D-aspartate (NMDA) receptors and may influence glutamatergic neurotransmission. The PRDM16 gene is involved in the development of brown fat, and its potential role in migraine is unknown. The TRPM8 gene encodes for a receptor on sensory neurons that is activated by cold temperatures and may be involved in neuropathic pain. Freilinger and colleagues[55] conducted a GWAS in 2326 persons with migraine without aura who were recruited from German and Dutch headache centers, and 4580 controls. They confirmed the associations with the rs11172113 (LRP1 gene) and rs10166942 (TRPM8 gene) markers noted in the Chasman study. They also found four additional potential loci that encode for MEF2D (myocyte enhancer factor 2D), TGFBR2 (transforming growth factor receptor 2), ASTN2 (neuronal protein astrotactin 1), and PHACTR1 (phosphatase and actin regulator 1) genes. The MRF2D gene is involved in the differentiation of neurons, and influences the number of excitatory synapses. The TGFBR2 gene regulates the

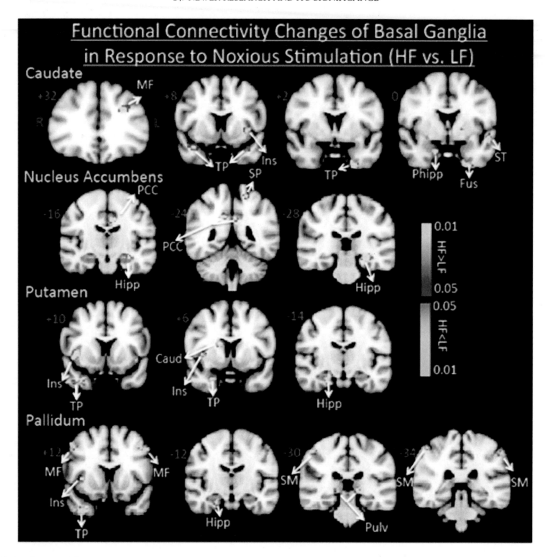

FIGURE 24.1 Functional connectivity contrast maps of the basal ganglia nuclei. Functional connectivity contrast maps of the basal ganglia nuclei during intermittent heat stimuli (pain threshold + 1°C on hand) in high frequency migraine patients vs. low frequency migraine patients. PCC: Posterior Cingulate Cortex, SM: SupraMarginal, SF: Superior Frontal, ST: Superior Temporal, SP: Superior Parietal, Ins: Insula, Hipp: Hippocampus, PHipp: Parahippocampus, Fus: Fusiform, Thal: Thalamus, Pulv: Pulvinar, TP: Temporal Pole, MF: Middle Frontal. Reproduced with permission from Maleki et al.[38].

production of the extracellular matrix, and mutations in this gene have been associated with migraine in families with aortic dissection.[56] The ASTN2 gene influences the migration of glial cells, while the PHACTRI gene controls synaptic formation and may play a role in endothelial function.

Linkage studies have been performed in large multi-generational families with high migraine prevalence. Lafreniere and colleagues[57] found a mutation in a KCNK18 gene on chromosome 10 that was linked to an autosomal dominant type of migraine with aura in one family. This gene encodes for the TWIK-related spinal cord potassium (TRESK) channel, which is located on trigeminal ganglion cells, and its overexpression decreases neurotransmission within trigeminal neurons.[58] A recent candidate gene study[59] performed in an Australian population did not find any of the three common polymorphisms of this gene to be associated with migraine with aura. Thus, this gene mutation may be a rare mutation that may account for sporadic cases of migraine with aura.

The above studies indicate that migraine is a poly-genic disorder. The gene mutations associated with familial hemiplegic migraine are completely different from more common subtypes of migraine in the general population. Familial hemiplegic migraine appears primarily to be a channelopathy, while more common types of migraine involve diverse mutations involving

TABLE 24.1 Gene Mutations, Genetic Loci, and Polymorphisms Associated with Familial Hemiplegic Migraine and More Common Variants of Migraine

Type of migraine	Gene mutation/polymorphism/genetic loci	Possible function
Familial hemiplegic migraine	CACNA1A	Encodes the $\alpha 1$ subunit of the P/Q type Ca^{2+} channel
	ATPA-2	$\alpha 2$ subunit of glial/neuronal sodium potassium pump
	SCNA1A	Neuronal Na 1.1 voltage-gated sodium channel
	PRRT2	Interacts with SNAP-25 proteins
Migraine overall	rs1835740 (close to MTDH and PGCP)	Both involved in glutamate homeostasis
	rs11172113 SNP (close to LRP1gene)	Involved with glutamatergic neurotransmission
	rs2651899 SNP (close to PRDM16 gene)	Involved with development of brown fat
	rs10166942 SNP (close to TRPM8 gene)	Cold receptor on sensory neurons
	rs 1050316, rs3790455 SNP's (close to MEF2D)	Influences number of excitatory synapses
	rs6478241 SNP (close to ASTN2 gene)	Influences migration of glial cells
	rs9349379 SNP(close to PHACTR1)	Controls synaptic activity and morphology
	Dopamine beta hydroxylase	Involved in norepinephrine synthesis
Migraine without aura	rs7640543 SNP (close to TGFBR2 gene)	Involved with production of extracellular matrix
Migraine with aura	TT variant of MTHFR gene	Vascular damage from increased homocysteine levels
	KCNK18 (encodes for TWIK K^+ channel)	Neurotransmission of trigeminal neurons

CACNA1, alpha-1 subunit calcium channel; ATPA2, ATPase, Na^+/K^+ transporting, alpha-2a polypeptide; SCNA1A, sodium channel, voltage-gated, type I, alpha subunit; PRRT2, proline-rich transmembrane protein 2; SNP, single nucleotide polymorphism; MTDH, astrocyte elevated gene 1; PGCP, plasma glutamate carboxypeptidase; LRP1, low density lipoprotein receptor related-protein 1; PRDM16, PR domain containing 16; TRPM8, transient receptor potential melastatin 8; MEF2D, monocyte enhancer factor 2D; ASTN2, neuronal protein astrotactin 1; PHACTR1, phosphatase and actin regulator 1; TGFBR2, transforming growth factor beta receptor 2; MTHFR, methylenetetrahydrofolate reductase.

homocysteine metabolism, glutameteric neurotransmission, TGF-β signaling pathways, synaptic and endothelial function, and TRPM8 receptors. Associations between genetic mutations and migraine may vary by ethnicity, race, and geographic location. No convincing data are available to suggest that the genes associated with migraine differ between men and women, but this has not been extensively studied. Table 24.1 includes a summary of the aforementioned gene mutations, genetic loci, and polymorphisms.

PHARMACOLOGICAL MODELS

A number of neuropeptides and other molecules have been studied that help elucidate the pathophysiology of migraine, including nitric oxide, calcitonin gene-related peptide (CGRP), and pituitary adenylate cyclase-activating peptide (PACAP-38). Pharmacological models of migraine have been developed in which these substances were infused intravenously in humans to determine if headaches or migraines were triggered.[60] These studies, as well as others linking these substances to migraine pathophysiology, will be discussed below.

Nitric Oxide

For many years, it has been observed that nitroglycerin triggers headache in susceptible persons.[61] Nitroglycerin is thought to trigger migraine through induction of nitric oxide because it is a nitric oxide donor.[62] Nitric oxide is liberated through the oxidation of L-arginine, which is catalyzed by three different subtypes of nitric oxide synthase (NOS) — endothelial (eNOS), neuronal (nNOS), and inducible (iNOS). eNOS is produced within the vascular endothelium, while nNOS is found in neurons within the peripheral and central nervous system. iNOS is located in neurons and glial cells, and can be induced by a number of stimuli affecting calcium signalling.[62] Administration of nitric oxide donors has been found to increase firing rates within the rat spinal trigeminal nucleus, and pretreatment with CGRP receptor antagonists can blunt this increased neuronal activity.[63,64] Another study[65] reported release of CGRP in the dura mater, after application of nitric oxide donors, promoting increases in meningeal blood flow. Therefore, nitric oxide has been postulated to be one of the molecules released during the cascade of neuronal events in the course of a migraine attack.

Pharmacologic infusions of nitroglycerin trigger headache in migraine as well as non-migraine patients.[66] Immediate headaches occurring 30 to 180 minutes after the start of the infusion developed in 80–90% of migraineurs and controls, but those with migraine reported greater pain severity with their headaches. In contrast to controls, migraineurs developed a delayed-type headache, 6–8 hours after the infusion, that had the characteristics of migraine without aura. Interestingly, auras are rarely (if ever) triggered by nitroglycerin, and only 50% of patients with migraine with aura will have the delayed headaches, which is far less than is experienced by those without aura.[67,68] Kruuse and colleagues[69] demonstrated that serum levels of CGRP do not increase 20 minutes after the start of a nitroglycerin infusion, suggesting that CGRP plays little role in triggering immediate-type headaches. Also, CGRP receptor antagonists do not prevent the headaches induced by nitroglycerin, thus confirming that these headaches are not CGRP dependent.[70] Other studies[71–73] have reported that vasodilation of the middle cerebral artery is induced with the nitroglycerin infusion, and its onset correlates with that of the immediate-type headache, but the vasodilatation abates prior to termination of the headache. Therefore, it is possible that MCA dilatation initiates immediate-type headaches, but it is not necessary for maintenance of these headaches.

NOS inhibitors have had mixed results in the abortive and preventative treatment of migraine. An early study[74] of a non-selective NOS inhibitor, L-NG methylarginine hydrochloride (546C88), reported relief from moderate to severe pain at 2 hours post-dosage in 66% of the 546C88 group as compared with 14% of the placebo group ($P < 0.05$). Blood pressure was elevated during the infusion, which is a known side effect of 546C88 as a result of inhibition of eNOS. More recent studies have focused on the more selective iNOS inhibitors, which have been found to be ineffective as abortive and preventative medications.[75,76] These failed results have led researchers to focus on nNOS inhibitors that have been shown to prevent allodynia in animal models of migraine.[77] Further studies are needed to determine if nNOS inhibitors have any benefit in the treatment of migraine.

CGRP

No other peptide has been studied as greatly in regard to the pathogenesis of migraine headache as CGRP. CGRP is synthesized within trigeminal afferents but can also be found in other parts of the trigeminal pain network, such as the PAG, thalamus, amygdala, and hypothalamus.[78,79] Thermal stimulation of the trigeminal ganglion in humans and cats leads to increases in CGRP in the extracranial circulation.[80] Serum levels of CGRP are elevated during ictal as compared with inter-ictal time periods during spontaneous migraine attacks. Also, inter-ictal levels in migraineurs are higher than in controls.[81,82] Infusions of CGRP trigger an immediate as well as a delayed attack that occurs 1–11 hours after its administration, which is a similar time-course to those attacks triggered by nitroglycerin.[83] CGRP levels are elevated in patients with chronic migraine as compared with controls during inter-ictal time periods, suggesting that CGRP levels could be a biomarker for chronic migraine.[84] Elevated CGRP levels may predict response to onabotulinum toxin A in those with chronic migraine.[85] Therefore, substantial evidence indicates that CGRP release is important to the pathogenesis of migraine headache.

The above data have led a number of researchers to target CGRP receptors in the abortive therapy of migraine.[86] The CGRP receptor is composed of three separate components: calcitonin receptor-like receptor (CRLR), receptor activity modifying protein 1 (RAMP1), and receptor component protein (RCP).[87–89] RAMP1 is necessary for expression of CRLR on the cell membrane, and RCP is important in activating secondary messenger systems. The CGRP receptor is found peripherally in the vascular smooth muscle and trigeminal ganglion cells, as well as centrally in the trigeminal nucleus caudalis, cerebellum, and cortex.[90,91] Thus, the CGRP receptor is located in many of the areas of the trigeminal pain network that are thought to be important in migraine pathophysiology.

Four different CGPR antagonists have been studied in the abortive therapy of migraine, including BIBN4096BS (olcegepant), MK-0974 (telcagepant), MK-3207, and BI 44370 TA. Olcegepant is administered intravenously, while telcagepant, MK-3207, and BI 44370 TA are given orally. All four have shown good efficacy in the abortive therapy of migraine in Phase II and III studies.[92–95] For example, 23.8% of those receiving telcagepant, 300 mg, achieved pain freedom at 2 hours as compared with 10.7% in the placebo group ($P < 0.001$ for the telcagepant–placebo comparison).[92] Another study[96] reported that telcagepant had equal efficacy to zolmitriptan in the abortive treatment of migraine, and fewer side effects. A very concerning adverse event reported in some of these trials is hepatotoxicity. Several studies[93,97,98] have demonstrated high liver function tests after administration of CGRP antagonists, with transaminase levels in the 1000 range. This side effect has tempered enthusiasm for this class of medication, and led Merck to postpone filing a new drug application for telcagepant.

The hepatotoxicity associated with CGRP antagonists has led some researchers to investigate the role of biologics (such as monoclonal antibodies) against CGRP itself or its receptors in the preventative treatment of migraine headache. According to Bigal,[99] monoclonal antibodies have advantages over CGRP antagonists in several ways. First, they are very "target specific," and therefore would be less likely to cause side effects from binding to other collateral sites. Second, they have long elimination half-lives, enabling them to be administered parenterally once a month or less. Third, they are not metabolized in the liver, and thus hepatotoxicity would be less likely with these drugs. Fourth, any side effects that would be experienced may be related to those resulting from targeting of the CGRP peptide or its receptor. Since the primary role of CGRP is vascular, there is concern that CGRP antagonists might produce hypertension or reduce the ability of the coronary arteries for stress- or ischemia-induced vasodilatation.

Three monoclonal antibodies targeting CGRP (ALD403, LY2951742, LBR-101) and one directed against the CGRP receptor (AMG 334) have been developed. The results of two double-blind placebo-controlled Phase II studies of monoclonal antibodies directed at CGRP in the prevention of migraine were presented at the American Academy of Neurology meeting in April 2014.[100,101] In the first study,[100] a single 1000-mg dose of ALD403 was administered intravenously, or the subject received matching placebo. The primary outcome measure was the mean change in migraine headache days at weeks 5 through 8 as compared with baseline. This study met this primary outcome measure, with migraine headache days decreasing by 5.6 days in the ALD403 group ($n = 81$) and 4.6 days in the placebo group ($n = 82$; $P = 0.05$). The proportion of patients with a 50% reduction in the frequency of migraine headache days was 60% and 33% in the treatment and placebo groups, respectively. In the second study, 217 patients received either LY2951742, 150 mg, or placebo on a biweekly basis.[101] The primary outcome measure was the change in migraine headache days at 12 weeks. The mean change in migraine headache days was significantly greater in the LY2951742 group than the placebo group (4.2 versus 3.0; $P < 0.003$). A 50% reduction in migraine headaches days was attained by 70% of the LY2951742 group and 45% of the placebo group (OR = 2.88; 95% CI: 1.78−4.69]). Both studies reported that adverse events were no different between the treatment group and placebo. These preliminary studies demonstrate proof of concept that monoclonal antibodies directed against CGRP may be effective in the prevention of migraine. Phase III trials have been planned for these promising new therapies for migraine headache.

PACAP-38

In recent years, PACAP (pituitary adenylate cyclase-activating polypeptide) is one of the more interesting neuropeptides to be studied in regard to the pathophysiology of migraine headache.[102] It belongs to the vasoactive intestinal polypeptide [VIP]/secretin/glucagon super-family, and has two forms: PACAP-27 and PACAP-38. PACAP-38 is primarily found in mammalian cells and is synthesized within trigeminal afferents, the trigeminal nucleus caudalis, the sphenopalatine ganglion, and parasympathetic neurons.[103−105] VIP and PACAP-38 share 68% structural homology. PACAP-38 functions as a neurotransmitter, but also acts as a hormone on the pituitary gland and vasodilates cerebral and dural arteries. PACAP-38 binds to $VPAC_1$, $VPAC_2$, and PAC_1 receptors, with the latter receptor binding only PACAP-38 and the two former binding both VIP and PACAP-38.[106]

Infusions of PACAP-38 produce an immediate-type headache in normal controls and in migraine patients.[107] These immediate-type headaches are of mild to moderate intensity, and occur within 10 to 50 minutes after the start of the infusion. In addition, 75% of the migraine patients also develop a "delayed" headache attack, meeting criteria for migraine without aura, that occurs 6 hours after the infusion.[108] These migraine headaches respond to sumatriptan. Infusions of VIP also trigger immediate-type headache in controls and migraine patients, but are less likely than PACAP-38 to produce the delayed migraine headaches occurring in only 18% of migraineurs.[108] Both VIP and PACAP-38 produce vasodilatation of the extracranial arteries (e.g., superficial temporal arteries, middle meningeal artery) 0 to 20 minutes after start of the infusion, but only PACAP-38 maintains the vasodilatation 2 hours later. However, the vasodilatation associated with PACAP-38 administration occurs bilaterally and is not more pronounced on the pain side as compared with the non-pain side of the head during a unilateral migraine attack. These data indicate that PACAP-38 produces a more long-lasting vasodilatation of extracranial arteries than VIP, which could partially explain its greater propensity to produce delayed-type migraine headaches. These data could also indicate an important role for the PAC_1 receptor in vasodilation, since only PACAP-38 binds to this receptor.

A recent study[109] reported that serum levels of PACAP-38 were increased during ictal time periods of a migraine attack as compared with inter-ictal time periods. This increase during ictal time periods was only observed in female migraineurs whose headaches were not related to the menstrual cycle. The author postulated a number of mechanisms through which PACAP-38 could trigger attacks of migraine, including direct sensitization of trigeminal afferents, mast cell degranulation, or its vasodilatory effects on the cranial vasculature.[109,110] Regardless of its mechanism of action, PACAP-38 and/or its receptors could represent a promising target for the future treatment of migraine headache.

TREATMENT

Medical devices represent one of the most important new treatment modalities for migraine headache. These include transcutaneous electrical and transmagnetic stimulation as well as occipital and vagal nerve stimulation. Each device applies an electrical current or magnetic field to specific nerves or to the brain itself, to either abort or prevent migraine headache. Some devices have been studied in patients with episodic migraine, while others have only been studied in those with chronic migraine. The studies supporting their use will be discussed below.

Supraorbital Transcutaneous Electrical Stimulation

Supraorbital transcutaneous electrical stimulation has been studied in the preventative treatment of migraine headache. The stimulator resembles a headband, and is placed in the frontal region of the head directly over the supratrochlear and supraorbital nerves (Figure 24.2). It is used for 20 minutes each day, and provides a transcutaneous electrical stimulus that depolarizes these nerves. Schoenen and colleagues[111] performed a randomized double-blind placebo-controlled trial of this device in 67 patients with episodic migraine. The frequency of migraine headache days per month was reduced by 29.7% as compared with baseline in those receiving supraorbital

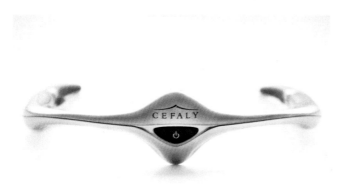

FIGURE 24.2 Supraorbital transcutaneous electrical stimulator. Reproduced with permission from Cephaly Technology Sprl.

transcutaneous electrical stimulation, while it increased by 4.9% in those receiving sham stimulation ($P = 0.054$ for the difference between groups in the intention-to-treat analysis). The proportion of subjects achieving a 50% or greater reduction in monthly migraine headache days was 38.2% in the stimulator group compared with 12.1% in the sham group ($P < 0.014$ for the difference between groups). The use of abortive medication as well as headache days per month were significantly less in the treatment as compared with the sham group. No side effects were reported in either of the study groups. Based on the results of this small clinical trial, this device has received approval from the United States Food and Drug Administration.

Occipital Nerve Stimulation

Three randomized controlled trials of occipital nerve stimulation (ONS) in the prevention of chronic migraine have been reported. Silberstein and colleagues[112] performed a randomized double-blind placebo-controlled study randomizing 268 subjects with chronic migraine to either ONS or sham ONS in a 2:1 treatment allocation. In addition, all subjects were required to have experienced occipital headaches, or pain originating from the neck or cervical spine. Only those subjects who had passed a trial period of ONS demonstrating pain reduction, and experiencing paresthesias in the distribution of the occipital nerve with the ONS, were eligible for inclusion in the trial. The primary outcome measure was a 50% reduction in a daily mean of a 100-mm visual analogue scale (VAS), which was attained by 17% of the ONS group and 13.5% of the sham group ($P = 0.55$). The ONS group had a significantly greater reduction in the frequency of headache days than sham ONS (-5.4 days per month versus -0.08 days per month; $P = 0.008$) and headache-related disability (-64 on the MIDAS scale in the ONS group versus -23 in the sham group; $P < 0.001$). Side effects in the ONS group included lead migration in 4.7%, persistent pain at the ONS site in 8.4%, infection in 2.8%, and device malfunction in 0.9%. Saper and colleagues[113] conducted a clinical trial in which 67 patients were randomized in a 2:1:1 allocation ratio into one of the following three groups: (1) adjustable stimulation (AS; $n = 33$) – patients had the ability to adjust the current, either up or down, depending upon response to therapy; (2) preset stimulation (PS; $n = 17$) – patients received 1 minute of stimulation per day, but had no ability to adjust settings; and (3) medical management (MM; $n = 17$) – patients received only medical management. The primary outcome measure was a 50% reduction in headache days with 39%, 6%, and 4.4% of the AS, PS, and

MM groups meeting this outcome measure, respectively (*P* values <0.03 for comparisons between AS versus PS, and AS versus MM groups). However, no primary outcome measure was chosen for this study, and thus these results must be considered exploratory. Lipton and colleagues[114] randomized 140 patients to ONS versus sham ONS in a 1:1 treatment allocation. This study failed to meet its primary outcome measure, which was the reduction in headache days per month. Therefore, results of ONS stimulation studies in chronic migraine have been mixed. Further study needs to be undertaken before ONS can be recommended as a therapy for refractory cases of chronic migraine.

Transcranial Magnetic Stimulation

Transcranial magnetic stimulation (TMS) has been studied in the abortive treatment of migraine headache. Lipton and colleagues[115] randomized 201 subjects who experienced attacks of migraine with aura to receive TMS of the occipital cortex or a sham procedure during the "aura phase" of the migraine attack. The stimulator was a portable device that was placed in the occipital region of the head and delivered two magnetic pulses to the area. Those subjects receiving TMS were more likely to achieve pain freedom at 2 hours (39% versus 22%; *P* = 0.018), and sustained pain freedom (29% versus 16%; *P* = 0.04). These data suggest that the likelihood of a migraine headache is decreased if one interrupts the aura with a magnetic field. It is possible that the magnetic field is terminating spreading cortical depression, which is postulated to the causative mechanism for the aura of migraine.

Two randomized trials of TMS of the prefrontal cortex as a preventative therapy for chronic migraine have been reported. Brighina and colleagues[116] randomized 11 patients with chronic migraine to either TMS (*n* = 6) or sham TMS (*n* = 5). Subjects received TMS or sham treatments every other day for 3 weeks, for a total of 12 treatments. Those receiving TMS had a significantly greater reduction in attack frequency, use of abortive medications, and the headache index (all *P* values <0.05). Conforto and colleagues[117] performed a randomized double-blinded parallel-group trial of TMS versus a sham procedure in 18 patients with chronic migraine. Subjects received 23 treatments with TMS or a sham procedure over 8 weeks. This study found that the frequency of headache days was significantly reduced in the sham-treated group compared with baseline, while no change was demonstrated in the TMS group. Therefore, conflicting results have been obtained regarding the role of TMS in the prevention of chronic migraine.

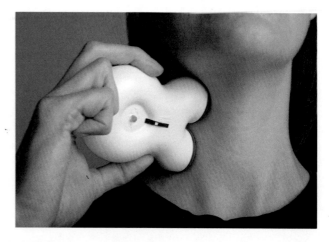

FIGURE 24.3 Transcutaneous vagal nerve stimulator. Reproduced with permission from ElectroCore, LLC.

Vagal Nerve Stimulation

Vagal nerve stimulation has been shown to provide analgesia in animal models of pain.[118–120] The mechanisms through which vagal nerve stimulation produces analgesia are unknown, but it was found to blunt an increase in extracellular glutamate in the trigeminal nucleus caudalis after a nitroglycerin infusion in allodynic rats.[121] Several case studies[122–124] have reported that an implantable vagal nerve stimulator placed for treatment of refractory depression or epilepsy also decreased the frequency of migraine in patients with both disorders. Recently, in a small open clinical trial, a transcutaneous vagal nerve stimulator has been developed and has been studied as an abortive medication for an acute migraine attack (Figure 24.3). Goadsby and colleagues[125] reported that 21% of patients with a moderate to severe migraine at baseline were pain free within 2 hours after use of the transcutaneous vagal nerve stimulator. No serious adverse events were reported, but one individual developed neck twitching and another complained of a raspy voice. In summary, vagal nerve stimulation shows promise in the treatment of migraine, but must be studied in large clinical trials before it can be recommended for the treatment of migraine headache.

CONCLUSIONS

Great strides have been made in recent years regarding research in the areas of epidemiology, neuroimaging, genetics, pathophysiology, and treatment of migraine headache. Epidemiological research has identified the demographics, comorbid medical disorders, and risk factors associated with chronic migraine, which may enable us to find ways to prevent and/or treat chronic migraine. Neuroimaging studies have found that the hypothalamus is activated early in a migraine attack during the prodrome. Therefore, the hypothalamus may very well supplant the dorsal pons as the "migraine generator." Other imaging studies suggest differences in gray and white matter, cortical thickness, and functional connectivity between the brains of migraineurs and controls. It is unclear if many of these changes occur as a direct result of migraine attacks or whether these changes precede the onset of migraine and simply predispose to it. Genetic studies have pinpointed the genetic loci and polymorphisms linked with more common forms of migraine, with and without aura. Identification of these loci will enable us to unlock the pathophysiology of migraine headache. Pharmacologic studies have confirmed the importance of nitric oxide, CGRP, and PACAP-38 in the pathophysiology of the migraine attack, and may provide targets for treatment interventions in the future. Medical devices offer promise in the treatment of patients with refractory chronic migraine. Therefore, the future is bright for migraine patients, as the abovementioned research will change our views on the pathophysiology of migraine headache and it subsequent treatment.

References

1. Buse DC, Manack AN, Fanning KM, Serrano D, Reed ML, Turkel CC, et al. Chronic migraine prevalence, disability, and sociodemographic factors: results from the American Migraine Prevalence and Prevention Study. *Headache*. 2012;52(10):1456−1470.
2. Blumenfeld AM, Varon SF, Wilcox TK, Buse DC, Kawata AK, Manack A, et al. Disability, HRQoL and resource use among chronic and episodic migraineurs: results from the International Burden of Migraine Study (IBMS). *Cephalalgia*. 2011;31(3):301−315.
3. Martin V, Pavlovic JM, Fanning KM, Buse DC, Serrano D, Reed ML, et al. The menopausal transition and its association with higher headache frequencies in women with migraine. Results of the American Migraine Prevalence and Prevention (AMPP) Study [abstract]. Paper presented at: American Headache Society, 2014; Los Angeles, CA.
4. Buse DC, Manack A, Serrano D, Turkel C, Lipton RB. Sociodemographic and comorbidity profiles of chronic migraine and episodic migraine sufferers. *J Neurol Neurosurg Psychiatry*. 2010;81(4):428−432.
5. Katsarava Z, Schneeweiss S, Kurth T, Kroener U, Fritsche G, Eikermann H, et al. Incidence and predictors for chronicity of headache in patients with episodic migraine. *Neurology*. 2004;62(5):788−790.
6. Zwart JA, Dyb G, Hagen K, Svebak S, Stovner LJ, Holmen J. Analgesic overuse among subjects with headache, neck, and low-back pain. *Neurology*. 2004;62(9):1540−1544.
7. Ashina S, Serrano D, Lipton RB, Maizels M, Manack AN, Turkel CC, et al. Depression and risk of transformation of episodic to chronic migraine. *J Headache Pain*. 2012;13(8): 615−624.
8. Aamodt AH, Stovner LJ, Langhammer A, Hagen K, Zwart JA. Is headache related to asthma, hay fever, and chronic bronchitis? The Head-HUNT Study. *Headache*. 2007;47(2):204−212.
9. Bigal ME, Lipton RB. Obesity is a risk factor for transformed migraine but not chronic tension-type headache. *Neurology*. 2006;67(2):252−257.
10. Scher AI, Lipton RB, Stewart WF. Habitual snoring as a risk factor for chronic daily headache. *Neurology*. 2003;60(8):1366−1368.
11. Scher AI, Stewart WF, Lipton RB. Caffeine as a risk factor for chronic daily headache: a population-based study. *Neurology*. 2004;63(11):2022−2027.
12. Scher AI, Stewart WF, Buse D, Krantz DS, Lipton RB. Major life changes before and after the onset of chronic daily headache: a population-based study. *Cephalalgia*. 2008;28(8):868−876.
13. Weiller C, May A, Limmroth V, Juptner M, Kaube H, Schayk RV, et al. Brain stem activation in spontaneous human migraine attacks. *Nat Med*. 1995;1(7):658−660.
14. Denuelle M, Fabre N, Payoux P, Chollet F, Geraud G. Hypothalamic activation in spontaneous migraine attacks. *Headache*. 2007;47(10):1418−1426.
15. Maniyar FH, Sprenger T, Monteith T, Schankin C, Goadsby PJ. Brain activations in the premonitory phase of nitroglycerin-triggered migraine attacks. *Brain*. 2014;137(Pt 1):232−241.
16. Stankewitz A, Aderjan D, Eippert F, May A. Trigeminal nociceptive transmission in migraineurs predicts migraine attacks. *J Neurosci*. 2011;31(6):1937−1943.
17. Eck J, Richter M, Straube T, Miltner WH, Weiss T. Affective brain regions are activated during the processing of pain-related words in migraine patients. *Pain*. 2011;152(5):1104−1113.
18. Valfre W, Rainero I, Bergui M, Pinessi L. Voxel-based morphometry reveals gray matter abnormalities in migraine. *Headache*. 2008;48(1):109−117.
19. Schmitz N, Admiraal-Behloul F, Arkink EB, Kruit MC, Schoonman J, Ferrari MD, et al. Attack frequency and disease duration as indicators for brain damage in migraine. *Headache*. 2008;48(7):1044−1055.
20. Kim JH, Suh SI, Seol HY, Oh K, Seo WK, Yu SW, et al. Regional grey matter changes in patients with migraine: a voxel-based morphometry study. *Cephalalgia*. 2008;28(6):598−604.
21. Schmidt-Wilcke T, Ganssbauer S, Neuner T, Bogdahn U, May A. Subtle gray matter changes between migraine patients and healthy controls. *Cephalalgia*. 2008;28(1):1−4.
22. Schmidt-Wilcke T, Leinisch E, Ganssbauer S, Draganski B, Bogdahn U, Altmeppen J, et al. Affective components and intensity of pain correlate with structural differences in gray matter in chronic back pain patients. *Pain*. 2006;125(1−2):89−97.
23. Robinson ME, Craggs JG, Price DD, Perlstein WM, Staud R. Gray matter volumes of pain-related brain areas are decreased in fibromyalgia syndrome. *J Pain*. 2011;12(4):436−443.
24. Rodriguez-Raecke R, Niemeier A, Ihle K, Ruether W, May A. Brain gray matter decrease in chronic pain is the consequence and not the cause of pain. *J Neurosci*. 2009;29(44):13746−13750.
25. DaSilva AF, Granziera C, Snyder J, Hadjikhani N. Thickening in the somatosensory cortex of patients with migraine. *Neurology*. 2007;69(21):1990−1995.
26. Erpelding N, Moayedi M, Davis KD. Cortical thickness correlates of pain and temperature sensitivity. *Pain*. 2012;153(8):1602−1609.
27. Kruit MC, van Buchem MA, Hofman PA, Bakkers JT, Terwindt GM, Ferrari MD, et al. Migraine as a risk factor for subclinical brain lesions. *JAMA*. 2004;291(4):427−434.
28. Palm-Meinders IH, Koppen H, Terwindt GM, Launer LJ, Konishi J, Moonen JM, et al. Structural brain changes in migraine. *JAMA*. 2012;308(18):1889−1897.
29. Trauninger A, Leel-Ossy E, Kamson DO, Poto L, Aradi M, Kover F, et al. Risk factors of migraine-related brain white

matter hyperintensities: an investigation of 186 patients. *J Headache Pain*. 2011;12(1).97–103.

30. Kruit MC, Thijs RD, Ferrari MD, Launer LJ, van Buchem MA, van Dijk JG. Syncope and orthostatic intolerance increase risk of brain lesions in migraineurs and controls. *Neurology*. 2013;80 (21):1958–1965.

31. Yamamoto Y, Craggs L, Baumann M, Kalimo H, Kalaria RN. Review: molecular genetics and pathology of hereditary small vessel diseases of the brain. *Neuropathol Appl Neurobiol*. 2011; 37(1):94–113.

32. Mainero C, Boshyan J, Hadjikhani N. Altered functional magnetic resonance imaging resting-state connectivity in periaqueductal gray networks in migraine. *Ann Neurol*. 2011;70(5):838–845.

33. Schwedt TJ, Larson-Prior L, Coalson RS, Nolan T, Mar S, Ances BM, et al. Allodynia and descending pain modulation in migraine: a resting state functional connectivity analysis. *Pain Med*. 2014;15(1):154–165.

34. Hadjikhani N, Ward N, Boshyan J, Napadow V, Maeda Y, Truini A, et al. The missing link: enhanced functional connectivity between amygdala and visceroceptive cortex in migraine. *Cephalalgia*. 2013;33(15):1264–1268.

35. Schwedt TJ, Schlaggar BL, Mar S, Nolan TR, Coalson RS, Nardos B, et al. Atypical resting-state functional connectivity of affective pain regions in chronic migraine. *Headache*. 2013; 53(5):737–751.

36. Yuan K, Qin W, Liu P, Zhao L, Yu D, Zhao L, et al. Reduced fractional anisotropy of corpus callosum modulates inter-hemispheric resting state functional connectivity in migraine patients without aura. *PLoS One*. 2012;7(9):e45476.

37. Yuan K, Zhao L, Cheng P, Yu D, Zhao L, Dong T, et al. Altered structure and resting-state functional connectivity of the basal ganglia in migraine patients without aura. *J Pain*. 2013;14(8):836–844.

38. Maleki N, Becerra L, Nutile L, Pendse G, Brawn J, Bigal M, et al. Migraine attacks the Basal Ganglia. *Mol Pain*. 2011;7:71.

39. Riant F, Roze E, Barbance C, Meneret A, Quyant-Marechal L, Lucas C, et al. PRRT2 mutations cause hemiplegic migraine. *Neurology*. 2012;79(21):2122–2124.

40. Gasparini CF, Sutherland HG, Griffiths LR. Studies on the pathophysiology and genetic basis of migraine. *Curr Genomics*. 2013;14(5):300–315.

41. Schurks M, Rist PM, Kurth T. STin2 VNTR polymorphism in the serotonin transporter gene and migraine: pooled and meta-analyses. *J Headache Pain*. 2010;11(4):317–326.

42. Peroutka SJ, Wilhoit T, Jones K. Clinical susceptibility to migraine with aura is modified by dopamine D2 receptor (DRD2) NcoI alleles. *Neurology*. 1997;49(1):201–206.

43. Fernandez F, Lea RA, Colson NJ, Bellis C, Quinlan S, Griffiths LR. Association between a 19 bp deletion polymorphism at the dopamine beta-hydroxylase (DBH) locus and migraine with aura. *J Neurol Sci*. 2006;251(1–2):118–123.

44. Fernandez F, Colson N, Quinlan S, MacMillan J, Lea RA, Griffiths LR. Association between migraine and a functional polymorphism at the dopamine beta-hydroxylase locus. *Neurogenetics*. 2009;10(3):199–208.

45. Rubino E, Ferrero M, Rainero I, Binello E, Vaula G, Pinessi L. Association of the C677T polymorphism in the MTHFR gene with migraine: a meta-analysis. *Cephalalgia*. 2009;29 (8):818–825.

46. Schurks M, Rist PM, Kurth T. MTHFR 677C > T and ACE D/I polymorphisms in migraine: a systematic review and meta-analysis. *Headache*. 2010;50(4):588–599.

47. Oterino A, Toriello M, Cayon A, Castillo J, Calas R, Alonson-Arranz A, et al. Multilocus analyses reveal involvement of the ESR1, ESR2, and FSHR genes in migraine. *Headache*. 2008;48 (10):1438–1450.

48. Colson NJ, Lea RA, Quinlan S, MacMillan J, Griffiths LR. Investigation of hormone receptor genes in migraine. *Neurogenetics*. 2005;6(1):17–23.

49. Rodriguez-Acevedo AJ, Maher BH, Lea RA, Benton M, Griffiths LR. Association of oestrogen-receptor gene (ESR1) polymorphisms with migraine in the large Norfolk Island pedigree. *Cephalalgia*. 2013;33(14):1139–1147.

50. Ghosh J, Pradhan S, Mittal B. Multilocus analysis of hormonal, neurotransmitter, inflammatory pathways and genome-wide associated variants in migraine susceptibility. *Eur J Neurol*. 2014; 21(7):1011–1020.

51. Corominas R, Ribases M, Cuenca-Leon E, Cormand B, Macaya A. Lack of association of hormone receptor polymorphisms with migraine. *Eur J Neurol*. 2009;16(3):413–415.

52. Kaunisto MA, Kallela M, Hamalainen E, Kilpikari R, Havanka H, Harno H, et al. Testing of variants of the MTHFR and ESR1 genes in 1798 Finnish individuals fails to confirm the association with migraine with aura. *Cephalalgia*. 2006;26(12):1462–1472.

53. Anttila V, Stefansson H, Kallela M, Todt U, Terwindt GM, Calafato MS, et al. Genome-wide association study of migraine implicates a common susceptibility variant on 8q22.1. *Nat Genet*. 2010;42(10):869–873.

54. Chasman DI, Schurks M, Anttila V, de Vries B, Schminke U, Launer LJ, et al. Genome-wide association study reveals three susceptibility loci for common migraine in the general population. *Nat Genet*. 2011;43(7):695–698.

55. Freilinger T, Anttila V, de Vries B, Malik R, Kallela M, Terwindt GM, et al. Genome-wide association analysis identifies susceptibility loci for migraine without aura. *Nat Genet*. 2012; 44(7):777–782.

56. Law C, Bunyan D, Castle B, Day L, Simpson I, Westwood G, et al. Clinical features in a family with an R460H mutation in transforming growth factor beta receptor 2 gene. *J Med Genet*. 2006;43(12):908–916.

57. Lafreniere RG, Cader MZ, Poulin JF, Andres-Enguix I, Simoneau M, Gupta N, et al. A dominant-negative mutation in the TRESK potassium channel is linked to familial migraine with aura. *Nat Med*. 2010;16(10):1157–1160.

58. Guo Z, Cao YQ. Over-expression of TRESK K(+) channels reduces the excitability of trigeminal ganglion nociceptors. *PLoS One*. 2014;9(1):e87029.

59. Maher BH, Taylor M, Stuart S, Okoliasanyl RK, Roy B, Sutherland HG, et al. Analysis of 3 common polymorphisms in the KCNK18 gene in an Australian Migraine case–control cohort. *Gene*. 2013;528(2):343–346.

60. Ashina M, Hansen JM, Olesen J. Pearls and pitfalls in human pharmacological models of migraine: 30 years' experience. *Cephalalgia*. 2013;33(8):540–553.

61. Tfelt-Hansen PC, Tfelt-Hansen J. Nitroglycerin headache and nitroglycerin-induced primary headaches from 1846 and onwards: a historical overview and an update. *Headache*. 2009;49(3):445–456.

62. Olesen J. The role of nitric oxide (NO) in migraine, tension-type headache and cluster headache. *Pharmacol Ther*. 2008;120 (2):157–171.

63. Koulchitsky S, Fischer MJ, Messlinger K. Calcitonin gene-related peptide receptor inhibition reduces neuronal activity induced by prolonged increase in nitric oxide in the rat spinal trigeminal nucleus. *Cephalalgia*. 2009;29(4):408–417.

64. Feistel S, Albrecht S, Messlinger K. The calcitonin gene-related peptide receptor antagonist MK-8825 decreases spinal trigeminal activity during nitroglycerin infusion. *J Headache Pain*. 2013;14(1):93.

65. Strecker T, Dux M, Messlinger K. Nitric oxide releases calcitonin gene-related peptide from rat dura mater encephali promoting increases in meningeal blood flow. *J Vasc Res*. 2002; 39(6):489–496.

66. Olesen J, Iversen HK, Thomsen LL. Nitric oxide supersensitivity: a possible molecular mechanism of migraine pain. *Neuroreport.* 1993;4(8):1027–1030.

67. Christiansen I, Thomsen LL, Daugaard D, Ulrich V, Olesen J. Glyceryl trinitrate induces attacks of migraine without aura in sufferers of migraine with aura. *Cephalalgia.* 1999;19(7):660–667 [discussion 626].

68. Afridi SK, Kaube H, Goadsby PJ. Glyceryl trinitrate triggers premonitory symptoms in migraineurs. *Pain.* 2004;110(3):675–680.

69. Kruuse C, Iversen HK, Jansen-Olesen I, Edvinsson L, Olesen J. Calcitonin gene-related peptide (CGRP) levels during glyceryl trinitrate (GTN)-induced headache in healthy volunteers. *Cephalalgia.* 2010;30(4):467–474.

70. Tvedskov JF, Tfelt-Hansen P, Petersen KA, Jensen LT, Olesen J. CGRP receptor antagonist olcegepant (BIBN4096BS) does not prevent glyceryl trinitrate-induced migraine. *Cephalalgia.* 2010;30(11):1346–1353.

71. Hansen JM, Pedersen D, Larsen VA, Sanchez-del-Rio M, Alvarez Linera JR, Olesen J, et al. Magnetic resonance angiography shows dilatation of the middle cerebral artery after infusion of glyceryl trinitrate in healthy volunteers. *Cephalalgia.* 2007;27(2):118–127.

72. Iversen HK, Holm S, Friberg L, Tfelt-Hansen P. Intracranial hemodynamics during intravenous infusion of glyceryl trinitrate. *J Headache Pain.* 2008;9(3):177–180.

73. Tegeler CH, Davidai G, Gengo FM, Knappertz VA, Trobst BT, Gabriel H, et al. Middle cerebral artery velocity correlates with nitroglycerin-induced headache onset. *J Neuroimaging.* 1996;6(2):81–86.

74. Lassen LH, Ashina M, Christiansen I, Ulrich V, Olesen J. Nitric oxide synthase inhibition in migraine. *Lancet.* 1997;349(9049):401–402.

75. Hoivik HO, Laurijssens BE, Harnisch LO, Twomey CK, Dixon RM, Kirkham AJ, et al. Lack of efficacy of the selective iNOS inhibitor GW274150 in prophylaxis of migraine headache. *Cephalalgia.* 2010;30(12):1458–1467.

76. Palmer J, Gulllard F, Laritjssens B, Wentz AL, Dixon R, Williams PM. A randomized, single-blind, placebo-controlled, adaptive clinical trial of GW274150, a selective iNOS inhibitor, in the treatment of acute migraine. *Cephalalgia.* 2009;29:124.

77. Annedi SC, Maddaford SP, Ramnauth J, Renton P, Rybak T, Silverman S, et al. Discovery of a potent, orally bioavailable and highly selective human neuronal nitric oxide synthase (nNOS) inhibitor, N-(1-(piperidin-4-yl)indolin-5-yl)thiophene-2-carboximidamide as a pre-clinical development candidate for the treatment of migraine. *Eur J Med Chem.* 2012;55:94–107.

78. Hokfelt T, Arvidsson U, Ceccatelli S, Cortes R, Cullheim S, Dagerlind A, et al. Calcitonin gene-related peptide in the brain, spinal cord, and some peripheral systems. *Ann NY Acad Sci.* 1992;657:119–134.

79. van Rossum D, Hanisch UK, Quirion R. Neuroanatomical localization, pharmacological characterization and functions of CGRP, related peptides and their receptors. *Neurosci Biobehav Rev.* 1997;21(5):649–678.

80. Goadsby PJ, Edvinsson L, Ekman R. Release of vasoactive peptides in the extracerebral circulation of humans and the cat during activation of the trigeminovascular system. *Ann Neurol.* 1988;23(2):193–196.

81. Goadsby PJ, Edvinsson L, Ekman R. Vasoactive peptide release in the extracerebral circulation of humans during migraine headache. *Ann Neurol.* 1990;28(2):183–187.

82. Ashina M, Bendtsen L, Jensen R, Schifter S, Olesen J. Evidence for increased plasma levels of calcitonin gene-related peptide in migraine outside of attacks. *Pain.* 2000;86(1–2):133–138.

83. Lassen LH, Haderslev PA, Jacobsen VB, Iversen HK, Sperling B, Olesen J. CGRP may play a causative role in migraine. *Cephalalgia.* 2002;22(1):54–61.

84. Cernuda-Morollon E, Larrosa D, Ramon C, Vega J, Martinez-Camblor P, Pascual J. Interictal increase of CGRP levels in peripheral blood as a biomarker for chronic migraine. *Neurology.* 2013;81(14):1191–1196.

85. Cernuda-Morollon E, Martinez-Camblor P, Ramon C, Larrosa D, Serrano-Pertierra E, Pascual J. CGRP and VIP levels as predictors of efficacy of onabotulinumtoxin type A in chronic migraine. *Headache.* 2014;54:987–995.

86. Hoffmann J, Goadsby PJ. New agents for acute treatment of migraine: CGRP receptor antagonists, iNOS inhibitors. *Curr Treat Options Neurol.* 2012;14(1):50–59.

87. Aiyar N, Rand K, Elshourbagy NA, Zeng Z, Adamou JE, Bergsma DJ, et al. A cDNA encoding the calcitonin gene-related peptide type 1 receptor. *J Biol Chem.* 1996;271(19):11325–11329.

88. McLatchie LM, Fraser NJ, Main MJ, Wise A, Brown J, Thompson N, et al. RAMPs regulate the transport and ligand specificity of the calcitonin-receptor-like receptor. *Nature.* 1998;393(6683):333–339.

89. Ma W, Chabot JG, Powell KJ, Jhamandas K, Dickerson IM, Quirion R. Localization and modulation of calcitonin gene-related peptide-receptor component protein-immunoreactive cells in the rat central and peripheral nervous systems. *Neuroscience.* 2003;120(3):677–694.

90. Hostetler ED, Joshi AD, Sanabria-Bohorquez S, Fan H, Zeng Z, Purcell M, et al. *In vivo* quantification of calcitonin gene-related peptide receptor occupancy by telcagepant in rhesus monkey and human brain using the positron emission tomography tracer [11C]MK-4232. *J Pharmacol Exp Ther.* 2013;347(2):478–486.

91. Oliver KR, Wainwright A, Edvinsson L, Pickard JD, Hill RG. Immunohistochemical localization of calcitonin receptor-like receptor and receptor activity-modifying proteins in the human cerebral vasculature. *J Cereb Blood Flow Metab.* 2002;22(5):620–629.

92. Connor KM, Shapiro RE, Diener HC, Lucas S, Kost J, Fan X, et al. Randomized, controlled trial of telcagepant for the acute treatment of migraine. *Neurology.* 2009;73(12):970–977.

93. Diener HC, Barbanti P, Dahlof C, Reuter U, Habeck J, Podhorna J. BI 44370 TA, an oral CGRP antagonist for the treatment of acute migraine attacks: results from a phase II study. *Cephalalgia.* 2011;31(5):573–584.

94. Olesen J, Diener HC, Husstedt IW, Goadsby PJ, Hall D, Meier U, et al. Calcitonin gene-related peptide receptor antagonist BIBN 4096 BS for the acute treatment of migraine. *N Engl J Med.* 2004;350(11):1104–1110.

95. Hewitt DJ, Aurora SK, Dodick DW, Goadsby PJ, Ge YJ, Bachman R, et al. Randomized controlled trial of the CGRP receptor antagonist MK-3207 in the acute treatment of migraine. *Cephalalgia.* 2011;31(6):712–722.

96. Ho TW, Ferrari MD, Dodick DW, Galet V, Kost J, Fan X, et al. Efficacy and tolerability of MK-0974 (telcagepant), a new oral antagonist of calcitonin gene-related peptide receptor, compared with zolmitriptan for acute migraine: a randomised, placebo-controlled, parallel-treatment trial. *Lancet.* 2008;372(9656):2115–2123.

97. Han TH, Blanchard RL, Palcza J, McCrea JB, Laethem T, Willson K, et al. Single- and multiple-dose pharmacokinetics and tolerability of telcagepant, an oral calcitonin gene-related peptide receptor antagonist, in adults. *J Clin Pharmacol.* 2010;50(12):1367–1376.

98. Salvatore CA, Moore EL, Calamari A, Cook JJ, Michener MS, O'Malley S, et al. Pharmacological properties of MK-3207, a potent and orally active calcitonin gene-related peptide receptor antagonist. *J Pharmacol Exp Ther.* 2010;333(1):152–160.

99. Bigal ME, Walter S. Monoclonal antibodies for migraine: preventing calcitonin gene-related peptide activity. *CNS Drugs*. 2014;28(5):389–399

100. Goadsby PJ, Dodick D, Silberstein SD, Lipton RB, Olesen J, Ashina M, et al. Safety and efficacy of ALD403, an antibody to calcitonin gene-related peptide, for the prevention of frequent episodic migraine: a randomised, double-blind, placebo-controlled, exploratory phase 2 trial. *Lancet Neurol*. 2014;13 (11):1100–1107.

101. Dodick D, Goadsby PJ, Spierings EL, Scherer J, Sweeney S, Grayzel D. Safety and efficacy of LY2951742, a monoclonal antibody to calcitonin gene-related peptide, for the prevention of migraine: a phase 2, randomised, double-blind, placebo-controlled study. *Lancet Neurol*. 2014;13(9):885–892.

102. Schytz HW, Olesen J, Ashina M. The PACAP receptor: a novel target for migraine treatment. *Neurotherapeutics*. 2010;7(2):191–196.

103. Tajti J, Uddman R, Moller S, Sundler F, Edvinsson L. Messenger molecules and receptor mRNA in the human trigeminal ganglion. *J Auton Nerv Syst*. 1999;76(2–3):176–183.

104. Uddman R, Tajti J, Moller S, Sundler F, Edvinsson L. Neuronal messengers and peptide receptors in the human sphenopalatine and otic ganglia. *Brain Res*. 1999;826(2):193–199.

105. Csati A, Tajti J, Kuris A, Tuka B, Edvinsson L, Warfvinge K. Distribution of vasoactive intestinal peptide, pituitary adenylate cyclase-activating peptide, nitric oxide synthase, and their receptors in human and rat sphenopalatine ganglion. *Neuroscience*. 2012;202:158–168.

106. Laburthe M, Couvineau A. Molecular pharmacology and structure of VPAC receptors for VIP and PACAP. *Regul Pept*. 2002;108(2–3):165–173.

107. Schytz HW, Birk S, Wienecke T, Kruuse C, Olesen J, Ashina M. PACAP38 induces migraine-like attacks in patients with migraine without aura. *Brain*. 2009;132(Pt 1):16–25.

108. Amin FM, Hougaard A, Schytz HW, Ashgar MS, Lundholm E, Parvaiz AI, et al. Investigation of the pathophysiological mechanisms of migraine attacks induced by pituitary adenylate cyclase-activating polypeptide-38. *Brain*. 2014;137(Pt 3): 779–794.

109. Tuka B, Helyes Z, Markovics A, Bagoly T, Szolcsanyi J, Szabo N, et al. Alterations in PACAP-38-like immunoreactivity in the plasma during ictal and interictal periods of migraine patients. *Cephalalgia*. 2013;33(13):1085–1095.

110. Baun M, Pedersen MH, Olesen J, Jansen-Olesen I. Dural mast cell degranulation is a putative mechanism for headache induced by PACAP-38. *Cephalalgia*. 2012;32(4):337–345.

111. Schoenen J, Vandersmissen B, Jeangette S, Herroelen L, Vandenheede M, Gerard P, et al. Migraine prevention with a supraorbital transcutaneous stimulator: a randomized controlled trial. *Neurology*. 2013;80(8):697–704.

112. Silberstein SD, Dodick DW, Saper J, Huh B, Slavin KV, Sharan A, et al. Safety and efficacy of peripheral nerve stimulation of the occipital nerves for the management of chronic migraine: results from a randomized, multicenter, double-blinded, controlled study. *Cephalalgia*. 2012;32(16):1165–1179.

113. Saper JR, Dodick DW, Silberstein SD, McCarville S, Sun M, Goadsby PJ. Occipital nerve stimulation for the treatment of intractable chronic migraine headache: ONSTIM feasibility study. *Cephalalgia*. 2011;31(3):271–285.

114. Lipton R, Goadsby PJ, Cady R. PRISM study: occipital nerve stimulation for treatment-refractory migraine. *Cephalalgia*. 2009;39:30 [abstract].

115. Lipton RB, Dodick DW, Silberstein SD, Saper JR, Aurora SK, Pearlman SH, et al. Single-pulse transcranial magnetic stimulation for acute treatment of migraine with aura: a randomised, double-blind, parallel-group, sham-controlled trial. *Lancet Neurol*. 2010;9 (4):373–380.

116. Brighina F, Piazza A, Vitello G, Aloisio A, Palermo A, Daniele O, et al. rTMS of the prefrontal cortex in the treatment of chronic migraine: a pilot study. *J Neurol Sci*. 2004;227(1):67–71.

117. Conforto AB, Amaro Jr. E, Goncalves AL, Mercante JP, Guendler VZ, Ferreira JR, et al. Randomized, proof-of-principle clinical trial of active transcranial magnetic stimulation in chronic migraine. *Cephalalgia*. 2014;34(6):464–472.

118. Randich A, Ren K, Gebhart GF. Electrical stimulation of cervical vagal afferents. II. Central relays for behavioral antinociception and arterial blood pressure decreases. *J Neurophysiol*. 1990;64 (4):1115–1124.

119. Ren K, Randich A, Gebhart GF. Vagal afferent modulation of spinal nociceptive transmission in the rat. *J Neurophysiol*. 1989; 62(2):401–415.

120. Aicher SA, Lewis SJ, Randich A. Antinociception produced by electrical stimulation of vagal afferents: independence of cervical and subdiaphragmatic branches. *Brain Res*. 1991;542(1):63–70.

121. Oshinsky ML, Murphy AL, Hekierski Jr H, Cooper M, Simon BJ. Noninvasive vagus nerve stimulation as treatment for trigeminal allodynia. *Pain*. 2014;155(5):1037–1042.

122. Mauskop A. Vagus nerve stimulation relieves chronic refractory migraine and cluster headaches. *Cephalalgia*. 2005;25(2):82–86.

123. Cecchini AP, Mea E, Tullo V, Currone M, Franzini A, Broggi G, et al. Vagus nerve stimulation in drug-resistant daily chronic migraine with depression: preliminary data. *Neurol Sci*. 2009;30 (suppl 1):S101–104.

124. Sadler RM, Purdy RA, Rahey S. Vagal nerve stimulation aborts migraine in patient with intractable epilepsy. *Cephalalgia*. 2002;22(6):482–484.

125. Goadsby P, Grosberg B, Mauskop A, Cady R, Simmons K. Effect of noninvasive vagus nerve stimulation on acute migraine: an open-label pilot study. *Cephalalgia*. 2014;34(12):986–993.

Index